RADAR
SYSTEM
ANALYSIS

The Artech House Radar Library

Radar System Analysis by David K. Barton

Introduction to Electronic Warfare by D. Curtis Schleher

Electronic Intelligence: The Analysis of Radar Signals by Richard G. Wiley

Electronic Intelligence: The Interception of Radar Signals by Richard G. Wiley

Principles of Secure Communication Systems by Don J. Torrieri

Multiple-Target Tracking with Radar Applications by Samuel S. Blackman

Solid-State Radar Transmitters by Edward D. Ostroff et al.

Logarithmic Amplification by Richard Smith Hughes

Radar Propagation at Low Altitudes by M.L. Meeks

Radar Cross Section by Eugene F. Knott, John F. Shaeffer, and Michael T. Tuley

Radar Anti-Jamming Techniques by M.V. Maksimov et al.

Introduction to Synthetic Array and Imaging Radars by S.A. Hovanessian

Radar Detection and Tracking Systems by S.A. Hovanessian

Radar System Design and Analysis by S.A. Hovanessian

Radar Signal Processing by Bernard Lewis, Frank Kretschmer, and Wesley Shelton

Radar Calculations Using the TI-59 Programmable Calculator by William A. Skillman

Radar Calculations Using Personal Computers by William A. Skillman

Techniques of Radar Reflectivity Measurement, Nicholas C. Currie, ed.

Monopulse Principles and Techniques by Samuel M. Sherman

Receiving Systems Design by Stephen J. Erst

Designing Control Systems by Olis Rubin

Advanced Mathematics for Practicing Engineers by Kurt Arbenz and Alfred Wohlhauser

Radar Reflectivity of Land and Sea by Maurice W. Long

High Resolution Radar Imaging by Dean L. Mensa

Introduction to Monopulse by Donald R. Rhodes

Probability and Information Theory, with Applications to Radar by P.M. Woodward

Radar Detection by J.V. DiFranco and W.L. Rubin

Synthetic Aperture Radar, John J. Kovaly, ed.

Infrared-to-Millimeter Wavelength Detectors, Frank R. Arams, ed.

Significant Phased Array Papers, R.C. Hansen, ed.

Handbook of Radar Measurement by David K. Barton and Harold R. Ward

Statistical Theory of Extended Radar Targets by R.V. Ostrovityanov and F.A. Basalov

Radar Technology, Eli Brookner, ed.

MTI Radar, D. Curtis Schleher, ed.

RADAR
SYSTEM
ANALYSIS

DAVID K. BARTON
CONSULTING SCIENTIST
RAYTHEON COMPANY
BEDFORD MA

Artech

This book is dedicated to the
two engineers who were responsible
for my entry into the
radar engineering field:

William F. Hoisington

and

Ozro M. Covington

Library of Congress Cataloging in Publication Data

Barton, David Knox, 1927-
  Radar system analysis.

  (The Artech radar library)
  Reprint of the ed. published by Prentice-Hall,
Englewood Cliffs, N. J., in series: Microwaves and
fields series and Prentice-Hall electrical engineering
series.
  Bibliography: p.
  Includes index.
  1. Radar.   I. Title
[TK6575.B365   1976]   621.3848   76-45811
ISBN 0-89006-043-6

Copyright © 1976.
ARTECH HOUSE, INC.
610 Washington Street
Dedham, Massachusetts 02026
Printed and bound in the United States of America
Library of Congress Catalog Card Number: 76-45811
Standard Book Number: 0-89006-043-6

10 9

Since the original publication of *Radar System Analysis* in 1964, there have been many new books on different aspects of radar. Most of these deal with detection theory, measurement, resolution, signal design, antennas, or similar specific subjects. Component and circuit technologies have changed and advanced in many areas, providing the radar designer with tools and techniques which were only on the horizon in the early 1960's. However, the fundamentals of system performance, and of synthesis and analysis of search and track radars, covered by this book, remain applicable to the new systems, whether they use mechanical or electronic beam steering, digital or analog processing, solid-state or thermionic devices.

Because the need for careful analysis remains as important as ever before, this new printing has been undertaken to provide copies to those engineers who

PREFACE

may have entered the radar system field since the last printing of the book. Several errors (both substantive and typographical) have been corrected, but the material remains otherwise in its original form.

DAVID K. BARTON
Lexington, Massachusetts

The author is indebted to Harlan Collar, head of the RCA Radar Systems Group at Moorestown, and to many members of that group, for their assistance and encouragement during the preparation of the manuscript. The contents of the book result from some eight years of study by the author and his associates at RCA, conducted largely under the guidance of S. M. Sherman. In particular, results have been drawn from work by Dr. Irving Kanter (in Chapter 13), and by Dr. Sidney Shucker (Chapters 9 and 16).

Members of the staff of the United States Naval Research Laboratory have been most helpful, both as sources of original material and illustrations, and in review of the manuscript. Lamont V. Blake provided much material and consultation in the areas of detection theory, search radar, and propagation effects. Dean D. Howard and John H. Dunn served in a similar role on tracking radar subjects.

ACKNOWLEDGMENTS

Professor Mischa Schwartz of Brooklyn Polytechnic Institute, during his review of the manuscript, has also contributed useful suggestions throughout the book.

CONTENTS

The development of radar during the first half of the twentieth century marked the greatest advance in methods of sensing remote objects since the telescope was invented in the year 1608. By providing its own, well-controlled source of illumination, the radar made possible not only detection but measurement of accurate radial distance to the target. The ability of radio waves to penetrate the atmosphere under all conditions of weather and the absence of strong ambient illumination in the frequency bands used by radar made possible much longer ranges and greater sensitivity than could be obtained in the visible portion of the electromagnetic spectrum. This chapter will discuss the basic limitations imposed upon radar target detection by the inevitable presence of random or "thermal" noise in the receiver. Future chapters will consider the effects of this noise on measurement of target position and velocity, as well as the sources and effects of other noise components such as background echoes and man-made interference.

THE THEORY OF TARGET DETECTION

1

1.1 RADAR ECHOES AND NOISE

Detection is the process by which the presence of the sought-after object, or target, is sensed in the presence of competing indications which arise from background radiation, undesired echoes, or noise generated in the radar receiver. In

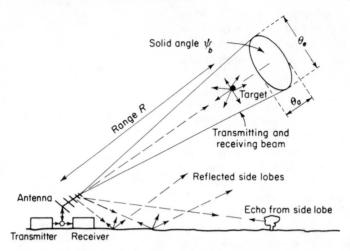

Figure 1.1   Typical radar-target geometry.

radar, the target is sensed by its reflection of radio-frequency energy originating in the radar's own transmitter.* The geometrical relationships in a typical radar case are shown in Fig. 1.1. The radar transmitter is connected to an antenna which illuminates the solid angle ψ_b often defined in terms of azimuth and elevation beamwidths θ_a and θ_e. The limits of the beam are not precisely defined in space, some of the energy going into side lobes which illuminate objects outside the intended coverage, or which reflect from the earth's surface to interfere with the direct rays at the target. The target at range R intercepts a small portion of the transmitted power and scatters it in various directions, returning some of it to the radar's receiving antenna. The exact calculations by which this returned signal strength may be found in a given case are covered in Chapter 4. The properties of the target which are important in detection calculations are described in Chapter 3. For the present, we shall merely assume that some portion of the transmitted signal, modified by the motion of the target, by target reflectivity characteristics, and by the transmission medium, reaches the receiver along with an assortment of undesired echoes and noise.

The Nature of Radar Echoes

The echo returned to the radar by a simple, point-source target will be an exact reproduction of the transmitted signal, shifted in time by an amount coresponding to the range delay, in frequency by the "Doppler

---

* The special case where the targets cooperate by responding with amplified signals through "beacons" or transponders will also be covered by this detection theory.

shift" owing to target radial velocity v_r, and in amplitude by the geometry of the radar-target situation. If the radar transmission consists of a train of rectangular pulses modulating a sinusoidal carrier, the echo will appear as a delayed train of pulses (during the time the target lies within the beam of the radar antenna). The delay time t_r of each pulse, relative to the corresponding transmitted pulse, will indicate the target range R, according to the relationship $t_r = 2R/c$, where c is the velocity of light. The center frequency of the received pulse will be shifted from the transmitted frequency f_t to a value $f_t + f_d = f_t(c + v_r)/(c - v_r)$, or approximately $f_d \cong 2f_t v_r/c$. The time at which the received pulse train reaches its maximum will indicate the point of closest approach to the center of the radar beam, as shown in Fig. 1.2(b). In the practical case, the echoes will deviate from the regular pattern shown because of interference between different reflecting surfaces on the target, variations in the propagation medium, and presence of multipath reflections. The average amplitude of the echoes may be reduced or enhanced by these factors, and the envelope of the returned echoes will generally be distorted and modulated as shown in

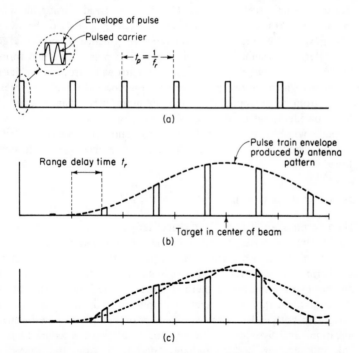

Figure 1.2 Typical echo pulse trains: (a) transmitter pulse train; (b) received pulse train; and (c) received pulse train from fluctuating target.

Fig. 1.2(c). Within this envelope, there will be corresponding changes in the range delay and Doppler shift of the echoes.

The properties by which the echo signals may be described include the amplitude of individual pulses, the average amplitude, the range delay and Doppler shift, the duration of the pulse train, and such other factors as width of individual pulses. For a given observation, the total energy returned to the radar by all pulses in the train may be used as an index of detectability and of measuring ability of the radar. As shown in the next section, there is an upper limit to the performance of the radar in target detection that is set entirely by the ratio of this total energy to the spectral density of noise in the receiver. Thus, the ability of a radar to detect targets can be shown to depend upon the *average* power of the transmitter, the *time* during which this power is brought to bear on the target, and the *geometry* of the radar-target situation, regardless of the type of signal modulation used in transmission. For the pulsed radar referred to earlier, the received energy may be represented as the product of received pulse power, pulse width, and number of pulses in the received train. Alternately, the average received power times the observation time t_o may be used to measure received energy. Other types of radar using modulated continuous-wave transmission will exchange energy equal to the average received power times the observation time. In practice, it makes a great deal of difference whether the received signal is pulsed or some form of *c-w,* in that the different signal processing systems will introduce inefficiency or loss factors which vary with the wave form used, and which may prevent the radar from even approaching the optimum performance level. The practical considerations of single-pulse signal-to-noise ratio, receiver bandwidth, pulse width or modulating wave form, and actual detection criteria will be discussed in more detail after the basic theory of detection is developed.

Sources of Noise in Radar

The first problem of radar is to detect targets against the natural background of random noise generated in the receiver and by "black body" radiation from the environment. This portion of the noise will be referred to as "thermal noise," whether it originates inside or outside the radar equipment. Typical sources of thermal noise are the "shot noise" of electron tubes used in the early stages of amplification in the receiver, noise introduced by crystal detectors or mixers prior to intermediate-frequency amplification, resistive loss in wave guide components connecting the antenna and receiver, and r-f radiation picked up from the atmosphere and the ground by the receiving antenna. These sources of noise are diagrammed in Fig. 1.3. In early radars, the performance of the receiver itself was so poor that the noise level could be calculated almost entirely

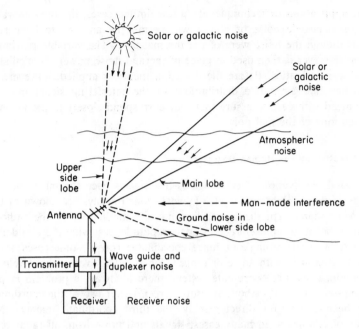

Figure 1.3 Sources of radar noise.

in terms of the receiver noise factor.* This noise level corresponded in typical receivers to the output of a resistive input termination at a temperature of 3000 to 5000° K. Compared to this temperature, the possible contribution of ambient temperature of the surroundings, near 290° K, was negligible. The attenuation of the atmosphere and of the r-f transmission line between antenna and receiver was important only in that it reduced the signal amplitude. More recently, as parametric amplifiers and masers† have brought receiver noise levels to the equivalent of a few degrees or a few hundred degrees Kelvin, the radiated noise coupled into the receiver by lossy elements in the r-f signal path has become an important component of radar input noise.

In later chapters, the effects of nonthermal noise from unwanted echoes, or clutter, will be discussed.‡ The characteristics of this type of noise are so variable that they cannot be included in a general discussion of detection theory, but must be considered in connection with specific

---

* See J. F. Reintjes and G. T. Coate, *Principles of Radar*, 3rd ed. (New York: McGraw-Hill Book Company, 1952), pp. 4–11.

† See, for instance, L. S. Nergaard, "Amplification—Modern Trends, Techniques and Problems," *RCA Review*, **21**, No. 4 (Dec. 1960), pp. 485–507.

‡ For a discussion of man-made interference, see Robert J. Schlesinger, *Principles of Electronic Warfare* (Englewood Cliffs, N. J.: Prentice-Hall, Inc., 1961).

radar applications or techniques. In a few limited cases, the random variation of electronic noise or clutter will permit the analysis to be carried
out as though the noise were of the thermal type. The variable portion of
the noise power is then used in place of thermal noise power in calculating
signal-to-noise ratios. Before the detection theory is applied, however, the
frequency and amplitude distributions of the interfering signal must be
scrutinized with care to assure that they correspond closely to the assumed
distributions of thermal noise.

Detection by the Human Operator

Visual observation of echo signals has always been the most common
means of detecting radar targets. Typical radar displays are shown in Fig.
1.4. Most search-type displays utilize intensity modulation of the cathode-
ray-tube beam to represent echo signal amplitude, reserving the two deflection axes to portray angle or range coordinates (or, in some cases, radial
velocity combined with angle or range). Since radar, even without velocity
resolution, is a three-coordinate system, there is always a problem in presenting data on a flat cathode-ray-tube screen. After two of the coordinates
have been chosen for direct display, the third coordinate remains as a
source of confusion. In many cases, signals and noise from all radar resolution elements of this unused coordinate must be "collapsed" or projected
onto the corresponding elements of the two-dimensional display. Thus, if
a series of narrow beams is stacked up vertically for elevation coverage in
a search radar, their outputs may be combined by video mixing and
presented on a conventional PPI display tube as the radar scans in azimuth
[Fig. 1.4(a)]. Each element of the resulting display will present the sum
of several noise voltages which were independently generated in the receivers of each beam. The detection loss from this type of signal processing
may partly offset the gain which was achieved by using separate, highly
directive patterns in receiving from different elevation angles. The use of
a single, broad beam to cover all elevation angles makes it possible to display the radar data without collapsing several receiver outputs, but sub-
stitutes a low-gain receiving antenna which will produce a lower signal-
to-noise ratio in the single receiver. The broad beam also precludes the
possibility of resolving two targets which have the same range and azimuth
but different elevations, which can be separated on supplementary displays when the stacked-beam antenna is used.

Displays intended for detection and measurement of targets within a
single antenna beam position (or with a slowly varying position) often
use a constant cathode-ray intensity and indicate signal amplitude by
modulating the beam from its regular position (see the A, J, K, and R
patterns of Fig. 1.4). The actual signal amplitude may be judged more

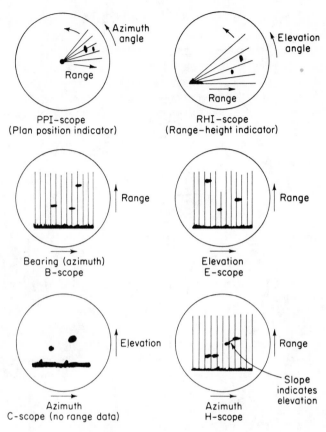

Figure 1.4 (a) Search radar displays.

accurately by the operator on these displays than on intensity-modulated types, but there remains only one deflection coordinate for indication of target position. This coordinate is usually used for range data, with both angular coordinates suppressed, time shared, or collapsed. Such displays are poorly suited to search of an extended area. For purposes of target detection, studies have shown that the two types are nearly equivalent when used properly.* The combination of the operator's perception and the persistence of the cathode-ray-tube screen provides integration of the signal over a period in the order of a second. Detection is based upon the ratio of received signal energy to noise power density, as measured by the

---

* See Lamont V. Blake, "Interim Report on Basic Pulse-Radar Maximum Range Calculation," Naval Res. Lab. Memo. Report 1106 (Nov. 1960).

See also Bibliography, Appendix B, for reference to more recent publications of related material.

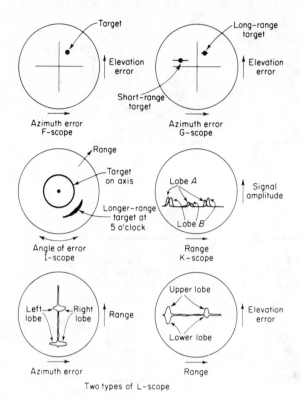

Two types of L-scope

Figure 1.4(b) Angle tracking displays.

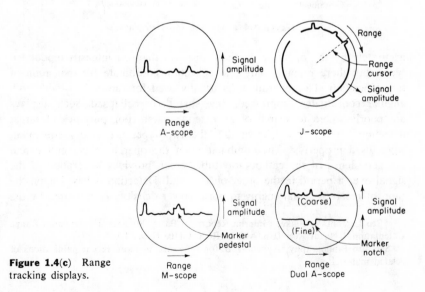

Figure 1.4(c) Range tracking displays.

total light output or the integrated deflection of the sweep from its normal pattern. Various attempts have been made to produce special displays with enhanced detection performance, by using nonlinear intensity or position modulation, but none has demonstrated any improvement over the original *A* or PPI displays. The comparison of experimental results, the basic displays being used with the optimum performance predicted by detection theory, shows that little room is left for improvement, and that future efforts along these lines cannot prove fruitful.

Automatic Target Detection

The uncertain and sometimes erratic performance of human operators as target-detection devices has led to development of many types of automatic detectors. These devices are seldom better than an alert and properly trained human, but they have the advantage of operating in a consistent and predictable fashion over extended periods of time. They also provide for more rapid transfer of data into the automatic computers which are often used with radar systems. Some of the more refined automatic detection schemes are able to adapt to a variety of input conditions, such as the presence of excess input noise, jamming, or clutter, without becoming saturated or signaling the presence of spurious targets.

A rudimentary target detector takes the form of a single "range gate" which samples the echoes and noise within a specific range delay interval extending over one or more pulse lengths (see Fig. 1.5). The receiver output at this range is stretched and averaged in a simple integrator, which may take the form of a single resistor and capacitor whose time constant is equal to several repetition periods. The output of the R-C integrator is applied to a biased diode or "threshold," which produces an output signal when the receiver output, averaged over the several repetition

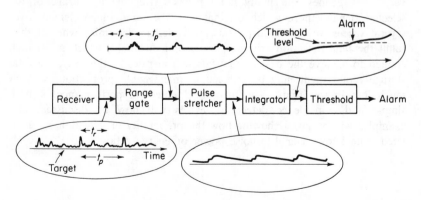

Figure 1.5 Simple automatic detector.

periods, exceeds a predetermined level. If the threshold voltage level is adjusted to lie well above the rms noise output of the receiver, false alarms caused by thermal noise may be held to any desired low rate. The use of high threshold settings will also result in failure to note the presence of actual targets when their signals are relatively weak, and hence the probability of detection will be a function both of signal-to-noise ratio and of threshold setting or false-alarm probability. The exact relationship between these parameters will be derived and discussed in Section 1.2.

The simple circuit described above may be used as an automatic alarm feature in a search radar where the operator may not be continuously alert, and where all targets of interest must pass through the chosen range interval. However, weak targets may pass through unnoticed unless the range of the automatic alarm gate is set well within the maximum detection range of the radar. The use of two or more range gates can overcome this limitation, when combined with a human operator who can take over after the automatic alarm has sounded. Fully automatic detection circuitry requires that essentially all range elements in the radar repetition period be instrumented with integrators and detection gates, and this requirement leads to considerable complication of equipment. It becomes very difficult to maintain all threshold settings at equal levels, and to hold them at the desired level with respect to varying receiver noise output. This has led to the development of a number of schemes for storing and integrating signals from all range elements, and for applying the integrated signals sequentially to a single threshold for automatic detection. The storage may take the form of acoustic delay lines or magnetic drums, or the signal and noise amplitudes may be reduced to digital form and stored in computer-style memory devices, such as cores or storage tubes. The digital approach is often used in a two-step or "coincidence" detection system. Here, signals are first screened by a threshold whose level is low enough to pass very weak echoes and many noise peaks. After digital storage of the selected signals, those which originated in the same range element over several successive repetition periods are added to determine whether they fulfill the final detection criterion, or second threshold, which is set high enough to achieve the desired low false-alarm rate. In this way, the total storage requirement may be reduced to manageable proportions without much loss in sensitivity. The theoretical detection performance of both simple and coincidence detection devices will now be discussed, and examples will be given showing how the probability of detection is calculated from known signal-to-noise power or energy ratios.

1.2 DETECTION OF A SINGLE RECEIVED PULSE

The foundation of radar detection theory was provided by the work of D. O. North at the RCA Laboratories in 1943, by S. O. Rice at the Bell Telephone Laboratories in 1944–45, and by J. I. Marcum and Peter Swerling of the RAND Corporation in 1947 to 1954. Their results will be used as a basis for discussing the probability of detection of a single sample of radar signal embedded in a background of random noise, and for extending this theory to the detection of groups of signal and noise pulses. More recent techniques in coincidence detection procedures will be covered by a summary of results originally derived by Mischa Schwartz,* who showed that such detection circuits can closely approach the ideal video integrator performance predicted by Marcum.

Results for a Single Sample of Noise Alone

One of the problems originally discussed by North concerned the probability that a single sample of random noise, or of sinusoidal signal added to random noise, would exceed a given threshold level after envelope detection. He assumed that the random noise was confined by a filter (such as an i-f amplifier) to a band of frequencies which was narrow compared to the center frequency of the filter. The resulting probability distribution of instantaneous voltage V applied to the envelope detector was defined by the Gaussian distribution

$$dP_v = \frac{1}{\sqrt{2\pi\psi_o}} \exp\left(\frac{-V^2}{2\psi_o}\right) dV \qquad \text{(1.1)}$$

Here, dP_v represents the probability that the instantaneous i-f amplitude lies between V and $V + dV$, and ψ_o represents the mean-square i-f output voltage of the noise alone, normalized to a circuit impedance of one ohm. The value of ψ_o is given by the effective input noise power (designated N in chapters to follow) times the power gain of the receiver.

After envelope detection, the video signal applied to the threshold is given by the Rayleigh distribution

$$dP_c = \frac{E_n}{\psi_o} \exp\left(\frac{-E_n^2}{2\psi_o}\right) dE_n \qquad (E_n \geq 0) \qquad \text{(1.2)}$$

The probability that the envelope of V lies between E_n and $E_n + dE_n$ is

---

* Schwartz's work was a continuation of studies which he started at Sperry Gyroscope Company in 1947. A group there, headed by E. Barlow, performed analyses which paralleled and agreed with those of Marcum.

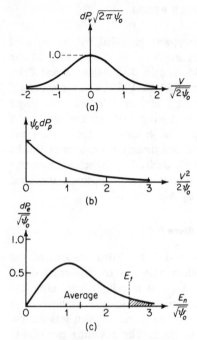

Figure 1.6 Probability distributions of noise: (a) Gaussian distribution of i-f noise voltage; (b) exponential distribution of i-f noise power; and (c) Rayleigh distribution of detected envelope.

denoted by dP_e, and it is assumed that the output of the detector follows the i-f envelope linearly without changing its amplitude. For other detector characteristics, the actual voltage applied to the threshold will be a single-valued function of E_n as given above, and appropriate values of threshold level can be found to give detection performance identical to that of the system using the lossless linear detector. Figure 1.6 shows the Gaussian distribution of i-f noise voltage V, the exponential distribution of i-f noise power, and the Rayleigh distribution of the detected envelope voltage E_n.

The probability p_n that a single sample of detected noise will exceed a given threshold level E_t is given by the area under the curve to the right of E_t in Fig. 1.6(c), which is

$$p_n = \int_{E_t}^{\infty} \frac{E_n}{\psi_o} \exp\left(\frac{-E_n^2}{2\psi_o}\right) dE_n$$

$$= \exp\left(\frac{-E_t^2}{2\psi_o}\right) \tag{1.3}$$

This relationship* permits us to calculate the setting of the threshold level E_t such that the desired "single-pulse false-alarm probability" p_n is obtained for a given mean-square noise power at the detector. An equivalent threshold level can be established for any arbitrary detector characteristic by finding the actual detector output voltage which corresponds to a peak i-f amplitude E_t, or a normalized short-term i-f power level $E_t^2/2$. As used here, the short-term period refers to the duration of a single noise sample, or approximately $1/B$ sec, where B is the i-f bandwidth. The value of ψ_o

---

* This is the same relationship given by E. M. Purcell on page 36 of L. N. Ridenour's *Radar System Engineering* (New York: McGraw-Hill Book Company, 1947). His short-term noise power P is equal to $E_t^2/2$, and his mean-square power P_o is our ψ_o.

is taken over a large number of such samples. A tabulation of threshold levels, in terms of both short-term i-f power and peak i-f voltage is given in Table 1.1 for various single-pulse false-alarm probabilities.

Table 1.1 FALSE ALARM PROBABILITY p_n VS. THRESHOLD SETTING
FOR CASE OF LINEAR SECOND DETECTOR

p_n	Normalized mean square threshold power $E_t^2/2\psi_o = \log(1/p_n)$	Normalized peak threshold voltage $E_t/\sqrt{\psi_o} = \sqrt{2\log(1/p_n)}$
1.0	0.0	0.0
0.5	0.69	1.17
0.2	1.61	1.79
0.1	2.30	2.15
0.05	2.99	2.45
0.02	3.91	2.80
0.01	4.61	3.04
0.005	5.30	3.26
0.002	6.22	3.49
0.001	6.91	3.71
0.0005	7.60	3.90
0.0002	8.52	4.13
0.0001	9.21	4.29
10^{-5}	11.5	4.80
10^{-6}	13.8	5.25
10^{-7}	16.1	5.69
10^{-8}	18.4	6.07
10^{-9}	20.7	6.43
10^{-10}	23.0	6.80
10^{-11}	25.3	7.11
10^{-12}	27.6	7.43
10^{-13}	29.9	7.73
10^{-14}	32.2	8.01
10^{-15}	34.5	8.31
10^{-16}	36.8	8.57

Results for One Sample of Signal-Plus-Noise

The presence of a steady sinusoidal i-f signal of peak amplitude E_s, added to the narrow-band noise, modifies the Rayleigh distribution at the detector output and produces the distributions shown in Fig. 1.7. The expression derived by North for the new distribution is

$$dP_e = \frac{E_n}{\psi_o} \exp\left(\frac{-E_n^2 - E_s^2}{2\psi_o}\right) I_o\left(\frac{E_n E_s}{\psi_o}\right) dE_n \qquad \textbf{(1.4)}$$

In the above expression, I_o is the Bessel function of the first kind with

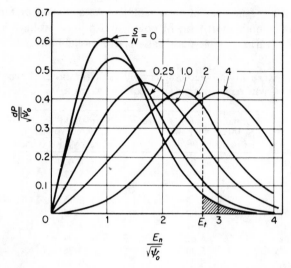

Figure 1.7 Probability distributions of envelope of signal plus noise.

imaginary argument, and E_n is now the amplitude of the envelope of signal-plus-noise. The quantity $E_s^2/2$ is the mean-square i-f signal voltage at the detector, equal to the received signal power (designated S in chapters to follow) multiplied by the power gain of the receiver. Hence, the ratio $E_s^2/2\psi_o$ is equal to the ratio S/N as calculated from the radar range equation. North showed that the maximum value of this ratio $E_s^2/2\psi_o$ could be equated to the ratio of the total received signal energy E to the noise power per unit band width, in the special case of the "matched filter" system. For instance, in the case of a single rectangular pulse of length τ and received power level S, the energy received is simply $E = S\tau$. The matched filter in this case has an amplitude vs. frequency response of the form (sin $x)/x$, where the parameter x represents $\pi\tau$ times the frequency measured relative to the carrier or center intermediate frequency. The equivalent noise bandwidth of this filter is exactly $B_t = 1/\tau$, so that the receiver noise level is given by $N = N_o B_t = N_o/\tau$. In this case, N_o is the thermal noise power per unit bandwidth measured over positive frequencies only. Thus, North's energy ratio E/N_o, which he designated λ, is exactly equal to the signal-to-noise ratio S/N for the matched-filter case with a single pulse.

For a train of pulses, the response of the matched filter follows the envelope of the (sin $x)/x$ curve, but consists of narrow response bands at intervals equal to the repetition rate of the pulses. The total noise bandwidth is reduced by the factor n, equal to the number of pulses observed in the train. The output signal-to-noise ratio $E_s^2/2\psi_o$ is now just n times the value for a single pulse. The same result can be found by considering that the total signal energy has increased to $E = nS\tau$, whereas the spectral density of noise remains unchanged.

More recent authors (Woodward, Siebert, Manasse, and Skolnik, for instance) have adopted the use of the energy ratio, but they use a value equal to twice North's λ. To avoid confusion with the symbols for wave length and range, this doubled energy ratio $R = 2\lambda = 2E/N_o$ will be represented by the script $\mathcal{R}$ throughout this book. Where pulsed radar is discussed, the expressions $2S/N$ or $2nS/N$ are preferred, since they represent the values actually observed at the output of the receiver, which is only approximately matched to the pulse spectrum.

The "single-pulse detection probability" p_d is given by the area under the distribution curve to the right of E_t in Fig. 1.7, which is

$$p_d = \int_{E_t}^{\infty} dP = \frac{1}{\psi_o} \int_{E_t}^{\infty} E_n \exp\left(\frac{-E_n^2 - E_s^2}{2\psi_o}\right) I_o\left(\frac{E_n E_s}{\psi_o}\right) dE_n \qquad \textbf{(1.5)}$$

An approximate evaluation of this integral was made by North* for values of p_d above 50 per cent, using the equation

$$p_d \cong \frac{1}{2}\left[1 + \text{erf}\left(\sqrt{\frac{1}{2} + \frac{E_s^2}{2\psi_o}} - \frac{E_t}{\sqrt{2\psi_o}}\right)\right]$$

$$= \frac{1}{2}\left[1 + \text{erf}\left(\sqrt{\frac{1}{2} + \frac{S}{N}} - \sqrt{\log\left(\frac{1}{p_n}\right)}\right)\right] \qquad \textbf{(1.6)}$$

The error function erf (x) is defined by

$$\text{erf}(x) = \frac{1}{\sqrt{\pi}} \int_{-x}^{x} e^{-y^2}\, dy$$

Thus the approximate value of p_d may be found directly from the known value of S/N and the required value of p_n or threshold setting, by using the tabulated values of the error function available in various mathematical handbooks.[†] Rice, using numerical integration methods derived by Bennett, evaluated Eq (1.5) to obtain a family of curves[‡] for the probability that the detected envelope would exceed the threshold level E_t. His plots were in terms of a parameter $(E_t - E_s)/\sqrt{\psi_o}$. To make them more convenient

---

* See Mischa Schwartz, *Information Transmission, Modulation and Noise* (New York: McGraw-Hill Book Company, 1959), p. 408. Schwartz uses a similar expression, omitting the term $\frac{1}{2}$ under the radical.

† Richard S. Burrington, *Handbook of Mathematical Tables and Formulas*, 3rd ed. (Sandusky, Ohio: Handbook Publishers, 1949), p. 273. Here the tabulation of $(1 + \alpha)/2$ gives p_d directly for $x/\sqrt{2} = \sqrt{\frac{1}{2} + (S/N)} - \sqrt{\log(1/p_n)}$.

‡ See S. O. Rice, "Mathematical Analysis of Random Noise, Parts III and IV," *Bell System Tech. Journal*, **24**, No. 1 (Jan. 1945), p. 103.

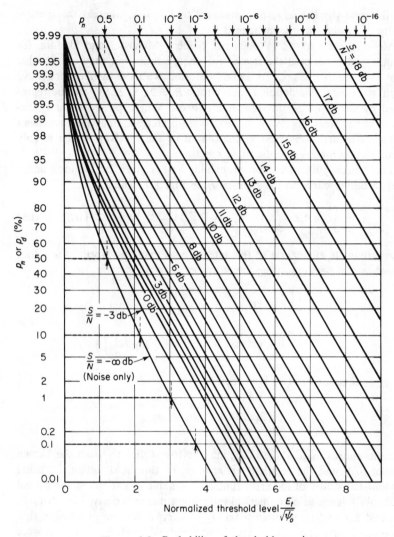

Figure 1.8 Probability of threshold crossing.

for radar analysis, they have been rescaled and replotted as in Fig. 1.8, to give the probability of threshold crossing p_n or p_d as a function of normalized threshold level $E_t/\sqrt{\psi_o}$ and signal-to-noise ratio. The curve for noise alone represents a plot of p_n versus threshold setting, whereas the remaining curves represent detection probability p_d for various signal-to-noise ratios. The threshold levels corresponding to different values of p_n are shown along the top scale of Fig. 1.8. When the threshold level is ex-

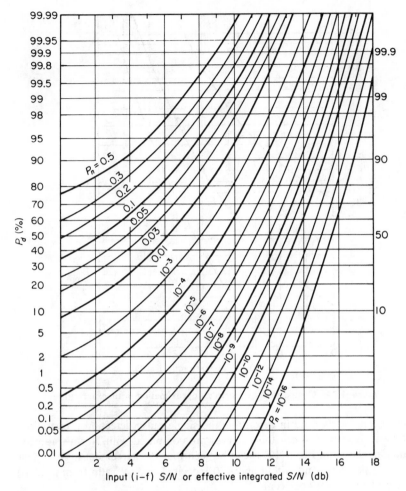

Figure 1.9 Probability of detection vs. S/N ratio.

pressed in terms of i-f noise voltage, as in this figure, the use of a lossless envelope detector is assumed. For any other detector characteristic, the corresponding threshold level will be the actual detector output for an instantaneous i-f power level E_t^2/ψ_o.

The rescaled Rice curves may now be converted to the form shown in Fig. 1.9, where the detection probability is plotted directly as a function of signal-to-noise ratio for fixed values of p_n corresponding to particular but unspecified threshold settings. In this form, the curves are entirely independent of the detector characteristics. These are the curves which

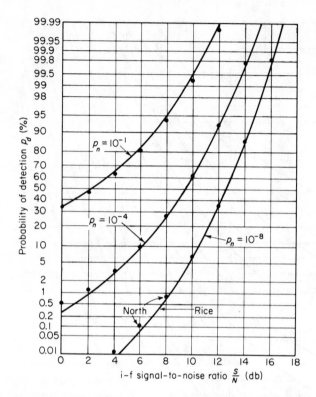

Figure 1.10 Comparison of North's approximation with exact curves.

have been widely used in describing the single-pulse detection probabilities of radar systems.* When these curves are compared with North's approximation, shown in Fig. 1.10, it can be seen that the agreement is good near $p_d = 50$ per cent, and within about 0.1 decibel in S/N ratio from $p_d = 10$ per cent to $p_d = 99$ per cent.

False-Alarm Time

We have defined the false-alarm probability p_n as the probability that a single sample of noise will exceed the detection threshold. In a receiver which operates continuously, with bandwidth B, there will be just B in-

---

* W. M. Hall, "Prediction of Pulse Radar Performance," *Proc. IRE*, **44**, No. 2 (Feb. 1956), pp. 224-31.

Mischa Schwartz, "A Coincidence Procedure for Signal Detection," *Trans. IRE*, Vol. IT-2, No. 4 (Dec. 1956), pp. 135-39.

W. M. Siebert, "A Radar Detection Philosophy," *Trans. IRE*, Vol. IT-2, No. 3 (Sept. 1956), pp. 204-21.

dependent samples of noise per second applied to the threshold, and the probability p_n will be related to the total number of alarms in a time t by the relationship

$$p_n = \lim_{t \to \infty} \frac{n_a}{Bt}$$

where n_a is the number of alarms. From this, we may define an "average false-alarm time" $\overline{t_{fa}}$ as

$$\overline{t_{fa}} \equiv \lim_{t \to \infty} \frac{t}{n_a} = \frac{1}{p_n B}$$

In cases where the receiver or detection circuit is turned off for a portion of the repetition period t_p, we may find the number of independent noise samples per repetition period as $\eta = t_g B$, where t_g is the portion of the period during which the noise reaches the threshold circuit. The number of false-alarm opportunities per second will be reduced from B to ηf_r, and the false-alarm time will be increased to

$$\overline{t_{fa}} = \frac{1}{p_n \eta f_r}$$

A slightly different definition of false-alarm time was used by Marcum.* He defined t_{fa} as the time in which the probability of *at least one* false alarm was equal to one-half. The difference in definitions was pointed out by Hollis,† who showed that Marcum's t_{fa} is about 69 per cent of the average false-alarm time $\overline{t_{fa}}$ defined above. He also tabulated the following probabilities of different numbers of false alarms occurring during each time interval (for low values of p_n)

Number of false alarms	Probability during $\overline{t_{fa}}$	Probability during $0.69\overline{t_{fa}}$
0	0.368	0.500
1	0.368	0.346
2	0.184	0.120
3	0.061	0.028
4	0.015	0.005

The "false-alarm number" used by Marcum, designated n_{fa} here (and n in the original paper), is given by

* J. I. Marcum, "A Statistical Theory of Target Detection by Pulsed Radar," RAND Corp. Research Memo, RM-754 (Dec. 1, 1947).

† Richard Hollis, "False Alarm Time in Pulse Radar," *Proc. IRE*, **42**, No. 7 (July 1954), p. 1189.

$$n_{\text{fa}} = \frac{1}{1.45 p_n}$$

This number represents the total number of opportunities for a false alarm during Marcum's t_{fa}. In our later discussion of integration and collapsing losses, we shall use a similar quantity defined by

$$n_f = \frac{1}{p_n} = \eta f_r \overline{t_{\text{fa}}}$$

In most cases, Marcum's results for a given value of n_{fa} will be identical to those for the same value of n_f, since the two quantities differ by only 45 per cent, and since the detection results are only slightly dependent upon the false-alarm number.

Matching Bandwidth to Pulse Shape

In his basic report, North considered the problem of matching the predetector filter response (normally that of the i-f amplifier) to the spectrum of the received signal. As stated earlier, the peak value of the ratio $E_s^2/2\psi_o$, which appears in Eqs. (1.5) and (1.6), can only be equated to the signal-to-noise ratio S/N for the case where a matched filter is used. The characteristic of this filter is a frequency response which is the complex conjugate of the spectrum of the echo signal received by the radar. When rectangular pulses are used, this ideal filter response is of the form $(\sin x)/x$, with an equivalent noise bandwidth $B = 1/\tau$. The resulting i-f signal envelope is an isosceles triangle whose base is 2τ in width and whose peak amplitude, normalized to the mean-square noise level ψ_o, is $E_s^2/2\psi_o$. If a rectangular filter is used (approximating the common, stagger-tuned i-f amplifier response, with flat top and steep sides), the optimum bandwidth derived by North is $B = 1.4/\tau$. Figure 1.11 shows his results for this type of filter, for the synchronously tuned filter, and for the single-pole filter, in terms of $U = E_s^2/2\psi_o$.

In the case of the synchronously tuned filter, it is assumed that the amplifier consists of n_i identical i-f stages, each with bandwidth B_1. The over-all bandwidth is given by

$$B \cong B_1 \sqrt{\frac{\log 2}{n_i}} \qquad\qquad \textbf{(1.7)}$$

The optimum value of noise bandwidth in this case is $B = 0.8/\tau$, which gives a value of $E_s^2/2\psi_o$ about one-half decibel lower than that of the matched filter. This loss, representing the reduction in peak value of signal-to-noise ratio relative to the matched filter, will be called the "matching

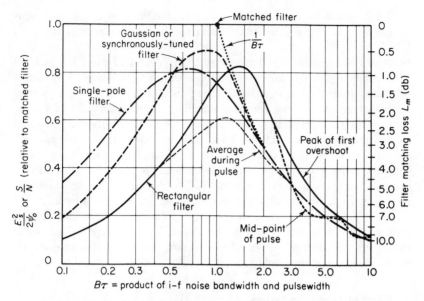

Figure 1.11 Performance of practical filters as compared to matched filter.

loss" and designated by L_m. (It is the reciprocal of North's parameter U.) We have added to Fig. 1.11 the curve derived by Schwartz* for a single-pole filter, plotted in terms of the effective i-f noise bandwidth. In this case, the optimum bandwidth is $0.63/\tau$, and the loss is 0.9 db.

In the figure, three curves are plotted for the case of the rectangular filter. The solid line gives the value at the time of the first overshoot of the pulse, which aproaches a level 0.7 db above the curve for $U = 1/B\tau$, for large values of $B\tau$. The curve given by North separates from this curve near $B\tau = 2.5$ and oscillates about the value $U = 1/B\tau$ as $B\tau$ is increased. This curve gives the value at the center of the output pulse. The third curve represents the average value during the period of length τ centered on the output pulse, and lies below the other curves in the region $0.4 < B\tau < 3.0$. This curve is more representative of the output viewed by a human operator on an oscilloscope display, since the eye tends to average the signal rather than to respond to a peak value. It is this curve, centered near $B\tau = 1.2$, which should be compared with the results of experiments in operator detection. It is interesting to note that the matched filter is not as efficient when the criterion is average output, its average being only 0.585 times its peak, compared to 0.613 for the rectangular filter with

* Mischa Schwartz, *Information Transmission, Modulation and Noise* (New York: McGraw-Hill Book Company, 1959), p. 289.

$B\tau = 1.2$. In the region of low $B\tau$, all curves tend toward $U = B\tau$, whereas at high $B\tau$ all but the first overshoot curve for the rectangular filter approach $U = 1/B\tau$.

It can be concluded that the detectability curves of Figs. 1.8 and 1.9 are directly applicable to threshold-type detection schemes, S/N ratios calculated from the radar equation being used, provided that a loss term L_m is included to account for filter matching. If the bandwidth of the receiver is $1.2/\tau$, the optimum for operator detection, the loss in peak S/N applied to a threshold is 0.9 db. This loss is accounted for, in this case, by using the value of S/N calculated from the actual bandwidth, but must be included as a separate factor when calculations are based on pulse energy. In cases where a synchronously tuned i-f filter is used, the noise bandwidth should be set to about $0.8/\tau$ and the threshold detection performance will be about 0.3 db better than for the rectangular filter. When detection is performed by a human operator, the loss will be near the 2.2 db value shown for the average output of the rectangular filter. This subject will be discussed more thoroughly in Chapter 6.

Effect of Target Fluctuation

The detection of fluctuating targets has been studied by Swerling* and by Schwartz,† whose results describe the performance of radar systems in detection of most actual targets. The so-called Rayleigh-distributed target has a cross-section distribution that is exponential.

$$dP_\sigma = \frac{1}{\sigma} \exp \frac{-\sigma}{\sigma} d\sigma \qquad (1.8)$$

The echo voltage follows the Rayleigh distribution. The usual measure of target size is the "average cross section" denoted by $\bar{\sigma}$. The corresponding average signal-to-noise ratio $\bar{S}/N$ is computed by using $\bar{\sigma}$ in the radar equation. The results obtained by Swerling for the single-pulse detection probability in this case are summarized in Fig. 1.12, where p_d is plotted against $\bar{S}/N$ for various false-alarm probabilities. Comparing this figure with the curves of Fig. 1.9 for steady signals, we see that the results are the same for $p_d = 33$ per cent, and that they remain closer to that p_d value at higher and lower $\bar{S}/N$ ratios than do the steady-signal curves. It becomes very difficult to achieve single-pulse detection probabilities above 90 per cent for the fluctuating target. On the other hand, the very low detection

---

* Peter Swerling, "Probability of Detection for Fluctuating Targets," RAND Corp. Research Memo RM-1217 (March 17, 1954).

† Mischa Schwartz, "Effects of Signal Fluctuation on the Detection of Pulse Signals in Noise," *Trans. IRE*, Vol. IT-2, No. 2 (June 1956), pp. 66-71.

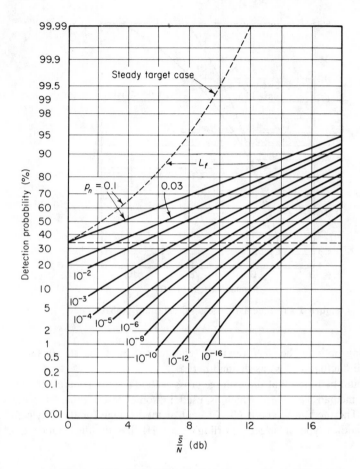

Figure 1.12 Probability of detection for fluctuating target.

probabilities which characterized the steady target at low S/N are enhanced considerably by the upward fluctuations of the Rayleigh target. A plot showing the difference between the two curves, for an arbitrary value of p_n, shows the "fluctuation loss" L_f, and is given in Fig. 1.13. This loss is the amount by which the average signal-to-noise ratio $\overline{S}/N$ must exceed the steady S/N to achieve a given probability of detection. It is applicable for a wide range of false-alarm probabilities centered on $p_n = 10^{-8}$. A similar curve, used by Hall,[*] showed as a "loss" the total change in signal-to-noise ratio required to reach detection probabilities other than 50 per

---

[*] W. M. Hall, "Prediction of Pulse Radar Performance," *Proc. IRE*, **44**, No. 2 (Feb. 1956), Fig. 7.

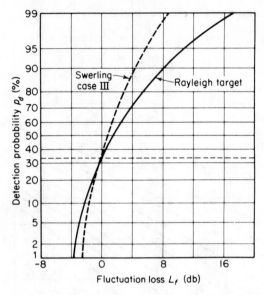

Figure 1.13 Fluctuation loss vs. detection probability.

cent, owing both to the S/N versus p_d characteristics of the steady target and to the fluctuation loss as defined here.

It should be noted that the average cross section $\bar{\sigma}$ is, in general, greater than the median value σ_{50} which is used in some descriptions of radar targets.* The median is about 69 per cent of the average for the Rayleigh target. Another type of target, Swerling's Case III, has the distribution defined by

$$dP_\sigma = \frac{4\sigma}{\bar{\sigma}^2}\exp\left(\frac{-2\sigma}{\bar{\sigma}}\right) d\sigma \qquad \textbf{(1.9)}$$

In this case, the fluctuation loss is somewhat less than that applicable to the Rayleigh target, as shown by the broken curve on Fig. 1.13. The median is 82 per cent of $\bar{\sigma}$ and the target may be described physically as a single large reflector combined with other, smaller objects which produce constructive and destructive interference with the main source. The Rayleigh target, on the other hand, is assumed to be composed of a large number of reflecting elements of comparable size (see Chapter 3).

* See Lamont V. Blake, "Interim Report on Basic Pulse-Radar Maximum Range Calculation," Naval Res. Lab. Memo Report 1106 (Nov. 1960).

Examples of Detection Calculation

As an initial example, assume that the threshold level is to be established at the output of an envelope detector of unknown characteristics, which is located at the output of an i-f amplifier with an output noise level of one volt rms. A false-alarm probability $p_n = 10^{-8}$ is desired. From Table 1.1 or Fig. 1.8, the threshold setting must be such that an instantaneous peak i-f voltage level of 6.07 v will just cause an alarm. This level may be established experimentally by substituting, in place of the i-f output, a noise-free sinusoidal signal with this peak amplitude, and adjusting the threshold at the detector output until it is just exceeded by the peak of the sinusoid. If a theoretical detector curve is available, the threshold level may be established as the detector output voltage for a peak input of 6.07 v.

To find the required S/N ratio for a given detection probability, the curves of Fig. 1.8 or Fig. 1.9 may be used. For example, assume that p_n is to remain at 10^{-8} as in the above example, and that p_d is to be 50 per cent. In Fig. 1.8, the threshold level $E_t/\psi_o = 6.07$ intersects with $p_d = 50$ per cent for an S/N ratio of 12.6 db. Use of Fig. 1.9 is more straightforward when there is no need to find the threshold setting: the curve for $p_n = 10^{-8}$ is followed to its intersection with $p_d = 50$ per cent, and the value $S/N = 12.6$ db is read directly from the abscissa. For the fluctuating target, Fig. 1.12 would be used in place of Fig. 1.9, and an average $\overline{S/N} = 14.3$ db would be found. The fluctuation loss L_f is thus $14.3 - 12.6 = 1.7$ db, which may also be read directly from Fig. 1.13 by using the curve for the Rayleigh target. Finally, in this single-pulse case, the average false-alarm time $\overline{t_{fa}}$ may be found as the reciprocal of $p_n B$, modified if necessary by the gating factor. For $B = 1.0$ mc, if it is assumed that 10 per cent of the total time between transmitted pulses is lost in recovery time and preparing for the next transmission, the false-alarm time in the above example would be the reciprocal of $0.9 \times 10^6 \times 10^{-8}$, or 111 sec.

1.3 INTEGRATION OF PULSE TRAINS

Only in unusual cases will the radar perform detection on the basis of a single received pulse, as just described. Instead, a train of several to several hundred pulses will be received from each target, and the entire train will be processed before it is decided whether or not a target is present. High detection probability can then be obtained even when the single-pulse or i-f S/N ratio is near or below unity. In an ideal processing arrangement, the energy from all the pulses would be added directly in a matched filter, prior to envelope detection. The result would be to achieve an effective, or integrated, value of signal-to-noise ratio $(S/N)_i$ equal to the single-pulse S/N multiplied by n, the total number of pulses in the train. This process is referred to as "coherent integration," because it requires that the phase of successive signals be consistent with that of a local reference sinusoid. The applications and limitations of the coherent techniques will be discussed in a later chapter on MTI and Doppler radar systems. For the present, it is sufficient to describe the resulting detection performance as equivalent to that of a single-pulse system with an i-f noise bandwidth $B_e = B/n \cong 1/n\tau$, where B is the bandwidth used for reception of the individual pulses of width τ. In calculating probability of detection, Figs. 1.8 and 1.9 are applicable, the effective value $(S/N)_i$ being used instead of S/N in the abscissa, and the false-alarm and detection probabilities for the pulse train being designated as P_n and P_d. The proper value of $(S/N)_i$ may be found equally well by using B_e in place of B in the radar equation, by assuming an equivalent pulse width $n\tau$, or by using the energy ratio computation with an observation time $t_o = n/f_r$ (see Chapter 4 for further discussion).

Video Integration

Practically all radars use video or "post-detection" integration to improve target detection capabilities without establishing a coherent reference for phase information. In the early literature on the subject[*] it was stated that the resulting power gain for purposes of detection was limited to $\sqrt{n}$ or even to $\sqrt[3]{n}$. However, the mathematical studies by Marcum and others showed that the video integrator could actually achieve a gain which approached n for cases where only a few pulses were to be processed, and which could be aproximated by $n^{0.8}$ over a wide range of conditions. If we define the integrated signal-to-noise ratio $(S/N)_i$ as that value which can be used directly in entering Figs. 1.9 and 1.12 to find target detectability, then the integration gain G_i may be written as

$$G_i \equiv \frac{(S/N)_i}{S/N} = n^\gamma \qquad \text{(1.10)}$$

The parameter γ, which represents integration efficiency, is itself a weak function of n or of S/N, with values lying between about 0.7 and 0.9 for most practical cases. A plot showing γ as a function of n, taken from

---

[*] E. M. Purcell, "The Radar Equation," Chap. 2 in *Radar System Engineering*, L. N. Ridenour, ed. (New York: McGraw-Hill Book Company, 1947).

D. O. North, "An Analysis of the Factors Which Determine Signal/Noise Discrimination in Pulsed Carrier Systems," RCA Labs. Tech. Report PTR-6C (June 25, 1943).

A. V. Haeff, "Minimum Detectable Radar Signal and Its Dependence Upon Parameters of Radar Systems," *Proc. IRE*, **34**, No. 11 (Nov. 1946), pp. 857-61.

Marcum's report,* is shown as Fig. 1.14. Only for very large n (corresponding to S/N less than unity) would γ approach a value of 0.5 as originally believed. The early results are explained by the introduction of other losses into the practical video integrator circuits and their output devices, including the human observer.

Loss Relative to Coherent Integration

An alternate way of describing the effectiveness of video integration is to express the "integration loss" L_i as representing the difference in effective power gain between the video integrator and what would have been achieved had coherent integration taken place. This loss, in decibels, is given by

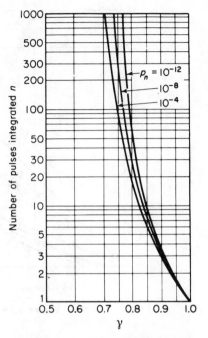

Figure 1.14 Integration efficiency vs. number of pulses integrated (from Marcum).

$$L_i = 10 \log_{10} n - 10 \log_{10} n^r = 10(1 - \gamma) \log_{10} n \qquad \textbf{(1.11)}$$

The radar equations using energy ratio or average power as parameters may be modified by inclusion of this loss to give the effective performance of practical systems which use video integration. A plot of L_i, calculated from Eq. (1.11) and Fig. 1.14 (with $p_n = 10^{-8}$) is shown in Fig. 1.15. Similar curves for specific values of P_n and P_d are given in Marcum's report. An average curve for L_i used by Hall† is shown as a dashed line in the figure, giving somewhat higher loss values than the curve based on Marcum, but the assumptions behind this curve are not stated.

The experimental curve used by Blake‡ for visual PPI detection also leads to values of L_i greater than those predicted by Eq. (1.11) at large

* J. I. Marcum, "A Statistical Theory of Target Detection by Pulsed Radar," RAND Corp. Research Memo, RM-754, Dec. 1, 1947, Marcum, Fig. 55. (The curves for $p_n = 10^{-12}$ and $p_n = 10^{-4}$, which were reversed in the original, are correct here.)

† W. M. Hall, "Prediction of Pulse Radar Performance," *Proc. IRE*, **44**, No. 2 (Feb. 1956), Fig. 6.

‡ Lamont V. Blake, "Interim Report on Basic Pulse-Radar Maximum Range Calculation," Naval Res. Lab. Memo Report 1106 (Nov. 1960), Fig. 3.

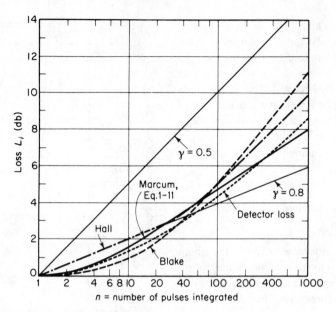

Figure 1.15   Integration loss.

n, but shows somewhat less loss for small n. This curve is also shown in the figure, along with two straight lines representing $\gamma = 0.5$ and $\gamma = 0.8$. The early experiments were conducted chiefly over the interval between $n = 10$ and $n = 1000$, where the slope of the curve parallels $\gamma = 0.5$. This explains why the presence of other, fixed, losses could be interpreted in terms of an integration gain of $\sqrt{n}$. In a more recent discussion,* the effects of integration and fluctuation loss are combined to arrive at an effective value $\gamma = 0.6$ for the case of fluctuating targets. This is a good approximation over quite a wide range of conditions.

It should be emphasized that the "integration loss" is only a relative reduction in performance, compared to what might have been achieved with coherent integration. It has been pointed out by Blake† that the loss is directly attributable to the detector which precedes the integrator. The detector loss‡ is negligible at high S/N ratios, but approaches the value of S/N itself when this ratio drops well below unity. The apparent dependence

* R. S. Raven, "The Calculation of Radar Detection Probability and Angular Resolution," Chap. 3 in *Airborne Radar* by Povejsil, Raven, and Waterman (Princeton, N.J.: D. Van Nostrand Company, Inc., 1961), p. 154.

† L. V. Blake, private communication.

‡ This is also referred to as "small-signal suppression loss" and results from the beating of noise against itself in the envelope detector. The output signal-to-noise ratio will vary as the square of S/N for $S/N \ll 1$.

of the parameter γ upon the number of pulses integrated may, therefore, be explained by the fact that most measurements and calculations are made under conditions which lead to an integrated value $(S/N)_i$ between 12 and 15 db, sufficient for reliable detection. In such cases, the input S/N ratio is near unity when 30 or 40 pulses are to be integrated, and this is where the knee of the curve for L_i is seen. To show that this explanation yields results which are consistent with Marcum's calculations and with measured results, curves for "integration loss" were prepared on the basis of a detector loss $C_a = (S + N)/S$, with a final $(S/N)_i$ equal to 13.5 db assumed. The average curve is plotted in Fig. 1.15, and is seen to agree with the curves derived as functions of n. As will be shown below, the loss may reasonably be described either as a function of n or of S/N at the detector.

As an example of calculation of integration gain, let us assume a system where 40 pulses of equal amplitude are added with uniform weighting to determine the presence or absence of a target in a given resolution element. For coherent integration, the enhancement in signal-to-noise ratio would be a factor of 40, or 16 db. Using Eq. (1.11) and a γ of 0.79 from Fig. 1.14, we find that the gain for video integration would be

$$G_i = (40)^{0.79} = 18, \quad \text{or } 12.6 \text{ db}$$

The exact gain, as mentioned earlier, would depend slightly upon P_n. The integration loss for this case would be $16 - 12.6 = 3.4$ db, which could also be read directly from Fig. 1.15 for $n = 40$. The alternate curves for integration loss when results of Hall or Blake or a constant $\gamma = 0.8$ are used differ by only a few tenths of one decibel for this case.

For the fluctuating target, assume that the single-pulse $\overline{S}/N$ is three decibels and $(S/N)_i = 15.6$ db. The probability of detection from Fig. 1.12 is 60 per cent, and the fluctuation loss L_f from Fig. 1.13 is 2.4 db. Using Raven's method, with $\gamma = 0.6$, we find that a combined loss caused by integration and fluctuation would be $0.4(16) = 6.4$ db. This compares with a combined loss of $3.4 + 2.4 = 5.8$ db, the methods described above being used. When we consider the approximate nature of the Rayleigh target assumption, this discrepancy is insignificant.

The fluctuation loss of Fig. 1.13 applies when the signals are correlated from pulse to pulse. Swerling and Schwartz also studied cases where fluctuating signals were uncorrelated from pulse to pulse. The loss for such signals is negligible for large values of n, and approaches the loss L_f shown in Fig. 1.13 when n approaches unity. Schwartz also considers the case of partially correlated signals, where the loss is intermediate between the values applicable to uncorrelated and fully correlated signals.

Collapsing Loss

Most radar integration and detection devices, including those which use cathode-ray tubes or range gates, are subject to "collapsing loss." This loss occurs when noise originating outside the radar resolution element which contains the signal is mixed, after envelope detection, with the signal and its associated noise. One typical example of this, which has been mentioned earlier, is the mixing at video of several receiver outputs, from beams at different elevation angles, prior to display on a PPI-scope. If the echo signal appears in only one beam at a time, the signal-to-noise ratio obtained in that receiver channel will be degraded at the output by the presence of noise contributed by the other receivers. A higher S/N ratio will be required in the signal channel to achieve a given performance, and this incremental increase in required signal power will define the collapsing loss L_c.

Marcum* calculated the value of this loss for various false-alarm and detection probabilities, and found that the loss was quite small for moderate amounts of extra noise. In a case where n samples of signal-plus-noise are received and mixed with m additional samples of noise in the integrator, Marcum's "collapsing ratio" ρ is given by

$$\rho \equiv \frac{m + n}{n} \qquad (1.12)$$

This represents the factor by which the number of independent noise samples applied to the integrator has been increased by the mixing or co-ordinate-collapsing process. It is apparent that the detection threshold will have to be raised if the false-alarm time is to be kept the same as it would have been if only n noise samples were received in association with the signal. The amount by which the threshold will have to be raised will depend not only on ρ but also on the number of independent opportunities for a false alarm to occur, after mixing and integration, in the false-alarm time $\overline{t_{\mathrm{fa}}}$.

$$n_{f'} = \frac{\eta f_r \overline{t_{\mathrm{fa}}}}{n} = \frac{n_f}{n} \qquad (1.13)$$

As before, η is the number of range resolution elements observed per sweep, and ηf_r represents the number of independent noise samples per second which leave a single receiver channel. The number of such samples occurring during the false-alarm time is $n_f = \eta f_r \overline{t_{\mathrm{fa}}}$, and $n_{f'}$ is the number of

* J. I. Marcum, "A Statistical Theory of Target Detection by Pulsed Radar: Mathematical Appendix," RAND Corp. Research Memo RM-753 (July 1, 1948), pp. 59-61.

independent noise samples appearing at the output of the integrator during this same period.

In cases such as video mixing of several receiver outputs, η remains constant regardless of the number of receivers, and hence $n_{f'}$ remains constant as ρ is varied. The corresponding collapsing loss, as calculated by Marcum, is shown in Fig. 1.16(a) for typical values of P_d and P_n. As in the case of integration loss, the collapsing loss varies only slightly with these probabilities. In the "constant $n_{f'}$" case, this loss is seen to approach

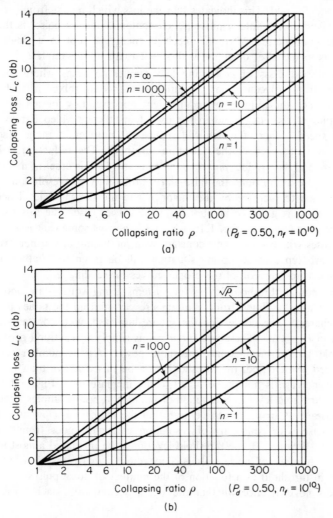

(a)

(b)

Figure 1.16 Collapsing loss vs. collapsing ratio. (a) Collapsing loss for constant $n_{f'}$. (b) Collapsing loss for constant $\rho n_{f'}$ (after Marcum).

$\sqrt{\rho}$ for very large values of n. Thus, for integration of 100 or more signal pulses, the loss would be 5 db when 10 receiver outputs were mixed, and the signal power would have to be increased by this amount to maintain the detection performance calculated for a single channel. In general, however, the collapsing loss will fall below $\sqrt{\rho}$, especially when only a few signal pulses are integrated. In fact, the collapsing loss curve for $n = 1$ will be found to correspond almost exactly to the integration loss curve of Fig. 1.15, the difference being due entirely to the difference in values of P_d and P_n used. This indicates that the loss which results from separation of received energy into n pulses, each contaminated by noise, is the same as that encountered when the received energy is kept in a single pulse, which is mixed after detection with $n - 1$ extra noise samples.

A second class of collapsing loss applies to cases where the noise samples from adjacent radar resolution elements are mixed in such a way as to reduce the number of independent opportunities for a false alarm. Examples of this are found in the use of insufficient video bandwidth or sweep speed, either of which will reduce the number of range resolution elements η observed at the detection threshold. The reduction is by the factor ρ, and the product ρn_f remains constant. In this case, in maintaining a given false-alarm time, the false-alarm probability per resolution element is permitted to increase by the factor ρ, and the threshold level need not be raised by as great an amount as in the earlier case. The loss for "constant ρn_f," is shown in Fig. 1.16(b), and is always somewhat less than for those cases where n_f remains constant with increasing ρ. Further examples of the two types of collapsing situation will be given in Chapter 4, with guidance as to selection of the proper curves for reading L_c.

The actual threshold levels for systems using square-law and linear detectors may be found from Marcum's report or from the tables prepared by Pachares.* The curves for the square-law case, with constant n_f, are shown in Fig. 1.17. The curve for $n = 1$ is simply a repetition of the data in Table 1.1 for mean-square threshold power. As the number of pulses integrated is increased, the threshold must be raised above that for $n = 1$, but the increase is not proportional to the increase in n. Figure 1.18 is a plot of threshold level as a function of number of noise samples integrated, for constant P_n or n_f. It can be seen that the increase is slow as n (or ρn in the case with collapsing of resolution elements) goes from unity to about 30. Beyond 30, the threshold level must be increased almost linearly with the number of noise pulses. Since these two figures show only the change in threshold as a function of the number of noise pulses integrated, it makes no difference whether the signal accompanies each noise pulse

---

* J. Pachares, "A Table of Bias Levels Useful in Radar Detection Problems," *Trans. IRE*, Vol. **IT-4**, No. 1 (March 1958), pp. 38–45.

(as in video integration of n pulses) or whether the signal appears only with some fraction of the noise pulses (collapsing loss).

The slope of the curves for threshold settings provides another explanation of the relatively small loss which accompanies video integration of a few pulses, or collapsing of several resolution elements. The threshold level is raised only by a factor of two when the number of noise samples goes from 1 to 10. In order to cross the threshold, the signal power (or mean-square signal voltage at the detector output, for the square-law case) need not be doubled, as the average noise level will contribute to reaching the threshold. As a result, the signal power need only be increased by about 1.7 db to maintain 50 per cent probability of detection in the presence of

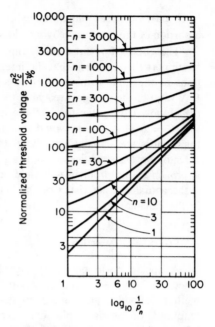

Figure 1.17 Variation in bias level with number of noise samples integrated (after Marcum).

ten noise samples. For very large numbers of noise samples, the threshold

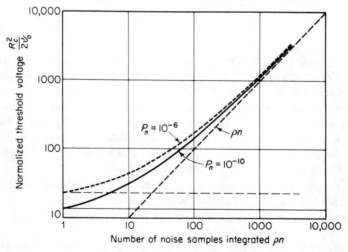

Figure 1.18 Variation of threshold level for constant P_n.

level must be raised more rapidly. The average noise level still makes up a portion of the increase, leaving a loss of about $\sqrt{\rho}$ or $5\log_{10}\rho$ in decibels.

The curves for threshold setting in the presence of noise alone tend to support the argument that the integration and collapsing losses are functions of the total number of noise samples integrated, rather than of the S/N ratio of the i-f signal reaching the envelope detector. At the same time, they follow closely the form of the detector (or small-signal suppression) loss $C_a = (S + N)/S$. It would therefore appear that the two explanations for the loss are actually equivalent statements of the effect of the envelope detector on the statistics of the integrated signal-plus noise voltage. Either approach to the problem will yield adequate results for detection analysis, if the actual characteristics of the detection system are taken into account in the calculations.

Coincidence Detection

A special form of binary signal processing known as "coincidence detection," or "sequential detection," can be used in place of the analog methods of video integration discussed above. In this binary scheme, the amplitude of the signal-plus-noise pulse for each range element of interest is applied to a threshold, and the number of resulting threshold crossings is stored for a period of n repetition periods. The number of crossings in a given range element is used as the basis for target detection. When at least k crossings take place out of n trials, a target is assumed to be present. The performance of such a detector was analyzed by Schwartz,* who found that the optimum value for k was approximately

$$k_o \cong 1.5\sqrt{n} \tag{1.14}$$

For this optimum setting, the system performs as a video integrator with $\gamma = 0.7$. For integration of 50 to 100 pulses, the loss relative to the optimum video integrator ($\gamma \cong 0.78$) is only a little over one decibel. This compares favorably with the performance of most practical video integrator circuits. Detailed calculations on the performance of this type of detection circuit are easily carried out for n less than 10, and these may be used to help evaluate the losses in search systems caused by beam shape, scan pattern, and such factors as will be discussed in later chapters.

The method of calculation is to start with the single-pulse probabilities of detection and false alarm and to assume that the same values of probability apply to each of the n pulses in the train. If the single-pulse prob-

---

* Mischa Schwartz, "A Coincidence Procedure for Signal Detection," *Trans. IRE*, Vol. IT-2, No. 4 (Dec. 1956), pp. 135-39.

ability of threshold crossing is p, then the probability of exactly j threshold crossings in n trials is given by the binomial distribution

$$P(j) = \frac{n!}{j!(n-j)!}p^j(1-p)^{n-j} \tag{1.15}$$

The probability that k or more pulses will cross the threshold and lead to an alarm will be the sum $P(k) + P(k+1) + \cdots + P(n)$. The required setting of the first threshold is calculated so that the single-pulse false-alarm probability p_n, substituted for p in Eq. (1.15), results in the desired system false-alarm probability P_n.

$$P_n = P_n(k) + P_n(k+1) + \cdots + P_n(n)$$
$$= \sum_{j=k}^{n} \frac{n!}{j!(n-j)!}p_n^j(1-p_n)^{n-j} \tag{1.16}$$

The system detection probability P_d for a given S/N is then calculated from the single-pulse p_d, as determined from Fig. 1.8 or Fig. 1.9, the same procedure being used as for system false-alarm probability.

$$P_d = \sum_{j=k}^{n} \frac{n!}{j!(n-j)!}p_d^j(1-p_d)^{n-j} \tag{1.17}$$

Examples of Coincidence Detection Calculation

An example will be given for processing of four pulses. The probabilities of exactly zero, one, two, three, and four threshold crossings when $n = 4$ are given by

$$P(0) = (1-p)^4$$
$$P(1) = 4p(1-p)^3$$
$$P(2) = 6p^2(1-p)^2$$
$$P(3) = 4p^3(1-p)$$
$$P(4) = p^4$$

The threshold criterion for an alarm, according to Eq. (1.13), should be $k_o = 3$, and when this is substituted in Eq. (1.16), we have

$$P_n = P_n(3) + P_n(4) = 4p_n^3(1-p_n) + p_n^4$$

For small values of P_n, an approximation for p_n may be found by setting $(1-p_n) \cong 1$ and $p_n^4 \ll 4p_n^3$, which gives

$$P_n \cong 4p_n^3$$

If the desired system false-alarm probability P_n is 10^{-8}, this leads to $p_n = 0.00136$, and the approximations used above are entirely justified. Figure 1.8 may now be used to find a normalized threshold setting $E_t/\sqrt{\psi_o} = 3.65$. The single-pulse detection probability as a function of S/N may now be read from Fig. 1.8 or Fig. 1.9 and used to calculate the system detection probability. Assume that we wish to find the value of S/N necessary to achieve $P_d = 90$ per cent. Using a trial and error procedure, we might start with two values of S/N and p_d, and find the corresponding values of P_d from Eq. (1.17). A convenient initial value for p_d is 50 per cent, for which S/N is 8.0 db, and P_d is

$$P_d = P_d(3) + P_d(4) = 4(0.5)^3(0.5) + (0.5)^4 = 0.31$$

For $p_d = 90$ per cent, S/N must be 10.7 db, and P_d is

$$P_d = 4(0.9)^3(0.1) + (0.9)^4 = 0.95$$

By comparing these results with Fig. 1.9, we find that the coincidence detection process has achieved a gain of 3.9 db with respect to the single-pulse case. This is equivalent to $\gamma = 0.65$. Either by interpolating or by starting from the single-pulse S/N of 14.2 db required to obtain $P_d = 90$ per cent (see Fig. 1.9) and using Eq. (1.10), we find the required S/N for the coincidence detection case to be 10.3 db.

In evaluating the effects of beam shape or scanning pattern for the four-pulse case, the method must take into account the variation in S/N and hence in p_d over the pulse train. The values of p_n will remain the same, since the noise level is not affected by the antenna beam shape. The system detection probability will be given by the following equations

$$P_d = P_d(3) + P_d(4)$$
$$P_d(3) = p_1 p_2 p_3 (1 - p_4) + p_1 p_2 p_4 (1 - p_3) + p_1 p_3 p_4 (1 - p_2)$$
$$+ p_2 p_3 p_4 (1 - p_1)$$
$$P_d(4) = p_1 p_2 p_3 p_4$$

Here, p_1, p_2, p_3, and p_4 represent the detection probabilities of the four individual pulses in the train, and these may be determined from the known beam shape and scan pattern. A further example using this method to compare different scan patterns will be given in a later chapter on search radar analysis.

Radar detection of targets is only the first step in a process which must include at least a range measurement if the device is to live up to its title as a *RA*dio *D*etection *A*nd *R*anging instrument. The general methods by which range and angular coordinates of the target are measured have not changed since World War II, and have been described in many of the references.* Only recently, however, has there been a concerted attempt to analyze with accuracy the fundamental limitations which govern these measurements. In this chapter we shall consider the basic processes used in radar measurement of target position and velocity, review the applicable theory, and see how it can be used to establish limits for operation of several types of practical radar system. In later chapters the same theory will be applied to a detailed analysis of search and tracking radar systems, and related to the practical equipment limitations to permit prediction of the performance to be expected from such systems in field operation.

THE THEORY OF
RADAR MEASUREMENT

2

2.1 MEASUREMENT OF RANGE AS TIME DELAY

The outstanting advantage of radar as compared to other means of

* See, for instance: L. N. Ridenour, ed., *Radar System Engineering* (New York: McGraw-Hill Book Company, 1947).

Merrill I. Skolnik, *Introduction to Radar Systems* (New York: McGraw-Hill Book Company, 1962).

target position measurement is its ability to make direct measurements of radial range in terms of the round-trip time delay of the transmitted signal. The techniques for measuring time delay have been developed to the point that new standards of length and time are necessary to prevent errors from arising in the basic reference system for long-range measurements. Given a recognizable feature in the transmitted wave form, and a signal which is reasonably strong relative to noise in the receiver, one can measure the delay between transmission and reception to the order of a nanosecond (10^{-9} sec, or a millimicrosecond) in many practical systems operating at ranges of thousands or millions of miles. In order to translate this delay into radial range with corresponding accuracy (about one-half foot), the velocity of light must be known or defined to within one part in 10^9 or better. This can be accomplished by changing the standard of length from the meter bar to the wave length of a particular atomic resonance line, and the standard of time to the period of the corresponding oscillation frequency. The vacuum velocity of light $c = f\lambda$ is thus defined to the precision of the atomic standard, which can be in the order of one part in 10^{11} over long periods of time.

Pulsed Ranging Systems

After a few early experiments using continuous waves for target detection, the pulsed system was adopted almost universally for accuracy and simplicity of range measurement. By operating the transmitter only during the brief interval of the pulse, the pulsed system frees the receiver from the potential sources of noise and competing signals that accompany continuous transmission from a nearby location. The sharp leading and trailing edges of the pulse wave form serve as ideal reference features for time delay measurement in wide-bandwidth systems, since the corresponding portions of the received signal are readily recognized. Much of the technology of radar has been concerned with optimizing the various components of the transmitter and receiver for processing pulsed signals without altering their shape or introducing competing noise energy. Recent developments in radar and information theory have provided the rigorous mathematical justification for long-standing practices in pulsed radar, as well as indicating new approaches to performance of accurate measurements.

From the earliest ranging systems to the present, the basic ranging process has consisted of matching the received signal against some sort of reference which is delayed by a known amount from the transmission. In search systems, the reference may be as coarse as a set of "range markers" appearing on the display scope. For more accuracy, a movable "strobe" marker may be used. In moving this marker into coincidence with the received signal, the operator turns a control which is coupled to an accurate range counter for visual or automatic range readout. In tracking systems, the corresponding techniques make use of manually positioned cursors or electronic markers on expanded ranging displays. Automatic tracking systems generate gating signals which sample the early and late portions of the received pulse and generate error signals to force the gates into coincidence with the center of gravity of the pulse. In some digital ranging circuits, the crossing of a threshold is sensed on both leading and trailing edges of the pulse, and the time delay is measured to the average of these two positions by using a counter and regularly spaced clock pulses. In this case, the reference is provided by the thresholds, which are analogous to the hairline cursors on an expanded display. In all cases, the basic process remains the same: matching of certain features of the received pulse against the reference marker, followed by accurate measurement of the time delay of the reference relative to the transmission.

Mathematically, the ranging process can be described as "correlation" of the received signal with the reference.* To illustrate this process, assume that the transmitted signal is stored for reference at the radar site. After insertion of a variable delay, the stored reference is used to multiply the received signal, the product being integrated over the entire time of observation of the target. When the delay has been adjusted to the range delay of the target, the two signals will match at all points, and the output, or "correlation function," will reach a maximum. A transmitted signal with random phase or amplitude characteristics will show low values of correlation function when the reference delay is set to any value other than the proper one. A pulse carrier signal, on the other hand, will exhibit the two periodic terms shown in Fig. 2.1, when a single target is present. One period will correspond to the pulse repetition frequency f_r, whereas the fine structure will repeat at the carrier frequency f_t. The fine structure is difficult to use in range measurement, since it is ambiguous at delay intervals equal to the r-f period, corresponding to a few centimeters in range. Furthermore, if the target is moving, the reference will have to be matched both to the range delay and to the Doppler frequency of the target if the correlation function is to reach its recognizable peak. For this reason, most practical ranging systems suppress the fine structure by envelope detection of the received signal prior to the multiplication process. The appropriate stored reference now consists of the delayed envelope of the transmitter pulse, represented by the range gate or strobe signal, which has been in use a good deal longer than the term "correlation." The envelope correlation function, shown in Fig. 2.1(d), represents the output of a single range gate matched to the pulse width of the signal, whereas the output of the "split-gate" circuit represents the derivative of this function.

* Robert J. Schlesinger, *Principles of Electronic Warfare* (Englewood Cliffs, N.J.: Prentice-Hall, Inc., 1961), pp. 38–42.

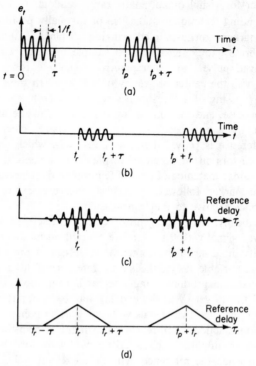

Figure 2.1 Correlation functions for pulsed transmission. (a) Transmitted signal $e_t = E_t \sin 2\pi f_t t$. (b) Delayed received signal. (c) Correlation function (complete). (d) Correlation function (envelope).

Effects of Noise

In the absence of noise, there would be no limit to the potential accuracy of the ranging systems discussed above. By expanding the scale of the display devices, amplifying the received signal, and stabilizing the thresholds or gates used in measurement, a specific point of the leading edge of the pulse, or the average of leading and trailing edge times, may be read to a very small fraction of the pulse width or rise time. The presence of noise introduces an uncertainty in the measurement, whose magnitude can be estimated in various ways. The error in measurement of a single pulse can be analyzed by assuming a bandwidth-limited rectangular pulse and noting the effect of noise upon the leading and trailing edges. If the bandwidth of the receiving system is B cycles per second, the rise time (and fall time) will be $\tau_e = 1/B$ sec.* This time is actually the time required

* Mischa Schwartz, *Information Transmission, Modulation and Noise* (New York: McGraw-Hill Book Company, 1959), pp. 43–47.

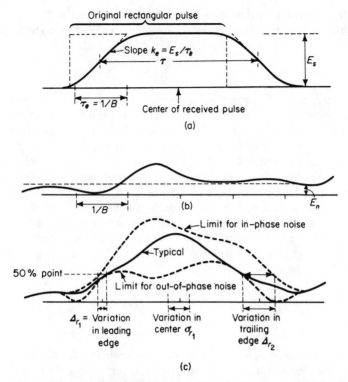

Figure 2.2 Range measurement on a bandwidth-limited rectangular pulse. (a) Pulse envelope after bandpass filter. (b) Noise envelope. (c) Noise-distorted pulse envelope.

to reach full pulse amplitude at the rate which is observed near the 50 per cent point on the pulse, where the voltage slope will be $k_e = E_s/\tau_e = E_s B$ volts per second (see Fig. 2.2). In most cases, the delay of the received pulse can be defined by averaging the leading and trailing edges as they pass this 50 per cent point, and the effect of a small additive noise term E_n will be to modify each of these crossing points by a delay $\Delta_r = E_n/k_e = E_n/(E_s B)$. For pulse widths τ greater than $1/B$, this noise will be independent from leading edge to trailing edge, so that the error in the average time delay will be (for $S/N \gg 1$)

$$\sigma_{r_1} = \frac{\Delta r}{\sqrt{2}} = \frac{1}{\sqrt{2}\, B \dfrac{E_s}{E_n}} = \frac{1}{2B\sqrt{\dfrac{S}{N}}} \qquad (2.1)$$

A more refined analysis is given by Woodward,* who expresses the

* P. M. Woodward, *Probability and Information Theory, with Applications to Radar* (New York: McGraw-Hill Book Company, 1955), p. 105.

standard deviation in time delay measurements in terms of a parameter β, which describes the spread of energy over the frequency spectrum.

$$\sigma_r = \frac{1}{\beta\sqrt{\mathcal{R}}}$$

The energy ratio $\mathcal{R}$ may be related to signal-to-noise ratio as described in the previous chapter, but the evaluation of β presents certain problems. Woodward defines β as the rms deviation of the power spectrum of the received signal relative to zero frequency, in units of radians per second. For rectangular pulses of width at least equal to $1/B$, after envelope detection has removed the carrier frequency, Skolnik* has shown that a good approximation of β can be found from $\beta^2 = 2B/\tau$. This approximation accounts for the fact that the pulse energy is not uniform over the passband when B exceeds $1/\tau$, so that the rms frequency deviation increases less rapidly than the bandwidth of the receiver alone. If we combine this expression for β with the relationship for the single-pulse energy ratio $\mathcal{R} = 2B\tau S/N$, we have

$$(\sigma_{r_1})_{\text{opt}} = \frac{1}{2B\sqrt{\dfrac{S}{N}}} \qquad (B\tau \geq 1) \tag{2.2}$$

It should be noted here that an increase in receiver bandwidth B will decrease the signal-to-noise ratio, so that σ_{r_1} is improved only by the square root of the bandwidth increase, rather than directly. In addition, the equation presumes that the approximate location of the pulse is already known, so that measurements are not made on noise alone, and that the rise and fall times of the transmitted pulse are short enough to produce energy throughout the receiver bandwidth B. If these conditions are met, the result agrees perfectly with Eq. (2.1), and shows that the ranging precision depends only upon the radar system bandwidth and the signal-to-noise power ratio.

Result of Prolonged Observation

We have discussed the precision of measurement on a single-pulse basis. Most pulsed radar systems will use the average of n separate readings taken over an observation interval t_o, where $n = f_r t_o$. The noise term in each of these readings will be independent of that in any other readings, and elementary statistics tells us that the standard deviation of the average will be $1/\sqrt{n}$ times that of each pulse.

---

* Merrill I. Skolnik, "Theoretical Accuracy of Radar Measurements," *Trans. IRE*, Vol. ANE-7, No. 4 (Dec. 1960), p. 125.

$$\sigma_r = \frac{1}{2B\sqrt{n(S/N)}} = \frac{1}{2B\sqrt{f_r t_o (S/N)}} \tag{2.3}$$

This equation shows the dependence of the ranging precision on an effective or integrated signal-to-noise ratio $n\,S/N$. This is similar to the integrated value $(S/N)_i$ used in detection calculations, and a similar detector loss will be present when the single-pulse S/N ratio is near or below unity. At low S/N, the assumption of small noise voltages and linear operation of the detectors and delay-measuring circuits will no longer be valid, and the errors in an actual system will increase above those predicted by the above equations. The operation of some of the common types of ranging circuits, such as the split-gate range tracker, will be discussed more fully in a later chapter, and the analysis of precision will be extended to S/N values near and below unity.

The form of the previous equations has been chosen deliberately to emphasize the S/N ratio of the individual pulse. It can readily be shown, however, that the results are the same as those given in the literature* which use the energy ratio $\mathcal{R} = 2E/N_o$. The single-pulse energy received by the radar is τS, and the total energy received during the observation time is $E = n\tau S = t_o f_r \tau S$. The noise power per unit bandwidth (if only the single-sided spectrum of positive frequencies is used) is $N_o = N/B$, so we may write

$$\frac{E}{N_o} = t_o f_r \tau B \frac{S}{N}$$

Substituting this in Eq. (2.3) leads to Skolnik's result.

$$\sigma_r = \left(\frac{\tau}{4BE/N_o} \right)^{1/2} \tag{2.4}$$

Range Resolution Limitations

In the above discussion of noise errors, it was assumed that only thermal noise, occurring during the pulse, contributed to measurement error. In the practical case, we must guard against the presence of additional target echoes, false echoes, or excess noise not considered in the theory given above. The property of pulse ranging systems that gives the high degree

* Merrill I. Skolnik, "Theoretical Accuracy of Radar Measurements," *Trans. IRE*, Vol. ANE-7, No. 4 (Dec. 1960), pp. 123–29.

W. M. Siebert, "A Radar Detection Philosophy," *Trans., IRE*, Vol. IT-2, No. 3 (Sept. 1956), pp. 204–21.

P. M. Woodward, *Probability and Information Theory, with Applications to Radar* (New York: McGraw-Hill Book Company, 1955).

of immunity from such errors is the range resolution provided by the near-rectangular shape of the transmitted pulse. Only signals and noise appearing within the time interval τ need be applied to the measuring circuits, once the pulse has been detected and its approximate position noted. Targets separated by a delay greater than τ can be completely resolved; that is, their positions may be measured separately with small interaction. The point is sometimes made that the presence and location of two targets only slightly displaced from each other can be recognized in the noise-free case by noting the different amplitude levels in the composite received signal. This "conditional resolution" is so dependent upon relative target amplitudes and over-all signal-to-noise ratio, however, that it is not a very useful characteristic in describing the performance of practical equipment.

Following Woodward's usage of the term "resolution,"* we shall consider that the range delay resolution is proportional to the reciprocal of the frequency span of the received signal, which depends upon the bandwidth of the entire system. The system bandwidth is roughly that of the element with the narrowest bandwidth in the complete system: the half-power width of the main lobe of the transmitted spectrum ($B_t = 1/\tau$), where a wider receiver bandwidth is used to avoid pulse distortion; or the receiver bandwidth B, where the transmitted pulse spectrum contains appreciable energy beyond this band. For the often encountered case of the matched system, the pulse width $\tau = 1/B$ will be an adequate measure of the resolution in range delay.

Another consideration closely related to resolution is the ambiguity of the measurement. In a conventional pulse radar, the range data has ambiguities at time delay intervals equal to the repetition period. Targets separated by approximately this distance cannot be resolved without introducing some variation into the otherwise regular pulse train. Where the "multiple-time-around" targets are at such a range that their echoes are weak or undetectable, this limitation may not be important, but in some systems the ambiguity can lead to severe measurement problems. This subject will be covered in more detail later, but it is mentioned here to emphasize that the figure for precision cannot always be derived or used without reservation.

Extension to C-W and FM Systems

In applying Woodward's theory to various forms of amplitude- and frequency-modulated ranging systems, two factors must be considered. First, we must decide whether the range measurement uses the "fine structure" of the transmitted spectrum, containing carrier phase information, or

* P. M. Woodward, *Probability and Information Theory, with Applications to Radar* (New York: McGraw-Hill Book Company, 1955), p. 116.

only the detected envelope from which this phase information has been removed. Having determined which of these two cases applies to the system under analysis, we must evaluate the rms frequency parameter β, and use this along with the energy ratio to find the limiting precision of range measurement. Second, we must consider the presence of possible ambiguities in the data caused by the modulation or the carrier, in order to see whether the data will be useful in the case of actual target measurements. If the fine structure contains ambiguities which are not resolvable on the basis of other data, we are limited to the use of envelope information alone in range measurement.

Consider first the case of an unmodulated c-w system. It is an accepted fact that such a system has no range measurement capability whatsoever, since there are no recognizable features by which the round-trip delay between radar and target can be measured. Actually, this is true only for the envelope of the signal, which has no frequency span and hence no resolution or precision. The carrier wave form consists of a sinusoid, which furnishes means of distinguishing different portions of the cycle, but no means of identifying which cycle at the transmitter produced the echo being observed. Although the echo is ambiguous at intervals of the r-f cycle, ranging is still possible if the target range delay is short compared to the period of the r-f cycle. This is not the usual case, it being more common to find thousands or millions of wave lengths between the radar and the target with double this number of ambiguities. This fact justifies the common judgment that ranging is impossible in the absence of carrier modulation.

The next step in c-w systems might be to add a second frequency slightly displaced from the first, as suggested by Hansen in an early work.* If we use the envelope information alone (in this case, the beat note between the two echo frequencies), the system will have ambiguities at range delays equal to one-half the period of this difference frequency, which can be made arbitrarily small to extend the unambiguous range. For a given frequency separation Δf, the rms frequency of the envelope will be $\beta = \pi \Delta f$ radians per second. The corresponding ranging precision can be related to the average signal-to-noise ratio $(S/N)_{av}$ and the observation time t_o as follows

$$\sigma_r = \frac{1}{\beta \sqrt{\mathcal{R}}} = \frac{1}{\pi \, \Delta f \sqrt{2 B t_o (S/N)_{av}}} \tag{2.5}$$

For a stable system with a nonaccelerating target, the bandwidth B can

---

* W. W. Hansen, "C-W Radar Systems," Chap. 5 in *Radar System Engineering*, L. N. Ridenour, ed. (New York: McGraw-Hill Book Company, 1947), pp. 139–43.

be narrowed to match the observation time $(B = 1/t_o)$, giving $\mathcal{R} = 2(S/N)_{av}$ and

$$\sigma_r = \frac{1}{\pi \, \Delta f \sqrt{2(S/N)_{av}}} = \frac{1}{4.5 \, \Delta f \sqrt{(S/N)_{av}}} \tag{2.6}$$

Obviously, the reduction in ambiguities accomplished by using small separations Δf is obtained only at the expense of degrading the range precision. The full-scale or unambiguous range corresponds to a delay time $t_p = 1/2\Delta f$ and the precision of measurement is about one-half of this value divided by the average signal-to-noise ratio. This would explain the failure of most practical systems to use the simple, dual-frequency approach to ranging, in spite of its other advantages. If carrier phase information were to be used along with the difference-frequency phase, the accuracy would be essentially the same as that of a pure c-w system at the same carrier frequency, and the same ambiguities would be present in the fine structure. To resolve these by using the envelope data, the range would have to be limited to a few tens of wave lengths of the carrier, and the frequency separation would have to be in the order of one-tenth to one-hundredth of the carrier frequency itself.

The addition of further sidebands to the basic c-w carrier can provide better resolution of ambiguities without compromising the ranging precision of the system. Appendix D discusses some of the possible modulation systems, in terms of spectra, envelope waveforms, and frequency vs. time functions. Many practical c-w radar configurations can be devised by using these basic modulation methods, but the linear-sweep fm system is the one most commonly exploited when c-w transmission is used. In this system, the sweep recurrence rate f_r is adjusted to equal $1/t_p$, where t_p is the maximum unambiguous range delay. The slope of the sweep, which determines the total frequency excursion over the time t_p, is set by the range resolution desired. For an excursion Δf, the range delay of the target can be resolved to $\tau' = 1/\Delta f$, and measured to a precision defined by

$$\sigma_r = \frac{\sqrt{3}}{\pi \, \Delta f \sqrt{2 \, (S/N)_{av}}} = \frac{1}{2.5 \, \Delta f \sqrt{(S/N)_{av}}} \tag{2.7}$$

Pulse Compression Radar

Although their resolution and theoretical limiting precision can be calculated from spectral distributions similar to those for c-w and fm systems, the various "pulse compression" schemes deserve special consideration. The initial approach to this technique was through linear modulation of frequency during the pulse, described as "chirp" by analogy to the audio transmission of the cricket.* The spectrum for this case is almost identical to that of the c-w system using linear fm, and the resolution and precision are expressed in the same form as would be obtained if the fm sweep were repeated at intervals equal to the pulse width without interruption of the transmission. The chirp transmission has a bandwidth $\Delta f \gg 1/\tau$, corresponding to Δf of the fm c-w system. The line spacing in the spectrum is f_r whether or not the transmitter is turned off between sweeps. The primary difference lies in the equipment used to generate, transmit, and process the signals, which in the chirp system resembles the conventional pulse radar much more nearly than it does the equipment of fm c-w systems. The relatively long "off" period of the transmitter in the chirp system also provides an excellent solution to the problem of isolation between transmitter and receiver, which plagues the c-w radar systems and severely limits their practical application. As an approximation, the range delay resolution and precision can be found by considering a conventional pulse radar with an effective pulse width $\tau' = 1/\Delta f = \tau/k_c$, where $k_c = \tau \Delta f$ is known as the "compression ratio" of the system (Appendix D, Fig. D.11).

Other systems of phase and frequency modulation during the pulse have been devised to overcome some of the limitations of the linear fm method. These include various modulating wave forms, both continuous and discrete (or digital, as they are sometimes called), which have the common property of producing broad distributions of energy in the frequency domain while permitting the transmission of long pulses or pulse trains at low peak power. Further descriptive data on these modulations appear in Appendix D. In each case, the range precision and resolution may be found by using appropriate values of the rms frequency β (based on the modulation envelope rather than on the carrier itself) in Woodward's equations. Equation (2.2) gives the limiting precision, whereas the resolution in delay is roughly $1/B_t$ or $1/\Delta f$, the reciprocal of the total spread of frequencies in the transmission.†

2.2 ANGLE MEASUREMENT

Radar is only one of the many types of instruments used to measure the angle of arrival of electromagnetic waves. The techniques used in radar for this purpose have much in common with radio direction-finding equip-

---

* J. R. Klauder, A. C. Price, S. Darlington, and W. J. Albertsheim, "The Theory and Design of Chirp Radars," *Bell System Tech. Journal*, **39**, No. 4 (July 1960), pp. 745–820.

Charles E. Cook, "Pulse Compression—Key to More Efficient Radar Transmission," *Proc. IRE*, **48**, No. 3 (March 1960), pp. 310–16.

† P. M. Woodward, *Probability and Information Theory, with Applications to Radar* (New York: McGraw-Hill Book Company, 1955), p. 118.

ment and with modern air-navigation systems. As in these devices, the radar measures the signal wave front arriving at the antenna, and indicates the direction normal to this wave front as being the "angle of arrival" of the signal. The wave front consists of a surface of equal phase. The basic process used in any of the several types of direction-finding antennas involves sampling the relative phase of the signal over a given receiving aperture, and seeking out a plane surface which best fits the actual equiphase surface. Signals may be taken from all elements of a continuous aperture, or from discrete points within the aperture. The seeking process may involve either simultaneous or sequential measurement, in each case an estimate being obtained of a plane over which the apparent variation in signal phase is minimized. Unlike the purely passive direction-finders, the radar may use its transmitting pattern to modulate the echo signal, producing variations in phase and amplitude of the received wave which depend upon the position of the antenna during transmission as well as reception. This can have both beneficial and troublesome effects on measurement of target direction.

Principles of Direction Finding

A point source of signals at relatively long range from the receiving antenna will radiate spherical waves into free space, which arrive at the antenna in the form of plane wave fronts. An elementary antenna such as a dipole or loop will provide maximum output when aligned with the wave front, and will show a sharp null when set at right angles to the wave front (polarization effects will be neglected for the present). By rotating such an antenna, the location of the lobes and nulls may be established, and comparison with a fixed (or "sense") antenna will permit resolution of the 180° ambiguity in the system. Simultaneous sampling of the outputs of loop and fixed antenna permits reading directly the angular error between the loop and the wave front, and this error signal may be used in a servo system to keep the antenna aligned with the wave. Alternatively, the wave may be sampled at four fixed locations around the sense antenna, and the direction of arrival may be measured by combining the four phase readings in a simple analog computer known as a "goniometer." This approach is used in the Adcock direction-finding system, and represents the forerunner of the modern "electronic scanning" or "phased array" radar antennas.

The elementary types of direction finder referred to above are subject to two major disadvantages. First, the gain of the antenna in such systems is very low, so that the range is severely limited in the high-frequency bands. In radar, where the frequency is very high and the echo signal is relatively weak, they find no application. In the second place, the wave arriving at the antenna cannot be considered as a simple spherical surface in free space, since there will inevitably be conducting and reflecting surfaces near the receiver or near enough to the propagation path to cause distortion of the wave front. Both these limitations are overcome to some extent by using directional antennas, which provide gain for the signal when properly pointed, and also discriminate against unwanted signal reflections.

Generation and Use of Narrow Beams

A high-gain antenna must sample the arriving wave front over a considerable aperture area, combining the signal components in proper phase relationships before feeding them to the receiver. This process results in the generation of a narrow angular beam from which the signals may be received with high gain. On each side of this beam are a series of nulls and side lobes, whose location and amplitude depend upon the relative weight given to signal samples received at different points over the aperture. Directional beams may be formed in many ways: by connecting in parallel a number of coplanar dipoles, by focusing energy from a large continuous aperture onto a single feed point with lenses or parabolic reflectors, by coupling to the wave over a long path parallel to its direction of travel (end-fire array), or by combinations of these methods. The basic properties of each of these antenna types are determined by the dimensions of the aperture (or effective aperture in the case of end-fire types), relative to the wave length of the signals being received, and the weighting given to energy received at different portions of this aperture. In one common case, where the sampling is continuous over a rectangular aperture and all elements are given equal weight in the combining process, the relationships among power gain, beamwidth at the one-way half-power points of the pattern, and aperture are related as follows*

$$G = \frac{4\pi A_r}{\lambda^2} = \frac{4\pi wh}{\lambda^2}\eta_a \cong \frac{4\pi}{\theta_a \theta_e} \qquad \text{(2.8)}$$

$$A_r = wh\eta_a \qquad \text{(2.9)}$$

$$\theta_a \cong 0.88\frac{\lambda}{w} \quad \text{radians}$$

$$\qquad \text{(2.10)}$$

$$\theta_e \cong 0.88\frac{\lambda}{h} \quad \text{radians}$$

Here, G is the power gain, referred to isotropic radiation; A_r is the effective

* Samuel Silver, *Microwave Antenna Theory and Design* (New York: McGraw-Hill Book Company, 1949), pp. 4 and 181.

receiving aperture; λ is the wave length of the signal; w and h are the width and height of the antenna, measured normal to the direction of arrival; and θ_a and θ_e are the horizontal and vertical beamwidths of the antenna. The efficiency factor η_a is included to account for variations in weighting (or illumination, to use the more usual antenna terminology) over the aperture. This factor will be near unity for the near-solid reflector or for closely spaced dipole elements. Circular apertures, apertures with tapered illumination, and other variations from the rectangular case given above will have similar relationships involving additional constants near unity.* In particular, the circular aperture of diameter D with edge illumination reduced to suppress side lobes will have an aperture efficiency near 60 per cent, a gain given by $G = 6D^2/\lambda^2$, and a beamwidth $\theta \cong 1.2\lambda/D$ radians in both axes.

Direction finding with high-gain antennas is usually carried out by comparing the outputs from two or more beam positions displaced about the direction of arrival of the signal. In the simplest configuration, a single beam is swept past the target, and the amplitude of the signal delivered to the receiver is observed as it rises to and falls from maximum amplitude (see Fig. 2.3). If the returned signal strength is constant and strong relative

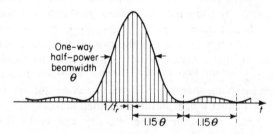

Figure 2.3 Signals received from scanning antenna.

to received and internal noise, the exact angle at which the maximum amplitude was reached can be taken as the angle of arrival in the coordinate where scanning takes place. This method is used in passive direction finding at VHF and above, and in many search radars which scan continuously in azimuth with broad vertical beams. Since the maximum of the beam may be quite broad and difficult to define, interpolation between the nulls or half-power points of the pattern will usually be preferred as a method of measurement. Location of the target in both angular coordinates is accomplished by scanning in azimuth with a pattern of many separate

* Samuel Silver, *Microwave Antenna Theory and Design* (New York: McGraw-Hill Book Company, 1949), p. 195.

beams, "stacked" in elevation, or by using a single beam which scans different elevation angles in sequence. Greater accuracy of measurement results when the beam, or beams, remains pointed near the target for prolonged periods of time, careful comparisons being made of the signals received from beam positions distributed around the actual angle of arrival. Refined means of tracking targets will be discussed in a later chapter, and only the general theory will be covered here.

Thermal Noise Limitations

The early descriptive texts on radar gave very little information on the precision to be expected from a given direction-finding technique. It was assumed that the azimuth angle of a target could be interpolated to within some fraction of the width of the echo on the PPI display, but little was known about the factors which limited the interpolation process. As the design trends led to larger and more expensive antennas to increase range and accuracy, studies were pressed into fundamental sources of measurement error, and new techniques were introduced to overcome these errors. In the tracking radar field especially, detailed studies were made as to the relative contributions of thermal noise from the receiver input, target scintillation and glint effects, and other error components. In the search radar field, a pioneer work by Swerling* at the RAND Corporation considered the limitations due to both thermal noise and certain types of target fluctuation. Using a Gaussian approximation of the beam shape, which fits very closely the actual characteristics of the main lobe but excludes side lobes, he arrived at the following expressions for limiting precision on a constant-amplitude target:

$$\text{For} \quad (S/N)_o \ll 1: \quad \sigma_{\min} = \frac{0.58\theta}{(S/N)_o\sqrt{n}} \qquad \textbf{(2.12)}$$

$$\text{For} \quad (S/N)_o \gg 1: \quad \sigma_{\min} = \frac{0.49\theta}{\sqrt{(S/N)_o n}} \qquad \textbf{(2.13)}$$

Swerling's results for the case of a rapidly fluctuating target, whose

* Swerling's symbols have here been translated into the consistent set of symbols used throughout this book. In his original paper, Swerling uses a beamwidth parameter β equal to the angle between the beam center and the point where the two-way power pattern drops to $1/e$ of its maximum value. Thus, his β is equal to 0.43θ, where θ is the usual one-way half-power beamwidth. Swerling's X_o is our $(S/N)_o$, and his N, the number of pulses received during beam motion through the angle 2β, is equal to $0.85n$, where n is the number of pulses received between half-power points on the one-way pattern. See Peter Swerling, "Maximum Angular Accuracy of a Pulsed Search Radar," *Proc. IRE*, **44**, No. 9 (Sept. 1956), pp. 1146-55.

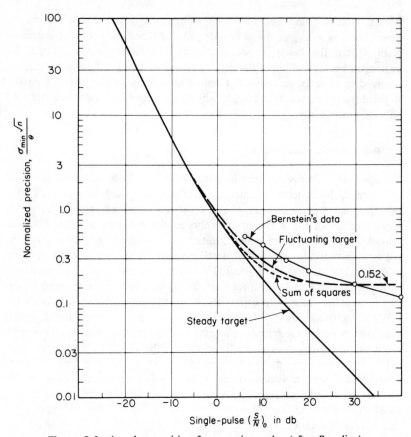

Figure 2.4   Angular precision for scanning radar (after Swerling).

amplitude is independent from each pulse to the next, are compared in Fig. 2.4 with the case of the steady target. As will be shown in a later chapter, the combined effects of thermal noise and target scintillation may be represented as the rms sum of two independent errors, one given by the thermal noise equations [Eq. (2.12) or Eq. (2.13) above] and the other by equations containing beamwidth, number of pulses observed, and target scintillation parameters. Also shown in Fig. 2.4 are points determined in an experimental study of a so-called "maximum likelihood estimator" which combined the errors due to thermal noise and target scintillation.*

More recent studies at the Massachusetts Institute of Technology

* R. Bernstein, "An Analysis of Angular Accuracy in Search Radar," *IRE Conv. Record* (1955), Part 5, p. 78.

Lincoln Laboratory* have applied the procedures developed by Woodward and Siebert in the analysis of range and velocity estimation to the case of angular measurements. An rms aperture length $\mathcal{L}$ is defined and used in the same way as the rms deviation of the spectrum in the case of range measurement. The basic thermal noise error, by analogy to the case of range error, becomes

$$\sigma_\theta = \frac{1}{\mathcal{L}\sqrt{\mathcal{R}}}$$

For a uniformly illuminated circular aperture of diameter D, the limiting precision in angle becomes

$$\sigma_\theta = \frac{2\lambda}{\pi\sqrt{\mathcal{R}}\,D} = \frac{0.62\theta}{\sqrt{\mathcal{R}}} \qquad \text{(2.14)}$$

The total energy received may be expressed in terms of the signal power, number of pulses, and beamshape loss L_p for the scanning case. Using Blake's loss figure of 1.6 db,† we find that the energy ratio becomes

$$\mathcal{R} = 2\left(\frac{S}{N}\right)_o n\frac{1}{L_p} = 1.39\,n\left(\frac{S}{N}\right)_o \qquad \text{(2.15)}$$

Equation (2.14) then reduces to

$$\sigma_\theta = \frac{0.53\theta}{\sqrt{n(S/N)_o}} \qquad \text{(2.16)}$$

It is interesting to note that the same equation with the same constants may be derived from the Lincoln Laboratory results for a uniformly illuminated rectangular aperture, if the error is expressed in terms of the beamwidth θ instead of aperture dimensions. The constant in the numerator of Eq. (2.14) (first part) becomes $\sqrt{3}$ instead of 2 when the aperture is rectangular, but the same reduction applies also to the beamwidth. We are justified, therefore, in assuming that Eq. (2.16) will hold for any

---

* Roger Manasse, "Range and Velocity Accuracy from Radar Measurements," Linçoln Lab. Group Report 312-26, Feb. 3, 1955, Astia Document AD 236, 236; "An Analysis of Angular Accuracies from Radar Measurements," Lincoln Lab. Group Report 32–24 (Dec. 6, 1955); "Summary of Theoretical Accuracy of Radar Measurements," Mitre Corp. Technical Series Report No. 2 (April 1, 1960), Astia Document 287, 563.

Merrill I. Skolnik, "Theoretical Accuracy of Radar Measurements," *Trans. IRE*, Vol. ANE-7, No. 4 (Dec. 1960), pp. 123–29.

† See Fig. 5.5.

type of aperture distribution if the received information is processed in the optimum fashion. The apparent discrepancy between the theoretical limit given by this equation and the results of Swerling for the simple scanning antenna are explained by the fact that Swerling's approach took advantage of the two-way pattern actually applicable to radar systems, where the echo signal is reduced by six decibels instead of three at the "half-power" point on the pattern. The two-way beam pattern is only 70 per cent as wide as the one-way pattern at the point where the echo power is one-half of the maximum value, and the angle error for the simple scanning case is some eight per cent below the theoretical limit for the one-way optimum estimator. A further improvement could be expected if a more refined measuring system were applied to the scanning radar beam.

Angular Resolution

All of the above analyses were carried out on the assumption that the target was known to be present within an angular sector roughly equal to the beamwidth. If false indications were received due to thermal noise appearing at some angle far removed from the actual location, the process of finding the rms error would lead to very large values of error compared to those predicted by the above equations. Readings could be obtained at angles as far as $180°$ from the true target. For this reason, the whole concept of measurement accuracy must be conditioned upon correct identification of a target in the first place. A high value of $\mathcal{R}$ or $n(S/N)_o$ will give a high probability of distinguishing the real target from thermal noise. Referring to Fig. 1.9 and the related discussion on effects of integration, we find that values of integrated signal-to-noise ratio between 10 and 50 are required to achieve reasonable probabilities of detection with low false-alarm rates. This implies that the limiting precision of angular measurement will be in the order of one-tenth the half-power beamwidth for most targets which can be distinguished against the background of thermal noise.

Strong signals and unwanted echoes or clutter will present more serious problems in angular resolution. Unlike the case of range measurement, where the transmitter can be turned off completely during the period between pulses, the antenna gain cannot be reduced to zero at points beyond the main lobe, except for discrete null points in the pattern. The side lobe and background response of the antenna covers the entire hemisphere, and reflections from the ground add more posibilities for unwanted signal reception. If the competing signals are of amplitude comparable to the desired signal, the basic resolution element in angle may be taken as the half-power beamwidth. Targets separated by this amount in either angular coordinate may be recognized as distinct targets, and measurements may be made on each with only small interaction, even though they may fall within the same range resolution element. However, a strong echo may appear in the side lobes of the antenna far from the main lobe, and it is difficult to define an angular region beyond which the desired target may be considered free of confusion. Angular resolution is thus conditioned upon relative amplitude to a greater extent than was true of range resolution, and small targets may be unresolvable even when separated by angles approaching 180° from very large targets. The two-way pattern characteristics of most radar antennas provide significant improvement in this respect when compared to one-way patterns, since the off-beam echoes will be attenuated by the square of the one-way pattern. This advantage cannot be applied to overcome jamming, where the unwanted signal depends only upon the receiving pattern of the radar, or to the constant-amplitude transponder which produces a fixed output for any interrogation above the threshold.

Multipath Propagation

A second serious limitation which must be faced before the measurement theory discussed earlier is applied is the inevitable presence of reflected signals from the earth's surface, which modify the free-space pattern of the antenna in the elevation plane. These effects are neither random, in the sense that they may be averaged out of observations taken over several pulses, nor predictable enough to be removed by calibration procedures. They depend a good deal upon the details of the surface contour underlying the radar beam, and the nature of the surface material. Most of the experimental tests of angular measurement theory are made in the azimuth plane over a relatively smooth surface, where the effects of multipath error are expected to be small. In the elevation plane, however, these effects are so pronounced that they have been depended upon to produce a lobing pattern for approximate measurement of target height.* The use of highly directive patterns in the elevation plane may reduce the amplitude of the reflected component to a small value when the targets are at relatively high elevation angles, but measurements of altitude on targets near the horizon are always impaired by surface reflections. A quantitative discussion of these problems will be reserved for a later chapter.

Discontinuous Antennas and Interferometers

Where high gain is not a primary requirement, but very narrow beams and high precision are desired, the antenna system may be constructed by using widely separated elements. An extreme example of this technique is

---

* L. N. Ridenour, ed., *Radar System Engineering* (New York: McGraw-Hill Book Company, 1947), pp. 184–87.

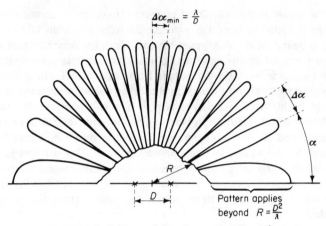

Figure 2.5   Pattern of interferometer, $D = 10\lambda$.

the interferometer, where the elements are placed many wave lengths apart over a long baseline. The pattern produced by a pair of elements at the ends of the baseline will consist of a series of nearly uniform lobes, punctuated by deep nulls, as in Fig. 2.5. The separation between adjacent lobes or nulls is given by

$$\Delta\alpha = \frac{\lambda}{D \sin \alpha}$$

Here, α is the angle between the baseline and the direction of the target, λ is the wave length, and D is the spacing between the elements. The rms aperture length, measured as before in units of 2π times the weighted distance from the center of radiation of the array, is just $\mathcal{L} = \pi D/\lambda$. The corresponding precision limit set by thermal noise is

$$\sigma_\alpha = \frac{\lambda}{\pi D \sin \alpha \sqrt{\mathcal{R}}} = \frac{\Delta\alpha}{\pi\sqrt{\mathcal{R}}} = \frac{0.27\,\Delta\alpha}{\sqrt{n(S/N)_o}} \qquad \textbf{(2.17)}$$

[The assumptions made here are the same as in Eq (2.16), where the target is assumed to move over the lobe with an average loss of 1.6 db in signal power.] It will be noted, as before, that the precision limit is related directly to the beamwidth, the lobe spacing $\Delta\alpha$ being exactly twice the half-power beamwidth of each lobe, and the constant in the numerator exactly half that which appeared in the equations using the beamwidth θ.

Although the precision of measurement appears higher for the two-element interferometer than for any other distribution of elements over the same aperture length D, this advantage is balanced by the presence of am-

biguities at intervals equal to $\Delta\alpha$. For this reason, the use of simple interferometers is normally limited to the measurement of angular rates, where
the approximate angle α is already known from other instruments. Where
the actual angle α must be measured by the interferometer, means must be
provided for resolving the ambiguities. This may be done by adding elements at intermediate distances along the baseline, or by using two or more
frequencies with the two end elements. Proper combination of data from
these supplementary measuring elements will usually permit the position
of a single target to be identified as lying within a particular lobe of the
detailed pattern set up by the end elements at a given frequency. However,
the presence of additional targets, unless they are resolved in range or frequency, may still introduce ambiguities, owing to the inability of the processing system to match the data gathered by different elements on a particular target. The use of many elements distributed in a random fashion
over the aperture, or of many frequencies with the two end elements, may
provide any desired probability of resolving these ambiguities. The detailed
analysis of such systems can be quite complex, especially when carried
out in both angular coordinates with targets at different ranges and Doppler
frequencies. A few of the practical system configurations will be discussed
in a later chapter on multistatic radar systems.

2.3 MEASUREMENT OF VELOCITY AS DOPPLER SHIFT

In addition to the usual three spherical coordinates of target position
(range and two angles), radar has the ability to measure directly the radial
velocity of the target, and to resolve targets on the basis of their differing
radial velocities. Signals reflected from a moving target whose radial velocity
component is v_r will show an increase in frequency, given by

$$f_d = f_t \frac{(c + v_r)}{(c - v_r)} - f_t \cong \frac{2f_t v_r}{c} = \frac{2 v_r}{\lambda_t} \qquad \textbf{(2.18)}$$

Here, f_t is the transmitted frequency, c the velocity of light, v_r the velocity
of the target towards the radar, and f_d the Doppler shift. Each element of
the transmitted spectrum will be shifted by an amount proportional to its
original frequency. The resulting effects on the ability of the radar to
measure and resolve differences in radial velocity will depend upon the
details of the transmitted spectrum and the duration of the transmission.

Radar Transmission Spectra

Spectra and wave forms characterizing the more common types of radar
transmission were discussed briefly in Section 2.1. We shall now consider

the effect of Doppler shift on these spectra. For simplicity, the discussion will be limited to cases where the deviation of the spectrum from its center frequency f_{to} is a small fraction of this frequency. The Doppler shift can then be viewed approximately as a simple translation of all components of the spectrum by a common amount $f_d = 2f_{to}v_r/c$. The received signal in the pure c-w case will consist of a single frequency located just f_d above f_{to}. For this case, a number of simple means is available for measuring the received frequency in a way that avoids all ambiguity in Doppler or velocity data. The picture will be much the same for other systems which use very few spectral lines in transmission. The amount of frequency translation f_d will be measurable by using any one of the lines or some combination of all lines, and it will be relatively easy to distinguish between the few lines upon reception. The practical problems for these systems lie in isolating the transmitter from the receiver so that low-amplitude echoes can be detected in that portion of the spectrum near the transmission.

In the case of the conventional pulse radar transmission, the spectral shift due to Doppler presents special problems. Normally, there will be hundreds of lines in the spectrum, with adjacent lines differing only slightly from each other in amplitude. The time separation between transmission and reception will serve to isolate the weak echoes from the high-power transmission, but the echoes may be shifted in frequency by an amount far greater than the separation between adjacent lines in the spectrum. The velocity indication will therefore be ambiguous at intervals corresponding to the repetition frequency (see Fig. 2.6). Two choices are open to the radar designer for measuring target velocity in this type of system. First, the shift of the entire spectral envelope may be measured, a discriminator whose bandwidth is

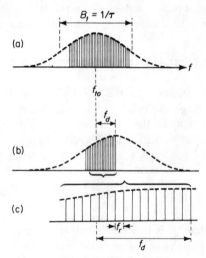

Figure 2.6 Spectra in pulsed radar system. (a) Transmitter spectrum. (b) Received spectrum. (c) Detail of received spectrum.

comparable to the width of the spectrum (approximately $1/\tau$ at the 40 per cent power points) being used. Alternatively, an attempt may be made to sense the shift of the individual lines in the spectrum, some supplementary source of data being used to resolve the ambiguities. We may estimate the limiting precision for either case, and apply the results to other types of transmission used in radar.

Shift of the Spectral Envelope

The spectrum of a single pulse received at the radar consists of a continuous band of energy, following generally the $(\sin x)/x$ form, with a main lobe of width $2/\tau$, measured to the nulls. Side lobes of relatively low level will appear on both sides of the main lobe, at intervals $1/\tau$. There will also be present a continuous noise spectrum whose amplitude will be approximately constant over the band occupied by the signal. By matching the i-f response of the radar to the width of the pulse spectrum, and passing the output through a discriminator whose response peaks are spaced $1/\tau$ cycles per second apart, the mean frequency of the received pulse may be measured. The discriminator characteristic may be approximated by a pair of idealized narrow-band filters whose bandwidth and separation are both $1/\tau$, as shown in Fig. 2.7(b). The noise output of each channel will be taken as E_n volts, and the noise after subtraction of the two channels will be $\sqrt{2}\, E_n$ since the two noise voltages may be considered essentially independent. The slope of the discriminator curve, shown in Fig. 2.7(c),

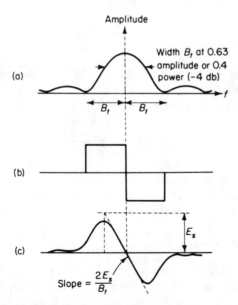

Figure 2.7 Idealized discriminator characteristics for spectrum envelope measurement. (a) Pulse spectrum $B_t = 1/\tau$. (b) Idealized discriminator filters. (c) Output vs. frequency off spectrum center.

is approximately $2E_s/B_t$, since a shift of the signal by an amount $B_t/2$ would produce a full output signal voltage E_s from one channel with only side lobe power in the other channel. Using this slope and the noise output $\sqrt{2}\,E_n$, we find the rms error in frequency measurement for a single pulse to be

$$\sigma_{f_1} = \frac{B_t}{2\sqrt{S/N}} = \frac{1}{2\tau\sqrt{S/N}} \qquad (2.19)$$

Where measurements are made on n pulses, separated in time so as to assure independence of the thermal noise components, the standard deviation will be $1/\sqrt{n}$ times as great.

$$\sigma_f = \frac{B_t}{2\sqrt{n(S/N)}} = \frac{1}{2\tau\sqrt{n(S/N)}} \qquad (2.20)$$

This result may be compared with Eq. (2.3) for range precision. The same form is observed, but with bandwidth B_t and pulse width τ interchanged.

In a more rigorous study, Manasse* has shown that the limiting precison of frequency measurement is dependent upon the rms time span of the signal and the energy ratio $\mathcal{R}$, the relationship having the same form as that for range or angle measurement.

$$\sigma_f = \frac{1}{\alpha\sqrt{\mathcal{R}}} \qquad (2.21)$$

The time span parameter α is defined as 2π times the rms duration of the signal, measured with respect to its center of energy. Thus, for a single rectangular pulse of duration τ, we will have $\alpha = \pi\tau/\sqrt{3}$, and

$$\sigma_{f_1} \cong \frac{1}{2.6\tau\sqrt{S/N}}$$

The discrepancy between the factor of two in the denominator of Eq. (2.20) and the 2.6 here may be attributed to the simplification in the assumed characteristics of the discriminator shown in Fig. 2.7(c).

Coherent Measurement

If the pulsed transmission is derived from a reference sinusoid of stable frequency and phase, so that the phase of the received signal remains

---

* Roger Manasse, "Summary of Theoretical Accuracy of Radar Measurements," Mitre Corp. Technical Series Report No. 2 (April 1, 1960), Astia Document AD 287, 563.

predictable from one pulse to the next, the energy will be found concentrated in discrete lines within the spectral envelope, rather than spread continuously over the transmission band. These lines will appear at intervals equal to the repetition frequency f_r, and will have a width at the 40 per cent power points given approximately by $1/t_o$, where t_o is the total time over which the target is illuminated by the radar. Such a transmission is termed "coherent" and provides an opportunity to make frequency measurements with much greater precision than that applied to the random-phase transmission of an unsynchronized oscillator. The time span of the received signal, for frequency measurement purposes, may now be taken as t_o rather than τ. Substituting this in Manasse's relationship gives us, for the coherent system

$$\sigma_{f_c} \cong \frac{1}{2.6 t_o \sqrt{n(S/N)}} \tag{2.23}$$

The precision is improved over the noncoherent value by the ratio t_o/τ, which may be in the order of a million, as well as by the factor $1/\sqrt{n}$, which accounts for the greater total energy processed as compared to the single-pulse case. Unfortunately, the ambiguities which result from the presence of the many similar lines in the pulse spectrum may lead to frequency errors in multiples of f_r unless some supplementary data is available to indicate the proper spectral line for measurement. This coarse designation may be provided, in some cases, by envelope discrimination. In other cases, differentiation of range data over periods of a few seconds can identify the proper interval in the spectrum. The use of two repetition rates in combination may also prove adequate to resolve the ambiguity. As in the case of the discontinuous antenna, however, the presence of multiple targets will introduce severe complications in the resolution process. Successful Doppler measuring systems have been built using higher than normal repetition rates to minimize the ambiguity problem, and a few highly ambiguous systems have been operated for tracking limited numbers of targets. These will be discussed at fuller length in the chapters on MTI search radars and Doppler trackers.

Frequency Resolution of Radar Transmissions

The ability to measure Doppler shift and other coordinates of the radar echo is dependent upon resolution of the echo signal from other competing signals in at least one of the four radar coordinates. In the frequency domain, resolution is dependent upon whether the signal is coherent or random in phase, and is conditioned upon solving the ambiguity problems referred to above. If the transmission is not coherent, or if the signal is rendered noncoherent by random phase modulation at the target or elsewhere in the system, the frequency resolution element is given by the width of the spectral envelope, or approximately by $1/\tau$ for pulsed transmissions. The envelope resolution for other types of transmission is given in Appendix D. The resolution of coherent signals is given basically by $1/t_o$, although only the pure c-w system provides this without any ambiguity. Woodward* shows that the total ambiguity in the domains of time and frequency will remain constant for all types of transmission, and can only be redistributed around the time-frequency plane by variation in the transmission wave form. He shows that the effective "area of ambiguity" in this plane is unity for all systems, although the product $\alpha\beta$ of the time and frequency parameters can be extended without limit to improve the precision of measurement. In cases where the region containing targets is limited, a system can be designed to minimize or eliminate ambiguity within this region by proper selection of the transmitted wave form, but this may not lead to a wave form which provides the desired measurement precision and other desired properties.

For example, we might assume that targets will be detected by the radar only if they lie within a time delay of one millisecond (equivalent to eighty nautical miles in range), and that the maximum target velocity is 1000 ft per sec. An unambiguous system can be designed which uses conventional pulse transmissions with a repetition frequency of 1000 pps. By selecting a transmitting frequency below 500 mc, the velocity ambiguities will be kept above the maximum target velocity when a coherent system is used. The pulse width is then open for the designer to use in optimizing the system, although it obviously must be kept to a small fraction of the one millisecond total delay time to avoid losing targets at short range. If only the spectral envelope is used for frequency resolution, a much higher transmitting frequency would have to be used to obtain useful discrimination against fixed targets. If we assume a maximum pulse width of ten microseconds, the spectrum would be 100 kc wide at the 40 per cent points. If the transmission were at a frequency of 35,000 mc, the Doppler shift corresponding to the maximum target velocity would be only 70 kc, and no useful velocity resolution would be obtained. Thus, at radar frequencies between 500 mc and the upper limit of the commonly used electromagnetic spectrum, it will prove impossible to avoid ambiguities between targets at different velocities and between targets and ground clutter, if pulsed transmissions are used.

By prolonging the period of observation, the precision of measurement may be improved without limit, except for practical equipment considerations. In the coherent system, the prolongation of the observation serves

* P. M. Woodward, *Probability and Information Theory, with Applications to Radar* (New York: McGraw-Hill Book Company, 1955), pp. 118–23.

also to reduce the width of the spectral lines (if it is assumed that the system is stable and that the targets cooperate by not introducing disturbing modulations of their own). The reduced line width brings about an apparent reduction in ambiguity areas within the region where targets may be detected. Actually, if the range of possible targets had not been limited arbitrarily to 80 mi, the prolonged observation time would have extended into space the region in which "multiple-time-around" echoes could be confused with desired targets, and the total area of ambiguity would remain constant as stated by Woodward.

A change from pulsed to fm c-w transmission can be made, but this would actually worsen the ambiguity problem by introducing crosstalk between range and velocity, as shown in Appendix D. The added area of ambiguity introduced by the fm transmission was formerly, in the pulse case, outside the region occupied by targets. The introduction of dual repetition rates and other forms of coding may extend the range at which completely ambiguous results are obtained, but such systems introduce new areas where there is some probability of making ambiguous readings, the total amount of ambiguity remaining constant.

The performance of a radar system in detection and measurement can be analyzed and predicted with confidence only when the target characteristics agree with their predicted values. Since a wide variety of targets may be encountered in use of radar systems in the field, we must devise target descriptions of sufficient latitude to **accommodate** wide variation in characteristics of individual targets at specific times. The importance of the signal-to-noise power ratio or energy ratio has been shown in the two preceding chapters, and procedures for calculating these ratios will be discussed in the next chapter. Knowledge of the "radar cross section" or backscattering coefficient of the target is essential in any such calculation. Although there are a few cases in which this coefficient is a constant, it will generally be found to vary considerably for each target as the aspect angle changes, or as internal motions of the target change its shape. These changes force us to use statistical methods to describe the radar target.

DESCRIPTION OF RADAR TARGETS

3

In measurement of target position and velocity, there are factors other than echo amplitude which are also of importance. The echo signal may consist of many components of energy scattered from points distributed over the surface of the target. The amplitude and phase of each component will vary as a function of time or aspect angle, and the interaction of these components may greatly affect the radar measurement process. An attempt will be made in this chapter to summarize all these important target characteristics and to describe them in ways which will permit accurate analysis of radar system performance.

3.1 DEFINITION OF RADAR CROSS SECTION

The first important characteristic of any radar target is a measure of its ability to reflect energy to the radar receiving antenna. The parameter used to describe this ability is known variously as the "radar cross section," the effective echoing area, or the backscattering coefficient.* This is defined as "4π times the ratio of reflected power per unit solid angle in the direction of the source to the power per unit area in the incident wave." Thus, if the target were to scatter power uniformly over all angles, its radar cross-section would be equal to the area from which power was extracted from the incident wave. Since the sphere has the ability to scatter isotropically, it is convenient to interpret the radar cross section in terms of an equivalent sphere.

Cross Section in Terms of a Sphere

A sphere whose radius a is large compared to the wave length will intercept the power contained in an area πa^2 of the incident wave, and will scatter this power uniformly over 4π steradians of solid angle. Its radar cross section, using the preceding definition, is therefore equal to its projected area πa^2. That portion of the surface which actually returns power to the radar is located close to the point where the wave first strikes the sphere, where the conducting surface lies parallel to the wave front. A physical definition of the radar cross section (for sufficiently large objects) would then be as follows: the radar cross section of a given target is the projected area of a conducting sphere which would produce the same signal as the target, if it were placed in the same position relative to the radar. For small targets, the equivalent sphere has a radius so small that a resonance phenomenon sets in (see below), and the physical definition becomes inadequate.

The cross section of a sphere is exactly one-fourth the total surface

---

* The *IRE Dictionary* gives "cross section" as a deprecated term referring to the equivalent echoing area of the target, but does not define "equivalent echoing area." Under "backscattering coefficient (echoing area)," p. 12, an appropriate definition is given, as used above. In this book, we shall follow common usage, where the term "radar cross section" persists, almost to the exclusion of the other synonymous terms.

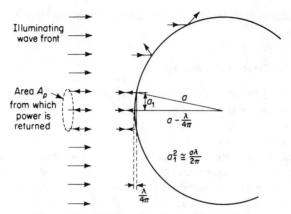

Figure 3.1 Effective flat-plate area of sphere.

area, and it can be shown* that the average radar cross section of any large object which consists of continuous, curved surfaces will be one-fourth of its total surface area. This relationship can be of great use in estimating the cross section of elongated or irregular bodies, provided that the ir-regularities are not in the form of sharp edges or structures which are resonant at the wave length being considered. Another physical interpreta-tion of cross section, which can be derived from the equivalent sphere, relates it to the flat-plate area A_p, which would provide an equal return signal if located at the target position and oriented normal to the radar beam. Since no portion of the sphere is actually flat, we shall take that portion which lies within a distance $\lambda/4\pi$ of the wave front when it just touches the front of the sphere (see Fig. 3.1). This is a circular disc of radius $a_1 = \sqrt{a\lambda/2\pi}$, whose area is $A_p = \pi a_1^2 = a\lambda/2$. If the nearly flat sur-face is considered to be an antenna, the power intercepted will be reradiated towards the radar with a gain $G = 4\pi A_p/\lambda^2$ [see Eq. (2.8)]. The resulting cross section, given by the product of gain times interception area,† will be

$$\sigma = A_p G = \frac{a\lambda}{2} \times \frac{4\pi a\lambda}{2\lambda^2} = \pi a^2 \qquad \textbf{(3.1)}$$

In some early articles on radar, this equivalent flat-plate area was used as a descriptive parameter for targets. It can be shown, in general, that the area lying within the distance $\lambda/4\pi$ of the wave front when it first makes contact with the target will give the flat-plate area A_p for the object, subject

---

* Donald E. Kerr, ed., *Propagation of Short Radio Waves* (New York: McGraw-Hill Book Company, 1951), p. 467.

† See the *IRE Dictionary*, p. 12.

to the limitation that the radius of curvature must be large compared to wave length.*

The Radar Target as an Antenna

The analysis based on equivalent flat-plate area demonstrates the use of antenna theory to describe the reradiation properties of targets. Whether the power which illuminates the scattering surfaces of the target originates at the radar or at some other point makes no difference in the resulting radiation pattern, so long as the proper phase relationships are established. Thus, a rectangular flat plate, oriented normal to the radar beam, can be equated to a uniformly illuminated aperture with constant phase of excitation, which produces a radiation lobe directed at the radar. The half-power width of the lobe will be $\theta = 0.88\lambda/L$ in that plane which includes the radar beam and the target dimension L. Side lobes, generally following the $(\sin x)/x$ amplitude pattern, will extend on each side of the main reflection lobe, at intervals $\Delta\theta = \lambda/L$. If the target is rotated through an angle α from its original position normal to the beam, the position of the main reradiation lobe will change by 2α, as in optical reflection. The excitation phase across the aperture will have changed, and the radiated beam will no longer be normal to the plate. Thus, the lobing structure of the target, when plotted with respect to target rotation angle relative to the radar beam, will change twice as rapidly as the pattern of the corresponding antenna with uniform, in-phase illumination. The basic lobe widths of the reflectivity pattern will be as follows

Half-power width of main lobe $\Delta = 0.44\lambda/L$ **(3.2)**

Main lobe between nulls $2\Delta_o = \lambda/L$ **(3.3)**

Side lobes between nulls $\Delta_o = \lambda/2L$ **(3.4)**

Detailed relationships for cross-section of some basic shapes have been derived by Kerr and others.† These show that the reflectivity patterns for rectangular plates follow the form

$$\sigma(\theta) = \sigma_o \left[\frac{\sin (kL \sin \theta)}{kL \sin \theta}\right]^2 \cos^2 \theta \quad \left(k = \frac{2\pi}{\lambda}\right) \qquad \textbf{(3.5)}$$

In the case of a circular plate, the factor $\sin (kL \sin \theta)$ in the numerator is replaced by the Bessel function $2J_1(2kL \sin \theta)$. In the case of the cyl-

---

* Roy C. Spencer, "Back Scattering from Conducting Spheres," Air Force Cambridge Research Lab. Report E5070 (April 1951), pp. 9–11.

† Donald E. Kerr, ed., *Propagation of Short Radio Waves* (New York: McGraw-Hill Book Company, 1951), pp. 445–69.

inder, the relationship is the same as for the rectangular plate, except that
the $\cos^2$ term is omitted. Figure 3.2 shows the similarity in patterns for the

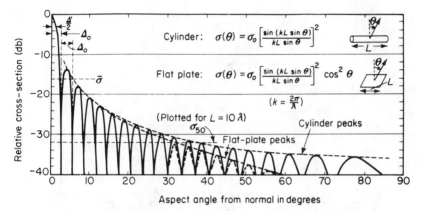

Figure 3.2 Reflectivity pattern for a cylinder and a flat plate.

rectangular plate and the cylinder when the object is many wave lengths
long. In such cases the lobe amplitudes drop to very low values before a
change in $\cos \theta$ appears. The circular disc has a similar pattern, with nulls
occurring at intervals $\Delta_o = 0.6\lambda/D$.

Cross Section of Simple Objects

A set of equations summarizing the maximum cross section values, lobe
widths, and total number of lobes presented by the target during a complete
rotation is shown in Table 3.1. Typical patterns of cross section vs. aspect
angle are shown in Fig. 3.3, to supplement the cylinder and flat-plate patterns of Fig. 3.2. All the equations are subject to the limitation that the
dimensions or radii of curvature must be large compared to wave length.
If an object is more complex than those listed in the table, its cross section
and reflectivity pattern may often be found by considering a combination
of these simple shapes which approximates the actual reflecting surface.
In the case of a finite cylinder, for instance, the actual pattern will consist
of a combination of the cylinder lobe structure, uniformly distributed
around the longitudinal axis and dependent upon the length L, with an
end-lobe structure caused by the flat plates at each end of the cylinder,
whose pattern will depend upon the diameter D. In the region between
side and end aspect, there will be two lobing periods, the relative amplitudes
depending upon the ratio of length to diameter and the aspect angle.

The behavior of cross section in the resonant region (where dimensions

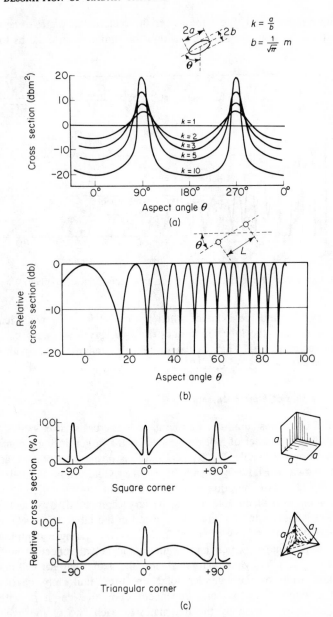

Figure 3.3 (a) Reflectivity of ellipsoid. (b) Reflectivity of dumbbell. (c) Reflectivity of corner reflector clusters.

Table 3.1 RADAR REFLECTIVITY CHARACTERISTICS OF SIMPLE BODIES

Object	σ_{max}	σ_{min}	Number of lobes	Major lobe width
Sphere	πa^2	πa^2	1	2π
Ellipsoid ($k = a/b$)	πa^2	$\dfrac{\pi b^2}{k^2}$	2	$\cong \dfrac{b}{a}$
Cylinder	$\dfrac{2\pi a L^2}{\lambda}$	null	$\dfrac{8L}{\lambda}$	$\dfrac{\lambda}{L}$
Flat plate	$\dfrac{4\pi A^2}{\lambda^2}$	null	$\dfrac{8L}{\lambda}$	$\dfrac{\lambda}{L}$
Dipole	$0.88\lambda^2$	null	2	$\dfrac{\pi}{2}$
Infinite cone (half-angle α)	$\dfrac{\lambda^2 \tan^4 \alpha}{16\pi}$	null		
Convex surface	$\pi a_1 a_2$			
Square corner reflector	$\dfrac{12\pi a^4}{\lambda^2}$		4	$\dfrac{\pi}{4}$
Triangular corner reflector	$\dfrac{4\pi a^4}{3\lambda^2}$		4	$\dfrac{\pi}{4}$

are in the order of the wave length) has been discussed in the literature.*
Curves for the sphere and the cylinder are shown in Fig. 3.4. Although
these are based upon theoretical calculations, there have been experimental
verifications in the case of the sphere.† Special cases of cross section that
varies with radar frequency are found in the dipole and in sharp-pointed
objects such as cones viewed from along the axis of symmetry. The resonance curve for the dipole, normalized to one square wave length at
the primary resonant frequency, is shown in Fig. 3.5. The particular dipole
on which these data were taken was extremely small in radius relative to
the wave length ($a = \lambda/300$). Kerr‡ gives the maximum value for the half-
wave dipole as $0.88\lambda^2$, in agreement with the test data shown here, and
the average value over all orientations and polarizations as $0.11\lambda^2$. The

---

* Donald E. Kerr, ed., *Propagation of Short Radio Waves* (New York: McGraw-Hill Book Company, 1951), pp. 445-54.
† S. B. Adler, "Pulsed Radar Measurement of Backscattering from Spheres," *RCA Review*, 23, No. 1 (March 1962).
‡ Donald E. Kerr, ed., *Propagation of Short Radio Waves* (New York: McGraw-Hill Book Company, 1951), p. 465.

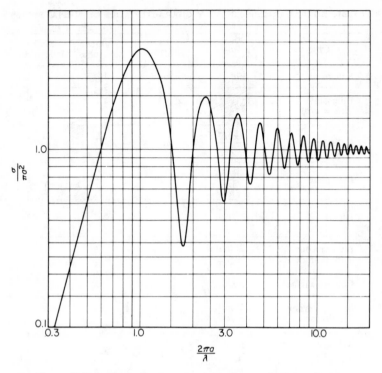

Figure 3.4(a) Normalized cross section for conducting spheres.

infinite cone, as listed in Table 3.1, has a cross section that varies as wave length squared. This is explained by the fact that the only discontinuity which is capable of returning energy is at the tip of the cone. This sharp tip may be replaced with a rounded point whose radius of curvature is some fraction of the wave length, leading to an effective area proportional to wave length squared. Where the cone is terminated in a flat plate or a matched section of a sphere,* larger cross-section values have been found, depending upon the nature of the discontinuity at the base of the conical surface. In some cases, the relationship $\sigma = 0.1a\lambda$ has been found valid, where a is the radius of the base of the cone. In particular, this appears to agree with the average of the curves given by Kennaugh and Moffatt for the cone of $15°$ half-angle matched to a sphere at the base. However, any relationship which expresses radar cross section as a function

---

* J. B. Keller, "Backscattering from a Finite Cone," *Trans. IRE*, Vol. **AP-8**, No. 2 (March 1960), pp. 175–82.

E. M. Kennaugh and D. L. Moffatt, "On the Axial Echo Area of the Cone-Sphere Shape," *Proc. IRE*, **50**, No. 2 (Feb. 1962), p. 199.

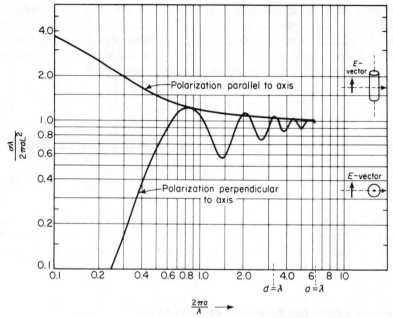

Figure 3.4(b) Normalized cross section for conducting cylinders.

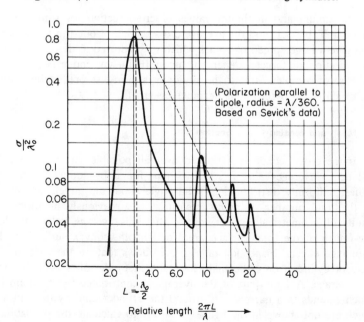

Figure 3.5 Cross section of dipole.

of wave length is dependent upon a number of assumptions, often unstated, relating to polarization, types of discontinuities, and other factors. Only in the so-called Rayleigh region, where all dimensions are small relative to the wave length, is there a clear frequency dependence for targets of all shapes. Small spheres follow the following relationship

$$\frac{\sigma}{\pi a^2} = 14,000 \left(\frac{a}{\lambda}\right)^4 \qquad (\lambda > 10a) \qquad \textbf{(3.6)}$$

Small circular discs of radius a follow the fourth-power law also, giving

$$\frac{\sigma}{\pi a^2} = 1125 \left(\frac{a}{\lambda}\right)^4 \qquad (\lambda > 10a) \qquad \textbf{(3.7)}$$

The lower cross section of the disc as compared to the sphere is to be expected, owing to the fact that the disc has less surface to interact with the incident wave. The average cross section of small objects may be used as an approximate measure of the volume of a conducting object.

3.2 DESCRIPTION OF COMPLEX TARGETS

The maximum and minimum values of cross section for simple objects, shown in Table 3.1, and the average cross section of one-fourth the surface area give a large spread of values which may be used to describe the radar cross section of targets, depending upon frequency and aspect angle. In the case of complex targets, the situation is even less predictable, and the use of statistical descriptions becomes essential.

Mean and Median Cross Section

If the cross section is measured over all possible aspect angles, there will be an average value $\bar{\sigma}$, which can be found by adding all measured values, uniformly distributed in solid angle about the target, and dividing by the total number of points measured. As previously mentioned, the results of geometrical optics indicate that this average value should be one-fourth of the total surface area for smooth, convex surfaces without resonance effects. Inspection of Fig. 3.2 shows that, in the case of a flat plate or cylinder, only the main lobe and first side lobes actually exceed this average. A large part of the average is contributed by the main lobe, which extends in a narrow belt around the cylinder, and by the end lobes, which are not shown in the figure. If we are to calculate the probability of detection for such an object, we must first find the probability that a given value of σ will occur, multiply by the probability of detection for that σ, and sum the result for all possible values of σ. This tedious process may be avoided by using the median cross section σ_{50} and restricting our calculations to the point at which a detection probability of 50 per cent is to be achieved. The median cross section is the value that is exceeded just 50 per cent of the time when the target is viewed over the range of aspect angles encountered in system operation. Although the procedure using σ_{50} does not give exactly the same result as the rigorous process referred to above, it can generally be assumed that the fluctuations above and below the median will balance in obtaining detection probabilities near 50 per cent. The use of the median has been adopted as a standard procedure by those responsible for calculations on Navy search radar systems,* and appears entirely adequate for most operational radar systems.

Amplitude Distributions

Accurate calculations of target detectability can be made when the complete amplitude distribution of target cross section is known. The distribution may be plotted either in terms of the probability that a given value of σ will be exceeded (or not exceeded), or as a probability that a given value will be observed (probability density). The simple targets of Table 3.1 have distribution functions which are difficult to express analytically, and which do not fit the commonly used models of the statistics texts. The following example, however, will illustrate how an oddly shaped distribution function can be used in detectability calculation, and will also show the accuracy of the calculation based upon median cross section.

Assume that a given radar at a given range achieves a probability of detection of 50 per cent for each scan when the target cross section has a value equal to the median value for the cylinder of Fig. 3.2. We wish to find the probability of detection for the cylinder at this range when it appears with random aspect angle. From Fig. 1.9, we find that the signal-to-noise ratio required for 50 per cent detection probability is about 12.5 db (for a false-alarm probability of 10^{-8}), and we assume that the radar achieves this value, after integration, on the median cross section σ_{50}, which is 32 db below the peak cross section observed at normal aspect to the cylinder. Figure 3.6 shows the distribution function for the particular cylinder whose reflectivity pattern was given in Fig. 3.2. Values of cross section near the median predominate, but there are regions where the peaks rise well above this, as well as nulls where the cross section approaches zero. From Fig. 1.9 we see that a drop in S/N of four or more decibels below that required for $P_d = 50$ per cent reduces the detection

* Lamont V. Blake, "Interim Report on Basic Pulse-Radar Maximum Range Calculation," Naval Res. Lab. Memo Report 1106 (Nov. 1960).

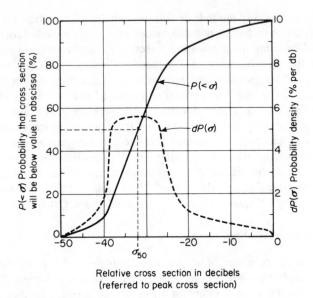

Figure 3.6 Distribution of cross section for cylinder $(L = 10\lambda)$. Distribution is taken in the plane containing the axis of the cylinder, not over sphere surrounding the object.

probability to near zero, so the portion of the distribution below -36 db may be considered to contribute nothing to target detection. Similarly, when the S/N rises four or more decibels above the value for $P_d = 50$ per cent, the detection probability approaches 100 per cent, and further increase may be ignored. All we must know of the distribution, therefore, is the total percentage above -28 db, that below -36 db and the approximate distribution between those values (all referred to the peak σ). Using one decibel steps, we make the following calculation:

Per cent distribution		P_d for this σ	Contribution to average P_d
Below -36 db:	27 %	0%	0.0%
-36 to -35 db:	5 %	2%	0.1%
-35 to -34 db:	5.5%	6%	0.3%
-34 to -33 db:	5.5%	16%	0.9%
-33 to -32 db:	5.5%	35%	1.9%
-32 to -31 db:	5.5%	60%	3.3%
-31 to -30 db:	5.5%	86%	4.6%
-30 to -29 db:	5 %	98%	5.0%
Above -29 db:	35.5%	100%	35.5%
Total	100 %		Average P_d: 51.6%

Thus, even for this target which departs radically from the Rayleigh distribution, the use of the median cross section yields results near $P_d = 50$ per cent, which are as good as the accuracy of the distribution data.

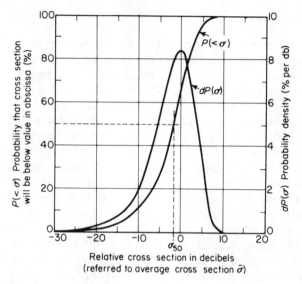

Figure 3.7 Exponential distribution (dP curve would be exponential if plotted on linear scale of σ).

If the target consists of many scattering elements of comparable size, with phase relationships which vary in a random fashion as a function of time, it can be shown that the distribution of cross section is exponential (Rayleigh amplitude distribution). As shown in Fig. 3.7, the median value for this case is approximately 69 per cent of the average $\bar{\sigma}$ used in defining the distribution [see Eq. (1.8)]. The fact that the echo signal voltage follows the Rayleigh distribution [as shown in Fig. 1.6(c) for detected thermal noise] leads to this type of target being called the "Rayleigh target." Common practice in radar system analysis is to assume a target of this type, with an appropriate value of $\bar{\sigma}$ unless the actual target is known to have some other specific reflectivity characteristic. This practice is entirely justified when microwave radars are used to observe large, complex objects over a variety of aspect angles, but the indiscriminate use of the Rayleigh distribution can lead to serious errors in other cases. When the wave length is long enough to approach the target dimension, it is far better to use measured or estimated distributions, even based on approximations of the target shapes to be observed. In cases like the cylinder discussed above, the substitution of a Rayleigh target with the same median cross section would not alter the situation near $P_d = 50$ per cent, but would lead to great differences where detection probabilities of 90 per cent or greater were to be achieved.

Frequency Spectra

The rate at which the target cross section changes can be of considerable importance both in calculation of detection probabilities and in estimation of measurement accuracy. If a search radar scans the area occupied by the target several times, with a moderate probability of detecting the target on each scan, it can achieve a very high cumulative probability of detection for a target that fluctuates significantly between successive scans. However, if the period of the fluctuation is long relative to the scan period, the same (possibly very low) value of cross section may be presented on every scan. The extra radar power necessary to overcome this problem is plotted as fluctuation loss in Fig. 1.13. In measuring azimuth with a scanning radar, the error for a steady target will depend largely on the signal-to-noise ratio, and strong signals make it possible to measure to a very small fraction of the antenna beamwidth. However, if the cross section changes appreciably during the period required to scan the beam across the target, the apparent azimuth of arrival, marked by the largest echo signal, may be shifted from the actual angle. This effect was shown in Fig. 2.4.

A complete description of target characteristics must include some specification of the periodicity of variation of cross section, as well as amplitude distribution. This may be done in several ways. If the aspect angle is changing in a regular fashion, the plots of cross section vs. aspect angle may be regarded as expressing cross section as a function of time, with the target rotation period t_a equated to 360° of aspect angle. For any target with many lobes in its pattern, the cross section will change appreciably over intervals of time much shorter than a complete rotation period. The total number of lobes per rotation was given in Table 3.1 for the simple shapes. The maximum lobing rate for elongated objects such as flat plates and cylinders is given by

$$f_{max} = \frac{\omega}{\Delta_o} = \frac{2\omega L}{\lambda} \tag{3.8}$$

Here, $\omega = 2\pi/t_a$ gives the rotational rate of the target in radians per second, Δ_o is the side lobe width from Eq. (3.4), L is the total length of the target, and λ is the wave length. The lobing rate will vary from zero to this maximum value as the aspect angle changes. Above this frequency, the spectral density of the echo signal strength will drop sharply, although the sharp nulls in the time function will produce some harmonics above f_{max}.

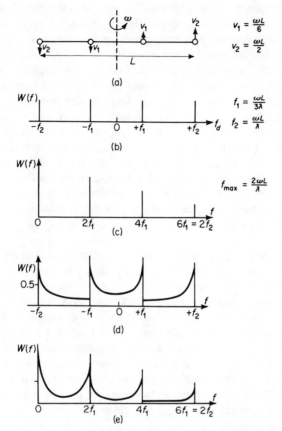

Figure 3.8 Synthesis of frequency spectrum for four-element target. (a) Spatial distribution of reflecting elements. (b) Doppler frequencies at normal aspect. (c) Doppler cross-mixing frequencies. (d) Distribution of Doppler frequencies for all aspects. (e) Distribution of cross-mixing frequencies for all aspects.

The reflectivity characteristics of some targets can be approximated by assuming a small number of scattering elements distributed in a known way over the total length or breadth of the target.* For a given target course or maneuver, the frequency spectrum of the cross section can be estimated in these cases by calculating the relative radial velocities of the various reflecting elements, converting these to Doppler frequencies, and finding the sum- and difference-frequency components that will be produced

* A. J. Stecca, N. V. O'Neal, and J. J. Freeman, "A Target Simulator," Naval Res. Lab. Report No. 4694 (Feb. 1956).

when the composite signal is passed through an envelope detector. The method of calculation is diagrammed in Fig. 3.8 for a target model consisting of four reflecting elements equally spaced across a target span L. The radial velocities encountered when the target span is normal to the radar beam are shown in Fig. 3.8(a). Four discrete Doppler frequencies will result, displaced symmetrically about the mean signal frequency, which is the transmitted frequency plus any Doppler frequency resulting from motion of the center of gravity of the target relative to the radar. After passing through the receiver and envelope detector, major components will appear at zero frequency, at f_{max} and at two intermediate values. If the power spectrum is measured over a period of time which includes several complete revolutions of the target, the discrete frequencies shown in Figs. 3.8(b) and 3.8(c) will spread to the forms shown in Figs. 3.8(d) and 3.8(e). In addition, of course, there will be relatively low-amplitude harmonics extending above f_{max} after the detection and cross-mixing process.

If actual recordings of cross section as a function of time are available, the frequency spectrum may be found by digital or analog computation. In one analog process,* the recording is made on magnetic tape, formed into a continuous loop, and played back repeatedly, at increased speed, into an audio spectrum analyser. A direct plot of output amplitude vs. frequency is obtained, which may be scaled to give the proper spectral density and frequency readings. In digital computation, the more usual procedure involves computation of the autocorrelation function for the cross-section data, and transformation of this into a frequency spectral density plot.† The autocorrelation function can also be used in place of the frequency spectral data as a descriptive function showing the time variation of cross section.

Commonly Used Mathematical Models

If the appropriate target model is a sphere, with constant cross section at all aspect angles, a single constant σ serves to specify the target. As a first step in introducing complexity into the target model, we might assume a sinusoidal variation of cross section at a frequency f_m, with a given fractional modulation m and a mean value $\bar{\sigma}$.

$$\sigma(t) = \bar{\sigma}(1 + m \sin 2\pi f_m t) \tag{3.9}$$

---

* Dean D. Howard, "Instrumentation for Recording and Analysis of Audio and Sub-Audio Noise," *IRE Conv. Record* (1958), Part 5, pp. 176–82.

† R. B. Blackman and J. W. Tukey, *The Measurement of Power Spectra from the Point of View of Communications Engineering* (New York: Dover Publications, Inc., 1958).

The above function resembles the pattern of an ellipsoid with small ellipticity, or the combined cross section of two targets, one of which is much larger than the other. A more common time function, representing the lobing encountered in elongated targets or combinations of two sources of comparable size, is given by a sine-squared function.

$$\sigma(t) = \sigma_{\max} \sin^2 (2\pi f_m t) \qquad (3.10)$$

The two-body or "dumbbell" target, which is analyzed in more detail in the next section, can be represented completely by the function

$$\sigma(t) = \sigma_{\max} \sin^2 \left(\frac{2L}{\lambda} \sin \omega t \right) \qquad (3.11)$$

Here, $\sigma_{\max}$ is four times the cross section of each of the two target elements, and the average value of σ is just twice the individual cross section. The lobes, of constant peak amplitude $\sigma_{\max}$, occur at a sinusoidally modulated rate as the two bodies rotate around the center of mass with a period $t_a = 2\pi/\omega$. The maximum lobing rate during the rotation is given by Eq. (3.8).

Targets of greater complexity are generally described in terms of their amplitude distributions and frequency spectra (or correlation times). The distribution most often used in the literature is the Rayleigh amplitude distribution, which leads to an exponential distribution of cross section described by

$$dP_\sigma = \frac{1}{\bar{\sigma}} \exp \left(\frac{-\sigma}{\bar{\sigma}} \right) d\sigma \qquad (\sigma \geq 0) \qquad (3.12)$$

In Swerling's classic work,* this was the distribution used for his "Case 1" and "Case 2" target models. The difference between the two models lies in the relationship between correlation time and scan period. In Case 1, the pulses received during a single scan are assumed to be equal, the amplitudes encountered in previous and succeeding scans being completely uncorrelated. The correlation time of the echo is thus considerably longer than the time-on-target for one scan, and considerably shorter than the time between scans. In terms of frequency spectra, this is equivalent to saying that most of the scintillation power lies well above the scan rate f_s and below a frequency f_b, which is defined by the ratio of scan speed to beamwidth. In Case 2, independence of echo amplitude from pulse to pulse is assumed, implying that most of the scintillation power is above the repetition rate in the spectrum.

---

* Peter Swerling, "Probability of Detection for Fluctuating Targets," RAND Corp. Research Memo RM-1217 (March 17, 1954).

Two other models used by Swerling follow the next amplitude distribution.

$$dP_\sigma = \frac{4\sigma}{\bar{\sigma}^2} \exp\left(\frac{-2\sigma}{\bar{\sigma}}\right) d\sigma \qquad (\sigma \geq 0) \tag{3.13}$$

Using the preceding distribution, Swerling's Case 3 had scan-to-scan fluctuation as in Case 1, and his Case 4 had the same pulse-to-pulse fluctuation as in Case 2. Physically, the two distributions can be considered to represent a large assemblage of equal reflecting elements, adding with random phase (Cases 1 and 2), or a single large reflector combined with a number of smaller ones (Cases 3 and 4). Kerr states that the "large number" of elements required to give an approximation of the Rayleigh distribution may be as small as four or five.* (In describing cross section, the Rayleigh distribution should not be confused with the Rayleigh scattering region, which refers to the case where the dimensions of the target are small with respect to wave length, and where the ratio of radar cross section to physical area varies as $1/\lambda^4$.)

More detailed assumptions as to frequency spectra than those used by Swerling are often necessary. In particular, scintillation power may extend down to the scan rate f_s, and may be continuous through the region near f_b. A white noise spectrum may be assumed to be as high as the repetition rate f_r, in which case the significant parameter is the spectral density, or mean-square fractional modulation per cycle per second of bandwidth. In other cases, spot frequency distributions corresponding to lobing frequencies of Eqs. (3.8) through (3.11) may also be used. Generally, however, the actual spectrum of cross-section variation may be fitted with a "Markoffian" or band-limited white noise spectrum, extending from zero frequency to infinity, but with steadily decreasing amplitude above a given frequency f_a.

$$W(f) = W_o \frac{f_a^2}{f^2 + f_a^2} \qquad (f \geq 0) \tag{3.14}$$

The frequency f_a is often referred to as the "half-power frequency" of the scintillation. Swerling's Cases 1 and 3 may be approximated by setting f_a one or two octaves below f_b, in those cases where f_b is much greater than f_s. His Cases 2 and 4 require that f_a be higher than the repetition rate f_r.

The total power in the Markoffian spectrum, which represents the variance of the cross section or the square of its standard deviation, is given by the integral of $W(f)$ from zero to infinity.

* Donald E. Kerr, ed., *Propagation of Short Radio Waves* (New York: McGraw-Hill Book Company, 1951), p. 554.

$$\sigma_a^2 = \int_0^\infty W(f)\,df = \frac{\pi W_o f_a}{2}$$

(The expression σ_a, designating the standard deviation of the cross section, is confusing but difficult to avoid once the statistical nature of cross section is recognized.) For the Rayleigh distribution, the standard deviation of cross section is equal to the mean value, and hence the relationship among the low-frequency density W_o, the half-power frequency f_a, and the mean cross section $\bar{\sigma}$ is, as follows

$$\bar{\sigma} = \sqrt{\frac{\pi W_o f_a}{2}} \tag{3.15}$$

For spectra of this type, the correlation time is given by $t_c = 1/(2\pi f_a)$. This time may be thought of as the time constant of the equivalent low-pass RC network that would produce the target spectrum when supplied with a white noise input. When two observations of the target are separated by this time interval, the cross section value for the second observation is only 37 per cent correlated with the first. After two or three such intervals, the cross section may be considered to be essentially independent of the old value.

3.3 SPATIAL DISTRIBUTION OF REFLECTIVITY

Radar measurement of target position and velocity is based upon the assumption that some point on the target may be defined as a position reference. Targets which are small with respect to the radar wave length present no problem in this regard, since they are seen as small point sources both in range and angle. Larger targets, however, show a noticeable shift of range, angle, and Doppler frequency as a function of aspect angle from which they are viewed. In order to predict the performance of a radar in its measurement function, we must be able to describe these target shifts, at least in a statistical sense, and to relate them to the measuring processes used by the radar.

Angular Glint

The terms "glint" and "scintillation" are often used interchangeably in the literature of radar* to describe both amplitude fluctuation and

---

* See, for instance, the *Dictionary of Guided Missiles and Space Flight*, pp. 285 and 547; the *IRE Dictionary*, pp. 127 and 145; *Air Force Manual AFM* 100-39, pp. 289 and 582; Nelson M. Cooke and John Markus, *Electronics and Nucleonics Dictionary* (New York: McGraw-Hill Book Company, 1960), pp. 195 and 417.

angular noise originating at the target. Where a distinction has been made between the two effects, the angular noise is consistently referred to as "glint" or "angle scintillation," whereas the fading or variation in signal strength is termed "amplitude scintillation" or "amplitude fluctuation."* Since, as we shall show in a later chapter, there are many sources of angle noise that do not originate with the target, the term "glint" will be used in this book to describe only the disturbance in apparent angle of arrival due to interference phenomena between reflecting elements of the target. "Scintillation" will be used only in reference to amplitude variation. This usage is consistent with the original derivation of both words: "glint" comes from the Middle English "glenten," meaning to turn aside or glance, "scintillation" is used in astronomy to describe the twinkling of stars, which is strictly an amplitude variation. The usage is also consistent with that of Rhodes† and with the definition of scintillation given in the Air Force publications for communications use.‡

It is sometimes erroneously assumed that glint represents merely the wandering of the center of reflection (or "radar center of gravity") over the extent of the target. As early as 1953, however, it was pointed out that the apparent radar center may lie outside the confines of the target a certain percentage of the time.§ For a linear distribution of reflecting elements along a rigid bar, the radar center lies beyond the ends of the bar approximately 13 per cent of the time, as the bar rotates around an axis perpendicular to its length and to the radar line of sight. It was also pointed out by Delano that the mean-square error in such a case is infinite, when a tracking system with unlimited bandwidth and dynamic range is

---

* Charles E. Brockner, "Angular Jitter in Conventional Conical-Scanning, Automatic-Tracking Radar Systems," *Proc. IRE*, **39**, No. 1 (Jan. 1951), pp. 51–55.

Richard H. Delano, "A Theory of Target Glint or Angular Scintillation in Radar Tracking," *Proc. IRE*, **41**, No. 12 (Dec. 1953), pp. 1778–84.

John E. Meade, "Target Considerations," Chap. 11 in *Guidance*, A. S. Locke, ed. (Princeton, N.J.: D. Van Nostrand Company, Inc., 1955).

Richard H. Delano and Irwin Pfeffer, "The Effect of AGC on Radar Tracking Noise," *Proc. IRE*, **44**, No. 6 (June 1956), pp. 801–10.

John H. Dunn, Dean D. Howard, and A. M. King, "The Phenomena of Scintillation Noise in Radar Tracking Systems," *Proc. IRE*, **47**, No. 5 (May 1959), pp. 855–63.

Dean D. Howard, "Radar Target Angular Scintillation in Tracking and Guidance Systems Based on Echo Signal Phase-Front Distortion, *Proc. NEC*, **15** (1959), pp. 840–49.

† Donald R. Rhodes, *Introduction to Monopulse* (New York: McGraw-Hill Book Company, 1959), pp. 4 and 11.

‡ AFM 100-39, p. 582.

§ Richard H. Delano, "A Theory of Target Glint or Angular Scintillation in Radar Tracking," *Proc. IRE*, **41**, No. 12 (Dec. 1953), p. 1781.

used. The special case of the two-element (or dumbbell-shaped) target is covered by several authors,* and will be discussed in some detail below.

The Two-Element Target

As an illustration of an extreme case of glint, consider a target which consists of two spheres or point sources, rigidly connected with a spacing L, and rotating around an axis perpendicular to L and to the radar line of sight. As the relative path lengths between the radar and the two elements vary, the two signals will alternately add and subtract, causing a series of maxima and minima in the signal seen by the radar. At the times of the maxima, when the two signals arrive at the radar in phase, the apparent angle of arrival will indicate the "center of gravity" of the two sources, as determined by the relative amplitudes of the echo signals. For all other phase angles, the wave will appear to originate from a point nearer the larger source. This type of target is used by Howard to demonstrate that glint can be understood in terms of distortion of the phase front received at the radar. By plotting the contours of equal phase radiating from the target, and comparing these with the concentric circles that would radiate from a single point source at the center of the dumbbell axis, an angular error can be defined in terms of the relative slope between the actual contours and the circles at the radar antenna, averaged over the extent of the aperture. Figure 3.9 shows such a plot for the case where the two sources are illuminated 180 deg out of phase. The angular error is given by the following

$$\delta' = \left(\frac{L \cos \psi}{2R}\right)\left[\frac{1 - k^2}{1 + k^2 + 2k \cos\left(\frac{4\pi L}{\lambda}\sin \psi\right)}\right] \qquad \textbf{(3.16)}$$

Here, the ratio of the two signal amplitudes is represented by k, the angle between the target axis and the radar wave front is ψ, and the quantity $(4\pi L \sin \psi)/\lambda = \phi$ represents the phase angle between the two echo signals at the radar.

Some of the properties of this target appear peculiar at first glance. When the second target is small relative to the first (small k), the presence

---

* John E. Meade, "Target Considerations," Chap. 11 in *Guidance*, A. S. Locke, ed. (Princeton, N.J.: D. Van Nostrand Company, Inc., 1955).

John H. Dunn, Dean D. Howard, and A. M. King, "The Phenomena of Scintillation Noise in Radar Tracking Systems," *Proc. IRE*, **47**, No. 5 (May 1959), pp. 855-63.

Dean D. Howard, "Radar Target Angular Scintillation in Tracking and Guidance Systems Based on Echo Signal Phase-Front Distortion," *Proc. NEC*, **15** (1959), pp. 840-49.

of the second target causes the apparent angle of arrival to oscillate about the actual direction to the first target, the deviation following the relationship

$$\delta' = \left(\frac{L\cos\psi}{R}\right)k\cos\phi = \theta_d k\cos\phi \qquad \text{(3.17)}$$

Here, θ_d gives the angle subtended by the target length L at the range R, and the other symbols are as defined for Eq. (3.16). The average indicated target position remains on the larger target, since the deviation is symmetrical. There is no way of determining on which side of the main target

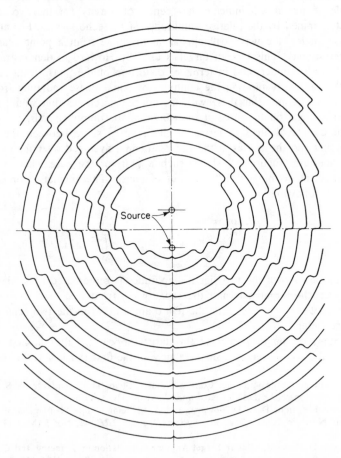

Figure 3.9 Phase front radiated from two-element target. (Courtesy D. D. Howard, Naval Res. Lab.)

the second target lies. As the second target increases in amplitude, the deviation follows the curves shown in Fig. 3.10. When the two elements are nearly equal in amplitude, the apparent angle of arrival lies midway between the two sources until the relative phase angle nears 180°, at which point a violent swing away from the smaller target takes place, exceeding the actual angular extent of the target by a factor $(1 + k)/2(1 - k)$. This factor goes to infinity for k equal to unity. For values of k above unity, the curves are repeated on the opposite side of the center line, the apparent position oscillating about the second target instead of the first.

The large glint errors predicted for the two-element target have been observed on modern radars with wide servo and automatic gain control bandwidths, leading to the belief that such errors are peculiar to the mono-

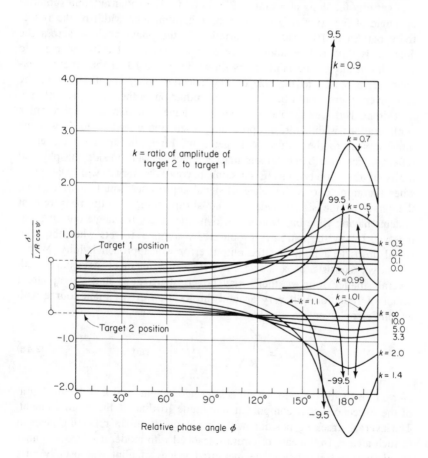

Figure 3.10 Normalized glint error for two-element target.

pulse processing methods. Actually, as Howard's analysis shows, the error will be the same for all types of angular measurement devices which evaluate the angle of arrival over a relatively small segment of the wave front. Limited dynamic range and speed of response of most practical direction-finding systems will tend to smooth out the error predicted by Eq. (3.16), and other errors may mask the glint term when it is so smoothed. For any given type of device, however, the glint error will remain proportional to the projected target length ($L \cos \psi$), and inversely proportional to the target range. If the receiving aperture subtends an appreciable angle at this range, the value of ϕ will vary over the aperture, tending to reduce the magnitude of the pointing error δ'. Thus, in a typical case, we might have an X-band radar ($\lambda = 3$ cm) with a five foot antenna diameter, tracking a target fifty feet long at a range of 100,000 ft. The radar antenna subtends an angle of 5×10^{-5} rad, whereas the minimum lobe width of the reflectivity pattern is 10^{-3} rad. The variation in the phase angle ϕ across the antenna is 0.05 rad or about three degrees, which is hardly sufficient to smooth out any of the glint errors shown in Fig. 3.10. If the antenna diameter is increased to the size of the reflection lobe, about 100 ft in the case given, there would be a significant reduction in the error encountered.

Although the extrapolation from two-element targets to more complex configurations is difficult, the preceding discussion allows us to draw some inferences as to the nature of target glint. First, the apparent center of reflection for distributed targets may at times lie well outside the physical limits of the target itself. Excursions beyond the target will take place when the amplitude of the reflected signal approaches a null, at which time the phase of the signal undergoes an abrupt reversal. For measurement systems having wide bandwidth and dynamic range, the angle output readings will follow the extremes produced by complex targets when the echo signal drops to a null, and the output error will approach infinity. Where the target consists of a single main reflecting element associated with several smaller sources, there may be no sharp nulls and the apparent center of radiation will vary around the location of the main source with an rms error given by

$$\sigma_g = \frac{1}{R}\sqrt{\frac{1}{2}\sum_i L_i^2 k_i^2} \qquad \text{rad} \qquad\qquad \textbf{(3.18)}$$

Here, L_i is the distance to the ith scattering element, and $k_i \ll 1$ is the ratio of the ith echo signal component amplitude to that of the main element. The average tracking position will remain on the main reflecting element in such a case. In the case of aircraft, tracked with moderate-sized antennas at microwave frequencies, the measured values of glint will usually vary from one-sixth to one-third of the maximum span of the target, depending upon the distribution of reflecting elements and the bandwidth of the angle-measuring circuits used.

Range Glint

In range measurement on an extended target, there will be a glint term very similar to the angular glint. In range, the glint error will bear a direct, linear relationship to the distribution of reflecting elements, and will be in-

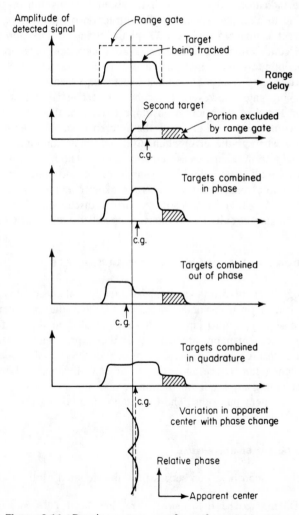

Figure 3.11 Ranging system waveforms for two-target case.

dependent of the range from radar to target. One experimental study*
has shown that the range glint is also about one-fourth of the extent of
the major scattering elements, projected on the radar line of sight. If the
two-element target is analyzed with respect to range glint, it will be seen
that the effect is almost identical to the angle case, although the range
glint will depend to a greater extent on the radar parameters and the circuits used for measurement. To illustrate this, Fig. 3.11 shows the wave
forms developed in the receiver and ranging system when the two elements
of the target are separated by a distance equivalent to one-half the pulse
width. The indicated range will oscillate about the position of the larger
target, owing to alternate addition and subtraction of the voltage from the
smaller target element as its relative phase changes. The normal range
tracking circuit, acting on video signals after envelope detection, cannot
develop an infinite error signal on the two-element target, even when the
two amplitudes and ranges are almost exactly equal and a null is created
in the tracking gate. However, if a coherent system is used to form the
range error signal at i-f prior to envelope detection, there will be a possibility of complete cancellation of the reference signal used to control
system gain, whereas the error channel signals may combine to produce
a finite signal into the range measurement channel. The result, after normalization in a circuit with unlimited dynamic range and speed of response,
will be an infinite range error which is the exact counterpart of the angle
glint error for $k = 1$ and $\phi = 180°$. A further discussion of the response
of various ranging circuits to glinting targets will be found in Chapter 12.

3.4 BEACONS AND ANTENNAS AS RADAR TARGETS

Radar range and performance can be greatly enhanced if the target
carries a beacon or transponder matched to the radar parameters. The
signal-to-noise ratio improvement will be calculated in the next chapter.
Equally important in many applications is the improvement in target definition for measurement purposes. In principle, the use of a beacon with
suitable antenna can reduce an extended target with range and angle glint,
and accompanying amplitude fluctuation, to a steady point-source for tracking or measurement purposes. The limitations in attaining this ideal situation will be reviewed briefly.

Beacon Output Characteristics

Once the beacon has been successfully interrogated by the radar trans-

* Dean D. Howard and B. L. Lewis, "Tracking Radar External Range Noise
Measurements and Analysis," Naval Res. Lab. Report 4602 (Aug. 31, 1955).

mitter, its reply is characterized by a relatively constant output power at a preset frequency. Environmental conditions may cause variation in both power output and frequency of the beacon, at rates which vary from slow drifts, caused by accumulation of heat, up to the frequencies encountered in vibration of some vehicles. Frequency shifts of the beacon transmitter will cause problems in acquisition by the radar, as well as possible measurement errors owing to reception of echo signals at the edge of the radar passband. The radar receiver bandwidth may have to be broadened from the value which is optimum from a signal-to-noise standpoint, unless the beacon transmitter can be stabilized to within a small fraction of the width of the pulse spectrum. If the interrogation of the beacon is not consistent, there may occur a phenomenon known as "skip-triggering," when some of the pulses are missing entirely in the response train. This and other less severe modulations mentioned before must be considered in analysis of the performance of radar-beacon systems. As a rule, however, they are less important than the signal-strength variations produced by beacon antenna pattern irregularities.

Beacon Antenna Patterns

In most beacon applications, the ideal antenna pattern is omnidirectional and lacking in nulls, and has the desired polarization in all directions. Although it is theoretically possible to generate an omnidirectional pattern, if polarization is permitted to vary over the sphere of coverage, it is impossible to generate controlled polarizations in all directions. In practice, it may not even be possible to generate an omnidirectional pattern in any polarization, and some compromise must be accepted. For tracking of aircraft by surface radar, a simple stub antenna, mounted below the craft, may be used to cover all azimuths around the aircraft, with essentially hemispherical coverage at somewhat reduced gain. The null directly beneath the craft is seldom important, but the absence of coverage in the upper hemisphere can be objectionable when the plane banks in turning. The same stub antenna or a small dipole mounted on the tip of the vertical stabilizer fin will cover most of both hemispheres, but with some shadowing below and in front of the plane, caused by the body and wing surfaces.

The expedient of using both top and bottom stub antennas, which might appear to solve the problem, leads to a complex pattern of lobes and nulls in directions near the horizontal plane, where both antennas are visible at the same time. This pattern will be essentially that of the interferometer, shown in Fig. 2.5. Since the nulls have almost infinite depth, the target may be entirely invisible when viewed from certain directions. Also, the glint characteristics will be those of the two-element target discussed earlier. As was shown, this can lead to some of the most extreme tracking errors encountered on any type of target. Accordingly, multiple beacon antennas should be used only in situations where the nulls are not encountered by the radar during important periods of tracking and detection, or where results from periods of observation between nulls may be extrapolated or averaged with the disturbed data. If multiple antennas must be used, the objective of creating a point-source target is usually lost during substantial portions of the track.

The above remarks can also be applied to missiles and satellites as radar targets. Unless the beacon antenna can be mounted on a fin or other projection to achieve near-omnidirectional performance, the addition of more antennas in parallel can be expected to result in serious lobing and signal modulation at the radar. When stable flight can be assured, it is possible to locate a single antenna to cover radars in a restricted solid angle, and even to provide gain over isotropic radiation. More generally, two or three antennas must bᵉ used to approximate the wide coverage required, and the intervening nulls between the main lobes will produce interference effects at the radar. Figure 3.12 shows the pattern of a typical three-element antenna system mounted on a cylindrical vehicle. Flush-mounted slot an-

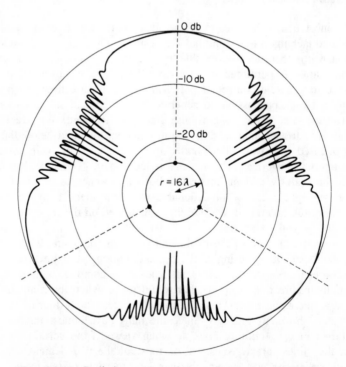

Figure 3.12 Radiation pattern of three-element beacon antenna.

tennas are often used to avoid aerodynamic effects and to withstand heating. By varying the angle of orientation of the slot or equivalent dipoles with respect to the axis, it is possible to achieve a certain amount of polarization diversity. Although the nulls will remain in the pattern measured with a specific received polarization, it may be possible to use variable polarization in the radar antenna to avoid the nulls over most of the aspect angles around the missile. For the configuration shown, nulls at the two ends are almost inevitable. Another means of avoiding nulls takes the form of phase modulation imposed on one or two of the antenna elements, causing the nulls to sweep rapidly past the angle at which the radar is located. In this case, if the relationship between radar repetition rate and phase-modulation rate is proper, it is possible to limit the duration of nulls to a small fraction of the smoothing time used in the radar servo or data processor.

Unwanted Illumination of the Target

In using the beacon to produce a point source for accurate angle and range tracking, another problem arises when the beacon antenna illuminates the surface of the extended target vehicle. Although this illumination and the resulting signal reflected to the radar may be many decibels below the main response, it appears as a multipath component with some delay and with a varying phase angle relative to the direct response. It also produces the very glint effects which were to have been eliminated by using the beacon. An indirect echo which is ten decibels below the direct ray and which is reflected by a flat surface separated by a distance L from the beacon antenna will produce the effect of a two-element target whose amplitude ratio is $k = 0.32$. Referring to Fig. 3.10, we see that this will lead to peak errors equivalent to about $L/2$ in linear distance at the target, or $L/2R$ in radians at the radar. The error will vary with aspect angle in the same way as the glint error for pure reflections, except that the phase variation will occur at half the rate, owing to the one-way path change which characterizes the noncoherent beacon.

Beacon Range Delay

Reflection of radar energy from a target occurs instantaneously when the wave front arrives at the target, but beacons require some time to detect and process the incoming signal before producing a response. The delay in response pulse will almost always be greater than the pulse width used, to avoid blocking the receiver before it has completed reception, and to prevent the beacon response from initiating another interrogation of its own receiver. Provision must be made in accurate ranging systems to correct for the expected beacon delay, and this delay must be held constant within narrow limits if serious range errors are to be avoided. In beacons using simple crystal-video receivers,* the response is initiated by the crossing of a fixed threshold by the video pulse. The threshold is placed sufficiently above the average noise level to hold the random noise triggering ("squitter rate") to a value well below the radar repetition rate. The detection curves of Figs. 1.8 and 1.9 may be used to calculate the threshold level and the triggering probabilities for noise and for signal-plus-noise samples. The variation in delay will be a function of video bandwidth and signal strength, and will also depend upon whether any special, nonlinear circuits or decoding devices (for multiple-pulse interrogation) have been included in the beacon to minimize delay errors. It is possible to hold variations in delay to a few millimicroseconds over a major part of the dynamic range of the input signal, range accuracies of a few yards thus being achieved. As the interrogation signal nears the noise level, there will appear a large variation in range delay, and with further reduction in signal the beacon will fail to respond.

Beacons using superheterodyne receivers may follow the same characteristics as the crystal-video models, with greater sensitivity and dynamic range, or they may be equipped with automatic gain control circuits to reduce the variation in delay to very small values. In principle, leading and trailing edges of the interrogation pulse may be averaged as in radar ranging systems, or the center of energy of the pulse may be tracked by a circuit which controls the beacon response time. However, this degree of complexity is seldom found in practical equipment, where reliability is at a premium. The most accurate ranging is accomplished by using repeater-type beacons, in which the signal is amplified at r-f or i-f without envelope detection. A detected video signal may be used to gate the output stage of the beacon, and a short delay is built into the beacon to permit reception of the entire radar pulse before response, but there are no variable threshold levels to introduce delays in the response. These beacons, in addition to reproducing the radar pulse on a cycle-by-cycle basis for coherent measurement, will faithfully reproduce the radar pulse width (and, in some cases, its possible irregular shape) and will only introduce distortion in the form of thermal noise if the interrogation signal is weak. Very few such beacons were available until the period of the 1960's, but the development of suitable amplifier-type klystron tubes and traveling-wave tubes has provided the basis for extensive development along these lines, and some solid-state designs have also become available.

Regardless of the type of beacon and antenna system chosen, the radiation patterns and delay characteristics must be evaluated and appropriate allowances made in calculating the nature of the signals returned to

* Arthur Roberts, ed., *Radar Beacons* (New York: McGraw-Hill Book Company, 1947).

the radar. The average signal level, amplitude distribution, frequency spectrum, and glint and delay terms can usually be estimated from measured patterns and interrogation tests run in the laboratory. Theoretical antenna plots and delay curves may also be used in the absence of other data. The rate of change of target aspect with respect to the radar line of sight is of great importance, as it will determine the rate at which antenna lobing will occur, and this in turn will establish the correlation time of the beacon signal for detection purposes, and the spectra of both angular and range delay errors. As will be shown in the next chapter, the beacon power output P_b invariably appears along with the beacon antenna gain G_b in equations giving signal power at the radar. We may, therefore, use the "effective radiated power" $P_e = P_b G_b$ in place of the two separate terms, with the understanding that this must be described as a variable quantity to account for antenna lobing and other modulations. Whether it is described as a function of time or as a statistical quantity will depend upon our knowledge of the antenna pattern and aspect angle of the vehicle. When these factors are taken into consideration, it may at first appear that the beacon has added only further trouble to an otherwise satisfactory radar system. However, the increase in range of operation on targets with small cross section cannot be achieved as economically in any other way. With an adequate understanding of the factors which limit radar-beacon performance, much of the difficulty can be eliminated and satisfactory performance can be obtained.

3.5 CHARACTERISTICS OF RADAR CLUTTER

Where fixed echoes result from discrete surfaces, such as buildings and well-defined land sources, their characteristics may be described in the terms already developed for complex targets. We shall be concerned here with clutter echoes which originate from a multitude of scattering elements distributed over the surface of the earth (as with ground or sea clutter) or throughout a volume of space (as with weather or chaff clutter). Such clutter echoes will generally fill at least one radar resolution element, and the amplitude of the echo will be proportional to the area or volume which lies within the element.

Amount of Clutter within Resolution Element

When a radar beam intersects a flat surface at an elevation* angle E,

---

* Strictly speaking, the "grazing angle" should be used instead of radar elevation (or depression) angle. At short range, where curvature of the earth may be neglected, no distinction need be made.

the area within the "effective beam angle" ψ_b may be identified as A_b (see Fig. 3.13a), given by

$$A_b = \frac{R^2 \psi_b}{\sin E} \tag{3.19}$$

If the beam were considered as a cone with elliptical cross section whose axes were θ_a and θ_e, then the solid angle within the beam would be $\pi \theta_a \theta_e / 4$. However, when the effect of the $(\sin x)/x$ or Gaussian beam pattern is considered, the effective solid angle is about half as great.* Thus we may consider the effective beamwidths to be $\theta_a / \sqrt{2}$ and $\theta_e / \sqrt{2}$, and the area of surface within the beam will be

$$A_b = \frac{\pi}{8} \times \theta_a \theta_e R^2 \times \frac{1}{\sin E} \tag{3.20}$$

In most cases, this area will be divided into two or more range resolution elements, each of which extends along the ground for a distance $\Delta R_g = \tau c / (2 \cos E)$. The area within the largest of these elements, near the center of the beam, will be

$$A_r = \frac{\theta_a R}{\sqrt{2}} \times \frac{\tau c}{2} \times \frac{1}{\cos E} \tag{3.21}$$

As shown in Fig. 3.13(a), this latter form will apply at long range and low elevation angles, the transition angle from Eq. (3.20) to Eq. (3.21) being given by

$$\tan E = \frac{\sqrt{2}\,\theta_e R}{\tau c} \tag{3.22}$$

The surface area contributing clutter will be denoted by A_c, and will be equal to A_b above this angle and A_r below it. For example, if the beamwidth θ_e is three degrees (0.052 rad), the pulse width τ three microseconds ($\tau c = 0.5$ mi), and the range R is ten miles, the transition is at an elevation angle $E = 55°$.

For volume clutter, the effective resolution element is defined by the dimensions $\theta_a / \sqrt{2}$, $\theta_e / \sqrt{2}$, and $\Delta R = \tau c / 2$, as shown in Fig. 3.13(b). The resulting volume is

---

* See Louis J. Battan, *Radar Meteorology* (Chicago: University of Chicago Press, 1959), p. 25.

J. R. Probert-Jones, "The Radar Equation in Meterology, *Proc. Ninth Weather Radar Conference* (Oct. 23–26, 1961).

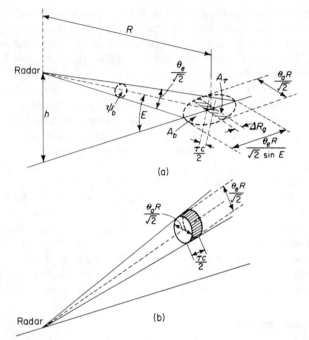

Figure 3.13 Geometry of clutter elements. (a) Surface clutter. (b) Volume clutter.

$$V_c = \psi_b R^2 \, \Delta R = \frac{\pi}{16} \theta_a \theta_e R^2 \tau c \qquad (3.23)$$

As in the case of surface clutter, the effective beam angle ψ_b may have to be increased to account for side-lobe energy. The increase may amount to a factor of 1.2 for low elevation angles, rising to 1.4 for vertical beams. In some cases, where the clutter fails to fill the entire beam, a reduction in effective area or volume must be made. The reduction factor may be calculated as the ratio of the total solid angle of the beam to that portion actually occupied by clutter, if the latter is known.

Amplitude of Clutter

Since radar clutter is considered to come from extended areas containing many scattering elements, it is described statistically in terms of average values and distributions about the average. The amplitude of clutter from a given resolution element can be described as the sum of a fixed and a random component, the ratio of powers (fixed/random) being designated

m^2. If the resolution element is dominated by large, rigid reflectors, m^2 will be very large. For such clutter sources as sea or weather return, m^2 will approach zero.* In the latter case, the amplitude of clutter will follow the Rayleigh distribution, and the power will follow the exponential distribution. These distributions were discussed in Section 1.2, in connection with thermal noise. Surface clutter, returned from land or sea, will be further described by the surface reflectivity, or average cross section per unit area of surface, designated by the dimensionless parameter $\sigma^0 = \bar{\sigma}/A_c$. A surface which, under given conditions, has a σ^0 of 0.1 will provide a radar cross section of one-tenth square meter for each square meter of surface within the radar beam. The average cross section within the radar resolution element will be given by $\bar{\sigma} = A_c\sigma^0$. In the case of volume scatterers, such as weather return, the corresponding parameter is the "radar reflectivity" η, expressed in units of cross section per unit volume and having the dimensions of (length)$^{-1}$. A typical rainstorm may have a radar reflectivity η equal to about 10^{-5} m^2/m^3 at X-band, so that a cross section of one square meter will correspond to a cube 45 m on a side. The average cross section within each radar resolution element in this case will be $\bar{\sigma} = V_c\eta$. More detailed data on the parameters σ^0 and η will be presented below.

Frequency Distribution

In addition to the amplitude characteristics of clutter, we are interested in the spectral distribution of the echo energy, since this will determine the ability of MTI systems to attenuate the clutter. In his pioneering work on this subject, Goldstein[†] plotted a number of power spectra and found that the spectral characteristics were identical at all radar wave lengths when plotted in terms of the parameter $f\lambda$, the product of fluctuation frequency times radar wave length. Barlow[‡] found that the spectra of almost all types of clutter could be represented by a Gaussian curve of the form

$$W(f) = W_o \exp\left(-a\frac{f^2}{f_t^2}\right) \qquad \text{(3.24)}$$

Here, W_o represents the power density of the clutter at zero fluctuation frequency, f_t is the radar operating frequency, and a is a dimensionless

---

* Herbert Goldstein, "Sea Echo," "The Origins of Echo Fluctuations," and "The Fluctuations of Clutter Echoes," Sects. 6.6–6.21 in *Propagation of Short Radio Waves*, D. E. Kerr, ed. (New York: McGraw-Hill Book Company, 1951), pp. 560–87.

† Herbert Goldstein, Ibid., pp. 573–86.

‡ Edward J. Barlow, "Doppler Radar," *Proc. IRE*, **37**, No. 4 (April 1949), p. 351. The equation describes the spectral density of clutter surrounding each transmitted spectral line, f representing the deviation from the line.

parameter which describes the relative stability of the clutter. Values of a given by Barlow range from 3.9×10^{19} for sparse woods on a calm day to 2.8×10^{15} for rain clouds. More recent discussions give the same spectrum in terms of an rms clutter frequency spread σ_c in cycles per second

$$W(f) = W_o \exp\left(-\frac{f^2}{2\sigma_c^2}\right) \tag{3.25}$$

The frequency spread is related to the rms velocity spread of the scattering elements σ_v by the familiar Doppler equation

$$\sigma_c = \frac{2\sigma_v}{\lambda} \tag{3.26}$$

By combining the above relationships, we find that Barlow's stability parameter a may be expressed as

$$a = \frac{f_i^2}{2\sigma_c^2} = \frac{c^2}{8\sigma_v^2} \tag{3.27}$$

Thus, we can interpret the parameter a physically as a function of the ratio of clutter velocity spread to the velocity of light. The use of σ_v to calculate the frequency spread σ_c directly from Eq. (3.26) is the preferred approach, since it preserves the physical significance of the various terms.

Typical values of clutter spread are shown in Table 3.3, which lists the values of $\sigma_c \lambda$, σ_v, and a obtained from various sources. More detailed data are given in the separate discussions on each type of clutter. The rms clutter spread σ_c is related to the half-power frequency of the clutter spectrum (used by Goldstein) through the relationship $\sigma_c = 0.85 f_{0.5}$, when a Gaussian spectrum is assumed.

Ground Clutter

Studies of ground clutter were reported by Goldstein, and results of a more recent experiment have been published by Grant and Yaplee.* Although there are many variables which enter into the process of backscattering from the ground, a few general statements can be made in summary of these studies. Goldstein found that the steady component of ground clutter could be attributed to all those scattering elements which move less than one-fourth wave length in the wind. Therefore, the ratio of steady to

* See also E. M. Sherwood and E. L. Ginzton, "Reflection Coefficients of Irregular Terrain at 10 Cm," *Proc. IRE*, **43**, No. 7 (July 1955), pp. 877-78.

Table 3.3 CHARACTERISTICS OF CLUTTER SPECTRA

Source of clutter	Wind speed (knots)	Ratio m^2	Barlow's a	$\sigma_c\lambda$ (cm/sec)	σ_v (ft/sec)	Reference
Sparse woods	(calm)		3.9×10^{19}	3.5	0.057	Barlow
Rocky terrain	10	30				Goldstein, p. 583
Wooded hills	10	5.2	7.2×10^{18}	8	0.13	// pp. 583–85
// //	20		2.3×10^{17}	45	0.74	Barlow
// //	25	0.8	9×10^{17}	23	0.38	Goldstein, pp. 583–85
// //	40	0	1.1×10^{17}	65	1.06	// //
Sea echo			2.4×10^{16}	140	2.3	Wiltse, et al., p. 226
// //		0	$(1–2)\times10^{16}$	165–205	2.5–3.3	Goldstein, pp. 580–81
// //	8–20		$(0.6–2.6)\times10^{16}$	100–220	1.5–3.5	Hicks, et al., p. 831
// //	(windy)		1.4×10^{16}	183	3.0	Barlow
Chaff	0–10	0	$(1.4–8)\times10^{16}$	75–180	1.2–3.0	Goldstein, p. 472
//	25	0	7×10^{15}	250	4.1	// //
//			10^{16}	215	3.5	Barlow
Rain clouds		0	$(0.7–3)\times10^{15}$	370–800	6–13	Goldstein, p. 576
// //			2.8×10^{15}	410	6.7	Barlow

Basic relationships: $\sigma_v = \dfrac{\sigma_c\lambda}{2}$ (cm/sec) $= \dfrac{\sigma_c\lambda}{61}$ (ft/sec) $\sigma_c = 0.85 f_{0.5}$

$$\sigma_v = \frac{c}{\sqrt{8a}} = \frac{10^9}{\sqrt{8a}} \text{ (ft/sec)} \qquad\qquad a = \frac{c^2}{8\sigma_v{}^2}$$

random component will be a function of wave length, the longer wave lengths showing the greater amount of steady power. The values of m^2 given in Table 3.3 were measured at a wave length of 9.2 cm. At this frequency, Goldstein found that wooded terrain exhibited very little fluctuation when the wind speed was below about fifteen miles per hour. Above 20 mph, the fluctuation increased sharply, indicating that the motion of small trees and branches had exceeded the $\lambda/4$ criterion. Although the critical wind velocity would be lower for shorter wave lengths, studies were not available to arrive at a definite relationship for fluctuation as a function of wind velocity and wave length. Above 50 mph wind speed, the steady component of clutter power had all but vanished, and the measured amplitude distributions followed the Rayleigh curve for an assembly of random scatterers. As shown in Table 3.3, the rms velocity of the scattering elements rises from a few hundredths of a foot per second, for sparse woods under calm conditions, to a little over one foot per second for gale winds. The ratio of wind speed to rms clutter velocity is in the order of 50 to 1.

The experiments of Grant and Yaplee were concerned with the magnitude of the ground clutter, measured by the reflectivity σ^0. Their results show that this parameter is almost independent of the depression (or elevation) angle E. Typical values of σ^0 for a surface covered with trees or tall weeds range from 0.0025 (or -26 db) at 3.2 cm wave length, to 0.03 (or -15 db) at 8.6 mm. These results are similar to the measured characteristics of the lunar surface as measured by radar astronomers, and are also applicable to rocky terrain. They may be used quite broadly whenever no specific data exist to justify other values. The trend towards lower magnitude of backscatter for longer wave lengths is indicative of the decreased roughness of the surface, when viewed by the longer waves.

Sea Clutter

Following Goldstein's thorough discussion of this subject, there have been several detailed studies on various characteristics of sea clutter. These have included the work at the Naval Research Laboratory by Schooley, Katzin, and Grant and Yaplee, a study made at Johns Hopkins University by Wiltse, Schlesinger, and Johnson; and a series of experiments reported by Hicks and his associates at the University of Illinois Coordinated Science Laboratory. The last of these produced a great deal of data on the spectral distribution of sea clutter, whereas the earlier results described primarily the magnitude of σ^0 as a function of various parameters.

Goldstein originated the use of the parameter σ^0 to describe clutter amplitude, and found that the value of σ^0 for sea clutter was dependent upon five factors:

1. The angle E which the incident ray makes with the horizontal (designated θ in the literature).
2. Radar wave length.
3. Radar polarization.
4. State of the sea, as affected by wind velocity.
5. Azimuth of the beam relative to the wave pattern on the sea (upwind, crosswind, or downwind).

Much of the early uncertainty as to the effects of wave length and angle of incidence was clarified in the analysis by Katzin, who showed that the characteristics of sea return changed sharply when the ray arrived at the surface with an angle less than a critical value E_c.

$$E_c = \sin^{-1}\left(\frac{h}{R_c}\right) \cong \frac{\lambda}{5H} \qquad \textbf{(3.28)}$$

Here, h represents the radar antenna height, H is the average height of the backscattering elements above the reflecting surface of the sea, and R_c is the transition range from "near zone" to "far zone" clutter characteristics. The height H is less than the physical height of the waves, in most cases. Katzin showed that the clutter power in the near zone, which was relatively unaffected by reflection phenomena, varied inversely as R^3. Within this zone, σ^0 varies inversely as λ. Beyond the transition range, the received power varies inversely as R^7, owing to the destructive interference caused by surface reflections. The scattering elements in this zone lie below the lowest lobe of the pattern formed by the interference phenomenon. Since the transition range is itself a function of wave length, the early attempts to arrive at a law relating σ^0 to wave length were subject to considerable difficulty, and diverse results were to be expected.

For rays within the near zone as defined above, Katzin found that a good explanation of sea clutter magnitude could be derived by considering the presence and distribution of flat "facets" at different angles on the sea surface. This explanation has since been expanded by Schooley* to explain the difference in magnitude of σ^0 measured in the upwind and downwind directions, which amounts to several decibels.

The effects of elevation (or depression) angle, wave length, and surface roughness are shown in Fig. 3.14, the upwind-downwind ratio being

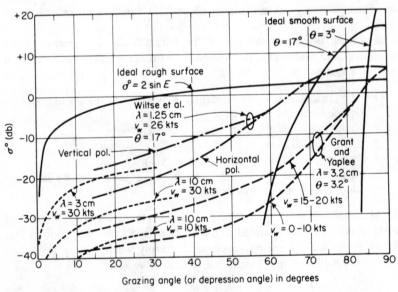

Figure 3.14 Reflectivity σ^0 for sea water at different wind speeds.

* Allen H. Schooley, "Upwind-Downwind Ratio of Radar Return Calculated from Facet Size Statistics of a Wind-Disturbed Water Surface," *Proc. IRE*, 50, No. 4 (April 1962), pp. 456–61.

neglected. The solid curves represent results which would be observed if the surface were ideally rough or smooth. These curves are based on an early study by Schooley,* who also gave experimental data showing agreement in the shape of the curves for low angles of elevation (three dotted curves in the lower left-hand corner of the figure). For normal sea conditions, the sea does not meet the criterion for perfect roughness at $\lambda = 3$ cm, but is rapidly approaching it. The fact that most clutter measurements have been carried out in the low-angle region from two to 20°, where the roughness criterion controls backscatter, tends to explain the inverse dependence of σ^0 upon wave length. Experimental curves for $\lambda = 1.25$ cm, taken from Wiltse's report, also tend to confirm Schooley's theory, when the effect of the "main lobe" near normal incidence is considered. The ideal smooth-surface curve for a beamwidth of 17° has been plotted from Schooley's equation. Since the actual surface is neither rough nor smooth, the amplitude of the main lobe will be reduced, and it will be broadened beyond its ideal width of 17° at the half-power points. Accordingly, the scattering caused by the surface roughness begins to appear at an elevation of about 55°, where the curves for vertical and horizontal polarizations begin to diverge. Below this angle, the curves tend to match the slope predicted by the rough-surface curve. Based upon Schooley's experimental data for $\lambda = 3$ cm, we would expect the magnitude of σ^0 for Wiltse's data to be somewhat greater, but the effect of wind direction and the inherent precision of the data do not permit exact comparisons.

The data from the report of Grant and Yaplee, taken at $\lambda = 3.2$ cm, are also plotted in Fig. 3.14, with the same sort of agreement. The ideal three deg beam pattern is broadened and reduced in amplitude, and the rough-surface reflection begins to show at an elevation of about 80°. In this case, since all runs were taken with horizontal polarization, the effect is marked by a divergence in data for different wind velocities. The effect of the main lobe extends down to about 55° elevation, where the curves begin to match the slope of the curve for ideal rough-surface reflection. The magnitude of σ^0 in the low-angle region is consistent with Wiltse's data.

The spectral distribution of sea-clutter energy, as measured by Hicks and his group, serves as an excellent indication of the physical origin of backscattered energy. Figure 3.15 shows the observed spectral spread of the clutter, in terms of both half-power width ($f_{0.5}$ in cps) and rms velocity

---

* Allen H. Schooley, "Some Limiting Cases of Radar Sea Clutter Noise," *Proc. IRE*, **44**, No. 8 (Aug. 1956), pp. 1043–47. He gave, for the smooth sea, the equation

$$\sigma^0 = \frac{4}{\theta^2} \exp\left[-\frac{3}{\theta^2}\left(E - \frac{\pi}{2}\right)\right]$$

where θ is beamwidth and E the elevation (or depression) angle, both in radians.

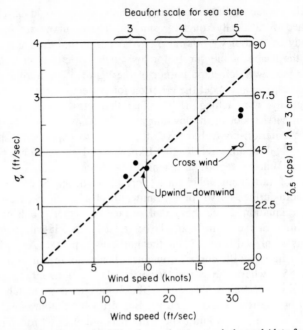

Figure 3.15 Spectral width of sea clutter vs. wind speed (data from Hicks, et al.).

spread (σ_v in ft/sec), plotted against the wind speed in knots. There is a rough proportionality between the wind speed and the rms clutter velocity, the ratio being in the order of eight to one. Analysis of the actual spectra showed the presence of return from whitecaps and spray, at the higher wind speeds. The spectral distribution of the echoes from wind-blown particles was centered about two knots downwind, and overlapped the spectrum of wave clutter, forming a continuous spectrum four or five knots wide. The width of this spectrum was almost independent of elevation angle, although there was a small increase in width (about 0.6 knots) as the elevation rose from 1° to 10°. The fact that the cross-wind measurement gave a smaller bandwidth is further indication that the scattering originates both from the wave surface and from blowing spray at high wind speeds. Hicks gives the half-power width as a function of "significant wave height" $H_{1/3}$ and mean wave period t_m.

$$v_{0.5} \cong \frac{11 H_{1/3}}{t_m}$$

The advantage of this expression is that it gives the dependence of clutter spread upon actual parameters of the sea state, rather than on wind speed. The latter is only an indication of the force being applied to the sea, and the sea state may depend upon other factors, such as tides and the duration of the wind condition. However, since the radar engineer will seldom have any way of determining the significant wave height in advance, the use of Fig. 3.15 or Table 3.3 will generally prove more convenient.

Weather Clutter

The discussion of weather clutter by Goldstein, et al.* provides an excellent background of the subject, and more recent work has led to fairly accurate relationships by which radar reflectivity of various types of precipitation can be estimated. The new relationships are so accurate, in fact, that the meteorologists are depending upon them to estimate the total rainfall and other conditions, as an aid in prediction of floods and similar events. Battan† gives the cross section of an individual droplet as

$$\sigma_i = \frac{\lambda^2}{\pi} \alpha^6 \left| \frac{m^2 - 1}{m^2 + 2} \right|^2 = \frac{\pi^5}{\lambda^4} |K|^2 d_i^6 \tag{3.29}$$

Here, α represents $2\pi a/\lambda$, where a is the radius of the droplet; m is the complex index of refraction, K is the ratio $(m^2 - 1)/(m^2 + 2)$, and d_i is the droplet diameter. The magnitude $|K|^2$ for microwave radar is about 0.93 for raindrops and 0.2 for ice crystals and snow. The reflectivity of a cloud of droplets is given by the summation over a unit volume of the individual drop cross sections σ_i.

$$\eta = \sum_{\substack{\text{unit} \\ \text{vol.}}} \sigma_i = \frac{\pi^5}{\lambda^4} |K|^2 \sum_{\substack{\text{unit} \\ \text{vol.}}} d_i^6 = \frac{\pi^5}{\lambda^4} |K|^2 Z \tag{3.30}$$

The quantity Z, representing the summation of d_i^6 over the unit volume, is called the reflectivity factor. It has been found that the usual distributions of droplet diameters in rain and snow clouds lead to the following relationships:

$$Z \cong 200 R^{1.6} \text{ mm}^6/\text{m}^3 \quad \text{(rain)} \tag{3.31}$$

$$Z \cong 2000 R^2 \text{ mm}^6/\text{m}^3 \quad \text{(snow)} \tag{3.32}$$

* Herbert Goldstein, Donald E. Kerr, and Arthur E. Bent, "Meteorological Echoes," Chap. 7 in *Propagation of Short Radio Waves*, D. E. Kerr, ed. (New York: McGraw-Hill Book Company, 1951), pp. 588–640.

† Louis J. Battan, *Radar Meteorology* (Chicago: University of Chicago Press, 1959), p. 27.

See also Goldstein, et al., p. 596.

Here the symbol R represents the precipitation rate in mm/hour of water. Table 3.4 gives the approximate values of R and Z for different types of weather. The maximum observed values of Z range up to 10^6 in the lower troposphere, and perhaps to 10^4 at altitudes above 30,000 ft. Figure 3.16 shows the radar reflectivity for the different conditions as a function of frequency.

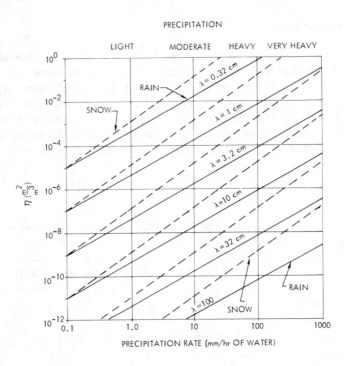

Figure 3.16 Radar reflectivity vs. precipitation rate.

The spectral spread of rain clouds is the most extreme of all types of clutter, as listed in Table 3.3. In addition to the spread of velocities within the cloud, the cloud mass will be in motion with the mean wind velocity. As a result of this, the use of conventional MTI, based upon a zero-velocity reference, is of limited value in reduction of weather clutter.

Chaff Clutter

The use of chaff was one of the earliest forms of radar countermeasure, and was particularly effective against the long-wave radars used in Europe during the major part of World War II. The theory pertaining to chaff

Table 3.4 REFLECTIVITY FACTOR FOR RAIN AND SNOW

Type of precipitation	Rate (mm/hr)	Reflectivity factor* Z (mm^6/m^3)
Light rain	0.5–5.0	60–2500
Moderate rain	5–25	2500–4×10^4
Heavy rain	25–125	4×10^4–5×10^5
Very heavy rain	125 and up	5×10^5 and up
Light snow	0.5–5.0	500–5×10^4
Moderate snow	5–25	5×10^4–1.2×10^6
Heavy snow	25–125	1.2×10^6–3×10^7

* *Note*: For Z in m^3, multiply figures in this column by 10^{-18} (maximum $Z \cong 10^{-12}$ m^3 results).

clutter is almost exactly that of weather clutter. The major difference lies in the fact that chaff dipoles can be cut to resonate at the radar frequency. Since the average cross section of a dipole with random orientation is $0.18\lambda^2$, and since the echoes may be considered to add noncoherently at the radar, the cross section of m dipoles is

$$\sigma = 0.18m\lambda^2 \qquad \text{(3.33)}$$

According to Schlesinger,* the total cross section of chaff weighing W pounds can be expressed as

$$\sigma \cong 3000\frac{W}{f_t} \text{ sq m} \qquad \text{(3.34)}$$

In the above, it is assumed that the chaff resonates at the radar frequency f_t (kmc), and that the dipoles are cut from material 0.01 in. wide and 0.001 in. thick. The distribution of chaff in the atmosphere during an aircraft operation is dependent upon the tactics used by the attacking force, but the total payload and the extent of the chaff cloud may be estimated to arrive at a figure for radar reflectivity η in m^2/m^3. As noted by Schlesinger, the chaff may have to be cut to match several different frequency bands used by different radars, with each chaff length covering a bandwidth of about $\pm$ 10 per cent. As a result, only a fraction of the total chaff payload will be available for jamming any particular radar.

The spectral spread of chaff will be approximately the same as precipitation in the same atmosphere. Table 3.3 shows generally smaller values for σ_v of chaff, since chaff is often used in a clear atmosphere without tur-

* Robert J. Schlesinger, *Principles of Electronic Warfare* (Englewood Cliffs, N.J.: Prentice-Hall, Inc., 1961), p. 130.

bulent winds. Rain, on the other hand, is usually accompanied by gusty winds. The ratio of wind speed to rms velocity spread will be approximately five to one for both chaff and rain clouds.

The radar equation is a basic relationship which permits the calculation of echo signal strength from known parameters of the radar transmitter, antenna, propagation path, and target. It may be extended to express the ratio of signal power to noise power in the radar receiver, and this ratio may then be used to calculate the performance to be expected of the radar as a detection or measurement device. If exact values of the parameters used in the radar equation are known, the equation will yield an exact value of received signal. In practice, however, knowledge of the radar-target parameters is imperfect and a "loss factor" is included to make the results of the equation agree with actual performance.

In this chapter, the basic relationships are derived and expressed in forms which lend themselves to easy and accurate computation. Definitions are given of the various parameters used, so that the equation may be applied in a precise fashion. The practical loss factors which enter into actual radar sys-

USING THE RADAR RANGE EQUATION

4

tems and observing methods are discussed, and means of estimating their magnitudes are described. Various forms of the equation, used in the literature, are compared and differences in application are described.

4.1 THE CLASSICAL RADAR EQUATION

The basic form of the radar equation has been derived in several previous works on radar.* The steps in the derivation will be repeated here to point out the exact definitions of the terms used and the assumptions which are implicit in the different forms of the equation. Assume first that the power of the radar transmitter, designated P_t, is radiated isotropically (i.e., uniformly distributed over the entire sphere surrounding the radar). Since the surface area of a sphere at a radius R from the radar is $4\pi R^2$, the uniform distribution of transmitted power over this surface will produce a power density given by

$$\left.\begin{array}{l}\text{Power per unit} \\ \text{area at range } R\end{array}\right\} = \frac{P_t}{4\pi R^2} \tag{4.1}$$

If a transmitting antenna of power gain G_t is now connected, the power density in the direction at which this gain applies will be increased to

$$\left.\begin{array}{l}\text{Power per unit area at range} \\ R \text{ in the given direction}\end{array}\right\} = \frac{P_t G_t}{4\pi R^2} \tag{4.2}$$

The above follows from Eq. (4.1) and the definition of power gain:† the directive gain of a transmitting antenna in a given direction, referred to an ideal isotropic radiator, is given by 4π times the ratio of the power radiated per unit solid angle in that direction to the total power delivered to the antenna.

The target at range R is illuminated by a wave whose power density is given by Eq (4.2), and the strength of the wave reflected towards the radar is

$$\left.\begin{array}{l}\text{Reflected power per unit solid} \\ \text{angle in direction of the radar}\end{array}\right\} = \frac{P_t G_t}{4\pi R^2} \times \frac{\sigma}{4\pi} \tag{4.3}$$

* See, for instance:

L. N. Ridenour, ed., *Radar System Engineering* (New York: McGraw-Hill Book Company, 1947), p. 21.

J. F. Reintjes and G. T. Coate, *Principles of Radar*, 3rd ed. (New York: McGraw-Hill Book Company, 1952), pp. 1–15.

Frederick E. Terman, *Electronic and Radio Engineering*, 4th ed. (New York: McGraw-Hill Book Company, 1955), p. 1016.

Donald J. Povejsil, Robert S. Raven, and Peter Waterman, *Airborne Radar* (Princeton, N.J.: D. Van Nostrand Company, Inc., 1961), p. 138.

† Samuel Silver, *Microwave Antenna Theory and Design* (New York: McGraw-Hill Book Company, 1949), p. 580.

The International Dictionary of Physics and Electronics, Walter C. Michels, senior ed. (Princeton, N.J.: D. Van Nostrand Company, Inc., 1956), p. 49.

The target cross section σ was defined in the previous chapter in such a way that the above equation follows from Eq. (4.2). The reflected wave arrives back at the radar with a power density given by $1/R^2$ times the power per unit solid angle reflected from the target.

$$\left.\begin{array}{l} \text{Reflected power per} \\ \text{unit area at receiver} \end{array}\right\} = \frac{P_t G_t}{4\pi R^2} \times \frac{\sigma}{4\pi} \times \frac{1}{R^2} \qquad (4.4)$$

The amount of power intercepted by the receiving antenna is this power density times the "effective aperture area" of the antenna, defined by

$$\left.\begin{array}{l} \text{Effective receiving} \\ \text{aperture} \end{array}\right\} = A_r = \frac{G_r \lambda^2}{4\pi} \qquad (4.5)$$

Here, G_r is the receiving power gain, and λ is the radar wave length.* The result, in terms of received signal power S, is given by

$$S = \frac{P_t G_t}{4\pi R^2} \times \frac{\sigma}{4\pi} \times \frac{1}{R^2} \times A_r = \frac{P_t G_t G_r \lambda^2 \sigma}{(4\pi)^3 R^4} \qquad (4.6)$$

This may be simplified if the same antenna is used for both transmitting and receiving, in which case the product $G_t G_r$ becomes G^2, giving the most common form of the radar range equation:

$$S = \frac{P_t G^2 \lambda^2 \sigma}{(4\pi)^3 R^4} \qquad (4.7)$$

Expressions for i-f Signal-to-Noise Ratio

Although Eq. (4.7) can be used directly to find the range at which the received signal power S is equal to some acceptable minimum value $S_{\min}$, as was done in some early works on radar,† it is preferable to introduce the concept of the "equivalent input noise level," designated N, to allow for the total system loss factor L, and to arrive at an equation for the signal-to-noise ratio in the i-f portion of the receiver. This ratio, S/N, is the starting point for most radar performance analyses, as indicated in the previous chapters, although for the reasons mentioned it falls short of measuring the total effectiveness of the radar.

* Frederick E. Terman, *Electronic and Radio Engineering*, 4th ed. (New York: McGraw-Hill Book Company, 1955), p. 899.

† L. N. Ridenour, ed. *Radar System Engineering* (New York: McGraw-Hill Book Company, 1947), p. 22.

J. F. Reintjes and G. T. Coate, *Principles of Radar*, 3rd ed. (New York: McGraw-Hill Book Company, 1952), pp. 1–16.

The equivalent input noise level can be found by considering first the definition of receiving system noise factor.*

$$\text{Noise factor} = \overline{NF} \equiv \frac{S/kTB}{(S/N)_o} \tag{4.8}$$

Here, S is the signal power as calculated from Eq. (4.7), k is Boltzmann's constant (1.38×10^{-23} w per cps per degree Kelvin), T is the absolute temperature of the signal source in degrees Kelvin, B is the equivalent noise bandwidth of the receiver in cps, and $(S/N)_o$ is the output signal-to-noise ratio of the receiver, measured prior to envelope detection but after i-f amplification. Present measurement standards† call for the noise factor of receivers to be measured with respect to an input termination at a reference temperature $T_o = 290°$ K. The noise factor equation may then be rewritten in the following form

$$\left(\frac{S}{N}\right)_o = \frac{S}{kT_oB\overline{NF}} \tag{4.9}$$

The system performance can most conveniently be expressed by assuming an ideal, noise-free receiver, with an input signal-to-noise ratio S/N representing the ratio of received signal power to equivalent input noise, such that $(S/N)_o = S/N$. For the case where the antenna may be considered to represent a source at the reference temperature, the equivalent input noise would be

$$N_r = kT_oB\overline{NF} \tag{4.10}$$

More generally, the antenna will have some other value of effective temperature designated T_a, and the equivalent input noise may be expressed in terms of an input temperature T_i which accounts for both antenna and receiver contributions to input noise.

$$N = k[T_o(\overline{NF} - 1) + T_a]B = kT_iB \tag{4.11}$$

Since the input temperature is an unfamiliar term in some discussions of radar, the "operating noise factor" $\overline{NF}_o$ may be used to describe system performance. This is defined by

$$\overline{NF}_o \equiv \frac{T_i}{T_o} = \overline{NF} - 1 + \frac{T_a}{T_o} \tag{4.12}$$

---

* S. N. Van Voorhis, *Microwave Receivers* (New York: McGraw-Hill Book Company, 1955), p. 2.

† Institute of Radio Engineers, *IRE Dictionary of Electronic Terms and Symbols* (1961), p. 96.

If the operating noise factor is substituted into Eq. (4.11), the equivalent input noise level N is given by

$$N = kT_oB\overline{NF}_o \qquad (4.13)$$

The operating noise factor will vary with environmental conditions such as sky temperature and elevation angle of the antenna, but its use is desirable because it describes the performance of the system directly, and may be compared on a decibel-for-decibel basis with other system parameters such as transmitter power and two-way antenna gain.

The combining of Eq. (4.7) with Eq. (4.13), with introduction of the total system loss factor L, results in one of the most useful forms of the radar range equation.

$$\frac{S}{N} = \frac{P_t G^2 \lambda^2 \sigma}{(4\pi)^3 R^4 k T_o B \overline{NF}_o L} \qquad (4.14)$$

Assumption Implicit in Equation for S/N

Since the signal-to-noise ratio found from Eq. (4.14) will be used extensively in analysis of radar performance, the assumptions on which this equation is based should be clearly understood:

1. The transmitter power P_t is radiated through the same antenna, with gain G, that is used in reception of the signal.
2. Departure from free-space propagation conditions must be accounted for by appropriate components of the loss factor L or by modification of the antenna gain G.
3. The transmitted power is assumed to lie within the bandwidth B of the receiver, after reflection from the target and propagation through the medium.
4. If P_t is the usual "peak transmitted power," which is actually the rms power during the pulse,* then S/N represents the ratio of mean-square signal voltage during the pulse to mean-square noise voltage, as measured at any point in the i-f portion of the receiver following the introduction of substantial gain and restriction of bandwidth to the value B.

This i-f or "single-pulse" S/N ratio does not measure the over-all effectiveness of the radar, but serves as an intermediate step in a number of further calculations. The effect of integration, as discussed in Chapters 1 and 2, enhances the performance of the system both for detection and

---

* Institute of Radio Engineers, *IRE Dictionary of Electronic Terms and Symbols* (1961), p. 103.

for measurement, and leads to results approaching what would be obtained if all the received energy were concentrated into a single pulse of power nS. However, the single-pulse S/N represents a quantity which can be calculated or measured more readily than the total received energy, and which can be visualized by the radar engineer as a starting point in system analysis. Several related quantities, often used in the literature on radar, may be found from the signal-to-noise ratio S/N. First, by setting S/N equal to unity and solving for range, we may find the reference range R_o, used as a normalizing factor in radar performance charts which give detectability or accuracy as a function of range for various target sizes.

$$R_o \equiv \begin{Bmatrix} \text{Range for} \\ \text{unity } S/N \end{Bmatrix} = \left[\frac{P_t G^2 \lambda^2 \sigma}{(4\pi)^3 k T_o B N F_o L} \right]^{1/4} \tag{4.15}$$

Alternate Forms of the Radar Equation

Different aspects of the radar problem are expressed more clearly by altering the form of Eq. (4.14) and introducing other radar parameters. For example, substitution of the effective receiving aperture area A_r, which was used in the derivation of the radar equation, demonstrates that the S/N ratio is independent of wave length for a given transmitting antenna gain and receiving aperture.

$$\frac{S}{N} = \frac{P_t G A_r \sigma}{(4\pi)^2 R^4 k T_o B N F_o L} \tag{4.16}$$

If this same aperture is used for transmitting as well, an inverse dependence upon λ^2 is shown.

$$\frac{S}{N} = \frac{P_t A_r^2 \sigma}{4\pi R^4 \lambda^2 k T_o B N F_o L} \tag{4.17}$$

More importantly, for the special case where the receiver is matched to the spectrum of the transmitted pulse of length τ, so that $B\tau = $ unity, the pulse energy $P_t \tau$ appears in the numerator as an important factor.

$$\frac{S}{N} = \frac{P_t \tau G^2 \lambda^2 \sigma}{(4\pi)^3 R^4 k T_o N F_o L} \tag{4.18}$$

Because of the more fundamental nature of the energy ratio, defined in Chapter 1 as the ratio of received signal energy to noise power density per unit bandwidth (see p. 14), this parameter $\mathcal{R}$ is frequently used in place of the i-f power ratio S/N. As was shown, the value of $\mathcal{R}$ for a single pulse is equal to twice the S/N ratio in the matched system, whereas the value for n pulses observed over a period t_o at a repetition rate f_r is

$$\mathcal{R} = \frac{2E}{N_o} = \frac{2S\tau n}{N/B} = 2t_o f_r \tau B \frac{S}{N} = 2n\frac{S}{N} \tag{4.19}$$

The radar equation in terms of energy ratio may then be written by combining Eq. (4.19) with Eq. (4.18)

$$\mathcal{R} = \frac{2P_t \tau f_r t_o G^2 \lambda^2 \sigma}{(4\pi)^3 R^4 k T_o \overline{NF_o} L} \tag{4.20}$$

This equation is simplified if the average transmitter power $P_{av} = P_t \tau f_r$ is used

$$\mathcal{R} = \frac{2P_{av} t_o G^2 \lambda^2 \sigma}{(4\pi)^3 R^4 k T_o \overline{NF_o} L} \tag{4.21}$$

In this form, it also shows that the energy ratio is completely independent of the type of modulation imposed on the transmission.

The energy ratio concept is advantageous in that it properly frees the analysis of system performance from arbitrary limitations of modulation wave form, and demonstrates a dependence only upon the total energy exchanged with the target and the noise density at the receiver input. It should be used with caution, however, for several reasons. First, in practical radar analysis problems, it is desirable, if not essential, to know the i-f S/N ratio, because this often determines whether a particular method of signal processing can be used, or whether the "detector loss" encountered will be excessive.* Second, as was mentioned earlier, it is the i-f S/N ratio which can be measured and observed most directly. Third, the confusion introduced by the factor of two and the difference between the single-sided and the double-sided noise spectral density may lead to errors in analysis unless the particular equation has been checked carefully to find which definition has been used in the derivation. Last, the fact that N_o has the physical dimensions of energy has led some writers to refer to it as the "noise energy per cycle" (per cycle of what?). It is actually defined as noise power per cycle per second of bandwidth. The only energy which is susceptible to measurement (in a calorimeter, for instance) is the product

* Wilbur B. Davenport, Jr. and William L. Root, *An Introduction to the Theory of Random Signals and Noise* (New York: McGraw-Hill Book Company, 1958), pp. 266-67. They refer to this loss as the "small-signal suppression effect."

of N_o times the receiver bandwidth B times the observation time t_o. It is seldom that the i-f bandwidth is matched to the observation time in a practical system.

Simplified Forms of the Radar Equation for Calculation

A major problem in use of the radar equation, when any of the forms given so far is used, is the performance of conversion of units to a common system, and the manipulation of the other constants and factors of ten which appear in the various terms. It has proved convenient to use a mixed system of units corresponding to those most often used in specifying and describing radar parameters. The equations are then converted to a logarithmic form, decibels being used, so that the result can be found by simple processes of addition and subtraction. In the first simplifying step, the consistent units of length required in all the previous equations are converted to a mixed system, in which wave length is expressed in centimeters, cross section in square meters, and range in nautical miles (6076 ft). When the resulting conversion constants are grouped with the constant terms $(4\pi)^3$ and kT_o, it is found that Eq. (4.14) assumes the following form

$$\frac{S}{N} = \frac{P_t G^2 \lambda^2 \sigma}{R^4 B N F_o L} \times \left(\frac{m}{cm}\right)^2 \times \left(\frac{nmi.}{m}\right)^4 \times \left(\frac{1}{4\pi}\right)^3 \times \frac{1}{290 \times 1.38 \times 10^{-23}}$$

$$= \frac{P_t G^2 \lambda^2 \sigma}{R^4 B N F_o L} \times 1.07 \tag{4.22}$$

If the usual approximation of 2000 yd = 1 nmi is used, the final constant reduces essentially to unity, the resulting simplified equation containing only four radar-target parameters in the numerator and four in the denominator.

The final simplification comes about when Eq. (4.22) is expressed in logarithmic form, with all parameters given in decibels with respect to the mixed system of units used above.

$$\left(\frac{S}{N}\right)_{db} = (P_t)_{dbw} + 2(G)_{db} + 2(\lambda)_{dbcm} + (\sigma)_{dbm^2} - 4(R)_{dbnmi}$$
$$- (B)_{dbcps} - (NF_o)_{db} - (L)_{db} \tag{4.23}$$

The use of the decibel form in expressing signal-to-noise ratios, power level, gain, cross section, noise factor, and loss is common practice in communications and electronic engineering. Extension to wave length (in decibels relative to one centimeter), bandwidth (referenced to one cycle per second) and range (referenced to one nautical mile) is a logical step.

In each case, the decibel quantity is calculated as $10 \log_{10}$ of the corresponding radar parameter, and is indicated by the original symbol in parentheses with the appropriate reference. Thus, a bandwidth of one megacycle (per second) would be written as $(B)_{\text{dbcps}} = +60$. The unique advantage of this type of calculation is that it permits answers to be found rapidly with minimum opportunity for arithmetic error. It is a pure coincidence that the use of the particular mixed system of units given above leads to elimination of all the conversion constants and physical constants in the radar equation.

The following expressions are the logarithmic forms corresponding to Eqs. (4.15), (4.18), (4.20), and (4.21).

For reference range R_o

$$4(R_o)_{\text{dbnmi}} = (P_t)_{\text{dbw}} + 2(G)_{\text{db}} + 2(\lambda)_{\text{dbcm}} + (\sigma)_{\text{dbm}^2}$$
$$- (B)_{\text{dbcps}} - (\overline{NF_o})_{\text{db}} - (L)_{\text{db}} \qquad \textbf{(4.24)}$$

For i-f S/N ratio (matched receiver)

$$\left(\frac{S}{N}\right)_{\text{db}} = (P_t)_{\text{dbw}} + (\tau)_{\text{dbs}} + 2(G)_{\text{db}} + 2(\lambda)_{\text{dbcm}} + (\sigma)_{\text{dbm}^2}$$
$$- 4(R)_{\text{dbnmi}} - (\overline{NF_o})_{\text{db}} - (L)_{\text{db}} \qquad \textbf{(4.25)}$$

Energy ratio in terms of peak power

$$(\mathscr{R})_{\text{db}} = (P_t)_{\text{dbw}} + (\tau)_{\text{dbs}} + (f_r)_{\text{dbcps}} + (t_o)_{\text{dbs}} + 2(G)_{\text{db}} + 2(\lambda)_{\text{dbcm}}$$
$$+ (\sigma)_{\text{dbm}^2} - 4(R)_{\text{dbnmi}} - (\overline{NF_o})_{\text{db}} - (L)_{\text{db}} + 3 \qquad \textbf{(4.26)}$$

Energy ratio in terms of average power

$$(\mathscr{R})_{\text{db}} = (P_{\text{av}})_{\text{dbw}} + (t_o)_{\text{dbs}} + 2(G)_{\text{db}} + 2(\lambda)_{\text{dbcm}} + (\sigma)_{\text{dbm}^2}$$
$$- 4(R)_{\text{dbnmi}} - (\overline{NF_o})_{\text{db}} - (L)_{\text{db}} + 3 \qquad \textbf{(4.27)}$$

In the above equations, the pulse width and observation time are expressed in decibels referred to one second. Similar forms can be derived for the equations using antenna aperture, with appropriate conversion constants expressed in decibels to account for the choice of the unit for area of the antenna, and to include the factor of 4π in Eq. (4.5).

Example of Range Calculation

As an example of the use of these equations, let us assume the following radar parameters

$$P_t = 1.0 \text{ Mw} = +60 \text{ dbw}$$
$$G = 44 \text{ db}$$
$$\lambda = 5.6 \text{ cm} = +7.5 \text{ dbcm}$$
$$B = 1.6 \text{ mc} = +62 \text{ dbcps}$$
$$\overline{NF_o} = 10 \text{ db}$$
$$L = 4 \text{ db}$$

To calculate the reference range R_o at which a zero decibel signal-to-noise ratio is achieved, let us assume a cross section $\sigma = 1.0 \text{ m}^2 = 0 \text{ dbm}^2$. Then, using Eq. (4.24), we obtain

$$4(R_o)_{\text{dbnmi}} = +60 + 88 + 15 + 0 - 62 - 10 - 4 = +87 \text{ dbnmi}$$
$$(R_o)_{\text{dbnmi}} = +21.75 \text{ dbnmi}$$
$$R_o = 150 \text{ nmi}$$

The same results would apply for a pulse width $\tau = 0.62 \ \mu\text{sec}$ in a matched system, if Eq. (4.25) is used. In this case also we can find that the range for $\mathcal{R}_1 = $ unity is greater by a factor $\sqrt[4]{2}$

$$4(R)_{\text{dbnmi}} = +87 + 3 = +90 \text{ dbnmi}$$
$$(R)_{\text{dbnmi}} = +22.5 \text{ dbnmi}$$
$$R = 178 \text{ nmi}$$

If a longer pulse were used without reduction in received bandwidth, the S/N ratio and R_o would remain the same, since Eqs. (4.23) and (4.24) are not dependent upon pulse width. However, the energy ratio would be increased in proportion to pulse width. This would indicate that an improvement in system performance should be possible by better matching. Steps in this direction would include reduction in i-f bandwidth, or possibly the use of a narrow video bandwidth after envelope detection. Unless such steps are taken, Eq. (4.25) cannot be used except as an indication of the potential value of $2S/N$ to be achieved by matching the system.

4.2 THE BEACON EQUATIONS

When the radar interrogates a beacon or "transponder" and receives a reply whose power is fixed by the beacon transmitter characteristics, separate calculations are necessary for the interrogation and response paths. The basic expression for the power available at the beacon receiver for interrogation is

$$S_b = \frac{P_t G G_b \lambda^2}{(4\pi)^2 R^2 L_t L_b} \tag{4.28}$$

Here, G_b is the beacon antenna gain in the direction of the radar; L_t is that part of the system loss which lies between the radar transmitter and the beacon antenna, and L_b is the loss between the beacon antenna and beacon receiver. The other symbols are as used in Eq. (4.7), and use of consistent units of length for λ and R is assumed. This equation may be derived by combining Eq. (4.2), for the power density incident at the beacon, with Eq. (4.5), for the aperture area of the beacon antenna, and adding the two loss factors.

On the response path, the power available at the radar receiver can be found by using a similar equation in which the beacon transmitter power P_b is substituted for P_t, and a return path loss L_r is substituted for L_t. The loss L_b is normally assumed to be the same for both interrogation and response, although two different values may be used if conditions dictate this. The signal-to-noise ratio in the radar i-f may then be expressed as

$$\frac{S}{N} = \frac{P_b G G_b \lambda^2}{(4\pi)^2 R^2 k T_o B N F_o L_r L_b} \tag{4.29}$$

The same simplifications and mixed system of units may also be used for the beacon case, yielding the following

Beacon interrogation signal power

$$(S_b)_{\text{dbw}} = (P_t)_{\text{dbw}} + (G)_{\text{db}} + (G_b)_{\text{db}} + 2(\lambda)_{\text{dbcm}} - 2(R)_{\text{dbnmi}} \tag{4.30}$$
$$- (L_t)_{\text{db}} - (L_b)_{\text{db}} - 127$$

Response S/N at radar

$$\left(\frac{S}{N}\right)_{\text{db}} = (P_b)_{\text{dbw}} + (G)_{\text{db}} + (G_b)_{\text{db}} + 2(\lambda)_{\text{dbcm}} - 2(R)_{\text{dbnmi}} \tag{4.31}$$
$$- (B)_{\text{dbcps}} - (\overline{NF_o})_{\text{db}} - (L_r)_{\text{db}} - (L_b)_{\text{db}} + 77$$

In radar countermeasures, the above equations are used to find the power available for an intercept receiver (analogous to S_b) or the jamming-to-noise ratio J/N at the radar receiver (analogous to S/N). Another relationship which is often used is the ratio of beacon (or jamming) power to echo power in the radar receiver.

$$\frac{J}{S} \quad \text{or} \quad \frac{S(\text{beacon})}{S(\text{echo})} = \frac{4\pi P_b G_b R^2 L_t}{P_t G \sigma L_b} \tag{4.32}$$

The above form uses consistent units of length, and is subject to the same simplifications used earlier to convert the radar and beacon equations to logarithmic form.

$$\left(\frac{J}{S}\right)_{db} = (P_b)_{dbw} + (G_b)_{db} + 2(R)_{dbnmi} + (L_t)_{db} - (L_b)_{db}$$

$$- (P_t)_{dbw} - (G)_{db} - (\sigma)_{dbm^2} + 77 \qquad \textbf{(4.33)}$$

As an example of using these equations, let us assume the same radar-target parameters given in the previous example, along with a beacon power of one watt peak, an omnidirectional beacon antenna, no beacon loss, and equal division of the four decibel system loss between L_t and L_r. At a range of 100 mi, we shall have

$$(S_b)_{dbw} = +60 + 44 + 15 + 0 - 40 - 2 - 127 = -50 \text{ dbw}$$

$$\left(\frac{S}{N}\right)_{db} = 0 + 44 + 0 + 15 - 40 - 60 - 10 - 2 + 77 = +24 \text{ db}$$

$$\frac{S(b)}{S(e)} = 0 + 0 + 40 + 2 - 2 - 60 - 44 - 0 + 77 = +13 \text{ db}$$

In a jamming case, the $+13$ db ratio would describe the J/S ratio if one watt of jamming power were emitted from the omnidirectional antenna within the bandwidth of the radar receiver, occupying the same interval of time at which the echo signal reflection took place. This would be the case if broadband noise with a power density of one watt per megacycle were emitted over the band used by the radar. The average jamming power required would then be given by one watt times the total tuning band of the radar, or that part of the band in which the radar was known to be operating.

4.3 EQUATIONS FOR BISTATIC RADAR SYSTEMS

In some cases, the radar transmitter and receiver may be widely separated, or may use different antennas for other reasons. By returning to the derivation of the radar equation, and modifying the assumptions to account for different transmitter and receiver sites, an equation for bistatic radar is developed.

$$\frac{S}{N} = \frac{P_t G_t G_r \lambda^2 \sigma_b}{(4\pi)^3 R_t^2 R_r^2 k T_o BN F_o L_t L_r} \qquad \textbf{(4.34)}$$

In the above equation, the subscripts designate the different parameters for the transmitting site and receiving site, and the cross section σ_b is the "bistatic cross section" defined in such a way that it accounts for the re-radiation in the direction of the receiver. Since the conversion constants are the same as those used in the basic equation, the logarithmic form may also be used here without the appearance of a constant term.

$$\left(\frac{S}{N}\right)_{db} = (P_t)_{dbw} + (G_t)_{db} + (G_r)_{db} + 2(\lambda)_{dbcm} + (\sigma_b)_{dbm^2}$$
$$- 2(R_t)_{dbnmi} - 2(R_r)_{dbnmi} - (B)_{dbcps} \qquad \textbf{(4.35)}$$
$$- (\overline{NF_o})_{db} - (L_t)_{db} - (L_r)_{db}$$

When either form is used, the equation gives the same results as would be obtained by introducing the geometrical mean values of antenna gain and range, over the two separate paths, into Eq. (4.14) or Eq. (4.23).

An important characteristic of bistatic radar is found when the angle between the transmitter and receiver paths approaches $180°$. In this "forward scatter" case, the bistatic cross section σ_b may greatly exceed the normal backscattering coefficient. This is due to the fact that the total power in the forward-scatter lobe is equal to that scattered over the remainder of the 4π steradians around the target.* If the receiving antenna can be placed within the narrow beam of the forward scatter, which is in the order of λ/L in width, the special benefits of the forward gain of the target may be realized, and the bistatic cross section will approach a value

$$\sigma_f = \frac{4\pi A^2}{\lambda^2}$$

This formula, derived originally by Siegel,[†] is the same as the flat-plate cross section measured normal to a plate whose area A is the same as the projected area of the target. A more detailed discussion of bistatic radar has appeared in the literature,[‡] showing the limitations of this type of system, and concluding that "the bistatic radar...cannot compete with monostatic radar in most radar system applications."

4.4 LOSS FACTORS FOR THE ON-AXIS TARGET

The several forms of the radar range equation [Eqs. (4.14) to (4.27)] include the basic radar-target parameters and a "system loss factor" de-

---

* See Section 16.1.

[†] K. M. Siegel, et al., "Bistatic Radar Cross Sections of Surfaces of Revolution," *Journal of Applied Physics*, **26**, No. 3 (March 1955), pp. 297-305.

K. M. Siegel, "Bistatic Radars and Forward Scattering," *Proc. 1958 Natl. Conf. on Aero. Electronics*, Dayton, Ohio, pp. 286-90.

[‡] Merrill I. Skolnik, "An Analysis of Bistatic Radar," *Trans. IRE*, ANE-8, No. 1 (March 1961), pp. 19-27.

signated by L. Several possible components of system loss have already been discussed as they arose in the derivation of the equations for detection probability in Chapter 1. These terms cover loss due to filter matching, fluctuation in target amplitude, integration of many pulses after envelope detection, and collapsing loss. These will be discussed in more detail here, and other sources of loss will be analyzed. For the present, the discussion will be limited to cases where the target is in the center of the radar beam. Losses which are peculiar to search or to tracking radars will be covered in later chapters on those subjects.

R-F Transmission Line Loss

The basic radar parameters of transmitter power, antenna gain, and receiving system noise factor (or noise temperature) are all influenced by transmission line loss. The radar equation given in the form of Eq. (4.14) is capable of providing exact results for free-space propagation conditions with no loss factor, provided that all these three basic radar parameters are measured at a single common point in the r-f line, or are referred to such a point. When this is done, any loss in the line between the transmitter and the reference point appears as a reduction in the transmitter power entered into the equation, and similarly for losses between the receiver or antenna, and the common point. It has long been the common practice, however, to specify and measure the transmitter power at the output of the final stage of the transmitter, regardless of what lies between this point and the antenna. A transmitter loss L_t must then be measured or calculated to account for attenuation in the transmission line, duplexer, and other components which carry the power to the antenna. In most practical radars, this loss will be in the order of one or two decibels, and may be estimated from charts or tables which give the loss in decibels per foot for the common types of coaxial line or wave guide at a given frequency. Typical radar duplexers introduce loses of about $\frac{1}{2}$ db in transmitter power.

Further losses may result from mismatch in the line (high voltage standing wave ratio or VSWR), or from the use of isolators (which protect the output tubes against arcing or breakdown from standing waves), from directional couplers used in monitoring output power, from variable attenuators, power splitters, rotary joints, switches, harmonic filters, and other specialized components in the transmission line. In the receiving line, the loss will be similar.* Those elements which are in the common path used for transmission and reception must be included both in L_t and L_r,

---

* Blake points out that the "available loss" should be used. This is the loss that would apply if all receiving circuits were matched in impedance. See Lamont V. Blake, "Antenna and Receiving-System Noise-Temperature Calculation," Naval Res. Lab. Report 5668 (Sept. 19, 1961).

unless they are applied as a reduction in antenna gain. In either case, they count double in the system loss, and also enter into the calculation of input temperature as discussed below.

Noise Factor

The operating noise factor defined in Eq. (4.12) may be measured or calculated at various points in the receiving system. If it is referred to the common r-f reference point used also for transmitter power and antenna gain, there need be no separate allowance for receiving loss L_r. More commonly, a "receiver" noise factor is specified and measured at the duplexer or between the duplexer and receiver input. In this case, the receiver noise factor must be increased by the amount of the loss factor L_r, or the L_r term must be included as a component of the system loss factor. In addition, for low-noise systems, the temperature contribution of L_r, as well as those of the antenna and the sky, must be calculated and used in finding total input temperature T_i and operating noise factor $\overline{NF_o}$. One procedure* starts with determination of the effective receiver temperature T_e using the relationship

$$T_e = T_o(\overline{NF} - 1) \qquad (4.36)$$

Here, T_o is taken as 290° K and $\overline{NF}$ is the conventional receiver noise factor, defined in Eq. (4.8). This noise factor may be referred to any point in the passive r-f line system preceding the receiver, depending upon where it is convenient to connect the measurement equipment. A second temperature contribution is then calculated for those portions of the r-f line which lie between the receiver reference point and the antenna.

$$T_r = T_o\left(1 - \frac{1}{L_r}\right) = \frac{T_o(L_r - 1)}{L_r} \qquad (4.37)$$

The loss factor L_r may include a gas-filled duplexer, characterized as a passive lossy element. If excess noise is generated by the gas discharge, it is added to T_r along with the loss temperature calculated above.

As a third temperature component, the antenna noise temperature T_a is calculated or measured, and referred to the receiver input as a slightly reduced term $T_a' = T_a/L_r$. Data from which T_a may be calculated are presented in a later chapter on propagation effects, and a series of plots for various frequencies and elevation angles is available in the literature.[†]

---

* Lamont V. Blake, "Interim Report on Basic Pulse-Radar Maximum Range Calculation," Naval Res. Lab. Memo Report 1106 (Nov. 1960), p. 155.

† Lamont V. Blake, "Antenna and Receiving-System Noise-Temperature Calculation," Naval Res. Lab. Report 5668 (Sept. 19, 1961).

The total input temperature T_i' referred to the receiver is the sum of the three terms

$$T_i' = T_a' + T_r + T_e = \frac{T_a}{L_r} + \frac{T_o(L_r - 1)}{L_r} + T_o(\overline{NF} - 1) \quad \text{(4.38)}$$

(The actual temperature of the transmission line should be used in the second term if different from T_o.) Using this receiver input temperature T_i' to calculate a value of operating noise factor referred to the receiver input requires that we include the loss term L_r also in the system loss L. An equivalent alternate procedure uses the total input temperature referred to the antenna terminal to calculate a system operating noise factor, in which case L_r need not be included as a separate component in system loss. In this case, the radar system input temperature T_i is found from

$$
\begin{aligned}
T_i &= T_a + T_o(L_r - 1) + L_r T_o(\overline{NF} - 1) \\
    &= T_a + T_o(L_r \overline{NF} - 1)
\end{aligned}
\quad \text{(4.39)}
$$

The alternate procedure is simpler, in that it requires only that the receiver noise factor be increased by L_r before being converted to a temperature term.

Antenna Losses

In deriving the radar equation, the "directive gain" G of the antenna was used to establish the transmitted power density at the target [Eq. (4.2)]. The received power was was determined by using the effective receiving aperture $A_r = G\lambda^2/4\pi$. Direct measurements of the antenna gain can seldom be made, and a gain figure is usually obtained by assuming an effective aperture equal to the physical aperture A times an antenna efficiency η_a, so that

$$G = \frac{4\pi A}{\lambda^2}\eta_a \quad \text{(4.40)}$$

Typical values of the efficiency factor η_a vary from 40 per cent to 70 per cent, depending upon antenna design. If an overly optimistic value of η_a is initially assumed, a compensating loss in antenna efficiency, designated L_n, must be included wherever G is found in the radar equations. This leads to a component L_n^2 in the system loss L. Another common relationship used to estimate gain is based upon the measured values of beamwidth in the two axes.

$$G = \frac{4\pi k}{\theta_a \theta_e} \qquad \text{(4.41)}$$

Here, θ_a and θ_e are the azimuth and elevation beamwidths, expressed in radians. If expressed in degrees, with k taken as 60 per cent, the expression becomes $25,000/\theta_a\theta_e$, an approximation often used when only the beamwidths are given.* When such an expression is used as the basis for calculating gain, an antenna loss term L_n may be found necessary to account for ohmic losses in the feed system, for excessive spillover loss, or for higher than usual side lobe or background levels in the antenna pattern. More accurate gain figures can be established by integrating the measured power density over 4π steradians of solid angle around the antenna, a procedure which accounts for all radiated power and reflection effects, whether resulting from side lobes, spillover, background illumination, or main lobe radiation. Ohmic losses are then the only contributor to the antenna loss L_n.

Atmospheric Attenuation and Propagation Factor

A later chapter will cover the absorption of radar signals by the atmosphere and ionosphere, and the effects of ground reflections, refraction, and ducting in modifying the free-space conditions assumed in derivation of the radar equation. The attenuation will be expressed in terms of a loss in decibels per mile of radar range through a given medium at a given frequency. A total loss L_a encountered in penetrating the entire atmosphere and ionosphere at various frequencies and elevation angles will also be plotted for the long-range radar case. For the beacon case, the loss in each direction will be one-half of the decibel value of L_a given for radar.

A more complex problem is the consideration of ground reflections, which are often included in the radar equation through the introduction of a "propagation factor" F in the radar range equation.† The effective antenna gain in the presence of reflections may be considered to be F^2 times the free-space gain, or the effective radar range may be found as F times the free-space value. If total reflection from the ground or sea surface is experienced, the maximum value of F will reach two, at points where the direct and reflected rays produce constructive interference. The radar range at such points is doubled, but complete nulls are formed at intermediate points by destructive interference. Since the F factor is quite variable with

* M. S. Wheeler, "Antennas and RF Components," Sects. 10-1–10-9 in *Airborne Radar*, by D. J. Povejsil, R. S. Raven, and P. Waterman (Princeton, N. J.: D. Van Nostrand Company, Inc., 1961), p. 522.

† Lamont V. Blake, "Interim Report on Basic Pulse-Radar Maximum Range Calculation," Naval Res. Lab. Memo Report 1106 (Nov. 1960).

operating conditions of the radar and with conditions of the surrounding surface, it may best be considered as a system loss component $L_x = 1/F^4$, and evaluated separately for conditions of interest. If averaged over a wide interval of elevation angles and surface conditions, the average value L_x will be found to be somewhat less than unity, indicating a small system gain caused by the equivalent image antenna contributed by the surface reflections.

Loss from Unmatched Receiver Bandwidth

The previous discussion of North's work (Section 1.2) has shown that the optimum value for the ratio of peak detected signal to rms noise can be achieved only when the receiver bandpass is matched to the transmitted spectrum. This condition can be approximated only with simple filters in the i-f amplifiers. For a rectangular pulse of width τ, the filter matching loss L_m (equal to the reciprocal of the ratio $E_s^2/2\psi_o$ as plotted in Fig. 1.11) has a minimum value of 1.2, or 0.8 db, when stagger-tuned i-f amplifiers are used with essentially rectangular bandpass. For synchronously tuned i-f amplifiers, the minimum loss was 1.1, or 0.4 db. In some cases, where the i-f bandwidth B is greater than the optimum value, a major portion of the matching loss L_m can be restored by reduction of the postdetection (or video) bandwidth to a value near half of the optimum i-f bandwidth. This will have the effect of substituting a collapsing loss (see below) for the filter matching loss which would be found from Fig. 1.11.

As an example of this loss evaluation, let us assume that a rectangular i-f filter is used, with a bandwidth of three megacycles, and that the pulse width is one microsecond. Figure 1.11 shows a matching loss of almost five decibels in average or mid-point S/N ratio, relative to the matched filter. This loss is about four decibels relative to the peak amplitude of the pulse output of the optimum rectangular filter ($B\tau = 1.4$). A portion of this loss can be regained if a low-pass video filter is used after the envelope detector. For instance, if the video bandwidth is 0.7 mc, we could obtain performance approaching that of the optimum rectangular filter, but falling below it by the value of the collapsing loss shown in Fig. 1.16(a). The collapsing ratio ρ is calculated from Eq. (4.44) (below), and found to be

$$\rho = \frac{1.4 + 3.0}{1.4} = 3.1$$

The corresponding value of L_c will be between 0.6 and 2.1 db, depending on the number of pulses n which are integrated to obtain detection. (We have used the "constant n_f" curve of Fig. 1.16(a) in this calculation, to arrive at a comparison with the case of matched i-f bandwidth, $B\tau = 1.4$.

The number of range resolution elements per sweep, denoted by η in Eq. (1.13), will be the same whether the i-f bandwidth is set at $1.4/\tau$, or the video bandwidth at $0.7/\tau$.)

The collapsing loss is seen to be less than the matching loss which would have been present with unrestricted video bandwidth, but there has been an irreversible loss owing to the presence of the extra noise power at the second detector. The use of excessive i-f bandwidth followed by matched video bandwidth may be considered equivalent to dividing each complete signal pulse into ρ separate, shorter pulses, each characterized by the S/N ratio which applied to the original pulse in the wide i-f bandwidth. These separate pulses, contaminated with independent noise samples, are then integrated by the video filter, and the integrated pulse groups may be further integrated over several repetition periods on a display tube. The result is the same whether the ρn pulses are equally spaced (as in video integration of a pulse train at high repetition rate), or whether they are integrated in n groups containing ρ adjacent pulses in each group. In either case, the loss incurred through the use of a wide i-f bandwidth is partially regained by integration of ρ times as many pulses (or portions of each pulse). In our example, the integration gain will be about four decibels, depending upon the number of complete pulses which was used for detection. If, for instance, ten of the one microsecond pulses are to be integrated, then the collapsing loss curve for $n = 10$ must be used, giving a loss of about 1.6 db for $\rho = 3.1$. As was shown in Chapter 1, the loss is greater for large n because the number of noise samples is increased by collapsing from n to ρn, rather than from one to ρ as in the case of a single-pulse detection scheme.

Gain Stability, Threshold Stability, and Operator Loss

The probabilities of detection and false alarm were calculated in terms of threshold settings at fixed levels above the rms noise voltage applied to the detector. In practice, some tolerance will have to be allowed, to accommodate drift in the threshold level and in the receiver gain level. Also, the input noise level may vary as a function of operating conditions. These drifts will cause the false-alarm rate to exceed the specified value unless the threshold is set initially above the calculated level. Drifts in the opposite direction will reduce the probability of detection, as will upward drifts in the threshold level. The resulting loss must be accounted for in the radar equation, if the signal-to-noise ratio found from this equation is to be used to find detection probability. This loss, denoted by L_g will depend upon the relative stability of gain, noise, and threshold levels, upon the time which elapses between adjustments and the accuracy of these adjustments, and on other characteristics of the specific circuits used in the radar. If a visual presentation, such as a PPI display, is used as the output device, it may be possible for an alert operator to readjust the receiver gain or his mental "threshold" judgment, to maintain a consistent level of performance. Whether or not this is done, there will be some "operator loss" in most practical operating situations. This loss, denoted by L_o, may amount to as little as 1 or 2 db, or may reach levels near 10 db when the operator is placed under poor conditions. Few systems will be able to avoid a loss of at least one decibel from the above factors, even when automatic equipment of considerable refinement is used.

Integration Loss and Integrator Circuit Loss

The theoretical minimum loss owing to integration of n pulses in an ideal video integrator was given as L_i in Fig. 1.15 and Eq. (1.11). To this loss must be added a circuit loss denoted by L_c, which depends upon how well the actual integrator circuit conforms to the uniform weighting function which was assumed in Marcum's analysis, and upon the possible introduction of excess noise in the circuits (e.g., electron tube shot noise, or noise caused by distortion or limiting in the integrator). In most cases, a continuously operating integrator with an exponential weighting function is used instead of the discontinuous or square response function which would correspond to the ideal. The integrator with exponential decay, such as the phosphor of the cathode-ray-tube display or the recirculating delay line, may introduce a loss of about one decibel relative to the optimum case for the steady target.* For fluctuating targets, the exponential characteristic appears to cause no more loss than does uniform weighting, although the usual fluctuation loss must still be considered. When the integrator stores the signals from n pulse repetition periods and then "dumps" the result to start accumulating the next n signals, there will also be a loss in the order of one decibel, owing to the fact that the signals from one target may be split between two integration cycles.

Collapsing Loss for Specific Cases

The general theory of collapsing loss was discussed in Chapter 1, and plots of L_c vs collapsing ratio ρ were presented in Fig. 1.16. Procedures for calculating the collapsing ratio will now be given, along with rules for determining which of the two curves (constant $n_{f'}$ or constant $\rho n_{f'}$) will apply to a given case. It will be recalled that the more serious loss is encountered when the number of independent opportunities for false alarms, denoted by $n_{f'}$, remains constant with varying ρ. When this number is reduced by

* Merrill I. Skolnik, *Introduction to Radar Systems* (New York: McGraw-Hill Book Company, 1962), p. 39.

the ratio ρ, the threshold may be reset slightly lower to regain a portion of the loss, and the reduced loss of Fig. 1.16b (constant $\rho n_{f'}$) will apply. These lower-loss cases will be discussed first.

1. *Cathode-Ray Sweep Speed.* Marcum* gives the collapsing ratio for display tubes as

$$\rho_1 = \frac{d + s\tau}{s\tau} \qquad\qquad \textbf{(4.42)}$$

Here, d is the diameter of the spot on the display screen, in millimeters or inches; s is the sweep speed in the same unit of length per second, and τ is the pulse width in seconds.

2. *Insufficient Video Bandwidth.* An effect similar to that of overlapping spot positions on the display may be caused by the use of video bandwidth B_v narrower than one-half the matched i-f bandwidth $B_t = 1/\tau$. The collapsing ratio for this case is†

$$\rho_2 = \frac{2B_v + 1/\tau}{2B_v} = \frac{2B_v + B_t}{2B_v} \qquad\qquad \textbf{(4.43)}$$

3. *Collapsing of Coordinates.* In using a two-dimensional display for three coordinates, the several independent resolution elements in the third, undisplayed coordinate are superimposed upon each other, or "collapsed." The collapsing ratio for this case is equal to the ratio of the total search interval covered during the integration time to the size of the corresponding resolution element.

$$\rho_3 = \frac{\omega_e t_v}{\theta_e} \quad \text{or} \quad \frac{\omega_a t_v}{\theta_a} \quad \text{or} \quad \frac{2\Delta R}{\tau c} \qquad\qquad \textbf{(4.44)}$$

Here, ω_e is the elevation scan speed in angular units per second; θ_e is the elevation beamwidth in the same angular units; t_v is the decay time of the integrator ($t_v = n/f_r$); ω_a is the azimuth scan speed; θ_a is the azimuth beamwidth; ΔR is the range search interval, and τ is the pulse width. For optimum results ($\rho = 1$), the integration time t_v should be set equal to the time-on-target $t_b = \theta/\omega$, in whichever angular coordinate is being scanned. The three

---

* J. I. Marcum, "A Statistical Theory of Target Detection by Pulsed Radar: Mathematical Appendix," RAND Corp. Research Memo RM-753 (July 1, 1948), p. 59.

† The expression $\rho = (B_{if} + B_v)/B_v$ appearing on p. 60 of Marcum's report and p. 228 of Hall's paper is in error when the conventional, single-sided video bandwidth is used.

cases given in Eq. (4.44) correspond to the use of a PPI or B-scope with elevation scan, the RHI or E-scope with azimuth scan, or the C-scope with targets in the range interval ΔR superimposed (see Fig. 1.4). When the integration time t_v exceeds the time-on-target in any coordinate, equations of this form apply. Where the entire search interval ΔE or ΔA is covered in less time than t_v, the terms $\omega_e t_v$ and $\omega_a t_v$ will be replaced by the corresponding search intervals ΔE and ΔA. In another case, where a narrow range gate is scanned over the range search interval ΔR, the collapsing ratio may be found as $\rho_3 = v_s t_v/\tau$, where v_s is the rate at which the gate is scanned, in units of delay time per second.

The three cases for ρ_1, ρ_2, and ρ_3 all correspond to "constant ρn_f," in Fig. 1.16b. No new noise samples are introduced into the system in these cases, but noise which was contained in resolution elements adjacent to the signal has been mixed with the signal and its associated noise sample. In the following cases, the increased collapsing ratio represents the introduction of new noise samples into the system, causing the larger loss shown in Fig. 1.16a.

4. *Excessive Bandwidth.* When the radar equation is being used in the forms of Eqs. (4.18), (4.20), or (4.21), or the equivalent forms based upon pulse energy or average power, the i-f bandwidth B does not appear directly. Loss caused by use of excessive i-f filter bandwidth may be accounted for by introducing a filter matching loss L_m, in cases where a threshold detector sensitive to peak amplitudes is used. However, as noted in the discussion of matching loss, the video bandwidth may be reduced to match the spectrum of the detected signal, a collapsing loss being substituted for the matching loss. The collapsing ratio in this case is the same as that calculated for insufficient video bandwidth, except that B replaces B_t.

$$\rho_4 = \frac{2B_v + B}{2B_v} \qquad\qquad \textbf{(4.45)}$$

Calculations made by using the radar equations in which B does appear directly, although more accurate for cases of unrestricted video bandwidth and peak detection, must be modified when B_v is reduced below $B/2$ to match the signal spectrum. An effective S/N ratio should be found, a bandwidth $B_r = 2B_v$ being used in place of the actual i-f bandwidth B. The S/N ratio so calculated is then reduced by the amount of the collapsing loss L_c found from ρ_4 and Fig. 1.16(a). The same procedure applies when the video noise is averaged by use of slow sweep speed or excessive spot size on the display.

5. *Mixing of Receiver Outputs.* In some radar systems, two or more receiver channels are mixed, and the resulting video signal displayed on a single cathode-ray tube or applied to a single threshold detector. If it is assumed that the target appears in only one receiver at a time, the collapsing ratio ρ_5 will be equal to the number of receivers whose outputs are mixed. This ratio may be modified by an appropriate gating factor when the receivers are not all operating continuously, or where gain control (such as sensitivity time control) is applied to one or more channels. This collapsing ratio applies equally well to cases where the receivers are attached to separate antenna feeds (as in stacked-beam antennas), or where parallel Doppler filter channels are fed into a single detection device.

6. *Excessive Gate Width.* There are cases where a fixed range gate is used to detect targets passing through a given range interval. Performance based upon the S/N ratio and the number of pulses integrated during target passage must be modified by a collapsing loss calculated from the ratio

$$\rho_6 = \frac{\tau_g}{\tau} \tag{4.46}$$

Here, τ_g is the width of the gate, and the collapsing ratio represents simply the number of possible pulse positions in the gate. The use of relatively long gates may permit integration of many more pulses on fast-moving targets as they pass through the gate, and the resulting gain will greatly exceed the collapsing loss.

In cases not specifically covered by the above discussion, the basic rules used to calculate the collapsing ratio and loss are as follows: If each signal sample is forced to compete with more than one noise sample, there will be a collapsing loss. In going from the ideal case ($\rho = 1$) to the actual case, if the number of independent opportunities for a false alarm in the output of the detection system is reduced by the factor ρ, then the curves of Fig. 1.16(b) (constant $\rho n_{f'}$) are used. If the number of such opportunities remains constant with increasing ρ, and the total number of noise samples is increased, then the curves of Fig. 1.16(a) (constant $n_{f'}$) are used. If the number of independent opportunities for a false alarm increases along with the increase in the number of noise samples, then the system bandwidth B (or the effective bandwidth B_c) has increased, and the value of S/N should be reduced in proportion to bandwidth.

In a given system, there may be two or more different sources of collapsing loss. In such cases, the over-all collapsing ratio should be calculated as the product of the separate ratios. Where the sources are all of the same type (constant $n_{f'}$ or constant $\rho n_{f'}$), the appropriate curve of Fig. 1.16 is used directly. Where they are of both types, loss values should be taken from both curves and the actual loss found by interpolation, the over-all collapsing ratio being used in each case.

Beacon Antenna Loss

An important cause of loss in radar-beacon systems has been discussed in a recent report by Pike.* He calls attention to the fact that beacon antennas are seldom matched to the polarization of the radar antenna, and that the resulting loss can cause complete failure of both interrogation and response paths, even when the beacon antenna pattern has a strong lobe in the direction of the radar (see Fig. 3.12, p. 92). In an attempt to avoid such large losses, radar antennas designed for beacon tracking may use circular polarization or some form of polarization diversity. When circular polarization is used with a linearly-polarized beacon antenna, there will be a constant loss of three decibels owing to "polarization mismatch" between the antennas. More recent designs for satellite tracking and communications have made use of "polarization tracking" antennas, in which the linearly-polarized ground antenna is kept aligned with the polarization of the signal received from the spacecraft. In this way, polarization loss may be avoided entirely, even for a spacecraft which changes its aspect angle and antenna polarization.

Total System Loss

The total system loss factor L is the product of the individual components. Therefore, in the logarithmic form, the decibel values of all losses are added to find the total decibel loss. In actual search or tracking radar, the losses for the "searchlighting" case, covered in this section, must be combined with other losses, as discussed in the following chapters.

* Beuhring W. Pike, "Power Transfer between Two Antennas, with Special Reference to Polarization," Pacific Missile Range, Pt. Mugu, Cal.: *Tech. Report* RMR-TR-63-1 (1963).

The search radar is assigned to cover a given volume of space, and is expected to detect and locate each target within this volume and those entering it within its operating period. The radar beam scans a given solid angle, while display equipment or automatic detection circuits inspect all assigned range elements, using some form of integration and thresholding process to discriminate between targets and random noise. The entire scan pattern is repeated at intervals, and the success of the system is measured by its ability to detect a target of given cross section within a given search time.

In this chapter, we shall use the basic radar equation, the relationships governing integration and detection probability, and a number of other simple equations to express the optimum performance which could be obtained with an ideal search radar. We shall then consider the effects of practical loss factors, applying the results of the ideal radar analysis to actual radar systems in such a way as to obtain a

THEORY OF THE
IDEAL SEARCH RADAR

5

quick estimate of the performance level to be expected. This approach will make it possible to compare search radar systems of radically different types against a single standard of performance, and to determine what factors contribute to superior performance of a given system in a particular situation.

In this discussion, target detection will be used as the sole criterion for search performance. Considerations of resolution, which are of great practical importance in search radar design, will be covered in subsequent chapters. The omission of resolution and accuracy from the following discussion will lead to an interesting result: the detection performance of a radar will be shown to be independent of the wave length used, and of the type of modulation imposed on the transmitted signal. The choice of these characteristics, along with scan rate and other radar parameters, will be governed by the requirements for quality of data on the targets, as will be shown later.

5.1 DERIVATION OF SEARCH RADAR EQUATION

In Chapter 1 we used the statistical theory of target detection to relate the probability of detection, the probability of false alarm, and the integrated signal-to-noise ratio $(S/N)_i$. Although the curves which express these relationships are difficult to reduce to simple mathematical form, we may use the graphs (Figs. 1.8 and 1.9) to find a value of $(S/N)_i$ that meets the requirements of any search assignment. Typical values range from 12 to 16 db (numerical ratios of 16 to 40), after allowance for all losses. We shall assume below that a suitable value of $(S/N)_i$ has been chosen, and proceed to consideration of the radar parameters necessary to achieve this on a target of given cross section.

Ideal Search Radar Assumptions

In order to define a limit to search radar performance, beyond which no actual system can be expected to perform, we shall make a series of assumptions which define the "ideal" search radar.* To the extent that a practical system fails to measure up to these assumptions, we may then attribute to it certain search system losses, which will permit the "ideal" equations to be applied to actual radars.

Assumption 1: Ideal Integration. It was shown in Chapter 1 that the integrated signal-to-noise ratio cannot exceed the value $n(S/N)$, which would have been obtained if all received energy had been gathered into a single pulse in a matched receiver. The ideal radar will be assumed to achieve this performance without any integration loss, regardless of how many pulses are actually exchanged with the target and how they are combined.

* This approach to search radar analysis originated with Barlow's group at the Sperry Gyroscope Company in 1948, where it was applied to a classified system.

Assumption 2: Uniform Search Pattern.* In calculating the number of pulses exchanged with the target during the search time t_s, we shall assume that the assigned solid angle ψ_s is scanned in a uniform fashion by the radar beam, whose included solid angle is ψ_b. The radar beam shape will be assumed to be rectangular, so that successive beam positions can be covered without overlaps or gaps. The number of beam positions to be searched will then be

$$n_s = \frac{\psi_s}{\psi_b} = \frac{A_m(\sin E_m - \sin E_o)}{\psi_b} = \frac{A_m(\sin E_m - \sin E_o)A}{\lambda^2} \qquad (5.1)$$

Here, A_m is the width of the azimuth sector to be covered; E_m and E_o are the upper and lower limits of the elevation sector; A is the antenna aperture, and λ is the wave length. For search at low elevation angles, where the sine is approximately equal to the angle, the term in parentheses may be replaced by $(E_m - E_o)$. All angles are in radians, and solid angles in steradians.

Assumption 3: Antenna Efficiency. The gain of the radar antenna will be assumed to be uniform over the rectangular beam of solid angle ψ_b, and zero outside the beam. The aperture will be assumed to be used with 100 per cent efficiency, so that the gain, beam angle, aperture, and wave length will be related as follows

$$G = \frac{4\pi}{\psi_b} = \frac{4\pi A}{\lambda^2} \qquad (5.2)$$

Assumption 4: Matched Receiver. The radar receiver will be assumed to be matched to the transmission spectrum, so that the pulse width τ, if a rectangular pulse is actually used, is equal to $1/B$. When a pulsed transmission at repetition rate f_r is used, the peak and average powers will be related by

$$P_{\mathrm{av}} = P_t \tau f_r = \frac{P_t f_r}{B} \qquad (5.3)$$

Modification of the Radar Range Equation

The radar range equation in the form of Eq. (4.17) will be taken as the starting point, expressing the single-pulse signal-to-noise ratio S/N at maximum range R_m.

---

* Some saving in power can be made by using nonuniform search patterns. See J. J. Bussgang and D. Middleton, "Optimum Sequential Detection of Signals in Noise, *Trans. IRE*, Vol. **IT-1**, No. 1 (Dec. 1955), pp. 5–18.

$$\frac{S}{N} = \frac{P_t A^2 \sigma}{4\pi R_m^4 \lambda^2 k T_o B \overline{NF_o} L} \tag{5.4}$$

Since target detectability will depend upon integrated signal-to-noise ratio, we shall multiply the above by the number of pulses n exchanged with the target. The total number of pulses transmitted during the search time t_s is simply $f_r t_s$, and these are divided equally between the n_s beam positions searched. Thus, we have

$$n = \frac{f_r t_s}{n_s} = \frac{f_r t_s \lambda^2}{A_m (\sin E_m - \sin E_o) A} \tag{5.5}$$

If we combine Eqs. (5.4) and (5.5), including a search loss factor L_s in place of the on-axis loss L, we arrive at the search radar equation (in consistent units)

$$\left(\frac{S}{N}\right)_i = n\left(\frac{S}{N}\right)$$

$$= \frac{P_t \tau f_r A t_s \sigma}{4\pi R_m^4 A_m (\sin E_m - \sin E_o) k T_o \overline{NF_o} L_s} \tag{5.6}$$

$$= \frac{P_{av} A t_s \sigma}{R_m^4 A_m (\sin E_m - \sin E_o)} \times \frac{1}{4\pi k T_o \overline{NF_o} L_s}$$

The equation has been divided, somewhat arbitrarily, into two terms, the first containing those radar-target parameters which have been given or are to be chosen by the system designer, the second containing the constants and loss terms.

Significance of the Search Radar Equation

The search radar equation states, in essence, that the performance of the radar (measured by the integrated signal-to-noise ratio) is dependent upon the average power transmitted, the aperture of the receiving antenna, the time available for search, and the given target cross section and search volume. The receiving noise factor and total system loss factors also enter into the equation, but these are subject to limits which are often beyond the control of the system designer, and they may be evaluated approximately before arriving at the specific type of radar to be used. The search radar equation could equally well have been derived from the standpoint of energy ratio, which was shown earlier to be equal to twice the integrated signal-to-noise ratio for a matched system without integration loss. In either case, the ideal limiting performance is independent of wave length, modulation wave form (which includes pulse width, repetition rate, or modulated c-w transmissions), and beamshape of the radar system. These factors will have their effects through the practical losses which are introduced in particular systems. and will also affect the validity of the assumptions made in deriving the equation. The number of beam positions and range resolution elements will have a small effect on the required value of $(S/N)_i$, in that the false-alarm time and required threshold level will vary with these factors, but the primary effect of these resolution parameters will be felt in the accuracy of the system for locating targets, and in its ability to overcome unwanted signals such as clutter and interference.

For a physical interpretation of the search radar equation, we may note that the product of average transmitter power and search time represents the energy output of the radar during the time interval for which $(S/N)_i$ is evaluated. The factor $R_m^2 A_m(\sin E_m - \sin E_o)$ in the denominator gives the area to be searched at the limit of coverage. Thus the first term of the equation represents the energy incident on an equivalent target sphere at maximum range, multiplied by a factor A/R_m^2. This last factor is the solid angle subtended at maximum range by the receiving aperture of the radar. The ratio $A/(4\pi R_m^2)$ gives the fraction of the target energy which is available to the radar receiver. It makes no difference what type or size of transmitting antenna is used, so long as the transmitted energy is distributed uniformly over the assigned search angle, when averaged over the search time, and so long as the backscattered energy reaching the radar aperture is collected efficiently. This means that the receiving beam or beams must be matched to the angles from which target energy arrives. The conventional means of insuring this is to use a common antenna for transmission and reception, and to scan this antenna slowly enough to permit each pulse to be received before the beam moves to the next position. Alternate approaches take the form of broad, fixed beams for both transmission and reception, and combinations of broad transmitting beams with many narrow receiving beams to achieve adequate receiving aperture. When a rapidly scanning beam is used, there may be a scanning loss, as will be discussed below.

5.2 ALTERNATE FORMS OF SEARCH RADAR EQUATION

A number of alternate forms of Eq. (5.6) will be useful in clarifying the capabilities of different systems under special conditions. For instance, if targets are known to pass through the search sector at a constant angular velocity $\omega_t = v_t/R_m$, the maximum allowable search time will be

$$t_s = \frac{E_m - E_o}{\omega_t} = (E_m - E_o)\frac{R_m}{v_t} \qquad \text{(for elevation motion)} \qquad \textbf{(5.7)}$$

or

$$t_s = \frac{A_m \cos E}{\omega_t} = \frac{A_m R_m \cos E}{v_t} \qquad \text{(for azimuth motion)} \qquad \textbf{(5.8)}$$

At elevation angles below about 30 deg, we may assume that the cosine of elevation is unity, giving (for the case of elevation motion)

$$\left(\frac{S}{N}\right)_i = \frac{P_{av} A \sigma}{R_m^3 v_t A_m} \times \frac{1}{4\pi k T_o N F_o L_s} \qquad \textbf{(5.9)}$$

For azimuth motion, the azimuth sector A_m in the denominator will be replaced by $(E_m - E_o)$. The above forms are useful in calculation of search system performance when a "fence" is to be erected against passage of a given class of targets.

For ease in calculation, a logarithmic form of Eq. (5.6) may be derived, the same procedures and units being used which were applied earlier to the basic radar equation.

$$\left(\frac{S}{N}\right)_i = (P_{av})_{dbw} + (A)_{dbft^2} + (t_s)_{dbs} + (\sigma)_{dbm^2}$$

$$- 4(R_m)_{dbnmi} - (A_m)_{dbr} - (\sin E_m - \sin E_o)_{db} \qquad \textbf{(5.10)}$$

$$- (\overline{NF}_o)_{db} - (L_s)_{db} + 52 \text{ db}$$

With the addition of three decibels, the above will yield the energy ratio $\mathcal{R}$. For Eq. (5.9), the logarithmic form is

$$\left(\frac{S}{N}\right)_i = (P_{av})_{dbw} + (A)_{dbft^2} + (\sigma)_{dbm^2} - 3(R_m)_{dbnmi}$$

$$- (v_t)_{dbmps} - (A_m)_{dbr} - (\overline{NF}_o)_{db} - (L_s)_{db} + 52 \qquad \textbf{(5.11)}$$

The reference units used here are the same as those applied in Chapter 4, with the aperture area A referred to one square foot, the azimuth sector A_m to one radian, and velocity v_t to one mile per second.

Examples Using the Search Radar Equation

As an initial example, assume that we wish to search an azimuth sector of 90° from the horizon up to 45° elevation, detecting targets of one square meter cross section out to 1000 mi range. An integrated signal-to-noise ratio of 14 db will provide the desired performance, which corresponds to 90 per cent detection probability at a very low false-alarm rate. Pending further evaluation, a search system loss of 16 db will be assumed to account for departures from the ideal search radar. As will be shown later, this amount of loss is almost inevitable in any practical system. Using a low-noise receiver design, we may take the operating noise factor as unity, corresponding to a total input temperature of 290° K. With the further assumption that the search time allotted for coverage of the volume is 10 sec, we can tabulate the known parameters of the radar and target as follows

$$\left(\frac{S}{N}\right)_i = 14 \text{ db}$$

$$t_s = 10 \text{ sec} = +10 \text{ dbs}$$

$$\sigma = 1 \text{ m}^2 = 0 \text{ dbm}^2$$

$$R_m = 1000 \text{ nmi} = +30 \text{ dbnmi}$$

$$A_m = \frac{\pi}{2} \text{ rad} = +2 \text{ dbr}$$

$$\sin E_m = 0.707 = -1.5 \text{ dbr}$$

$$\overline{NF_o} = 0 \text{ db}$$

$$L_s = 16 \text{ db}$$

Solving for average power and aperture, we find that $(P_{av}) + (A) = 88.5$ db. The requirements of the system could be satisfied by such combinations as the following, provided a specific configuration were used which would agree with the loss assumptions

$$P_{av} = 50 \text{ kw} = 47 \text{ dbw}$$

$$A = 14,000 \text{ ft}^2 = 41.5 \text{ dbft}^2 \qquad (120 \times 120 \text{ ft})$$

or

$$P_{av} = 500 \text{ kw} = 57 \text{ dbw}$$

$$A = 1400 \text{ ft}^2 = 31.5 \text{ dbft}^2 \qquad (38 \times 38 \text{ ft})$$

This type of calculation is of considerable value in finding the approximate magnitude of the system design problem. The particular manner in which the average power and aperture might be used will depend upon many considerations, which will be covered in the succeeding chapters.

As a second example, assume that the results obtained above dictate a radar of such size that it cannot be used in the over-all system under consideration. An alternate solution to the search problem is needed, which will necessarily involve reduced performance, more search time, less range, or a smaller search angle. The tradeoff for search time is obvious. If 100 sec can be allotted, the product of average power and aperture may be reduced by a factor of ten, with significant savings in cost and size. However, a high-speed missile could penetrate the volume of coverage by some 400 mi before detection, and the average range at which targets would be detected might be reduced substantially below the 1000 mi figure. The range at which 90 per cent of the targets would have been detected would be near 600 mi, in the case of incoming missiles.

Another approach to the search problem might take the form of a narrow beam scanning at low elevation angle, so that all targets would rise through the restricted coverage angle. If the maximum target velocity is four miles per second, and the remaining parameters are the same as those listed above, Eq. (5.11) may be used to find an appropriate power-aperture product $(P_{av}) + (A) = +76$ db. The system performance requirements have now been reduced by more than 12 db, as compared to the original requirements. Whether this sort of solution can be used will, of course, depend upon the geometry of the possible target paths and whether a low-elevation beam will cover all possible approaches to the area under protection of the radar system. One interesting result of this analysis is the indication that the detectability is not dependent upon the beamwidth in the direction of target motion. The use of a broader beam provides more search time and permits the same energy density at the target. The beam can be extended all the way to the original 45° elevation angle without any increase in power requirements. As noted above, however, the additional search time may not be available if targets are approaching at high speed. This might dictate the concentration of energy near the horizon to achieve a high probability of detection before the target has much opportunity for penetration. These considerations will be discussed in more detail in later chapters, but they serve to indicate some of the uses and limitations of the search radar equation.

5.3 ESTIMATING SEARCH LOSSES

The search radar system will be subject to all the losses discussed in Chapter 4 for the case of the on-axis target, and will also have loss terms which are due to the fact that the beam scans past the target. When using the search radar equation, we must also include loss terms which account for the failure of the actual radar to live up to the ideal model assumed in deriving that equation. Referring back to the four major assumptions of Section 5.1, we find that portions of this loss have already been covered for the on-axis case. The so-called integration loss L_i, resulting from detection prior to signal integration, was discussed in Chapter 1, as was the filter matching loss L_m. The antenna efficiency loss L_n was discussed in Chapter 4, but an additional component must be added to this loss when the search radar equation is used, and this will be discussed below. New loss components caused by beam shape, overlap or gaps in scan coverage, and possible motion of the beam during the range delay time must be estimated and included in L_s. These loss components are also quite sensitive to the particular radar configuration used, although it will be shown that there are minimum values achievable for each loss and that the total of all losses is surprisingly insensitive to changes in system design.

Scan Distribution Loss

The scan distribution loss L_d describes the reduction in performance, relative to the ideal search radar, which results from the failure to integrate into one signal all energy returned from a target during the search time t_s. In deriving the search radar equation, we found the total number of beam positions n_s that were included within the solid angle assigned for search, and calculated the total number of pulses exchanged with each target as $n_t = f_r t_s / n_s$. If this number of pulses is actually exchanged on a single scan past the target, and integrated as a group, there will be no scan distribution loss. Generally, however, the beam will be scanned more rapidly, and will pass the target location two or more times during the search time t_s. On each scan, the number of pulses to be integrated will be some fraction of n_t, and the integrated signal-to-noise ratio *per scan* will be less than the value calculated for integration of n_t pulses. The fact that there are now two or more independent opportunities to detect the target will compensate partially for the reduced integration gain, but there will remain a drop in system performance described by the loss L_d. The loss may be calculated as shown in the following example.

Assume that a search radar with adjustable scanning rate is used, and that a relatively large number of pulses per beamwidth would be obtained at the lowest rate, where the search volume is scanned only once in the given search time. By speeding up the scan, the radar could be made to complete two, three, or more scans in this period, but with fewer pulses per beamwidth in each individual scan. The over-all probability of detecting the target on at least one of the j scans would be given by the cumulative probability P_c, which is

$$P_c = 1 - (1 - P_d)^j \qquad \textbf{(5.12)}$$

The probability of detecting the target on each scan is P_d, and the probability of failing to detect is $1 - P_d$. Hence, the probability of failing on all j scans is $(1 - P_d)^j$, if the successive scans are assumed to give independent results, and the cumulative probability of detection is one minus the probability of total failure. In Table 5.1 we have shown the required single-scan detection probability P_d for a cumulative probability of 90 per

Table 5.1 CALCULATION OF SCAN DISTRIBUTION LOSS FOR CUMULATIVE
DETECTION PROBABILITY OF 90 PER CENT

Number of scans j	Required P_d (%)	Required P_n	Required $(S/N)_i$ (db)	Relative integration gain (db)	Relative S/N (db)	Scan distribution loss L_d (db)
1	90	10^{-8}	14.3	10	4.3	0
2	68.5	5×10^{-9}	13.5	7	6.5	2.2
3	53.5	3×10^{-9}	13	5.2	7.8	3.5
4	44	2.5×10^{-9}	12.7	4	8.7	4.4
5	37	2×10^{-9}	12.5	3	9.5	5.2
6	32	1.7×10^{-9}	12.3	2.2	10.1	5.8
7	28	1.4×10^{-9}	12.1	1.5	10.6	6.3
8	25	1.2×10^{-9}	12	1	11	6.7
9	22.5	1.1×10^{-9}	12	0.5	11.5	7.2
10	20.6	10^{-9}	11.9	0	11.9	7.6
100	2.3	10^{-10}	10.5	-10	20.5	16.2

cent on j scans, where j varies from one to ten. The single-scan false-alarm
probability P_n is also shown, in order to achieve a false-alarm probability
of 10^{-8} for the entire period of search. The required value of integrated
signal-to-noise ratio $(S/N)_i$ can be found from Fig. 1.9, and the available
integration gain will be considered proportional to the number of pulses per
scan (integration loss will be considered separately). It is apparent that
the decrease in integration gain exceeds the reduction in required $(S/N)_i$
as the number of scans is increased. As a result, the single-pulse signal-to-
noise ratio S/N must be increased to hold P_c constant, and the last column
of the table represents this increase, defined as the scan distribution loss.
The entry for $j = 100$ shows that the loss increases very rapidly for large
numbers of scans, the required $(S/N)_i$ ratio dropping slowly. If this loss
were the only determining factor, it would be advisable to scan the volume
only once, all pulses being grouped into a single integrated signal for detection, or to integrate over several scans.

When the target fluctuates in amplitude between scans, the fluctuation
loss L_f, shown in Fig. 1.13, must be applied in determining the required
value of $(S/N)_i$. The loss L_f is a rapidly increasing function of single-scan
detection probability P_d, and for this reason the use of a single scan to
achieve a high detection probability carries a heavy penalty. The fluctuation
loss drops to zero (in decibels) when $P_d = 33$ per cent, and will provide
some gain relative to steady signals when P_d drops below this value. Figure
5.1 shows the combined effects of losses from video integration, scan distribution, and fluctuation, plotted as a function of the number of scans for
$P_c = 90$ per cent, and compares the steady target with the fluctuating target.

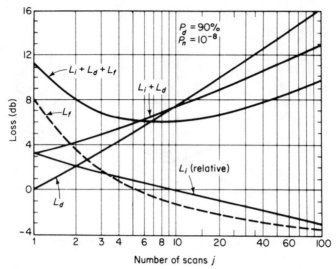

Figure 5.1 Scan distribution loss and related losses vs. number of scans.

For targets whose characteristics may vary between steady signal and Rayleigh fluctuation, the optimum number of scans will be near $j = 4$, under the conditions assumed.

The scan distribution loss can be shown to be relatively independent of the detection probability. Over a range of values from $P_c = 50$ per cent to 99 per cent, the loss for $j = 10$ varies less than one-half decibel from the 7.6 db given in Table 5.1. The loss increases when the detection probability decreases. The fluctuation loss, however, varies rapidly with changing detection probability, as shown in Fig. 5.2. The curves shown are valid for total pulses n_t between about forty and a few thousand, the values of integration loss being slightly greater for large numbers of pulses and slightly less for small numbers. When the total number of pulses is reduced to approach the number of scans, there will enter into consideration a significant increase in losses due to beam shape and scanning, as will be shown below.

Storage of signal information from scan to scan can lead to reduced scan distribution loss, since integration can then be performed over a larger fraction of the total number of received pulses. Such storage can be provided on a cathode-ray-tube display in the case of rapid scans, or by a digital process which accumulates the results of several successive scans. The output signal will now be initiated only when a target has been detected in the same resolution element in k out of j successive scans. Table 5.2 shows the way in which this procedure may be analyzed, and

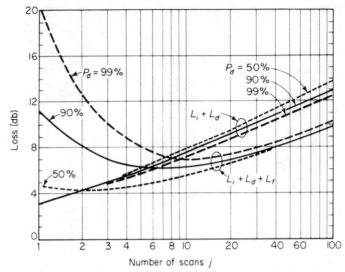

Figure 5.2   Effect of detection probability on combined losses.

Fig. 5.3 gives the results for j between one and ten. Although the combined integration and scan distribution losses are now almost independent of j for the steady-target case, the use of scan-to-scan storage leads to greater fluctuation losses when the single-scan P_d exceeds 33 per cent. In

Table 5.2 CALCULATION OF SCAN DISTRIBUTION LOSS FOR OVER-ALL DETECTION
PROBABILITY OF 90 PER CENT WITH SCAN-TO-SCAN STORAGE

Number of scans	Detections required	Required P_d	Required P_n	Required $(S/N)_i$	Relative integration gain	Relative S/N	Scan distribution loss L_d
j	k	(%)		(db)	(db)	(db)	(db)
1	1	90	10^{-8}	14.3	10	4.3	0
2	2	95	10^{-4}	12.3	7	5.3	1
3	2	89	6×10^{-5}	11.9	5.2	6.7	2.4
3	3	96.5	2×10^{-3}	11.4	5.2	6.2	1.9
4	2	68	4×10^{-5}	10.8	4	6.8	2.5
4	3	85	1.6×10^{-3}	10.1	4	6.1	1.8
4	4	97.4	1×10^{-2}	10.8	4	6.8	2.5
6	2	51	2.6×10^{-5}	10.1	2.2	7.9	3.6
6	3	67	8×10^{-4}	9.2	2.2	7	2.7
6	4	80	5×10^{-3}	8.9	2.2	6.7	2.4
6	6	98.25	0.0465	9.9	2.2	7.7	3.4
10	3	45	4×10^{-4}	8.2	0	8.2	3.9
10	5	64	8×10^{-3}	7.5	0	7.5	3.2
10	10	98.95	0.16	9.1	0	9.1	4.8

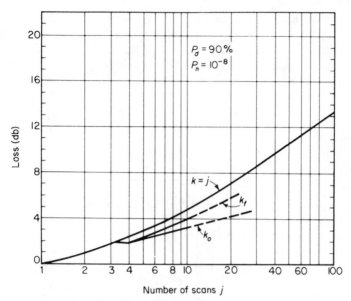

Figure 5.3 Scan distribution loss for systems with storage.

addition, it may require a considerable amount of extra equipment to store and process the data received over a period of several scans, and properly to correlate the results from a given resolution element. As a result, the simple system using independent detection on each scan is preferred in most cases. Whichever procedure is used, there will be a practical limit to the performance of the system which is well below that of the ideal radar discussed earlier. The total loss from scan distribution, fluctuation, and integration will be at least 6 db, unless 10 or more scans are stored and correlated to arrive at each detection decision. The losses due to beam shape and scanning will add further to this 6 db minimum, as will be shown below, and will force the use of two or more pulses per scan in an efficient system.

Figure 5.3 illustrates curves for scan distribution loss, using three different detection criteria in the scan-to-scan correlation process: $k = j$ requires detection on all scans; k_o corresponds to the optimum coincidence detection criterion derived by Schwartz [see Eq. (1.14)]; k_f is the optimum for fluctuating targets. Values of k_o and k_f are compared in Table 5.3, showing that the effect of target fluctuation is to favor the simple, single-scan detection criterion when less than five scans are to be processed. Figure 5.4 shows the combined effects of fluctuation, integration, and scan distribution losses, plotted as a function of the number of scans. The curve for $k = 1$, taken from Fig. 5.1, permits us to compare the advantage of

Table 5.3 OPTIMUM CRITERIA FOR COINCIDENCE DETECTION
USING SCAN-TO-SCAN STORAGE

Number of scans j	Optimum k_o for steady target [see Eq. (1.14)]	Optimum k_f for fluctuating target
1	1	1
2	2.1	1
3	2.6	1
4	3	1
6	3.7	2
10	4.7	3

using scan-to-scan storage or correlation for the various cases. These curves
may all be calculated by using the theory of coincidence detection, as
described by Schwartz* and discussed in Chapter 1.

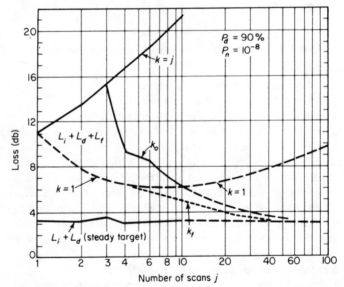

Figure 5.4 Combined losses for systems with scan-to-scan storage.

Beam Shape Loss

In deriving the search radar equation for the ideal radar, the beam was
assumed to be rectangular in cross section, with uniform gain over the

---

* Mischa Schwartz, "A Coincidence Procedure for Signal Detection," *Trans. IRE*,
Vol. **IT-2**, No. 4 (Dec. 1956), pp. 135–39.

rectangle. The actual beam will be elliptical in cross section, and will follow the $(\sin x)/x$ shape in its one-way voltage pattern. The beam shape loss L_p represents the additional signal strength required to make up for the reduced amplitudes of pulses received when the beam axis moves off the target. One loss will apply when the beam is scanned in one coordinate only, with the axis passing directly over the target (as in the case of a scanning fan beam which extends well beyond the target in the stationary coordinate). When the target is permitted to depart from the beam axis in both coordinates, or where a two-dimensional scan is used, a second beam shape loss is incurred. The loss for both these cases has been calculated by Blake* for the condition where many pulses are exchanged during the motion of the beam across the target. His value for the one-dimensional case was $L_p = 1.6$ db, and this has been verified for a large range of problems, both video integration and optimum coincidence detection being used. When integration is used, the beam shape loss will remain constant for as few as two pulses per beamwidth. When the optimum coincidence detection procedure is used, L_p will remain constant at about 1.6 db for $P_d = 50$ per cent for any number of pulses per beamwidth. Figure 5.5 shows the variation in beam shape loss for other conditions. It is interesting to note that this loss is only 1.1 db for the system which depends upon the cumulative probability of detection rather than on coincidence detection, offsetting in part the scan distribution loss for this case. When the detection probability exceeds 50 per cent, the beam shape loss increases with decreasing numbers of pulses per scan (or of scans per target passage through the beam), reaching 3.3 db for $P_d = 90$ per cent and 4.3 db for $P_d = 99$ per cent in the one-hit case. Those detection procedures which require more than the optimum number of hits per beamwidth will lead to larger beam shape losses, as shown in the figure.

As with Blake's results, the losses shown in Fig. 5.5 must be applied to each of two coordinates unless the center of the beam crosses directly over the target on each scan. Thus, when a pencil beam is scanned over a two-dimensional surface to locate the target, or where the target passes through a narrow search sector in which the beam is scanning in one coordinate, the decibel losses must be calculated for each coordinate and added. As an example, let us assume that a $1°$ pencil beam is scanned over a $20°$ azimuth sector at fixed elevation, a scan rate of $10°$ per sec and a repetition rate of 300 pulses per sec being used. If video integration of about 30 pulses is used (Blake would recommend integration of 25 pulses in the steady-signal case), the beam shape loss in the azimuth coordinate would be 1.6 db. If the target rises through the one degree elevation sector at a rate of $\frac{1}{10}°$ per sec, the radar will pass the target five times, between half-power points

* Lamont V. Blake, "The Effective Number of Pulses per Beamwidth for a Scanning Radar," *Proc. IRE*, **41**, No. 6 (June 1953), pp. 770–74.

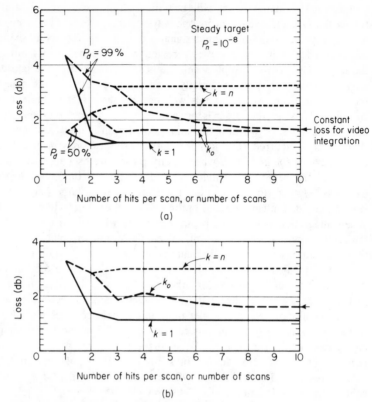

Figure 5.5 Beamshape loss vs. number of pulses or scans. (a) Beam-shape loss for $P_d = 50$ per cent, 99 per cent. (b) Beamshape loss for $P_d = 90$ per cent.

in the one-way elevation pattern. This would cause an additional 1.6 db loss if a system were devised to integrate all 150 pulses received during the five scans. The additional loss would be only 1.1 db if the cumulative detection probability were used as a criterion, and 2 db if a coincidence detection scheme were used to arrive at a final detection probability of 90 per cent or higher.

The beam shape loss interacts with the scan distribution loss when the number of hits per beamwidth is reduced below about six. The increasing beam-shape loss for use of a single scan ($j = 1$) will more than offset the advantage obtained by minimizing the scan distribution loss, even for the steady-target case. The interaction with fluctuation loss is less pronounced. Calculations for a wide variety of cases, coincidence detection being used, show an increase of 0.5 to 1.0 db in required S/N ratio for the fluctuating target, beyond that found in adding L_f and L_p from separate calculations. As would be expected, the increase of 1 db applies to the case of high detection probability, where the fluctuation loss is already high. When the uncertainty in the fluctuation loss for actual (non-Rayleigh) targets is considered, the small increase in combined loss can probably be ignored in most practical cases.

Scanning Loss

One more loss which depends upon the number of pulses received per beamwidth is the scanning loss, defined as the loss component due to motion of the beam between transmission and reception.* As in the case of beam shape loss, the scanning loss L_j will be constant for cases where more than six pulses are exchanged per beamwidth, but its value will be approximately zero for this case. For more rapid scanning, the product of transmitting gain G_t and receiving gain G_r must be used for each pulse or portion of continuous transmission in place of the factor G^2 in the radar equation. The relative positions of the beam at the time of transmission and reception must be considered, and the product $G_t G_r$ will always lie below G^2. Figure 5.6 shows the shape of the gain patterns for the product, normalized to equal peak amplitude. Figure 5.7 shows (dashed line) the loss corresponding to the drop in this peak amplitude, and the total loss, taking into account also the narrowing of the product pattern for high scan rates. This loss increases sharply when the beam moves more than one-half its width during the range delay time. It should be noted that the loss is expressed in terms of beam motion during the range delay time, $t_r\omega/\theta$, rather than in pulses per beamwidth $n = f_r\theta/\omega$. This is because the use of high, ambiguous repetition rates can increase the number of pulses per beamwidth without reducing scanning loss. It makes no difference how many pulses are in the medium between the radar and the target, if the beam has moved to a new location before they are received. The full amount of the scanning loss L_j appears for the most distant targets, which

---

* The term *scanning loss* has been employed historically to indicate various losses encountered as a result of beam motion, including those resulting from integration of fewer pulses when compared to the search-lighting radar, and the beam shape loss described above. Nelson M. Cooke and John Markus in *Electronics and Nucleonics Dictionary* (New York: McGraw-Hill Book Company, 1960), p. 416, define scanning loss as "the reduction in sensitivity when scanning across a radar target as compared with the sensitivity obtained when the radar beam is directed constantly at the target. Scanning loss is due to the change in antenna position during the interval in which the signal travels from the antenna to the target and back." Our usage here corresponds to the latter part of this definition, and excludes the effects of the limited number of pulses available for integration, the variable amplitudes of these pulses during the scan, and possible collapsing losses.

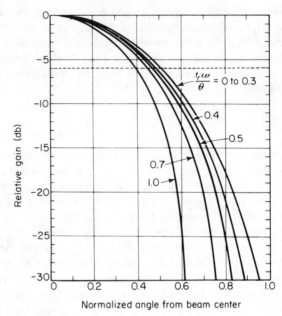

Figure 5.6 Effective beam patterns for rapid scans (Two-way radar case).

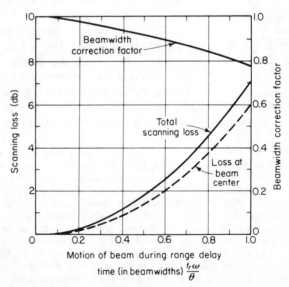

Figure 5.7 Scanning loss vs. scan speed.

is exactly where the radar needs its maximum gain. At short range, the motion of the beam between transmission and reception will be very slight, and little or no loss is incurred. By offsetting the receiving beam from the transmitting beam enough to compensate for the motion during the target delay time, it is possible to reduce scanning loss for targets at a given range. The equipment complexity involved in such a scheme has not encouraged this approach to reduction of loss, especially when the other search losses are found to increase with very rapid scans.

Gate and Filter Spacing Losses

Just as there are losses caused by beam shape in the two angular coordinates, the use of range gates and Doppler filters will introduce losses owing to spacing in the time-frequency plane. These losses will also depend upon the degree of overlap between adjacent channels, and the speed of scanning in these two coordinates if such scanning is used. Fortunately, the shape of time gates and filter passbands may be controlled much more easily than beam shape, and many overlapping channels may be used with-

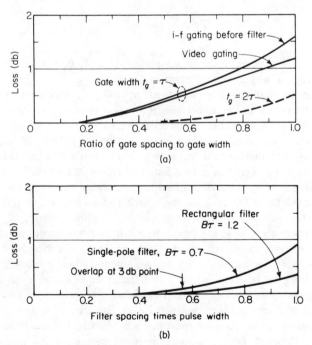

Figure 5.8 Gate or filter spacing loss. (a) Gate spacing loss. (b) Filter spacing loss.

out encountering the economic limits which affect antennas. The opportunity for unlimited amplification at intermediate frequency, prior to gating and filtering, also permits the use of any desired number of channels with overlapping coverage and without degradation in signal-to-noise ratio. As a result, there is no theoretical minimum value of loss from this source, and no practical limit except that set by the problems of size, weight, complexity, and cost of the i-f processing equipment. Figure 5.8 shows the spacing loss L_e for different combinations of rectangular and $(\sin x)/x$ wave forms and frequency spectra. It should be noted that the systems which are closely matched (minimum L_m from Fig. 1.11) require closer spacing of range and Doppler channels to maintain their high efficiency. When a relatively wide rectangular filter is used to receive a rectangular pulse, considerable detuning of the receiver (or Doppler shift of the signal) is permitted without much loss in signal-to-noise ratio.

Antenna Efficiency Loss

The search radar equation was derived by using the assumption of 100 per cent aperture efficiency, such that

$$G = \frac{4\pi}{\psi_b} = \frac{4\pi A}{\lambda^2}$$

The aperture A appears in the final equation [Eq. (5.6)] as a first-power term, whereas it was a squared term in the initial equation [Eq. (5.4)]. The number of beam positions searched, $n_s = \psi_s/\psi_b$, was used to reduce A to the first power in the numerator. Reduction in antenna efficiency will therefore appear as a decrease in the effective value of A, with an implicit increase in the beam angle ψ_b and in the number of pulses per target. In practice, however, much of the lost power gain is distributed into side lobes and spillover loss rather than in broadening of the main lobe of the antenna. The resulting loss in integrated signal-to-noise ratio will be greater than that calculated by reducing A in the search radar equation to reflect the actual aperture efficiency. In a typical case, where an efficiency of 50 per cent is assumed, the one-way gain will be cut to one-half its optimum value (for a loss of three decibels) and the beam angle ψ_b will increase by only 50 per cent, for an extra loss of 1.5 db. An antenna efficiency loss $L_n = 4.5$ db would have to be used in the search radar equation to describe the result, whereas a loss of six decibels would be used in those equations where A^2 appears. If there had been no increase in side lobe or spillover losses, then the loss in A^2 would have remained six decibels, but the beam angle would have doubled, providing twice as many pulses for integration and a net loss of only three decibels in search performance.

Minimum Search Losses

The foregoing discussion makes it clear that there is no single optimum design for a search radar, and that a balance must be achieved in each case between the basic parameters of average power and aperture, on the one hand, and the loss components owing to various scan patterns and detection procedures, on the other. With regard to the assumptions used in deriving the performance of the ideal search radar, there will be some minimum loss values which cannot be avoided in any practical system. These may be summarized as follows

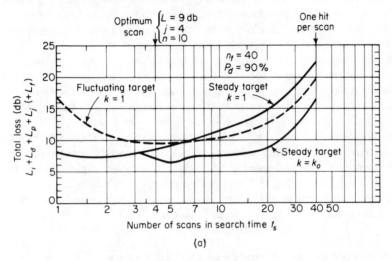

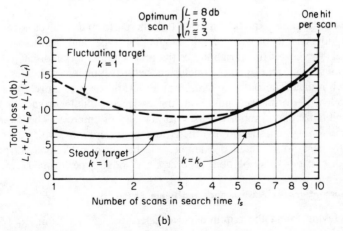

Figure 5.9 Combined search losses vs. number of scans. (a) $n_t = 40$. (b) $n_t = 10$.

1. Antenna efficiency loss L_n	3.0 db
2. Matching loss L_m	1.0 db
3. Combined $L_i + L_d + L_p + L_j + L_f$	9.0 db
Minimum combined loss	13.0 db

The above figures are based on an antenna gain efficiency of 63 per cent (2 db loss in gain relative to ideal), and a detection probability of 90 per cent for the complete search process. There must be at least nine pulses per target position, divided into three scans with three hits per scan, and with uncorrelated signal amplitude over the scan period. Figure 5.9 shows how the combined losses vary with the number of scans for two cases, one with many pulses available per target position and one with relatively few pulses. Although the steady-target case permits the loss to be minimized by using scan-to-scan storage and correlation, the more realistic case of the fluctuating target shows this storage to be of little value. The total loss increases quite sharply for fewer than two pulses per scan or two scans per target position, so at least four pulses per target position should be allowed, and nine or more are advisable. The actual loss for any particular type of scan may be found by referring to Figs. 5.1 through 5.8. Where the overall probability of detection must be greater than 90 per cent, the optimum number of scans will increase slightly above the number indicated in Fig. 5.9, and the minimum loss will be greater. Where the target fluctuates at such a low rate that its amplitudes remain correlated over the entire search time, an even higher loss will be incurred.

5.4 USE OF LOSS ESTIMATES IN SYSTEM DESIGN

The search radar equation showed that the performance of an ideal system was independent of wave length, modulation type, and beam shape of the radar. In view of the variability of the practical loss terms discussed above, we must now modify this conclusion before applying it to design of a search radar. The radar system, in order to approach as closely as possible the ideal performance, must scan the assigned volume in such a way as to minimize the combined losses for targets of interest. The restrictions placed on system design by these loss considerations will be reviewed briefly here. More detailed examples will be given in Chapter 8, following the discussions of resolution, accuracy, and related performance factors.

Limitations on Wave Length and Beam Shape

Having found the required power-aperture product from Eq (5.6) or one of the equivalent forms, the system designer has wide latitude in choice of transmitter power and aperture areas. Once the aperture area has been chosen, however, the range of frequencies over which optimum operation can be obtained may be quite limited. In the earlier example, it was found that antennas in the 40 to 100 ft size class were necessary. The maximum range-delay time for targets at 1000 mi is about 12.5 msec, making available a total of 800 exchanges with targets at maximum range during the ten sec search time. (If pulsed transmissions are used, this would be expressed as 800 transmitted pulses to cover the assigned volume, but the same considerations apply if other modulation types are substituted.) The assigned solid angle is about one steradian. Thus, if a single search beam is to be used with eight pulses or exchanges per beam position, there must be no more than 100 beam positions, and the solid angle of the beam must include at least 0.01 steradian (i.e., the width of a circular beam must be at least 0.128 rad or 7.5°). This will limit the choice of radar wave length to four feet or greater when the smaller antenna is used with 500 kw of average power, and the larger antenna with 50 kw of power can be used only with wave lengths greater than ten feet. The corresponding upper frequency limits of 250 or 100 mc might or might not be suitable for radars meeting the other requirements of the system.

The limitation on wave length may be overcome to some extent by using more than one search beam (see Fig. 5.10). According to the theory, the transmitted energy may be distributed over the assigned solid angle in a uniform fashion by using any one of several techniques, so long as the receiving aperture is able to collect the reflected energy efficiently. In the limiting case, fixed beams may be established to cover the entire solid angle for reception. The transmitted power may be shared between the many feed points which correspond to these beams, or may be radiated from a single, smaller aperture which illuminates the entire search volume simultaneously. The total number of pulses available per beam position would then approach 800 for the example cited above, unless the repetition rate were reduced to limit the integration loss. In this approach to the system, it would be important to assure that the full receiving aperture be available for the entire search period for each beam position, requiring that many separate feed points and receivers be used. Although this might prove inconvenient or expensive, the possible range of operating frequencies would be increased as the square root of the number of beams.

At the other end of the frequency scale, it would obviously be inefficient to use a beamwidth which exceeded the assigned solid angle. In practice, the transmitted energy can be distributed uniformly only if the beamwidth of the transmitting antenna is much narrower in both coordinates than the assigned search angle. This is because of radiation of appreciable energy beyond the half-power contour of the beam, and the practical impossibility of generating beam shapes which fall off rapidly from a broad maximum. Similar considerations apply to the receiving pattern.

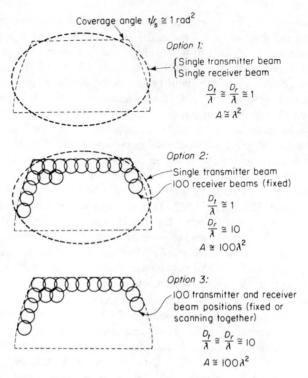

Coverage angle $\psi_s \cong 1\,\text{rad}^2$

Option 1:
Single transmitter beam
Single receiver beam

$$\frac{D_t}{\lambda} \cong \frac{D_r}{\lambda} \cong 1$$

$$A \cong \lambda^2$$

Option 2:
Single transmitter beam
100 receiver beams (fixed)

$$\frac{D_t}{\lambda} \cong 1$$

$$\frac{D_r}{\lambda} \cong 10$$

$$A \cong 100\lambda^2$$

Option 3:
100 transmitter and receiver beam positions (fixed or scanning together)

$$\frac{D_t}{\lambda} \cong \frac{D_r}{\lambda} \cong 10$$

$$A \cong 100\lambda^2$$

Figure 5.10 Optional means of providing search coverage.

Limitations on Modulation

The use of pulsed transmissions, assumed in many of the derivations above, illustrates the problems of optimizing beamwidth and search patterns. As was noted in the discussion of scanning loss, the increase in pulse repetition rate above the unambiguous value cannot overcome the loss encountered when too few hits are available on each target. Furthermore, when video integration is used, the integration loss will be increased by use of the higher repetition rate. The frequent use of high repetition rates in search radar is explained by considerations other than search losses (see Chapter 7). Similarly, the use of various forms of modulated c-w transmission cannot assist in overcoming limitations in scan rate, even though it may be possible to reduce integration loss below the level encountered in video integration of pulses. Any type of modulation may be used in an optimum system, so long as the receiver remains matched and the integration loss is kept within reasonable limits. Where a scanning beam is used, an upper limit is placed on the time available for coherent or predetector integration, which cannot exceed the time required for the beam to pass the target. This will limit the integration efficiency of the system in a way which is described by the scan distribution loss. The only improvement in detection performance that can be expected from coherent integration, using pulse or c-w transmission, is the elimination of part of the integration loss component shown in Figs. 5.1, etc. For optimum scan patterns, this improvement amounts to less than two decibels.

In summary, it may be stated that the wide variety of practical search radar systems which have been proposed and used over the years has been dictated more by considerations of resolution, accuracy, and data rate than by basic differences in detection efficiency. The ability to resolve in Doppler frequency, as in MTI, has been of particular importance. The theoretical limit to detection performance against a thermal noise background, covered by this chapter, is applicable to all types of systems, and may be used to arrive at a first approximation to a search radar design meeting any set of requirements. It may also be used to check rapidly on the reasonableness of a proposed system, even in the absence of detailed data as to its method of operation. If its power-aperture product is insufficient for the search job, no amount of data processing can help.

The preceding chapter established limits to the performance of search radar systems, when they operate against a background of thermal noise. Now we shall consider some of the additional factors which govern the design of a search radar system, and see how the freedom in choice of system parameters is narrowed by the requirements for resolution and measurement over specific volumes of coverage. To a large extent, these new factors will be governed by the types and locations of the targets to be found, and by the use to which the radar data will be put. We shall discuss in a qualitative way the nature of the more common radar search problems, and the characteristics of radars applied to their solution. The use of radar coverage charts will be discussed, and antenna limitations explored from the systems viewpoint. Finally, some of the techniques used in practical detection and position-measurement systems will be described.

DETECTION AND MEASUREMENT IN SEARCH RADAR 6

6.1 APPLICATION OF RADAR TO SEARCH PROBLEMS

A search radar system, by definition, is intended to display targets as soon as possible after they enter the assigned coverage volume.* Before we attempt to discuss in detail

* *IRE Dictionary of Electronic Terms and Symbols* (1961), p. 127. We prefer the use of volume rather than area to define coverage.

the factors which control the performance of such a radar, we should consider the different types of targets and the appropriate coverage volumes to which the radar may be assigned.

Description of Search Problems

The earliest attempts to apply the principles of radar were concerned with the detection of ships and aircraft, equipment on the ground or on shipboard being used. These same problems still account for the major portion of radar equipment in use, and the equipment is now split between ground, ship, and airborne types. The characteristics of ground and ship radars are very similar, although the size and weight of the shipboard equipment are somewhat more restricted. Airborne radar is subject to much greater restriction in size, weight, and input power available. The type and location of the targets for which the radar is intended to search has a greater effect on the equipment. Search of the sea surface can be carried out with very narrow (pencil) beams, and the primary source of competing noise is the background of sea clutter. If the search is directed upwards, to find aircraft or missiles, the coverage will usually be extended to a large elevation sector, and thermal noise will be an important consideration. The general characteristics of the several major classes of search problem will be discussed below, as an introduction to the problems of coverage, detection, and measurement.

Sea Surveillance

The first proposals for use of radar were those of the German engineer Christian Hulsmeyer and of Marconi, who were concerned with problems of collision avoidance and navigation for ships. Many thousands of small radars are now in use for this purpose on ships and small craft. Land-based harbor-surveillance radars are installed in many critical locations, and data may be relayed to small boats through conventional radio and television circuits. It was the military application of surface surveillance for location of naval targets that led to much of the early development effort in radar, and paved the way for the more peaceful applications to navigation. The need for location of surface craft has given way to the antisubmarine applications of radar as a subject for new development efforts, but the Navy is still largely dependent upon surface-surveillance radar for the conduct of its operations.

The common characteristics of radars for sea surveillance have been noted: relatively short range and narrow beams, with provision for minimizing sea-clutter returns. The range of operation is limited almost entirely by geometry rather than by radar sensitivity. The limit imposed by line of sight is given by

$$R_h = 1.23(\sqrt{h} + \sqrt{h_t})$$ (6.1)

where R is in nautical miles and the radar height h and target height h_t are in feet. For a shipboard radar, with its antenna on top of a mast 100 ft above sea level, the range to a surface target will be only 12 nmi when line of sight is lost. The range may increase up to twice this figure if the target has comparable height, but even 20 to 30 mi is a relatively short range compared to most air-search radars. There will, of course, be some transmission of energy beyond the line-of-sight limit, but the echo signal strength beyond R_h will fall off much more rapidly with range than it will in free space. A further consequence of the limited range is that most sea-search radars use comparatively high repetition rates, low powers, and small antennas.

Land Surveillance and Mapping

The early airborne radars were able to present maplike pictures of the surrounding land surface on a PPI display. They could be used for navigation, in much the same way as direct visual observations, and they made possible the location of large land targets for blind bombing through clouds or at night. Since the altitude at which the radar could be operated was great, the short-range limitation which applied to surface-based radar was overcome, and the relatively large radar cross section of land targets made it possible to operate at ranges of 100 or 200 mi. Broad fan beams were used to provide coverage extending from a point nearly under the aircraft to the horizon. In attempting to locate targets of tactical interest, it was necessary to improve the resolution of airborne radar in all three position coordinates, and often in Doppler frequency as well. The size limitation placed on airborne antennas led to use of very short wave lengths for the sake of narrow beamwidths, and more recently to the development of the "synthetic aperture" type of radar,* in which observations taken at different points on the aircraft's flight path are combined coherently to achieve the effect of an extended antenna.

Land surveillance by ground radar, used occasionally in World War II, has become a standard technique in army units. Since they are limited to short range by line-of-sight considerations, the major problems are rejection of ground clutter and achievement of some degree of angle resolution while portable antennas are used. Such practical problems as achieving light weight, low power consumption, and silent operation are also of major importance for radar to be used in the front lines.

---

* L. J. Cutrona, W. E. Vivian, E. N. Leith, and G. O. Hall, "A High-Resolution Radar Combat-Surveillance System," *Trans. IRE*, Vol. **MIL-5**, No. 2 (April 1961), pp. 127–31.

Air Search

The category of air-search radar includes many of the common applications, extending from early warning and area surveillance to weather radar and mortar-location equipment. In historical sequence, the first application was in early warning and ground control of interceptor aircraft. The early-warning radars emphasize long-range coverage at low elevation angles, being limited in range by aircraft altitude and line-of-sight paths. Area-surveillance radar, used in continuous plotting of target position for intercept control or navigation, will be characterized by shorter range (generally 20 to 100 mi) and extended vertical coverage (up to 30° or even 60°). Height-finding radar is used in area surveillance as an adjunct to the main search radar, or sometimes as an integral feature of the main radar. Height finding is needed when the vertical coverage of the main radar extends over many thousands of feet in target altitude. It also provides a third dimension in target resolution and measurement. A radar that has built-in height-finding capability is known as a "3-D" radar. Similar types of equipment are used on land and shipboard, and airborne counterparts exist for all but the largest surface radars.

The weather radar developed from unintentional viewing of weather clutter on surveillance-radar displays. Similar in design to the radars used in searching for aircraft, the weather radar is optimized for locating and measuring clouds, precipitation, or turbulent conditions in the atmosphere.

The specialized field of mortar- and artillery-location radar grew from the accidental observation of mortar shells by antiaircraft radar during World War II. It was found that a tracking radar could follow the projectiles, and that the track could be extrapolated back to the firing point and used to direct counterbattery fire. Search-type radar was later applied to this problem, using a pair of scanning beams to measure two points on the rising part of the trajectory. Relatively short-range and low-elevation coverage is required, and both clutter and thermal noise are important factors.

Missile and Satellite Surveillance

With the coming of the space age, a family of radars was developed for missile and satellite detection. These systems are analogous to the air-search radars, but the minimum and maximum ranges are greatly extended. The special limitations on motion of objects along ballistic paths make possible a number of compromises in the volume of coverage assigned to these radars, and clutter problems are greatly reduced by limiting the coverage to regions outside the earth's atmosphere. In some cases, it is possible to provide satisfactory search by using a narrow sector or "fan," through which space objects must pass to enter the protected volume. In the case of satellites, where detection could take place on any one of several orbits, a single such fan near the equator will eventually detect a satellite at any orbital inclination. The very large powers and apertures required for satisfactory search beyond the atmosphere have limited such radars primarily to land sites.

6.2 COVERAGE OF THE SEARCH VOLUME

Types of Scan

The size and shape of the volume of coverage will vary widely for the different applications discussed above, but a few basic types of scan will provide coverage for all cases. The most common scans are

1. Horizon scan, for detection of surface targets [Fig. 6.1(a)].
2. Continuous 360 azimuth scan at low elevation angle, for long-range detection of aircraft or missiles [Fig. 6.1(b)].
3. As above, but restricted to an azimuth sector less than 360.
4. Fixed-elevation or fixed-azimuth scans over limited sectors, for detection of targets entering a defined volume from a given direction, or for refinement of measurements as in height-finding [Fig. 6.1(c)].
5. Two-dimensional sector scans of limited extent, as used in acquisition by tracking radar (see Chapter 14) or in surveillance of a specified approach corridor [Fig. 6.1(d) and (e)].
6. Continuous coverage extending from the horizon to a specified altitude, for aircraft surveillance [Fig. 6.1(f)].
7. As above, but inverted for surface search from an airborne radar [Fig. 6.1(g)].

As shown in the figure, the beams used in the first five types of scan are usually simple pencil or fan beams scanned in one coordinate. The width of the beam in a direction normal to the scan is matched to the assigned coverage angle or to the target velocity in that direction, so as to assure two or more opportunities to see each target as it passes through the search volume. When such a beam is scanned in two coordinates, as in Fig. 6.1(d) or (e), the sector must be restricted to permit coverage within a reasonable time period.

The last two types of scan shown in the figure pose special problems in antenna design since the antenna pattern should be matched to the coverage volume for maximum efficiency. For short ranges, where the curvature of the earth is slight, the appropriate pattern follows the "cosecant-squared" form, shown in Fig. 6.1(f) and (g). This pattern is formed

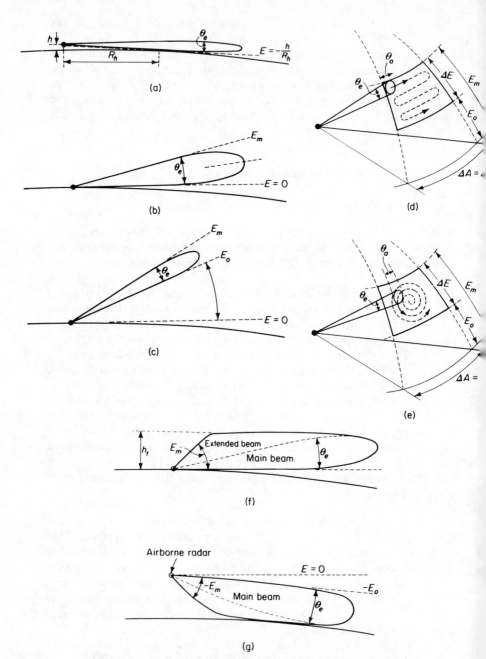

Figure 6.1 Basic types of search scan. (a) Horizon scan. (b) Long-range fan beam. (c) Fixed-evaluation scan. (d) Raster scan. (e) Spiral scan. (f) Cosecant-squared beam for air search. (g) Cosecant-squared beam for surface search from airborne radar.

by diverting a relatively small amount of energy from the main pencil or fan beam into an extended fan above (or below) the main beam. The effect is to maintain a constant echo signal strength for a target which moves along a line of constant altitude in the region of extended coverage. Here, the antenna gain function will vary with elevation angle according to the cosecant squared or R^2/h_t^2. The G^2 term in the numerator of the radar equation will cancel the R^4 term in the denominator [see Eq. (4.7), etc.], providing the desired constant signal strength. Means of obtaining the cosecant-squared pattern will be discussed below.

Line-of-Sight Limit

All radars which operate above 100 mc in frequency are subject to the limitations of line-of-sight ray propagation. Although this limit may prove beneficial in certain cases, furnishing a mask to reduce unwanted ground or sea clutter, the failure in low-altitude coverage imposed by earth's curvature is one of the major limitations in search radar systems. It has been overcome only in the trans-horizon radars which use ionospheric reflection to achieve propagation beyond the optical line of sight. The normal vertical coverage obtained by radars operating at VHF and above is shown by coverage charts such as Figs. 6.2 and 6.3. The first chart uses an expanded vertical scale to emphasize long-range, low-elevation

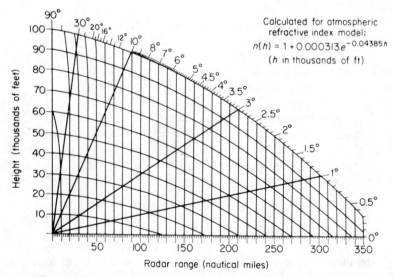

Figure 6.2 Vertical-coverage chart for aircraft. (Courtesy L. V. Blake, Naval Res. Lab.)

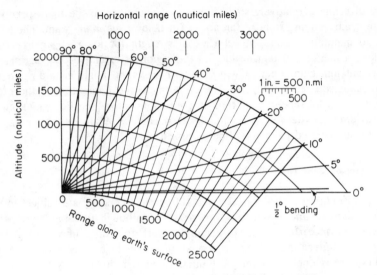

Figure 6.3   Vertical-coverage chart for missiles and satellites.

coverage as used in aircraft detection. The second uses equal scale-factors, and is appropriate for missile and satellite problems. Both include corrections for bending of the rays in the troposphere, although this effect is barely perceptible in the second chart, and rays are plotted as straight lines.

The effect of ionospheric bending cannot be reduced to such simplified form, since it is dependent upon the operating frequency of the radar and on variable characteristics of the ionosphere. However, the maximum bending under normal conditions is only about $\frac{1}{8}°$ at 200 mc, decreasing as the square of frequency, and the effect may usually be ignored.

The vertical-coverage pattern of the radar is represented by a plot of the one-way voltage gain pattern of the antenna on either of the charts shown. Such a plot is shown in Fig. 8.2 for a simple radar. The scale factor of the plot in range will depend upon the target and the required signal-to-noise ratio or detection probability for which the coverage is to be given. A family of coverage patterns may be plotted for different targets and detection criteria.

Effects of Surface Reflections

The effects of surface reflections on search-radar coverage have been known since early in World War II, but their importance is still overlooked by many who test and use radar in the field. Complaints about "holes" in the vertical antenna pattern of a search radar are generally traced to these fundamental reflection phenomena. Most search-radar antennas illuminate the surrounding ground or sea surface with appreciable power levels. Part of this power is scattered over a wide angle, including some backscattering to the radar in the form of surface clutter. The remainder is either absorbed by the surface or reflected upwards at an angle equal to the angle of incidence, where it combines with the direct ray to form an interference pattern. The result may be described in terms of an "image" antenna, shown in Fig. 6.4. The image antenna is assumed to be coupled to the real antenna through an attenuator representing the reflection loss at the surface, and through a variable phase shifter. It is pointed downwards at an angle equal to the upward tilt in the pattern of the real antenna. If the range R to the target is large compared to the height h of the antenna above the surface, the rays arriving at the real and image antennas may be considered parallel, and the added length Δ_r in the path to the image antenna is given by

$$\Delta_r = 2h \sin E_t = \frac{2hh_t}{R} \tag{6.2}$$

Here, E_t is the elevation angle of the target, h_t is its altitude above the reflecting surface, and R is the range. The phase of the reflected signal will differ from that of the direct ray by $\psi + 2\pi\Delta_r/\lambda$ rad, where ψ represents the phase shift associated with surface reflection. The low-angle pattern of interference will be similar to that shown in Fig. 2.5 for the interferometer, with lobes spaced by an angular interval.

$$\Delta E = \frac{\lambda}{2h} \tag{6.3}$$

At higher angles, the amplitude of the reflected ray will be reduced by the directivity of the antenna pattern and the reduced reflection coefficient of the surface, producing the lobing pattern shown in Fig. 6.4(b). The low-angle lobes, if total reflection from the surface is assumed, will extend the radar coverage to twice the free-space range of the radar, and the nulls will be of infinite depth. If the main lobe of the antenna is tilted upwards, the amplitude of the lobing will fall off rapidly with elevation, as the direct ray arrives near the axis of the beam and the reflected ray moves further from the axis of the image antenna pattern. Not until the surface illumination is some 20 db below the maximum gain of the direct ray will the lobing be reduced to negligible amplitude, however. For this reason, ground-reflection lobes are almost invariably a serious problem at elevation angles less than the width of the main lobe, and their effects often extend well

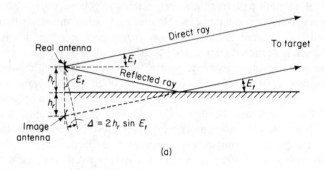

(a)

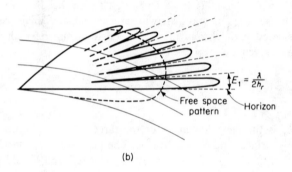

(b)

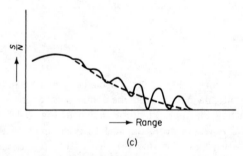

(c)

Figure 6.4 Effects of surface reflections. (a) Geometry of reflection.
(b) Lobing structure. (c) Signal strength vs. range.

into the high-altitude coverage of the radar. The signal strength from a
constant-altitude target will appear as shown in Fig. 6.4(c), the lobing rate
depending upon the target velocity as it moves through the interference
pattern.

Airborne radars are also influenced by surface reflections, although the geometry may differ radically from that shown for the ground radar. The vertical-coverage chart of Fig. 6.2 may be used as a basis for construction of a chart for airborne radars, as shown in Fig. 6.5. The free-space pattern of the airborne antenna is drawn, and possible reflecting paths are located as shown to indicate strong, medium, and weak reflections. Since the target range can no longer be considered much greater than the range at which the reflections take place, the lobing pattern cannot be considered as a simple modification of the angular coverage of the antenna. Within the indicated regions of strong and medium reflections, the lobes are so narrow that targets of high radial velocity pass through them very rapidly, with a resulting fluctuation of the signal amplitude, superimposed upon the fluctuation of target cross section. The effect of this combined fluctuation may be treated by the same type of statistical analysis used for the cross-section fluctuation when it occurs by itself.

The problem of ground reflections may be reduced by using narrow beams in a fixed elevation sector, a stacked-beam system, or a single narrow beam which scans in elevation. Only the lowest beam position, grazing the surface, is then affected to any great extent, although side lobes of upper beams may receive reflection. In the stacked-beam configuration there may still be variations in signal amplitude as the target moves from one beam to the next, but deep nulls can be avoided. On the other hand, the possible doubling of detection range which accompanies severe lobing

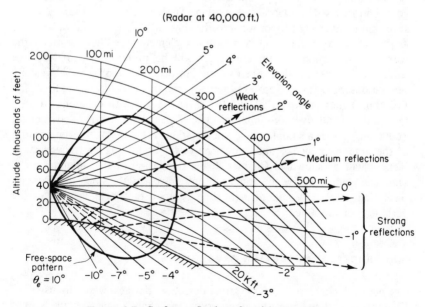

Figure 6.5 Surface reflections for airborne radar.

may be of great value when the lowest part of the main lobe falls in an area of important long-range coverage, and when periodic nulls are tolerable. In airborne applications, the limited vertical dimension of the antenna leaves little choice as to the beamwidth or beam shape, and the resulting wide beams are generally subject to the type of lobing shown in Fig. 6.5.

Antenna Limitations

The upper limit to antenna size may be set either by mechanical considerations or by the need to scan the assigned volume with some minimum time-on-target. It was established in Chapter 5 that a scanning beam should not move more than about one-third beamwidth in the range-delay time of the most distant target, if excessive scanning loss is to be avoided. Also, the data rate (or scan rate) must be held above some minimum value which assures two or more scans past the target before it leaves the scanned volume, if the beam shape loss is to be kept at a reasonable level. The use of higher scan rates may be dictated by system requirements for more frequent position measurements. These considerations encourage the use of broad beams, whereas the need for high S/N ratio and good resolution presses the design in the direction of narrow beams. The generation of simultaneous narrow-beam clusters provides a partial solution in some systems where the additional complexity can be tolerated.

The time-on-target requirement set by scanning loss may be less stringent than that set by Doppler resolution. The width of the observed target spectrum is inversely proportional to the time-on-target and the radar wave length. If a particular velocity resolution element is called for, then the corresponding minimum time-on-target will be proportional to the wave length of the transmission. For example, in order to resolve radial velocity separations of 1 ft per sec, a system operating at a wave length of 1 ft (1000 mc) must sample the echo for at least $\frac{1}{2}$ sec, with coherent processing of the signal over this interval. The system with a 2 in. wave length (6000 mc) needs a sample of $\frac{1}{12}$ sec duration. As a result, the beamwidth of the antenna may be made much narrower at the higher frequency. In fact, as will be shown in Chapter 7, the dimension of the antenna in the scanning plane may be the same for any wave length, since the Doppler resolution in velocity is dependent directly upon the linear velocity of the periphery of the antenna.

6.3 PRACTICAL DETECTION PROCEDURES

The general characteristics of target-detection devices were discussed in Chapter 1, and the use of cathode-ray tubes, range-gated channels, and video integrators was mentioned. In actual search-radar systems, the choice of detection procedure is limited by the data rate required, the traffic-handling capacity, the characteristics of the device which will receive the radar data, and the size, weight, and complexity permitted in the radar installation. The most frequent choice has always been the cathode-ray-tube integrator with a human operator as the detection and measurement system. The characteristics and limitations of this and other systems will be reviewed briefly here.

Cathode-Ray-Tube Integrator

The experiments described by Haeff, Payne-Scott, Lawson, and others have established that an alert operator, when not overburdened with too large an area to search, can achieve a detection performance level nearly as good as that predicted by Marcum for the optimum video integrator (see Section 1.3). Exact comparisons are difficult to make, owing to the uncertainty as to the operator's "false-alarm rate," but a minimum loss of about 2 db relative to the ideal "matched-filter" system will be derived below. The operator can also make measurements of range and angle on the display, can remember the past history of tracks and extrapolate them through regions of lost signal, and can relay the measurements by voice or by electromechanical aids to remote points where the data is to be used. His ability to adapt to changes in system noise level, presence of clutter, and target maneuver also leads to the choice of a human operator in many applications. Let us consider some of the factors that may detract from optimum operator performance, and assign loss terms to describe the departure of the operator from optimum video integration. Extensive experiments on the performance of human operators with cathode-ray-tube displays have been reported by Lawson and Uhlenbeck, Haeff, Payne-Scott, and Ashby, Josephson, and Sydoriak. Some of the losses found in these analyses have already been described in connection with the development of basic detection theory, but the findings on human operators will be summarized here.

Two of the most critical loss factors related to the use of video displays are dependent upon the i-f and the video, or postdetection, bandwidth of the system. In Fig. 6.6(a), the matching-loss curves which apply to automatic detection are shown for peak and average outputs of a rectangular filter. The curve for the average output of a two-pole filter is also shown, and it may be assumed that the characteristics of the usual radar i-f amplifier will lie somewhere between these two. Whereas the operation of the automatic detection circuit may depend upon the peak output of the filter, the human operator will see an average of the output, and will resolve signals whose duration is that of the pulse width. For this reason, the solid line marked "average observed power" has been drawn as a compromise between the average-output curves, and indicates the output power ratio when averaged over the pulse width τ for a near-rectangular filter response.

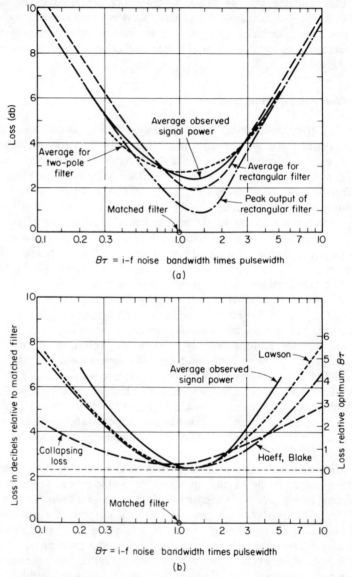

Figure 6.6 (a) Comparison of filter matching losses for peak and average signal. (b) Comparison of matching-loss curves for oscilloscope detection against observed power and collapsing loss.

The minimum loss of 2.3 db relative to the peak output of a matched filter occurs near $B\tau = 1.2$, which agrees with experimental results.

In Fig. 6.6(b), this compromise curve is compared with the observed losses given by Lawson and Uhlenbeck and by Haeff (whose results are used by Blake in his analysis of search radar). The zero decibel reference for the oscilloscope observations has been shifted to account for the minimum loss of 2.3 db at $B\tau = 1.2$. To confirm the amount of the minimum loss, calculations were made using the collapsing-loss theory of Marcum, with two collapsing ratios computed from Eqs. (4.43) and (4.45). An optimum postdetection bandwidth $B_v = 0.6/\tau$ was assumed, and the integration of many pulses was assumed in entering Fig. 1.16 to find the losses. The minimum combined collapsing loss from the two sources was found to be 2.6 db at $B\tau = 1.0$. A comparison of these curves indicates that

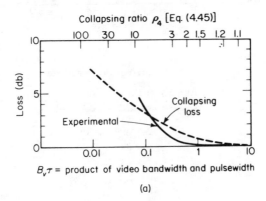

(a)

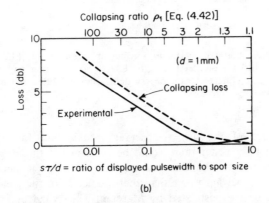

(b)

Figure 6.7 Effects of video bandwidth and sweep speed. (a) Loss owing to insufficient video bandwidth. (b) Loss owing to nonoptimum sweep speed. (Matched i-f bandwidth assumed in both cases.)

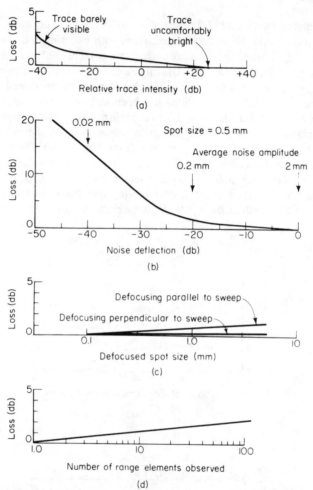

Figure 6.8 Losses in detection for human operator. (a) Effect of A-scope intensity on detection. (b) Effect of deflection amplitude (A-scope). (c) Effect of cathode ray defocusing. (d) Effect of observing large-range intervals.

the performance of the display-observer combination for nonoptimum matching is slightly better than would be predicted if the average signal-to-noise power ratio at the filter output were used. At the same time, the performance is somewhat worse than that predicted for the ideal integrator and threshold device postulated by Marcum. The broad optimum region near $B_T = 1.2$ is properly predicted by the average filter-output curve. The minimum loss of 2.3 db for the display relative to the matched filter and peak-sensitive detector must be considered in establishing the detection and

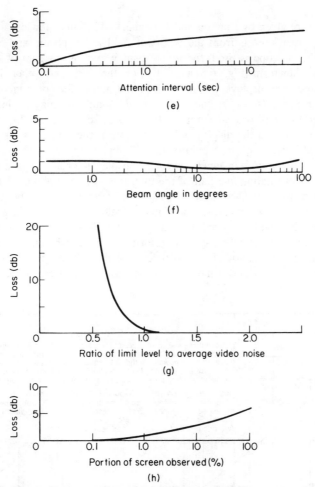

Figure 6.8 cont. (e) Effect of extended attention interval. (f) Effect of beam angle (PPI or similar display). (g) Effect of video limiting (PPI or other intensity-modulated display). (h) Effect of observing large screen area.

false-alarm probabilities for the human operator, and in comparing these with the curves for automatic detection.

When the i-f bandwidth has been set near the optimum value ($B\tau = 1.2$), the loss from insufficient video bandwidth or sweep speed will be as shown in Fig. 6.7. These losses were also covered by the collapsing-loss theory in Chapter 4, and corresponding values of L_c are plotted along with the experimental results. A series of relatively small loss terms is shown in Fig. 6.8(a)–(h). With the exception of the first curve, all these results were given by Lawson and Uhlenbeck. The curve for loss as a function of screen area is from the report of Ashby, Josephson, and Sydoriak. When the A-scope is used, the total "display loss" is the sum of the components shown in Fig. 6.8(a)–(e). For the PPI and other intensity-modulated area displays, this loss will be the sum of the components shown in Fig. 6.8(c), (f), (g), and (h). An additional loss may be incurred if the operator is distracted from his function of scanning the display. If he is able to observe the display only a fraction of the time, this loss may be described in terms of an "operator factor" applied to the single-scan detection probability. For instance, when the display is observed in a series of short glances totaling one-half of the total time, the probability of detection computed from signal-to-noise and loss considerations should be multiplied by one-half. In other cases, where the operator observes the display continuously but with reduced efficiency owing to some environmental or mental factor, his effect on system performance is described more accurately by an added loss factor of one to several decibels. The operator will obviously give his best performance when he has been alerted to the probable

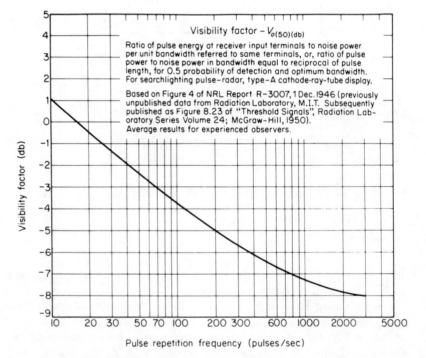

Figure 6.9 Visibility factor for A-scope display. (Courtesy L. V. Blake, Naval Res. Lab.)

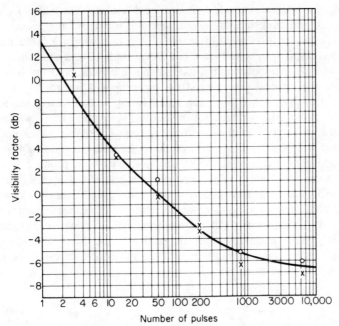

Figure 6.10　Visibility factor for PPI display. (Courtesy L. V. Blake, Naval Res. Lab.)

presence of a target, preferably within a particular region on the display. He may then be expected to match the performance described by Fig. 6.9 for the A-scope, or Fig. 6.10 for the PPI or similar display. Both these curves are used by Blake to describe the requirement for a detection probability of 50 per cent, after accounting for other losses such as matching of bandwidth. These curves express the integration capability of the operator-display combination, and their use (as an alternate procedure to that described in Chapter 1) will be discussed in Chapter 8.

Video Integrators

The early experiments with video integration circuits led to disappointment, largely because it was not realized that the CRT display and human eye were already integrating with excellent efficiency. When the video integrator was placed in series with the normal display, the result was often to reduce the effectiveness of integration by introducing circuit noise, nonlinearity, or gain instability. For some applications, however, video integrator circuits have proven valuable. One such application overcomes random pulse interference and provides a relatively clear display when the normal PPI integration would be rendered useless. The video from the receiver is subjected to limiting at a level slightly above the average noise level. The limiter is followed by a video integrator, which adds the number of pulses normally received over one beamwidth of antenna motion (see Fig. 6.11). A threshold circuit follows the integrator, and the bias level is set so that several pulses must be superimposed in order to produce an output to the display tube. The reception of a single strong pulse from a nearby radar will not produce an output, since the limiter will reduce its level to a fraction of the threshold setting. A train of pulses not synchronized with the repetition rate of the radar and integrator will be distributed over different range elements, and none will produce a visible output. Occasional noise pulses and random pulses will appear, but only as a random background against which the desired echoes will remain visible. With such an integrator, the detectability of echoes against a background of thermal noise will remain roughly the same as with normal CRT integration, depending upon the setting of the threshold and the stability of the integrator. In no case will the theoretical performance of the optimum integrator be exceeded.

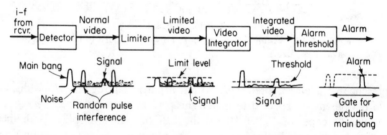

Figure 6.11   Use of video integrator to eliminate random pulses.

Another application for the video integrator is in the use of severely collapsed displays such as the C-scope, where many hundreds of range elements may be superimposed on a single area of the CRT screen. By inserting some type of integrator and threshold prior to the intensity-modulating input of the display tube, the collapsing loss may be eliminated. The same consideration applies when it is desired to use a single automatic detection threshold to cover a large number of range elements.

Several methods are available for performing video integration on a continuous basis (Fig. 6.12). Recirculating delay lines of the type originally developed for MTI cancellation systems may be used to sum the video output of a normal envelope detector, rather than to take successive differences between outputs of an MTI receiver. Magnetic drums or discs may be used in a similar fashion, the speed of the device being locked to the

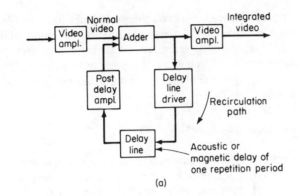

(a)

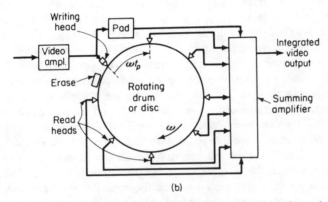

(b)

Figure 6.12 Two video integration methods. (a) Integration using recirculation loop. (b) Integration using magnetic storage.

radar repetition rate or vice versa. When the magnetic method is used, a number of reading heads may be located at intervals around the drum such that they are separated in time by an amount equal to the repetition period. The outputs of the several heads may be added linearly or in an appropriate weighting circuit. By erasing the oldest piece of recorded data just prior to its reaching the writing head, the problem of recirculation gain stability is completely overcome. When the disc or drum has only enough resolution to permit a single set of data to be carried around its periphery, it is necessary to read the old data just prior to reception of the next pulse from the same range delay, to erase the the old data, and then to add the old data to the new prior to writing the sum onto the erased part of the track. The gain of the loop must be held within narrow tolerances if oscillation or inefficient integration is to be avoided. Similar processes are possible when electrostatic storage tubes based on cathode-ray-tube designs instead of magnetic tracks are used. Photographic techniques have also been used, with rapid-processing systems to make available the output data within seconds of the time of signal reception. In all these processes, resolution in range must be retained if collapsing losses are to be avoided. Other losses will depend entirely upon the ability to maintain constant gain and freedom from circuit noise.

Automatic Target Detection

A rudimentary device for automatic detection was described in Chapter 1. To use such systems in search radar, it is generally necessary to cover the entire range of the radar with detection gates and thresholds, and to provide automatic transmission of range and angle data when a target has been detected. Both analog and digital techniques are available for such detection systems, the principal problem being to provide the necessary number of channels without undue cost, complexity, or unreliability. When each range element is instrumented with its own gate and threshold, the theoretical detection performance can be achieved without collapsing loss. By overlapping the gate circuits and making them somewhat longer than the pulse width, the sum of the gate spacing loss and the collapsing loss may be held to a value below one decibel without doubling or tripling the number of channels. Some type of automatic gain control or "automatic video noise level" control will generally be needed to maintain a constant false-alarm rate as the receiver and other circuits vary in gain. Automatic threshold level adjustment may also be required. As an alternative, in an analog integration system, the outputs of all integrator channels may be scanned at a high rate and compared with a single threshold reference. When an output occurs, the position of the scanning switch or sampling device serves to indicate the range of the target.

In digital detection schemes, the coincidence or "double-threshold" procedure is usually used to avoid the necessity of storing thousands of noise samples received during several repetition periods. Only signal-plus-noise samples which exceed the relatively low first threshold are stored for comparison with succeeding repetition periods. A digital counter may be assigned to each range element to accumulate the number of crossings of the first threshold over some fixed interval (see Fig. 6.13), or the fact of a threshold crossing in a given channel may be stored in some central memory device along with information as to the range at which it occured. In either case, some means must be provided to destroy or forget information after it has been in storage longer than the desired integration time. Thus, in Fig. 6.13, a "dump" control signal resets the counter to zero after some fixed number of repetition periods. As shown in Chapter 1, the use of the coincidence detection procedure involves some loss relative

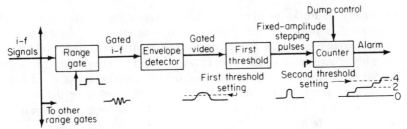

Figure 6.13 Automatic detection system using coincidence method.

to optimum video integration. The loss may be reduced by storing more data to indicate the approximate amplitude of signals which cross the first threshold (e.g., by encoding the signal-plus-noise samples in four or eight discrete amplitude levels). This will require the storage of signals and noise whose amplitudes are lower than those which would be required to pass the threshold in a simple coincidence system, and the resulting complexity and increased storage capacity can be justified only if the value of restoring the one or two decibels of integration loss is very high.

6.4 MULTIPLE-TARGET MEASUREMENT AND TRACKING

As was noted earlier, the function of the search radar includes both target detection and position measurement. The accuracy of measurement need not be as high as in most tracking radar systems, but the range and at least one angular coordinate must be found and made available for use outside the radar. The required accuracy, data rate, and resolution will depend upon the use to be made of the data, and these factors will have great influence on the choice of radar parameters.

System Requirements for Position Data

In air-defense radar systems, search radars are used for early warning, alerting of specific defensive or offensive systems, ground-controlled interception, direct control of defensive missiles, and designation to tracking radars. In each case, the ability of the search radar to provide data on many targets is of prime importance. Accuracy figures may vary from the order of one mile to some fraction of this figure. Resolution in range is usually a fraction of one mile; the angular resolution often exceeds this by a large factor, being set by the maximum usable antenna size. Azimuth beamwidths of about one degree are typical of many search radars, although much better resolution would be used if permitted by the data rate and time-on-target requirements. In order to maintain identification and continuity of tracks on the many targets appearing in the radar search volume, data rates of several scans per minute are used. Very much the same considerations apply to radar used in air traffic control, both in the vicinity of airports and en route between airports. The closer the targets approach to regions of critical importance (interception, in air defense; landing approach, in air traffic control), the better must be the accuracy and resolution. Also, as the rate of target maneuver increases, so must the data rate of the radar.

In using search radar data on aircraft, the PPI type of display serves as a primary measuring device. Occasional readings of altitude may be gathered from a related height-finding radar or from radio communications (in the case of cooperative targets). A series of range and azimuth readings, with successive points separated by a few seconds in time, can be formed into a continuous track on the radar screen or on separate plotting boards, and these tracks will serve a variety of needs. The same information may be handled automatically in a computer and used to derive control signals for navigation, guidance, and designation to narrow-beam tracking radars. The target position may be determined by reference to a grid of marker scales on the display, giving polar or rectangular coordinates relative to the radar, or by reference to features of the surrounding land, observed by the radar or superimposed on the display. The relative positions of two or more targets may also be read directly. Improved accuracy and resolution on the display are obtained by expansion of small segments of the coverage to fill the entire display screen, so long as other displays and operators are available to maintain surveillance over the remainder of the assigned volume. Since there is no limit to the number of video outputs which can be derived from a single radar receiver, as many as twenty displays are sometimes attached to a single radar for simultaneous tracking of many targets by human operators. Even when automatic equipment is used for measurement and tracking, a number of displays is needed to monitor the operation, assign new targets, and edit the data being fed to the computers.

The Human Tracker

The human observer constitutes one of the most flexible and reliable means of measuring and tracking targets, when he is used within his traffic capacity. Using controllable azimuth and range cursor scales on a long-persistence CRT screen, he may perform interpolation within the radar resolution element, measuring both quantities to perhaps 10 per cent of the resolution element. In range, his accuracy will often be limited more by the resolution of the display than by the radar pulse width, and it is here that the expanded display will prove especially valuable in a system designed for high accuracy.

In a radar-controlled intercept operation, the radar observer may make near-simultaneous measurements on two targets. Where only moderate accuracy is needed, he may control the interception by talking the controlled aircraft onto the pursuit course. Better accuracy and smoother control result when the operation of the radar observer's cursor scales is coupled automatically through a computer to generate continuous commands for the interceptor. In either case, once the track has commenced, the operator is alert to the presence of both targets, anticipates their future positions, and can perform his function on signals considerably weaker than those required for initial detection. His most serious problems arise when target fluctuation, clutter, or interference obscures one or both targets completely for more than one scan period. In this case, the errors in his estimated target positions grow geometrically, approaching such proportions that he must scan a considerable area of the display to reacquire the targets. During the period when no data are received, the command signals generated by his cursor-scale motions may force the interceptor seriously off course and make later interception impossible.

In air traffic control, similar operations are performed by human operators in giving navigation commands to pilots. Here, the major problem may be in identification of aircraft under observation, owing to the high traffic density in some areas. Aircraft flying at different altitudes are superimposed on the PPI display, and may perform similar maneuvers to stay on assigned airways. If an adequate period of time and good communications are given, positive identification can be assured by execution of specific maneuvers under ground command. The use of radar beacons, to provide both identification and altitude reports on a continuous basis, is now relieving much of the burden on traffic controllers and pilots alike, making possible rapid and reliable control of high-performance aircraft.

Automatic Track-while-scan Devices

Analog and digital computers are used in place of or in addition to human operators in some tracking systems. When the analog device is used, its input is derived from range and azimuth gate circuits, which sample the radar video output only at those times when the selected target echo appears (Fig. 6.14). Having been placed initially on the target with approximately matched rates in both range and azimuth, the gates move under the control of the computer to extrapolate the target's course in rectangular coordinates and to measure any departure of the target from the predicted position. The range and azimuth error signals are fed to the computer on each scan to correct the extrapolation rates. In one system, the computer stores position and rate data for a single scan, modifying this data as a result of comparison with measured position. On the nth scan

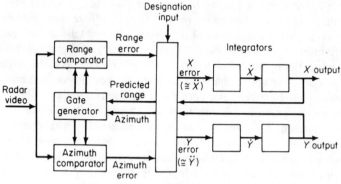

Figure 6.14   Typical track-while-scan channel.

across a target, the measured coordinate X is compared with the predicted value $\overline{X}_n$ formed from prior tracking, and new estimates of position and velocity for the next period are formed as follows

$$\overline{X}_{n+1} = \overline{X}_n + \beta(X - \overline{X}_n) + \Delta t \dot{X}_n$$

$$\dot{X}_{n+1} = \dot{X}_n + \alpha(X - \overline{X}_n)$$

Values of the "damping factors" α and β for position and velocity may be varied from about 0.05 to 1.0, depending on the quality of the data and the duration of the prior track. These factors will determine the response of the system to noise and to changes in target velocity. Adaptive processes for minimizing rms error in a similar system have been described,* based on sampling of the error $(X - \overline{X}_n)$ to determine whether noise or lag error is predominant. The size of the tracking gates may also have to be varied to optimize the system, and provisions must be made for carrying tracks through one or more scan periods in which no signal is obtained.

As an example, let us assume that a track has been established on an aircraft flying in a straight line, and crossing 30 mi from the radar. On each scan, the radar measurement is carried out with an accuracy approaching the thermal noise limit (see Chapter 2)

$$\sigma_r = \frac{c/2}{2B\sqrt{n(S/N)}} \simeq \frac{\tau(c/2)}{2\sqrt{n(S/N)}} \tag{6.4}$$

---

* A. H. Benner and R. Drenick, "An Adaptive Servo System," *IRE Conv. Record* (1955), Part 4, pp. 8–14.

$$\sigma_\theta = \frac{0.53\theta}{\sqrt{n(S/N)}} \tag{6.5}$$

If the radar has a pulse width of three microseconds and a beamwidth of one degree, the target which can be detected reliably (integrated signal-to-noise ratio of 15 db) will be located with an accuracy equal to about one-tenth of the resolution-element dimensions. Thus, σ_r will be about 150 ft, and σ_θ will be $\frac{1}{10}°$ or 1.7 mrad. At a range of 30 mi, the error in the direction normal to the beam will be 300 ft. A typical data rate would be six scans per minute, which would permit the target to move 6000 ft between scans, when traveling at a speed of 350 knots. After smoothing over about four scans, the predicted position would be within the error limits of the radar measurement, for the target with no acceleration, and the tracking error would vary in a random manner about the true position, with an rms error perhaps half that of the individual radar observation (see Chapter 13).

If the target in this example were capable of maneuvering with one g acceleration (32 ft per sec^2), the error between scans would approach a peak value given by

$$\Delta_p = \tfrac{1}{2}a(\Delta t)^2 = 16(10)^2 = 1600 \text{ ft} \tag{6.6}$$

Thus, if the maneuver is started immediately after the radar scans over the target, the aircraft could be outside the limits of the range gate by the time of the next scan. Since the track-while-scan computer must smooth over more than one scan, the error may easily be four times as great for this target. The tracking gates must be made wide enough to measure targets which deviate appreciably from the expected position, and this will increase the level of the tracking noise and increase the possibility of including unwanted targets in the gate. A balance must, therefore, be reached between thermal noise, lag errors, and the possibility of target loss caused by insufficient integration when rapid scan is used. In addition, since most targets fluctuate in amplitude enough to cause loss of measurement on some scans, there must be enough extra measurements to carry a track through occasional scans with no signal.

When a digital computer is used, similar problems arise, but there is a greater possibility of using a flexible program with data stored over much broader intervals around the extrapolated point. By storing the range and azimuth of all detected targets, and selecting those which correspond to the extrapolated target positions, it is possible to track targets with much greater maneuverability. In those cases where no target appears at the expected position, the computer may be programmed to search its memory for targets just outside the expected resolution element, the search expanding until a suitable target is found. There remains, of course, the possibility of confusing two targets when they pass close together with simultaneous maneuvers.

Using the theory developed for tracking radars (see Chapter 9), we may characterize the track-while-scan device by an effective bandwidth β_n, from which both thermal noise and lag errors are computed. The upper limit of this bandwidth is $\nu_s/2$ or $j/2t_s$, where $\nu_s = j/t_s$ represents the number of complete scans per second executed by the radar. The number of individual radar measurements averaged to arrive at the thermal noise component of system error will be $2\nu_s/\beta_n$, and the output error will be equal to the radar error divided by the square root of this ratio. Thus, in our example, the averaging of four scans by the computer, with a ten second scan period, would lead to a tracking bandwidth of $\frac{1}{20}$ cps, and an error of one-half the individual radar measurement. The acceleration error constant of such a system is approximately given by

$$K_a \doteq 2.5\beta_n^2 = 0.00625 \text{ per sec}^2 \tag{6.7}$$

The acceleration lag Δ_p will be equal to the target acceleration divided by K_a, or about 160 ft of error per foot per second of acceleration. By reducing the smoothing time to cover only two scans (maximum β_n), the value of K_a is increased by a factor of four, and the lag error will be only 40 ft per ft per sec^2 of target acceleration, or 1300 ft for a constant maneuver at one g. The initial transient value may exceed this, as was shown in the earlier calculation.

Height-Finding and GCA

A common form of height-finding is a special case of track-while-scan, in which a rapid elevation scan is used at the azimuth of the designated target. The range is also designated to the height-finder, leaving a minimum chance of confusion between targets or loss of detection. With scan periods of one or two seconds, the equivalent bandwidth β_n of the height-finding system will usually be between 0.1 and 0.5 cps. Unless the target is maneuvering at a great rate, there will be no appreciable lag error during the measurement. On the other hand, the interval between measurements may be extended well beyond the scan period of the main search radar, as the height-finder is shared between many targets at different azimuths. The human tracker or computer must call for height data at a rate consistent with the target's capability for vertical maneuvers, as determined by Eqs. (6.6) and (6.7) for lag error. The suitability of this type of height-finder is based upon the fact that most aircraft are severely restricted in their ability to maneuver in altitude. This permits tracks to be maintained with only occasional height readings. In cases where height data must be obtained at the same rate as azimuth and range, the use of a 3-D radar is indicated.

The stacked-beam height finder operates in much the same manner as the monopulse tracking radar, comparing the simultaneous outputs of two or more beam positions to interpolate the position of the target. Since the extremely high accuracy of the tracking radar is not usually required, the stacked-beam system may use video amplitude comparison, rather than comparing the amplitude or phase of signals in the r-f or i-f portions of the receiving system. In the video comparison system, the gains of the receivers must be closely matched to make the measurements accurate. The elevation measurement circuits may operate on all targets detected, or only on those covered by range and azimuth gates of a track-while-scan system. Interpolation to about one-tenth of the elevation beamwidth is commonly obtained over most of the elevation sector. By using narrow beams in the low-elevation region and broader beams in the higher regions, the desired high accuracy of altitude measurement may be obtained at long ranges without using an unreasonable number of channels. Another advantage of the stacked-beam system in search is that the ground clutter is limited to the lowest beam or two. The upper beams may be used without MTI or with noncoherent MTI to cancel weather or chaff echoes.

The scanning type of GCA radar operates in a manner similar to the nodding-beam height finder discussed above, but it scans in both azimuth and elevation, using two separate fan beams. A skilled operator, observing the dual display (a combination B and E scope is used with special markers) can close the control loop to the aircraft by voice commands at intervals of a few seconds. The bandwidth of such a system is in the order of $\frac{1}{10}$ cps, and there have been cases where this has proven inadequate for high-performance aircraft. The target in this case, represented by the airport runway, is fixed, and the system need eliminate only the errors due to wind effects, operator and pilot error, and initial transients caused by the change in the aircraft flight path as it enters the GCA pattern. With the coming of high-speed jet aircraft and restricted approach patterns, the capabilities of the human operators in both the GCA station and the aircraft are strained near the breaking point. For approaches under very poor weather conditions, the use of automatic GCA, with either tracking radar or fast-response track-while-scan computers, becomes necessary. Errors are reduced in such systems by increasing the accuracy and rate of the radar measurements, and by eliminating the human errors introduced in initiating and following the commands.

Programmed Scan and Tracking

A final type of measurement system to be described represents a combination of search, height-finding and tracking techniques, using a very agile radar beam under command of a computer. This type of radar is made possible by the development of rapid-scan antennas, which may perform discontinuous scans over extended regions. The conventional height-finder, operated in conjunction with a search radar, was an approach to this technique, in that the energy transmitted by the complex of radars could be distributed in a flexible fashion by control of the height-finder. However, in this case there was a more or less uniform spreading of search radar energy over the entire hemisphere, and the low-powered height-finder could modify this by only a small amount. In the extreme form of programmed scan, it is possible to vary the energy distribution from uniform search over the hemisphere to concentration on a single beam position, excluding all others. Such a radar can serve as a tracking radar or a 3-D area surveillance radar, and can adjust its scan rates to match the distribution of targets. It will not be necessary to analyze the operation of such a radar separately, since all its functions are covered by the theory applied to search and tracking radars of the conventional type.

The introduction of coherent signal-processing systems, in which the phase of the echo signal is used along with its amplitude, provides the radar with an effective fourth dimension for resolution and measurement. This fourth dimension is the Doppler frequency shift of the signal, equivalent to radial velocity of the target. In search radar, the additional dimension is used almost entirely to eliminate unwanted clutter from the background, selecting as targets only those objects which move with some minimum velocity relative to the radar or to the fixed background. The MTI, or moving-target indicator, is defined as a device which limits the display of radar information primarily to moving targets.* A more advanced type of system is the pulsed-Doppler radar, defined as a pulsed radar system which utilizes the Doppler effect for obtaining information about the target, not including simple resolution from fixed targets.† Several types of MTI will be described in this chapter, and the factors which limit the performance of the com-

MTI TECHNIQUES IN SEARCH RADAR

7

mon types of system will be analyzed. Advanced techniques in MTI systems will be reviewed briefly.

7.1 APPLICATIONS OF THE DOPPLER PRINCIPLE

Some of the earliest experiments in

* *IRE Dictionary of Electronic Terms and Symbols*, 1961, p. 93.
† Ibid., p. 112.

radar depended entirely upon the Doppler shift of reflected c-w waves to indicate the presence of moving targets against the background of fixed targets. These early systems had no range resolution and very little angular resolution, so the appearance of detectable energy at a resolvable frequency, adjacent to the frequency of transmission, was necessary for target detection. As pulsed radar came into experimental use, the importance of Doppler shift was reduced, while efforts went into increasing the sensitivity and the resolution of radar systems in the three spatial coordinates. Only later, when the power and sensitivity of pulsed radar had risen to the point where thermal noise lay below the level of the fixed-target background over a large part of the range sweep, was it found necessary to undertake the development of MTI techniques for use with pulsed radar systems.

Calculation of Doppler Shift

The moving target which travels at a velocity v_t will have a radial component given by $v_r = v_t \cos \alpha$, where α is the angle between the vector v_t and the radar beam. Referring to Eq. (2.18), the reflected signal for outbound targets ($v_r > 0$) will be lower in frequency than the transmission by a Doppler shift whose absolute value f_d is given by

$$|f_d| = f_t\left(1 - \frac{c - v_r}{c + v_r}\right) \cong \frac{2v_r f_t}{c} = \frac{2v_r}{\lambda} \text{ cps} \qquad \textbf{(7.1)}$$

In the above equation, consistent units of length must be used in expressing $c, v_r,$ and λ. The same relationship may appear in the following forms

$$f_d \cong \frac{102v_r}{\lambda} \text{ cps} \qquad (v_r \text{ in knots}, \lambda \text{ in cm})$$

$$f_d \cong \frac{89v_r}{\lambda} \text{ cps} \qquad (v_r \text{ in mph}, \lambda \text{ in cm}) \qquad \textbf{(7.2)}$$

$$f_d \cong \frac{61v_r}{\lambda} \text{ cps} \qquad (v_r \text{ in ft/sec}, \lambda \text{ in cm})$$

In an MTI system, the radar must be made to distinguish between reflections from fixed targets, whose spectrum will duplicate that of the transmitter, and those from moving targets, whose spectrum will be shifted by the amount f_d expressed above. (The special case of the radar system on a moving platform, where both fixed and moving targets are shifted, will be discussed later.)

Early Developments in MTI

Although the need for MTI was recognized early in World War II, a number of factors delayed its introduction until the end of the war. The basic problem in applying MTI to a conventional pulsed radar is that the width of the transmitted spectrum greatly exceeds the amount of Doppler shift produced by most targets. For example, a typical search radar may operate at a frequency of 1300 mc ($\lambda = 23$ cm), using a pulse width of 6 μsec. An aircraft moving at 300 knots will produce a maximum Doppler shift of 1330 cps. The transmitted spectrum extends to a width of 333 kc between the first nulls, with several spectral side lobes beyond this, so that a region several hundred times as wide as the Doppler shift is covered by components of the transmitted energy. In order to provide frequency resolution of moving targets under these circumstances, the transmitter must be kept as stable as possible in both repetition rate and transmitted frequency, and a suitable phase reference must be maintained for use in detection of the echo signals. When this is done, the received energy spectrum will consist of a number of narrow lines, as shown in Fig. 7.1. These lines, appearing at intervals equal to the repetition rate of the radar, represent the Fourier components of the repetitive bursts of r-f energy, as seen by the

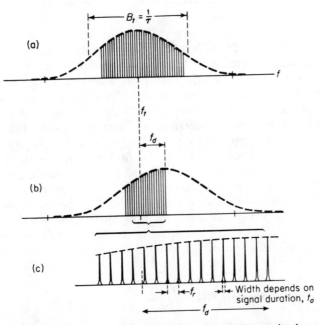

Figure 7.1 Spectra in pulsed radar system. (a) Transmitted spectrum. (b) Received spectrum. (c) Detail of received spectrum.

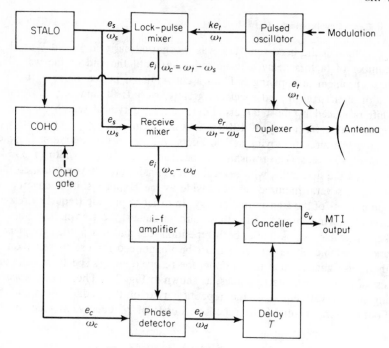

Figure 7.2 Pulsed-oscillator MTI.

receiver. Development of the first MTI systems required the establishment of an adequate phase reference for use in the receiver, and of the special filter needed to reject the energy from fixed targets, appearing in spectral lines corresponding to zero Doppler shift. Since most radars had been forced into the microwave portion of the spectrum by use of electronic jamming and by antenna considerations, the transmitters were predominantly of the magnetron type. These pulsed oscillators cannot be kept locked to a phase-stable reference signal, so the first problem was to develop a system which could be locked in phase with the magnetron at each transmission, and still preserve a phase reference over many pulse repetition periods. The type of system shown in Fig. 7.2 was developed during World War II, and remains in use today for MTI radars. The principles of its operation will be described, and some of the basic terminology of MTI will be defined before we go on to the more recent refinements in MTI techniques.

Simple Pulsed-Oscillator MTI

The type of system represented by Fig. 7.2 is one form of "coherent MTI," in which the phase reference is maintained within the radar itself.

This is as opposed to the "noncoherent MTI," in which the moving target is detected with reference to the phase of signals from fixed targets at the same range from the radar.* Since the transmitter used is a magnetron pulsed oscillator, whose starting phase is random from one repetition period to the next, the local phase reference must be maintained in two oscillators, the "STALO" and the "COHO." The STALO, or stable local oscillator, provides a continuous reference signal of high phase stability at r-f, and is used as the local oscillator for the superheterodyne receiver. The COHO, or coherent local oscillator, monitors the phase of each transmitted pulse relative to the STALO, and stores this phase information until all echoes resulting from this pulse have been received. The COHO operates at i-f, and is locked by an i-f pulse derived from mixing the transmitter pulse with the STALO signal in the "lock-pulse mixer." Having been locked in phase by this pulse, it continues oscillating at the i-f for one repetition period, after which it is momentarily interrupted and then re-locked to the next transmission. Echo signals are received as in the conventional radar and converted to i-f by mixing with the STALO. After i-f amplification, the phase of the signal and the COHO are compared in the phase detector, whose output can be considered an i-f signal centered at zero frequency (sometimes known as "bipolar video").

The output of the phase detector will appear as shown in Fig. 7.3(a). The fixed targets form a stationary pattern over the period required for the antenna to move through one beamwidth. The moving targets oscillate in phase relative to the baseline of the trace or to the fixed targets upon which they are superimposed. In most practical MTI systems, the signal will be limited in the i-f amplifier, so the bipolar video output to the canceller will depend only upon the phase of the signal relative to that of the COHO. In the canceller, the bipolar video signals are stored for exactly one repetition period, after which they are subtracted from the undelayed output of the phase detector. The canceller output will appear as shown in Fig. 7.3 (b), with the moving targets producing outputs which vary at the Doppler frequency. For use on intensity-modulated displays, this output is passed through a full-wave rectifier, from which it emerges as shown in Fig. 7.3(c).

The entire MTI system operates at a sampling rate given by the repetition frequency, and so the maximum apparent Doppler frequency at the output will be equal to one-half the radar prf. Targets whose real Doppler shifts are higher than this will appear with output frequencies equal to the difference between the Doppler frequency and the nearest harmonic of

---

* The *IRE Dictionary of Electronic Terms and Symbols*, p. 26, defines a coherent system as one in which the signal output is obtained by demodulating the received signal after mixing with a local signal having a fixed phase relationship to that of the transmitted signal to permit the use of the information carried by the phase of the received signal.

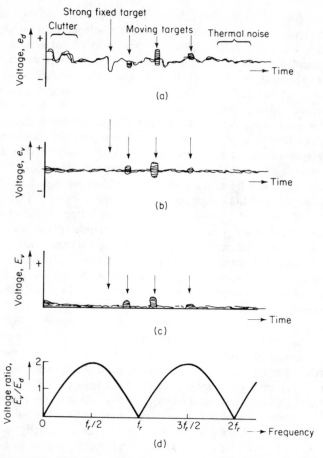

Figure 7.3 Waveforms and response in coherent MTI. (a) Phase detector output. (b) Canceller output. (c) Rectified MTI video. (d) MTI response vs. Doppler frequency.

the prf. When the Doppler shift is equal to the prf or one of its harmonics, the apparent output frequency will be zero, and cancellation will occur as for fixed targets. This leads to the presence of "blind speeds" in the MTI response, corresponding to velocities given by

$$v_{bj} = j\frac{f_r\lambda}{2} \qquad j = \pm0, 1, 2, \ldots \qquad (7.3)$$

The response of the phase detector and canceller to targets of different Doppler shifts is plotted in Fig. 7.3(d), and takes the form of a rectified sine wave with nulls at each blind speed. This type of response is not optimum for cancellation of actual clutter, as will be shown later, but it will serve as a basis for comparing the effectiveness of the more refined MTI filters to be described.

Pulsed-Amplifier MTI

It was realized from the start that the phase-locked COHO, operating with the pulsed oscillator of the magnetron type, was the source of much practical difficulty in operation of MTI systems. The performance in most cases was limited by the phase instability inherent in this technique. A more desirable MTI configuration, shown in Fig. 7.4, uses two continu-

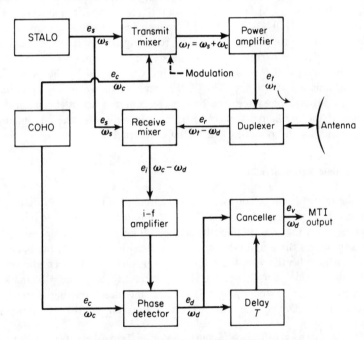

Figure 7.4 Pulsed-amplifier MTI.

ously operating oscillators as STALO and COHO, deriving the transmission frequency from the sum of the two reference frequencies. A high-power pulsed amplifier is then used in place of the magnetron oscillator in the transmitter. It was noted at the end of World War II that such a fully coherent system "can be used at frequencies of a few hundred megacycles per second but is not practical at microwave frequencies in the absence of suitable power amplifiers."* During the 1950's, a series of high-power klystron amplifiers was developed, providing stable outputs of mega-watt power levels in all the normally used radar bands. It was soon estab-lished that such amplifier tubes, driven by intermediate amplifier chains made up of lower-power klystrons or traveling-wave tubes, could preserve the phase stability of the best crystal or atomic oscillators, eliminating the transmitter as a limiting factor in MTI performance.

The operation of the pulsed-amplifier system is almost identical to that described above for the pulsed oscillator, and the wave forms and response curves are identical. The only difference, other than the greater inherent stability, is in the direction of flow of the COHO signal. In the pulsed-amplifier system, this signal is sent to the mixer which replaces the locking-pulsed mixer of the pulsed-oscillator system. This transmit mixer combines the COHO and STALO to produce an r-f reference frequency equal to the sum of the COHO and STALO frequencies, and this is amplified and pulse-modulated for transmission. The resulting transmitted spectrum con-sists of a series of narrow lines located exactly at harmonics of the repeti-tion frequency, and extending over a band determined by the pulse shape and transmission frequency. This is the same as shown in Fig. 7.1, but in the pulsed-oscillator system the lines were centered about the natural fre-quency of the oscillator, and could be defined only after the r-f signal was compared with the phase-locked COHO reference.

Dynamic Range of MTI

The dynamic ranges of the canceller circuits and of the PPI display are severely limited, requiring some form of compression prior to the phase detector. This may be supplied by hard limiting in the i-f amplifier, which also suppresses the amplitude variations caused by motion of the beam and variation in reflectivity of the clutter background. It has been shown that such limiting does not change the shape or relative amplitudes of the spec-tral components caused by signal, clutter, and noise passing through the i-f amplifier.† During the time in which the sum of the i-f voltages of

---

* A. G. Emslie and R. A. McConnell, "Moving Target Indication," Chap. 16 in *Radar System Engineering*, L. N. Ridenour, ed. (New York: McGraw-Hill Book Company, 1947), p. 630.

† It has been shown that the S/N ratio at the output of an ideal limiter varies from $\pi/4$ to two times the input S/N, as the latter quantity varies from zero to infinity. The gain of two at high S/N is the result of suppression of the in-phase component of noise, which would also have been suppressed by the action of the phase-sensitive detector following the limiter. Hence, the effective S/N and signal-to-clutter ratios will remain essentially the same whether or not the limiter is in-cluded in the system. See W. B. Davenport, Jr., "Signal-to-Noise Ratios in Band-Pass Limiters," *Jour. Appl. Phys.*, Vol. 24 (June 1953), pp. 720–27.

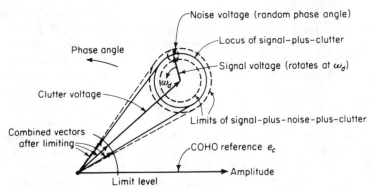

Figure 7.5 Vector diagram of received MTI signals for a given range element.

these components is above the limit level, the excess power is converted to harmonics of the i-f, which are suppressed in the narrow bandpass of the succeeding stages. As a result, when the clutter echo is very strong there will be considerable reduction in the amplitude of signal and thermal noise components leaving the receiver.

A vector diagram showing this action is given in Fig. 7.5. Within a given range resolution element, at a given position of the beam, the clutter can be represented by a large vector at some arbitrary phase angle relative to the COHO reference. A moving target at the same range will appear as a rotating vector added to the clutter component. The rotation rate will be equal to the Doppler frequency. Added to both will be a thermal noise component whose phase and amplitude is random from pulse to pulse. After passing through the limiter and following i-f stages, the total vector will be reduced to some standard amplitude, but its phase will remain the same as it was prior to limiting. The output of the phase detector will be proportional to the cosine of the angle between this vector and the COHO reference, and will have variable components representing both random noise and the Doppler-shifted signal. For a given power ratio of clutter to noise at the receiver input, the same ratio will apply to the rms output of the phase detector, since half of the power of each component will appear in quadrature with the reference. When the clutter is in phase with the reference, the output clutter-to-noise ratio for that range element will be doubled. The signal-to-noise ratio and signal-to-clutter ratio will also preserve their rms values in passing through the limiting i-f amplifier and phase detector.

The action of the canceller is that of a linear filter whose response is described by Fig. 7.3(d). In the ideal case, the clutter components will appear at zero frequency and at integral multiples of f_r, and will be eliminated completely. The noise will pass through the filter with a power gain of two, corresponding to the rms difference of two independent samples of noise which appear in each output sample. Signal response will vary from zero at the blind speeds to a power gain of four at the frequencies of maximum response. The mean-square signal gain for targets distributed uniformly over all frequencies will also be two, so there will be no change in the ability of the MTI radar to detect the average signal in thermal noise, when compared to the conventional radar. Nor will the appearance of thermal noise at the canceller output, after full-wave rectification, be any different from that of the normal PPI display, except in the presence of strong clutter. The clutter will reduce the level of both noise and targets, as described earlier, and will contribute, in practice, a fluctuating random component whose appearance is similar to that of thermal noise.

Describing MTI Performance

The measure of performance of an MTI system is its ability to improve the detectability of moving targets against a background of strong clutter. The property of an MTI radar which can best be defined and measured is its "cancellation ratio," which measures the reduction in amplitude of fixed targets as they pass through the system. The IRE definition of cancellation ratio is, as follows:

> "In a radar MTI system, the ratio of a fixed target signal voltage after cancellation to the voltage of the same target without MTI cancellation."*

Since it is not always possible to turn the MTI canceller off and on for testing, the Navy Bureau of Ships in 1955 gave a definition in terms of a test procedure which eliminates the ambiguity otherwise arising from the factor of two in the response curve of Fig. 7.3(d). This test procedure reads as follows:

> "Cancellation tests on the radar shall be made by observing the ratio of the uncancelled pulse residue amplitudes of simulated fixed targets at the range desired on the A-scope, entirely under MTI operation. Since the A-scope cannot be switched into the cancellation unit of the radar prior to the point of cancellation, the reference level of the uncancelled pulse shall be established by means of a selectable output which transmits a train of simulated target pulses on *alternate pulse periods only*. This output shall establish a desired signal to noise ratio, wherein the noise measured is that of the MTI receiver."†

---

* *IRE Dictionary of Electronic Terms and Symbols*, p. 18.
† Military Specification, Radar Signal Simulator SM-65 ( )/UP, MIL-15293 B(SHIPS), March 28, 1955, with Amendment 1, Sept. 24, 1955.

Thus the reference response level corresponds to the level of one shown on Fig. 7.3(d), and not to the peak response of the MTI system for the optimum target. Normal practice is to compare peak amplitudes of signal and residue, or correspondingly rms values as estimated visually on the A-scope during the test.

Another measure of improvement in detectability is the "subclutter visibility" of the moving target under given clutter conditions. There is no IRE definition of this term, but the military source used above defines it as "the ratio of peak moving target signal to peak fixed target signal existing at the input when the two signals are equal in amplitude at the output." In a properly designed system, this should be one-half the cancellation ratio, since the peak moving target signal will be passed with a voltage gain of two relative to the reference response level.* However, the non-linear operation of circuits in the phase detector or canceller may act to suppress the moving-target output relative to fixed targets, and the sub-clutter visibility test may be invoked to check for this effect. Unfortunately, the term "visibility" implies some variation between operators, and no consistent standard has been applied. An earlier definition† requires that the average moving-target signal amplitude should be at least three times the rms clutter fluctuation in order to be "visible" on the PPI display, which would lead to quite different standards of performance than the definition quoted above. Since there is no single accepted definition for subclutter visibility, and since the cancellation ratio will describe the effects of all recognized system design parameters of the MTI radar, we shall use only cancellation ratio in the following discussions. Later, this will be compared with the "improvement factor" defined by Steinberg, which accounts for the average gain of the canceller for signals spread uniformly over the frequency interval between blind speeds.

7.2 INTERNAL STABILITY REQUIREMENTS FOR SIMPLE MTI

The simple system described above contains all the elements of any MTI radar, and may be used as an example for calculation of component stability requirements. These requirements may be derived either in terms of spectral densities of various spurious signal terms, or as simple functions of time superimposed upon the ideal wave forms which have been assumed in describing MTI operation. We shall start with the simple mathematics of

---

* Note that both cancellation ratio and subclutter visibility are given by voltage ratios less than unity.

† A. G. Emslie and R. A. McConnell, "Moving Target Indication," Chap. 16 in *Radar System Engineering*, L. N. Ridenour, ed. (New York: McGraw-Hill Book Company, 1947), pp. 652-53.

the time functions, applied to the pulsed-amplifier system, and extend the analysis to the locking-pulse operation of the pulsed-oscillator system.

Analysis of Ideal Single-Delay MTI

The output of the single-delay canceller shown in Fig. 7.4 may be stated as a function of the following quantities:

1. Radar frequency f_t.
2. Target velocity v_r.
3. Target range R, or equivalent range delay t_r.
4. Pulse repetition frequency f_r, or repetition period $t_p = 1/f_r$.
5. Phase constants for the STALO, COHO, and transmitter.

We shall begin by writing the expressions for the outputs of the two reference oscillators and the transmitter. Instantaneous voltages will be written with lower-case letters, e.g., $e(t)$, to show voltage as a function of time. The instantaneous phase terms will be similarly written, e.g., $\theta(t)$, to show phase as a time function.

STALO output

$$e_s(t) = E_s \cos \theta_s(t) \tag{7.4}$$

$$\theta_s(t) = \omega_s t + \phi_s(t) \tag{7.5}$$

$$E_s = \text{constant}$$

COHO output

$$e_c(t) = E_c \cos \theta_c(t) \tag{7.6}$$

$$\theta_c(t) = \omega_c t + \phi_c(t) \tag{7.7}$$

E_c = constant during all significant portions of the repetition period

Transmitter output

$$e_t(t) = E_t \cos \theta_t(t) \tag{7.8}$$

$$\theta_t(t) = \omega_t t + \phi_t(t) \tag{7.9}$$

$$\omega_t = \omega_s + \omega_c \tag{7.10}$$

$$\phi_t(t) = \phi_s(t) + \phi_c(t) + \phi_o(t) \tag{7.11}$$

E_t = constant for intervals $jt_p < t < jt_p + \tau$, where $j = 0$, 1, 2, ... for successive repetition periods.

E_t = zero at all other times

The terms ϕ_s, ϕ_c, and ϕ_t represent phase settings of the STALO, COHO, and transmitter, relative to their reference sine waves at ω_s, etc. These phase settings initially will be considered as constants and then be permitted to vary as the effects of various instabilities are explored. At any given time $t = jt_p + t_d$ between the jth and the $(j + 1)$th transmissions, the transmitter will be off and the r-f pulse will be passing a point at a range ct_d from the radar, with a phase angle given by $\phi_t(t - t_d)$ or $\phi_t(jt_p)$ relative to the reference sine wave at the transmitting frequency ω_t.

A signal reflected from a target of arbitrary cross section at a range R from the radar will return to the receiver as a delayed and attenuated r-f pulse. If we consider, for the present, a small, isolated target with a range delay t_r, the echo signal wave form can be expressed as

Received echo signal

$$e_r(t) = E_r \cos \theta_r(t) \tag{7.12}$$

$$\theta_r(t) = \theta_t(t - t_r) = (t - t_r)\omega_t + \phi_t(t - t_r) \tag{7.13}$$

E_r = constant depending upon the radar equation, over intervals $jt_p + t_r < t < jt_p + t_r + \tau$

E_r = zero at all other times

The range delay time t_r is given by

$$t_r = \frac{2R}{c} = \frac{2}{c}(R_i + v_r t) = t_i + \frac{2v_r t}{c} \tag{7.14}$$

In the above, the radar observations are assumed to begin with some initial range R_i at a time $t = 0$, and the velocity v_r is taken as positive for outbound targets. The effects of the second-order Doppler shift and time delay are ignored. Also, the effects of all fixed time delays within the radar are included in t_i. The Doppler frequency of the target will be given by Eq. (7.1), and the corresponding radian frequency is

$$\omega_d = \frac{2v_r \omega_t}{c} \tag{7.15}$$

We may now combine Eqs. (7.5), (7.12), and (7.14), expressing the received signal as

$$e_r(t) = E_r \cos \left[(\omega_t - \omega_d)t - \omega_t t_i + \phi_s(t - t_r) \right. \\ \left. + \phi_c(t - t_r) + \phi_o(t - t_r) \right] \tag{7.16}$$

Upon reception, this signal is mixed with the STALO reference, the difference being selected by the i-f amplifier. The resulting i-f signal is

$$e_i = E_i \cos \theta_i(t) \tag{7.17}$$

$$\begin{aligned}
\theta_i(t) &= \theta_r(t) - \theta_s(t) \\
&= (\omega_c - \omega_d)t - \omega_t t_i + \phi_s(t - t_r) - \phi_s(t) \\
&\quad + \phi_c(t - t_r) + \phi_o(t - t_r)
\end{aligned} \tag{7.18}$$

It can be seen that the i-f signal appears at a frequency $\omega_c - \omega_d$, and has a phase term $\omega_t t_i$ proportional to the round-trip path length to the target, as well as four phase terms originating in the radar. After amplification and limiting, which adjusts the magnitude of E_i to some standard value, the i-f signal is mixed with the COHO reference in the phase detector, and low-pass filtering selects the difference-frequency components which give the zero-frequency i-f or bipolar video.

Bipolar video signal

$$e_d(t) = E_d \cos \theta_d(t) \tag{7.19}$$

$$\begin{aligned}
\theta_d(t) &= \theta_i(t) - \theta_c(t) \\
&= -\omega_d t - \omega_t t_i + \phi_s(t - t_r) - \phi_s(t) \\
&\quad + \phi_c(t - t_r) - \phi_c(t) + \phi_o(t - t_r)
\end{aligned} \tag{7.20}$$

E_d = constant representing the peak output of the phase detector for signals in phase with the COHO.

E_d = zero except when a received signal E_r exists.

The final step in the MTI process is that of cancellation, which involves subtraction of a stored value of e_d from the value taken exactly one repetition period later.

Cancelled video signal

$$\begin{aligned}
e_v &= e_d(t) - e_d(t - t_p) \\
&= E_d[\cos \theta_d(t) - \cos \theta_d(t - t_p)]
\end{aligned} \tag{7.21}$$

In the ideal radar, all the internal phase terms may be taken as constant, and the cancelled video output becomes

$$\begin{aligned}
e_v &= E_d[\cos(\omega_d t + \omega_t t_i) - \cos(\omega_d t + \omega_t t_i - \omega_d t_p)] \\
&= 2E_d\left[\sin \omega_d\left(t - \frac{t_p}{2} + \psi \right) \sin \frac{\omega_d t_p}{2} \right]
\end{aligned} \tag{7.22}$$

This output defines a sinusoid at ω_d, with a peak amplitude given by

$$E_v = 2E_d \left| \sin \frac{\omega_d t_p}{2} \right| = 2E_d \left| \sin \frac{\omega_d}{2f_r} \right| \tag{7.23}$$

This is the response previously plotted (Fig. 7.3d).

Effects of Radar Instabilities

Let us now consider the effect of the several internal radar phase terms which enter into the equation for cancelled video and derive expressions for the amplitude of the uncancelled residue appearing when these phase terms vary as a function of time. For simplicity, we shall assume that the target itself is fixed. Equation (7.21) then gives us, for small variations in phase during a repetition period, the following expression for uncancelled residue.

Amplitude of residue

$$E_v = E_d[\cos(\theta_d + \Delta\theta_d) - \cos\theta_d] \tag{7.24}$$
$$= -E_d \Delta\theta_d \sin\theta_d$$
$$\Delta\theta_d = \theta_d(t) - \theta_d(t - t_p) \tag{7.25}$$

Fixed clutter signals will appear at all possible values of phase angle θ_d, so the peak amplitude of the residue will be simply

$$(E_v)_{\text{peak}} = E_d \Delta\theta_d$$

Using the definition of cancellation ratio given earlier, we see that the peak voltage corresponding to the reference level for uncancelled video will be equal to E_d, and the cancellation ratio C will be given by the peak value of the sinusoidal phase instability $\Delta\theta_d$ in radians, or $\sqrt{2}$ times the rms phase instability. The complete expression for this phase instability is obtained by combining Eqs. (7.20) and (7.25).

Phase instability

$$\Delta\theta_d = \phi_s(t - t_r) - \phi_s(t) - \phi_s(t - t_r - t_p) + \phi_s(t - t_p)$$
$$+ \phi_c(t - t_r) - \phi_c(t) - \phi_c(t - t_r - t_p) + \phi_c(t - t_p) \tag{7.26}$$
$$+ \phi_o(t - t_r) - \phi_o(t - t_r - t_p)$$

or

$$\Delta\theta_d = \Delta\theta_s + \Delta\theta_c + \Delta\theta_o$$

The total phase instability is thus the sum of three terms originating in the STALO, the COHO, and the transmitter, respectively. The STALO and

COHO terms have exactly the same form, each containing four instantaneous values of oscillator phase error corresponding to the times of transmission and reception of the two successive pulses. Let us consider the
STALO phase term, and investigate the effects of different types of phase
error.

STALO phase error

$$\Delta\theta_s = \phi_s(t - t_r) - \phi_s(t) - \phi_s(t - t_r - t_p) + \phi_s(t - t_p) \qquad \text{(7.27)}$$

By inspection, we see that the residue will be zero for any of the following
STALO phase functions

$$\phi_s(t) = \text{constant}$$
$$\phi_s(t) = \phi_s(t - t_r)$$
$$\phi_s(t) = \phi_s(t - t_p)$$
$$\phi_s(t) = \Delta\omega_s t \quad \text{(constant tuning error of STALO)}$$

However, if the STALO drifts in frequency at a constant rate, we may
express its frequency and phase error as

$$\omega_s(t) = \omega_{si} + A_s t$$
$$\theta_s(t) = \omega_{si} t + \frac{A_s t^2}{2}$$

There will result a residue given by

$$\Delta\theta_s = \frac{A_s}{2}[(t - t_r)^2 - t^2 - (t - t_r - t_p)^2 + (t - t_p)^2]$$

$$= -A_s t_r t_p = -\frac{2\pi t_r \dot{f}_s}{f_r} \qquad \text{(7.28)}$$

In the above, $\dot{f}_s$ is the rate of change of reference frequency in cps per
second, equal to $A_s/2\pi$. The same result is given by Emslie and McConnell.*

Sinusoidal Modulation of Oscillators

If the reference oscillator is subject to a sinusoidal phase modulation,

---

* A. G. Emslie and R. A. McConnell, "Moving Target Indication," Chap. 16 in
Radar System Engineering, L. N. Ridenour, ed. (New York: McGraw-Hill Book
Company, 1947), p. 639.

the residue will depend upon both the amplitude and the frequency of the modulation. Consider a phase error given by

$$\phi_s(t) = A_p \sin \omega_m t \qquad (7.29)$$

Substituting this in Eq. (7.27), we obtain

$$\Delta\theta_s = A_p[\sin \omega_m(t - t_r) - \sin \omega_m t - \sin \omega_m(t - t_r - t_p)$$
$$+ \sin \omega_m(t - t_p)] \qquad (7.30)$$
$$= 4A_p \sin \frac{\omega_m t_p}{2} \sin \frac{\omega_m t_r}{2} \sin \frac{\omega_m(2t - t_r - t_p)}{2}$$

The final term represents a variation in residue at the modulating frequency ω_m, and may be set equal to unity to find the peak value of the residue.

$$C = (\Delta\theta_s)_{\text{peak}} = 4A_p \left| \sin \frac{\omega_m t_p}{2} \sin \frac{\omega_m t_r}{2} \right|$$
$$= 4A_p \left| \sin \left(\pi \frac{f_m}{f_r} \right) \sin (\pi f_m t_r) \right| \qquad (7.31)$$

Here, $f_m = \omega_m/2\pi$ is the frequency of the phase modulation. For frequency modulation, the peak deviation in reference oscillator frequency will be $f_p = f_m A_p$, and this may be substituted into the above equation to find the residue in terms of peak frequency deviation instead of peak phase excursion.

From the above, it can be seen that the phase or frequency modulation component of the reference oscillator which occurs at an integral multiple of the repetition frequency will produce no output residue. Also, if the modulation happens to appear at a frequency $f_m = 1/t_r$ or at some multiple of this frequency, there will be no residue for clutter appearing at the corresponding range $R = t_r c/2$. The worst situation is encountered at maximum range, where $t_r = t_p$. In this case, when modulation is at an odd multiple of one-half the repetition frequency, the residue will reach its maximum value.

$$C_{\max} = (\Delta\theta_s)_{\max} = 4A_p$$

For frequency modulation with constant peak deviation, the maxima occur at somewhat lower frequencies, the first being at $f_m = 0.37f_r$ instead of at $0.5f_r$. This is due to the fact that the peak phase excursion must drop steadily as modulation frequency rises, to keep a constant frequency deviation. Figure 7.6(a) shows the residue, normalized to one degree of peak phase modulation A_p. The peak values correspond to $C = 4/57.3$,

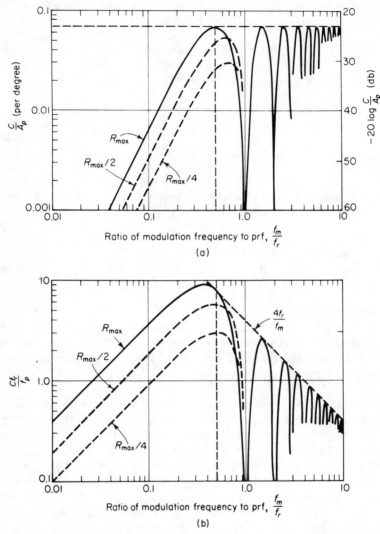

Figure 7.6   (a) Cancellation ratio for phase modulation. (b) Cancellation for frequency modulation.

representing the conversion from degrees to radians. In the low-frequency region there are three separate curves for targets at different ranges. Figure 7.6(b) shows the residue for the frequency-modulated case, normalized by the ratio of prf to peak frequency deviation f_p. Both sets of curves are equally applicable to STALO and COHO modulation, but the percentage of stability required of the STALO will be more stringent than that for the COHO, due to the much lower operating frequency of the latter. For example, the residue from a one degree phase modulation would have a peak value of $0.07E_d$ when the modulating frequency is $f_r/2$, $3f_r/2$, etc. If cancellation of fixed targets 40 db below their original level is required, the amplitude of the phase modulation would have to be held to $0.14°$ or less at these frequencies. When expressed in terms of peak frequency deviation, the deviation would have to be below 0.1 per cent of the repetition rate for 40 db cancellation. In a typical radar operating at 1000 pps, this would permit a deviation of only one cps. If the radar frequency is 3000 mc, the corresponding STALO stability would be three parts in 10^{10}, whereas the COHO stability would be only three parts in 10^8 for a 30 mc i-f system.

Transmitter Stability Criteria

The third part of Eq. (7.26) gives the transmitter contribution to the output phase error.

$$\Delta\theta_o = \phi_o(t - t_r) - \phi_o(t - t_r - t_p) \tag{7.32}$$

This term will be zero only if ϕ_o is constant or periodic at an integral multiple of the repetition frequency. In the often encountered case of sinusoidal modulation from power-supply variation or mechanical vibration, we will have a phase modulation of the form

$$\phi_o(t) = A_t \sin \omega_m t \tag{7.33}$$

leading to

$$\Delta\theta_o = A_t[\sin \omega_m(t - t_r) - \sin \omega_m(t - t_r - t_p)]$$

This reduces to the form

$$\Delta\theta_o = 2A_t \sin \frac{\omega_m t_p}{2} \cos \omega_m \left(t - t_r - \frac{t_p}{2}\right) \tag{7.34}$$

This describes a sinusoidal residue at ω_m, with a peak amplitude given by

$$C = (\Delta\theta_o)_{\text{peak}} = 2A_t \left| \sin \frac{\omega_m t_p}{2} \right| = 2A_t \left| \sin \left(\pi \frac{f_m}{f_r}\right) \right| \tag{7.35}$$

This response is of the same form as that of moving targets, plotted in Fig. 7.3(d). The effect of the transmitter instability does not depend upon target range but only upon repetition rate relative to the modulation frequency. The same result may be expressed in terms of peak transmitter frequency deviation f_p by substituting for A_t the ratio f_p / f_m.

Description of Locking-Pulse Process

In a pulsed-oscillator system, the STALO output will remain the same as that given above, but the COHO output will no longer be expressed as a continuous function. We shall, therefore, define an artificial COHO reference wave form which will provide a continuous reference phase $\theta_r(t) = \omega_c t$. The COHO output will superimpose on this reference a phase variation $\phi_c(t)$, defined as a continuous function of time and carrying the usual instabilities caused by drift and modulation, plus a discontinuous term $\phi_L(j)$ representing the phase jump due to the locking pulse on the jth transmission. As a result, we shall have

$$\theta_c(t) = \omega_c t + \phi_c(t) + \phi_L(j) \tag{7.36}$$

On each transmission, $\phi_L(j)$ will assume a value determined by the transmitter starting phase, and this value will be stored by the COHO for one repetition period. The phase jump due to the locking pulse will be the amount necessary to bring $\theta_c(j_{t_p})$ into agreement with the locking pulse phase at the time of transmission.

Transmitter output phase

$$\theta_t(t) = \omega_t t + \phi_o(t)^* \tag{7.37}$$

This output phase omits the COHO and STALO phase terms, since the transmitter now establishes its own phase. The received and i-f signals will be given by Eqs. (7.16) and (7.18), the terms $\phi_s(t - t_r)$ and $\phi_c(t - t_r)$ again being omitted. The locking pulse phase is

$$
\begin{aligned}
\theta_L(jt_p) &= \theta_t(jt_p) - \theta_s(jt_p) \\
&= \omega_c jt_p + \phi_o(jt_p) - \phi_s(jt_p)
\end{aligned}
\tag{7.38}
$$

Thus, the locking action forces the COHO to assume a phase such that

$$\phi_c(jt_p) + \phi_L(j) = \phi_o(jt_p) - \phi_s(jt_p)$$

or

$$\phi_L(j) = \phi_o(jt_p) - \phi_s(jt_p) - \phi_c(jt_p) \tag{7.39}$$

This is the value required to make $\theta_c(jt_p)$ equal to $\theta_L(jt)$. At a time t_r after

* See Eq. (7.9).

jt_p, when the echo from the range $R = t_r c/2$ is being received, the COHO phase will be

$$\theta_c(t) = \omega_c t + \phi_c(t) + \phi_o(t - t_r) - \phi_s(t - t_r) - \phi_c(t - t_r) \quad \textbf{(7.40)}$$

Subtracting this phase from the phase of e_i, we obtain the phase at the output of the phase detector.

$$\theta_d(t) = -\omega_d t - \omega_i t_i + \phi_s(t - t_r) - \phi_s(t) + \phi_c(t - t_r) - \phi_c(t) \quad \textbf{(7.41)}$$

This is exactly the same as Eq. (7.20), except that the transmitter phase error term ϕ_o has been cancelled out by the action of the locking pulse. As a result, the analysis of the pulsed-amplifier MTI may be applied to either case, so far as the reference oscillator errors are concerned.

In principle, the locking-pulse method can be applied to remove the effects of transmitter instability on any system. Practical considerations make this undesirable, since there are new sources of instability in the action of the locking-pulse circuit. First, the problem of designing a highly stable reference oscillator is made much more difficult when its output phase must be capable of rapid locking during the transmitted pulse. Stability implies a very high-Q circuit, whereas rapid phase shift in response to the locking pulse implies a low-Q circuit. Second, the locking process may not be accurate enough to achieve the desired level of residue. A new COHO phase error component may be added if the locking action fails to achieve perfect matching of the phase θ_c to θ_L by the end of the transmitted pulse. Such a phase error, which is random from pulse to pulse, may best be described by a standard deviation σ_L in radians for each pulse. The fractional residue after cancellation will be given by the difference between two successive locking errors, or

$$(\Delta\theta_d)_{\text{rms}} = \sqrt{2}\ \sigma_L$$

or

$$C = 2\sigma_L \quad \textbf{(7.42)}$$

To achieve a residue of one per cent from this source would require locking to an accuracy of 0.007 rad or 0.4° on each pulse. This tends to explain why the phase-locked systems cannot reach the performance levels of the fully coherent, amplifier-type systems.

Timing Stability Criteria

In addition to the uncancelled residue produced by instabilities in the transmitter and reference oscillators, there will be components produced by timing errors in the system. When the relative delay between successive pulses from a discrete target, arriving at the cancellation point in the receiver, differs by an amount Δ_t, the canceller output will contain a signal whose energy is $(2\Delta_t/\tau)^2$ times the energy of the uncancelled pulse. If this energy lies within the bandpass of the succeeding video amplifiers, it will appear on the cathode-ray-tube indicator. The cancellation ratio will then be limited to the value $2\Delta_t/\tau$. If the delay variation is expressed as an rms value σ_t which is independent from pulse to pulse, the rms value of Δ_t is $\sqrt{2}$ times σ_t. Similar results will apply when the clutter extends over a considerable range interval, and when the variation appears in pulse width instead of in delay.

When the timing stability of the canceller is considered, the wave form at the phase detector output is of importance. The equations given above for transmitter instability are applicable to the usual pulse shapes, which are trapezoidal or Gaussian in shape. However, when the receiver is not properly tuned, it is possible to produce an output to the canceller which is not at zero frequency i-f, but is superimposed on a carrier frequency Δf equal to the receiver tuning error. One or more cycles at this frequency may appear as the output pulse, increasing the residue which appears as a result of timing error. In order to assure that the output video will not contain more than a fraction of one cycle of this carrier during each pulse width, the tuning error must be held within the limit

$$\Delta f < \frac{1}{\pi \tau} \tag{7.43}$$

This requirement is no more stringent than the usual tolerance imposed by the restricted i-f bandpass in noncoherent radar receivers, where the tuning should be within a small fraction of the bandwidth $B = 1/\tau$.

7.3   RESPONSE OF MTI SYSTEMS TO CLUTTER

The response of the simple (single-delay) MTI circuit to clutter which is not absolutely fixed may be calculated by referring to the tabulation of clutter characteristics in Chapter 3 (see Table 3.3). We shall give the response in terms of cancellation ratio, and then determine what improvements are needed in the cancellation filter to achieve an acceptable level of performance.

Cancellation Using Single-Delay MTI

The frequency response of the single-delay canceller was shown in

Fig. 7.3(d), and may be described in terms of a power gain function of the form

$$G(f) = 4 \sin^2\left(\pi \frac{f}{f_r}\right) \qquad (7.44)$$

Here, f represents the deviation of the signal frequency from the nearest harmonic of the repetition frequency contained in the i-f spectrum. When expressed on the same basis, the clutter spectrum of Eq. (3.25) may be used directly.

$$W(f) = W_o \exp\left(-\frac{f^2}{2\sigma_c^2}\right)$$

Both the canceller response and the clutter spectrum will be repeated at intervals equal to the repetition frequency, as will any signal received through echoing of the transmitted pulse. Accordingly, the performance of the MTI system may be evaluated within any frequency interval of width f_r near the center of the receiver bandpass. Whatever cancellation and signal-to-clutter ratios are determined for such an interval will apply to the entire output of the MTI system, which includes the sum of signal and clutter components throughout the receiver bandpass.

The clutter power within one such interval of the spectrum will enter the canceller at a level given by

$$P_{ic} = \int_{-f_r/2}^{f_r/2} W(f)\,df \cong W_o \int_{-\infty}^{\infty} \exp\left(-\frac{f^2}{2\sigma_c^2}\right) df = \sqrt{2\pi}\, W_o \sigma_c \quad (7.45)$$

The approximations involved in setting the limits of integration at infinity in Eq. (7.45) are valid when the clutter spread $\sigma_c \ll f_r$, which must be the case for successful MTI action. The input clutter power may be normalized to unity for each interval f_r by setting the low-frequency spectral density W_o equal to the value $1/(\sqrt{2\pi}\sigma_c)$. In order to find the residual clutter power after cancellation, we integrate the product $G(f)W(f)$ over this same interval.

$$P_{oc} = 4W_o \int_{-f_r/2}^{f_r/2} \sin^2\left(\frac{\pi f}{f_r}\right) \exp\left(-\frac{f^2}{2\sigma_c^2}\right) df$$

$$\cong \frac{4\pi^2 W_o}{f_r^2} \int_{-\infty}^{\infty} f^2 \exp\left(-\frac{f^2}{2\sigma_c^2}\right) df = \frac{4\pi^2 W_o}{f_r^2} \sqrt{2\pi}\,\sigma_c^3 \qquad (7.46)$$

The voltage cancellation ratio, as defined earlier, is equal to the square root of the ratio P_{oc}/P_{ic}, or

$$C = 2\pi \frac{\sigma_c}{f_r} = \frac{4\pi\sigma_v}{f_r \lambda} \tag{7.47}$$

Note that the cancellation ratio is a number less than unity, in accordance with the definition. Expressed in decibels, we would obtain a negative number.

$$(C)_{\mathrm{db}} = 20 \log_{10} C = 20 \log_{10} \left(2\pi \frac{\sigma_c}{f_r} \right) \tag{7.48}$$

In most discussions, the absolute magnitude of the cancellation ratio in decibels is used (e.g., the system with a voltage residue of one per cent and a ratio P_{oc}/P_{ic} of 10^{-4} will be described as having a cancellation ratio of 40 db, rather than minus 40 db). Such a usage can be clarified, as it is in one early analysis of clutter fluctuations, by using the term "clutter attenuation" for the negative of Eq. (7.48).*

Another important measure of MTI performance is the improvement in signal-to-clutter ratio which is obtained by inserting the canceller circuit. Steinberg† considers the relative ratios of signal to clutter before and after the canceller circuit, taking the signal as the average over all possible target velocities. Since these velocities may cover many intervals f_r in the spectrum, he considers the signal energy as being distributed uniformly over this interval, and obtains an average signal power gain $\bar{G} = 2$. On this basis, the "improvement factor" I will be

$$I = \frac{\bar{G}}{C^2} = 2 \left(\frac{f_r}{2\pi\sigma_c} \right)^2 \tag{7.49}$$

The cancellation ratio attainable for various types of clutter, a simple MTI system with no internal instabilities being used, is plotted in Fig. 7.7. The improvement factor I will be 3 db greater than the magnitude of the cancellation ratio shown.

---

* R. S. Grisetti, M. M. Santa, and G. M. F. Kirkpatrick, "Effect of Internal Fluctuations and Scanning on Clutter Attenuation," *Trans. IRE*, Vol. **ANE-2**, No. 1 (March 1955), p. 38. When the approximation applied above is valid ($\sigma_c \ll f_r$), their first equation reduces to the form

$$\text{decibel attenuation} = 10 \log_{10} \left[\frac{2}{a} \left(\frac{\pi f_t}{f_r} \right)^2 \right]$$

Using the relationship between Barlow's stability parameter a and the rms clutter spread σ_v, we will obtain the negative of Eq. (7.48) above.

† Bernard D. Steinberg, "Signal Enhancement by Linear Filtering in Pulse Radar," Lectures 20–22 of special summer course, University of Pennsylvania (June 1961), Eq. 21.14.

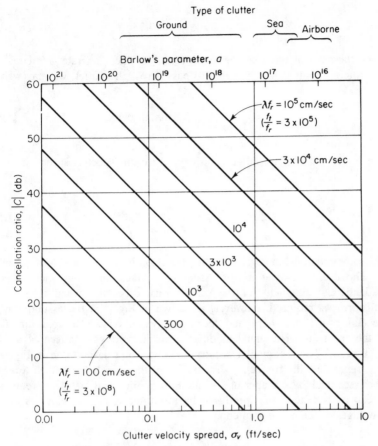

Figure 7.7 Maximum cancellation ratio for simple MTI circuit.

Effect of Antenna Motion

When the radar beam is scanned over a broad search sector, by mechanical rotation of the antenna or by other means, the spectral spread of the received clutter signals is increased. The increase can be attributed to the amplitude modulation of the echo signal by the two-way beam pattern, or, equivalently, to the introduction of an added velocity component by the motion of the antenna. Steinberg* expresses the rms width of the spectral component due to scanning as

---

* Bernard D. Steinberg, *Ibid.*, Eq. 20.3. His beamwidth parameter $\Delta\theta$, equal to twice the standard deviation of the Gaussian beamwidth, is equal to 0.84 times the usual one-way half-power beamwidth θ_a.

$$\sigma_c = \frac{\omega}{\sqrt{2}\,\pi\,\Delta\theta} = \frac{\omega}{3.78\theta_a} \qquad \text{(7.50)}$$

Since the ratio of the antenna beamwidth θ_a to the scan rate ω is equal to the time t_o required for the beam to pass across the target, we may write

$$\sigma_c = \frac{1}{3.78t_o} \qquad \text{(Gaussian beam)} \qquad \text{(7.51)}$$

The limitation in cancellation ratio caused by this spectral spread is given by

$$C_{\min} = \frac{2\pi\sigma_c}{f_r} = \frac{2\pi}{3.78f_r t_o} = \frac{1.66}{n} \qquad \text{(7.52)}$$

The same relationship is given in the discussion by Emslie and McConnell.*

The usual justification for assuming a Gaussian beam shape is the mathematical simplicity of the calculations, especially when the use of transforms is required. However, as shown in Fig. 7.8, the assumption of the Gaussian beam shape and spectrum gives a very close approximation of the spectrum of a signal modulated by the two-way pattern of a (sin x)/x beam. The rectangular spectrum, shown as a dashed line in Fig. 7.8 (a), represents the frequency distribution of the signal incident on a target when scanned by an aperture with uniform illumination. The frequency spread Δf is exactly that produced by the differing radial velocities at the two edges of the moving antenna.

$$\Delta f = \frac{D\omega}{\lambda} = \frac{2v_p}{\lambda} = \frac{\omega}{1.15\theta_a} \qquad \text{(7.53)}$$

The peripheral velocity $v_p = \omega D/2$ is directed towards the target at one edge of the antenna and away from it at the other end, and the uniform weighting of illumination across the aperture provides the rectangular spectrum shown. The corresponding one-way beam voltage wave form is of the (sin x)/x form, with nulls spaced in time by $2\lambda/\omega D$ or λ/v_p sec. In order to find the voltage spectrum of the returned echo signal, we must find the transform of the two-way beam voltage wave form. This is triangular in shape, extending from $-\Delta f$ to $+\Delta f$, and its square gives the power

---

* A. G. Emslie and R. A. McConnell, "Moving Target Indication," Chap. 16 in *Radar System Engineering*, L. N. Ridenour, ed. (New York: McGraw-Hill Book Company, 1947), p. 646.

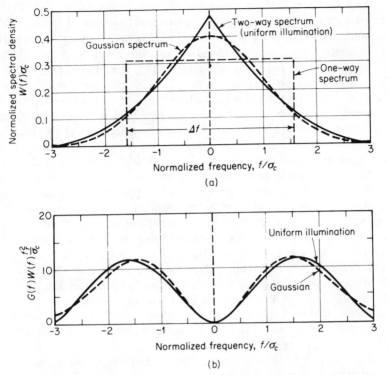

Figure 7.8 Effect of scanning on cancellation. (a) Clutter spectra owing to antenna scanning. (b) Spectra of clutter residue after cancellation.

spectrum plotted for the two-way signal in Fig. 7.8(a).* The rms spread of this two-way spectrum is given by

$$\sigma_c = \frac{\Delta f}{\sqrt{10}} = \frac{1}{3.65 t_o} \tag{7.54}$$

For this case, the limiting cancellation ratio is

$$C_{\min} = \frac{2\pi\sigma_c}{f_r} = \frac{2\pi}{3.65 f_r t_o} = \frac{1.73}{n} \tag{7.55}$$

(The spectrum of the residue is shown in Fig. 7.8b.)

---

* Samuel Silver, *Microwave Antenna Theory and Design* (New York: McGraw-Hill Book Company, 1949), Table 6-1, p. 187. He gives the relationship between the triangular aperture illumination and the $(\sin x)^2/x^2$ voltage pattern of the beam.

As was shown in the previous section, the average power gain of the canceller for moving targets provides a factor of two increase in the signal-to-clutter power ratio, or $\sqrt{2}$ increase in voltage ratio. When this is applied to the cancellation ratios calculated above, they agree with the "clutter attenuation" factors appearing in the literature.* This agreement is because the factor of two is equivalent to that appearing in these other papers, where it is said that the additional factor of two results from the separation of clutter residue into equal phase and amplitude components. Since most MTI systems limit the i-f signal prior to phase detection and cancellation, it appears more accurate to ascribe the factor of two to the gain in response to the average signal.

The above derivation explains physically the relationship between the peripheral velocity of the antenna and the Doppler spread of the clutter signal, which has been claimed to represent an accidental relationship.† This velocity serves to generate the spectrum of the incident radiation on the target, and to broaden this spectrum further upon reception, as shown above. The spreading of the echo signal into a triangular (voltage) spectrum is explained by the fact that all parts of the rectangular echo spectrum are incident upon all parts of the moving antenna upon reception, generating spectral components out to twice the one-way spread but at reduced amplitude. The final spectrum may be calculated also by convolution of the rectangular spectrum with itself, or by transformation of the $(\sin x)^2/x^2$ voltage wave form received.

The rms velocity which describes the antenna component of the clutter spectrum is

$$\sigma_{va} = \frac{\lambda}{2}\sigma_c = \frac{v_p}{\sqrt{10}} = \frac{\lambda \Delta f}{2\sqrt{10}} = \frac{\lambda \omega}{7.3\theta_a} \qquad \textbf{(7.56)}$$

When a tapered illumination is used, the results are almost exactly the same when expressed in terms of the beamwidth θ_a or the time-on-target t_o. The antenna may then be considered as equivalent to a somewhat smaller antenna with uniform illumination, scanning at the same angular rate but

* L. N. Ridenour, ed., *Radar System Engineering* (New York: McGraw-Hill Book Company, 1947), p. 646.

Merrill I. Skolnik, *Introduction to Radar Systems* (New York: McGraw-Hill Book Company, 1962), p. 150.

R. S. Grisetti, M. M. Santa, and G. M. Kirkpatrick, "Effect of Internal Fluctuations and Scanning on Clutter Attenuation," *Trans. IRE*, Vol. **ANE-2**, No. 1 (March 1955), p. 39.

F. R. Dickey, Jr., "Theoretical Performance of Airborne Moving Target Indicators," *Trans. IRE*, Vol. **PGAE-8**, No. 2 (June 1953), pp. 12–23.

† Merrill I. Skolnik, *op. cit.*, p. 79.

having a smaller peripheral velocity. This is true because the weighting of the aperture by varied illumination has the same effect on beamwidth as on clutter spectra. It will be recalled that the same type of relationship applied in Chapter 2 when the limiting accuracy of angular measurement was related to the beamwidth and the "rms aperture width" determined by the weighting of illumination.

In applying the derivation to antennas where the scanning is accomplished by other than mechanical means, it is necessary to consider merely the velocity of phase change across the aperture. The beam can be scanned only by altering this phase front, and the signal spectrum is the same whether the phase change is brought about by motion of the reflector, by physical motion of phase-shifter elements, or by electrical modulation of phase.

The spectral or velocity spread due to scanning is combined in an rms fashion with other components, such as those due to internal motion of clutter. The complete spectrum will have a broader Gaussian shape, with a variance given by the sum of the variances of the individual components.

$$\sigma_{vt}^2 = \sigma_{v1}^2 + \sigma_{v2}^2 + \cdots$$

Although the Gaussian aproximation may not have been perfect for the scanning component alone, the total received clutter spectrum, composed of components due to internal clutter and radar instabilities as well as antenna motion, will be very closely approximated by a Gaussian curve with the above variance.

Effect of Radar-Clutter Velocity

Another source of spectral spread in received ground clutter echoes is the relative velocity between the radar and the clutter mass. This can result from motion of the radar relative to the ground, or from drifting of a cloud of reflectors with the average velocity of the wind. When a very narrow beam is used, the average velocity will cause a simple displacement of the entire clutter spectrum away from zero frequency. However, when the beam is broader than some minimum value, the clutter at the edges of the beam will have a slightly different radial velocity from that in the center. The velocity difference will be proportional to the average velocity $\bar{v}_r$ and to the off-axis angle $\Delta\theta$.

$$\Delta v_r = \bar{v}_r \, \Delta\theta = \bar{v} \, \Delta\theta \sin \alpha \qquad (7.57)$$

Here, α is the angle between the velocity vector and the center of the beam. Since the edges of the beam are not sharply defined, it may be represented again by a Gaussian function whose rms value is proportional to the half-power beamwidth: $\sigma_\theta = \theta/12.36$ (see Appendix C). The corresponding velocity spread will have an rms value given by

$$\sigma_{ra} = \sigma_\theta \bar{v} \sin \alpha = \frac{\theta}{2.36} \bar{v} \sin \alpha \tag{7.58}$$

In order to express the result in terms of a frequency spread, we multiply by the factor $2/\lambda$ to get

$$\sigma_{fa} = \frac{\theta \bar{v}}{1.18\lambda} \sin \alpha \cong \frac{\bar{v}}{w} \sin \alpha \tag{7.59}$$

This last form of the equation is based on the approximation $\theta \cong 1.2\lambda/w$, and shows that the frequency spread owing to finite beamwidth is independent of radar wave length when an antenna of given size is used. As in the case of spectral spreading from antenna rotation, the component due to radar-clutter velocity should be added in an rms fashion to other components in order to arrive at a total rms width for calculation of cancellation ratio. Figure 7.7 may be used to read limiting cancellation ratio, the total spectral width σ_{rt} being substituted for the clutter width σ_c in entering the chart.

As an example showing the relative importance of the various factors, consider the case of chaff or weather clutter, where the internal fluctuation contributes about four feet per second to the total spectral width. If the search antenna is 40 ft wide and scans at six revolutions per minute ($\omega = 0.63$ rad per sec), the rms spread σ_{va} will be about 7.5 ft per sec. If the average wind speed aloft is 40 ft per sec and the beamwidth θ_a is 1.5° (0.027 rad), the spread caused by motion of the air mass across the beam will be less than $\frac{1}{2}$ ft per sec. Similarly, for a vertical beamwidth of 20° (0.35 rad), the spread owing to motion of the mass towards the radar will be about 1 ft per sec. These last two factors will become quite important when the radar is airborne, since in such a case the velocity of the aircraft may be hundreds of feet per second and the beamwidth several degrees. In this event, as when any appreciable average velocity exists between the radar and the clutter, the simple type of MTI described earlier must be modified to cancel the nonzero average frequency which results.

7.4 REFINEMENTS IN MTI

The earliest MTI systems, which used either the pulsed-oscillator or the pulsed-amplifier configuration with a single-delay canceller, have been discussed above, and certain limitations in performance have been estab-

lished. Since one of the most serious limitations is imposed by the characteristics of the clutter signals, the improvements introduced by more stable
transmitters and oscillators have not made it possible to obtain satisfactory
MTI operation in certain environments. A number of further refinements
have been developed which overcome these limitations to some degree, and
which extend the application of MTI into many regions where the simple
systems could not provide any measure of useful performance.

Improvement in MTI Filter Response

The sharp rejection notch provided by the single-delay canceller is not
adequate in cases where the clutter signal spreads beyond about one per
cent of the radar repetition frequency. Two techniques were developed to
overcome this limitation. In the first, a second canceller is placed in series
with the one shown in Figs. 7.2 and 7.4. The frequency response of the
system will then be the square of that shown in Fig. 7.3(d) and expressed
by Eq. (7.23). We may analyze the double-delay canceller system in the
same manner as was used for the simple system, starting with the equation
for the power gain of the canceller.

$$G(f) = 16 \sin^4\left(\pi \frac{f}{f_r}\right) \tag{7.60}$$

When this is multiplied by the Gaussian clutter spectrum of Eq. (3.25),
and integrated over an interval equal to the repetition frequency, the power
of the clutter residue is found to be

$$P_{oc} = \frac{16\pi^4 W_o}{f_r^4} \int_{-f_{r/2}}^{f_{r/2}} f^4 \exp\left(-\frac{f^2}{2\sigma_c^2}\right) df = \frac{16\pi^4}{f_r^4} W_o \sqrt{2\pi}\, 3\sigma_c^5 \tag{7.61}$$

As in the single-delay case, the cancellation ratio is the square root of the
ratio of P_{oc} to P_{ic} [see Eq. (7.45)], or

$$C = \sqrt{3}\left(\frac{2\pi\sigma_c}{f_r}\right)^2 \tag{7.62}$$

Now, however, the average signal gain $\bar{G}$, measured over all frequencies, is
equal to six, providing an improvement in signal-to-clutter power ratio of

$$I = \frac{\bar{G}}{C^2} = 2\left(\frac{f_r}{2\pi\sigma_c}\right)^4 \tag{7.63}$$

The cancellation ratio is now a function of the square of the ratio σ_c/f_r, and the improvement factor in power ratio of signal to clutter is a function of the fourth power of this frequency ratio. In cases where only a small amount of cancellation was provided by the single-delay system, the double-delay filter will provide a much greater improvement. On the other hand, the width of the rejection notches around each blind speed will be greater, and targets flying near the blind speeds will be lost more easily in the noise.

The addition of feedback around the double-delay canceller permits further adjustment of the shape of the filter passband, as shown in Fig. 7.9. The width of the notch may be reduced, while at the same time the cancellation characteristics of the double-delay system in the immediate vicinity of the blind speed may be, to a large extent, preserved.* If it is assumed that adequate stability can be achieved within the two canceller loops, the limiting factor in establishing the response curve will now be the almost total loss of target response when the target approaches the blind speed.

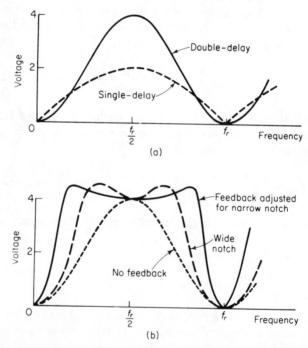

Figure 7.9 Comparison of canceller response curves. (a) Response of double-delay canceller. (b) Response of double-delay canceller with and without feedback.

* W. D. White and A. E. Ruvin, "Recent Advances in the Synthesis of Comb Filters," *IRE Conv. Record* (1957), Part 2, pp. 186–200.

The same limitation applies when the filter response is obtained through the use of a multipole filter in the range-gated system to be described below.

When the number of range elements to be processed is not excessive, MTI filtering may be accomplished through the use of simple bandpass filters following each range gate (see Fig. 7.10). These filters may be operated at any desired center frequency, if the range-gated signals are heterodyned to that frequency with a suitable coherent oscillator. Since the range resolution has been established by the range gate, the original signal bandwidth need no longer be maintained, and the new center frequency may be less than the bandwidth of the original signal, so long as it remains well above the repetition frequency of the radar. Operating at frequencies of ten to a few hundred kilocycles per second, filters may be designed which properly match the width of the lines in the signal spectrum, determined by the time-on-target during radar scan. Such a system, shown in Fig. 7.10, provides coherent integration over this time interval, and permits measurement and resolution of target velocity. It may, therefore, be considered a pulsed-Doppler radar system, rather than just an MTI system.

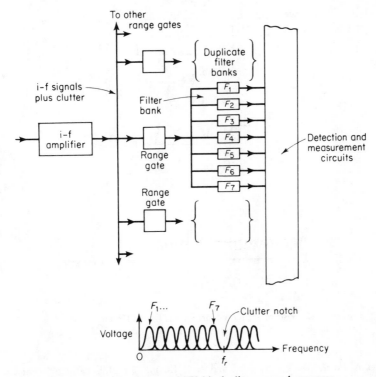

Figure 7.10 Range-gated MTI block diagram and response.

When only MTI operation is needed, a single complex filter may follow each range gate, designed to pass all frequencies except those near the frequency of the clutter. Response curves similar to those shown in Fig. 7.9 are easily obtained, but no coherent integration or velocity resolution between different targets is provided.

Overcoming Blind Speeds with Staggered PRF

An MTI system which is designed to minimize the loss of targets owing to blind speeds has been described.* In this system, every second transmitter pulse is delayed by an interval Δt which is small compared to the repetition interval t_p. As shown in Fig. 7.11, the response of this type of circuit falls below the maximum for a system with uniform repetition rate, but the nulls at some blind speeds are largely eliminated. When the ratio of t_p to Δt is very large, the blind speeds are absent near the center of the response curve, but deep nulls appear at blind speeds near the two ends of

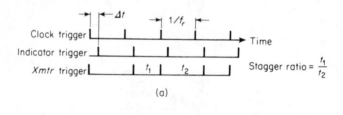

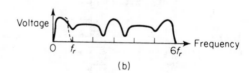

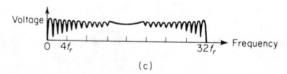

(after Fowler)

Figure 7.11 Use of staggered prf in MTI. (a) Timing of triggers. (b) Response for 5/7 stagger ratio. (c) Response for 63/65 stagger ratio.

* C. A. Fowler, A. P. Uzzo, Jr., and A. E. Ruvin, "Signal Processing Techniques for Surveillance Radar Sets," *Trans. IRE*, Vol. **MIL-5**, No. 2 (April 1961), p. 104.

the response. Staggered prf may be used in conjunction with any one of the canceller techniques described earlier, so long as the timing of the system is restored before the signals are fed into the range gates or delay lines.

Compensation for Average Velocity

When the clutter mass is moving relative to the radar, the rejection notch should be centered at a frequency different from zero. This problem arises in shipborne radar, airborne radar, or when the clutter is being carried by the wind. One approach to this problem involves variation of the COHO frequency as a function of platform velocity and azimuth angle during the scan. The rejection notch remains fixed in absolute frequency, but the clutter spectrum is kept centered in the notch by the COHO action. Control of the COHO offset frequency may be derived from an external velocity-measuring device and azimuth resolver, or by holding the average output frequency of the short-range clutter within the fixed notch, afc techniques being used in the radar i-f system. A second approach uses the average clutter signal over a small range interval to control the frequency of a supplementary COHO, cancelling any signal that extends over an interval larger than a few pulse widths in range, regardless of its radial velocity. A moving target that extends for less than one pulse width will remain visible against either clutter or noise background, as in the simple MTI. This system is closely related to the noncoherent MTI described below.

Noncoherent MTI

The noncoherent MTI system depends upon the presence of continuous clutter echoes to provide a coherent reference for detection of moving targets. Since the reference is not maintained within the radar itself, the radar operation is described as noncoherent. This system is sensitive only to those targets which have a radial velocity component relative to other targets at the same range, and provides no output for isolated targets against a background of thermal noise. The i-f signals are detected in a normal envelope detection circuit, which is followed by the canceller system. To avoid saturation in the canceller on strong echoes, i-f amplifiers of the logarithmic-response type are normally used. When the output of the detector is processed by one of the canceller devices described above, the Doppler signal representing the velocity difference in echo signals will appear at the output. Since this type of MTI circuit has no output for isolated targets, its use must be restricted to regions where the clutter background covers the desired search area completely. Outside such regions, normal video or output of a coherent MTI receiver must be made available for display.

One last type of MTI for pulsed radar will be mentioned here, since it is also noncoherent in its operation. This is the "area MTI" system, in which Doppler shift is ignored in favor of motion of the target echo envelope in range or angle on the display. A simple means of exploiting this technique is the so-called "true motion" display used in marine radar, in which the center of the scan on the PPI is displaced to conform to the motion of the ship which carries the radar. By using a display tube with very long persistence, the signal "blips" from moving targets will appear to have "tails" following them, indicating their previous locations and their velocities. Applied to land-based radar, the long-persistence display alone will provide the degree of memory necessary for indication of moving targets. In harbor surveillance, the land areas show up in recognizable shapes, outlining the channels or water areas under observation. Fixed markers such as buoys, piers, and breakwaters give constant, sharply defined echoes. Moving ships are identified by their characteristic tails, whose extent is an indication of speed and whose direction indicates the heading of the ship. Although there is no electronic suppression of the fixed targets, as required by the definition of MTI, the human eye and brain are able to perform some of the filtering required for picking out the moving targets, and the need for complex radar circuits is largely eliminated.

True area MTI employs an electronic storage device to suppress the fixed targets before they are sent to the display tube. By storing a complete scan in a high-resolution electrostatic pattern, and comparing this with the video signals arriving on the next scan, the area MTI can provide reasonable cancellation of fixed targets. As with the true-motion display, the moving target must travel through an appreciable fraction of a resolution element in either range or angle to appear as a moving target at the output of such a system. The response is not limited to targets with radial velocity components, and there are no blind speeds other than zero. This form of MTI has not found wide use, primarily because it requires a radar of very high resolution to be sensitive to targets with moderate speeds and because the resolution, dynamic range, and stability of storage tubes do not permit the degree of cancellation needed in many cases.

Pulsed-Doppler Search Radar

The definition of a pulsed-Doppler radar distinguishes it from MTI only in that it uses the Doppler resolution for purposes other than simple rejection of fixed targets. In practice, most pulsed-Doppler radars have evolved into forms which are quite distinct from the conventional radars which have been described. This is especially true of radars used for search, where the restricted time-on-target does not permit the removing of Doppler ambiguities by lengthy processing of the radar data. The need to reduce the number of such ambiguities, therefore, leads to the use of higher repetition frequencies than would otherwise be found desirable, and to the use of special techniques to overcome the resulting range ambiguities.

The range-gated approach to signal processing (see Fig. 7.10) lends itself to pulsed-Doppler applications, since the number of range elements in a high-prf system may be small and the number of filters may be large. Each Doppler channel of each range gate constitutes a separate channel for detection, and the relative outputs of adjacent channels may be compared to perform both range and velocity measurements by interpolation. The same performance can, of course, be achieved with continuous memory devices which store i-f signals for several repetition periods and deliver multiple Doppler outputs through special filter circuits.

The performance of pulsed-Doppler radar for detection has been described and compared with c-w and conventional pulsed systems.* The details of the analysis will not be repeated here, but the results may be stated simply. The analysis of search radar performance using the approach of Chapter 5 may be applied directly to pulsed-Doppler radars, since it was essentially independent of the type of modulation used. Calculations using the single-pulse signal-to-noise ratio may still be made, and the effects of coherent integration may be described in terms of a reduction in effective i-f bandwidth by a factor n equal to the number of pulses integrated in the Doppler filters. This number of pulses is given by the ratio of the repetition rate to the noise bandwidth of the Doppler filter. The resulting predetection signal-to-noise ratio is the same whether the calculation starts from the peak power and total i-f bandwidth, or from average power and total bandwidth of the Doppler filter responses, repeated over the several lines which make up the i-f signal spectrum.

Radar systems using long pulses are capable of significant Doppler resolution within a single pulse. In such a case, the pulsed-Doppler system takes the form of a number of parallel receiver channels separated by frequency increments equal to the reciprocal of pulse duration. Range gating, if used, may now take place either before or after the Doppler filters, since the range resolution is limited by the pulse width rather than by filter bandwidth. As in the case of the coherent system using range gating and Doppler filtering, each range and Doppler channel constitutes a separate detection channel, and interpolation may be used to make measurements with considerable accuracy. Where the pulse is used with frequency or phase modulation to improve range resolution, it is still possible to resolve Doppler shifts to an increment equal to the reciprocal of total pulse duration, but simple filters are no longer adequate for this purpose. A series

---

* J. J. Bussgang, P. Nesbeda, and H. Safran, "A Unified Analysis of Range Performance of CW, Pulse, and Pulse Doppler Radar," *Proc. IRE*, **47**, No. 10 (Oct. 1959), pp. 1753–62.

of special "compression filters" may be matched to the possible spread of Doppler frequencies, and each such filter will furnish an output consisting of target signals within its Doppler passband, resolved in range to an interval given by the reciprocal of the transmitted bandwidth.

The theory of search radar operation, discussed in the preceding three chapters, has been available in various forms for many years. It has been applied in different ways to arrive at a great variety of search radar designs for different purposes. In some cases, two or more radars designed for the same purpose take such different forms that they bear little resemblance to each other, and the relative merits of the different designs are the subject of great controversy. This is due to the large number of factors which enter into system performance, and the many possible compromises which can be made in arriving at the design for a radar. Each radar development is governed by a set of operating requirements or specifications, stated or unstated, and the compromises are made in attempting to satisfy most of the requirements without exceeding a restricted development or production budget.

As new components and techniques become available, the radar designer is given more latitude in his choice of radar parameters, and the

SYNTHESIS AND ANALYSIS
OF SEARCH RADAR SYSTEMS

8

need for compromise between requirements is reduced. Simultaneously, the operating requirements may be made more stringent, forcing a new round of parameter adjustments and compromises. It is for this reason that we have seen the trend in search radar frequencies shift from the HF and VHF bands used early in World War II to the microwave equipment developed by the Massachusetts Institute of Technology Radiation Laboratory, back to VHF and UHF in the late 1950's, and again toward microwaves in recent equipment. There is no "correct" design for a given search radar performance. At a particular time, there may be a design which represents the best compromise between the stated requirements and the available components, but there will often be two or three different designs which compete for this position.

In order to approach an optimum design for a given application, the first step is the specification of realistic performance goals. The several factors which enter into this step will be discussed below. After this, one must synthesize a radar, or perhaps two or three competing designs, keeping within the bounds of available components, funds, and time. Finally, when a radar has been designed or constructed, its performance must be analyzed from a theoretical viewpoint, to guide its successful application in the field and to determine what factors require emphasis in experimental evaluation. These analytical steps will be covered in this chapter, and a later chapter will discuss the experimental techniques used in field evaluation.

8.1 PERFORMANCE SPECIFICATIONS FOR SEARCH RADAR

The outcome of a radar development or modification program is determined to a large extent by the adequacy of the specifications which guide it. Although it is impossible to prepare complete specifications which will have the same meaning for everyone in the radar field, it is desirable to approach this ideal as closely as permitted by time and the precision of the language. This means that the basic radar terms must be defined and understood, and that all key performance goals must be defined using these terms. The following discussion is intended to clarify the major problems in search radar performance specification, and to relate these problems to the theory covered by the previous chapters of this book.

Table 8.1 shows the subjects which are covered in a typical system performance specification for a search radar.* The general function of the radar will be stated, and this will be followed by a number of detailed requirements which will govern the design and determine its feasibility and cost. If the detailed requirements are to be realistic, leading to a practical radar design, they must be checked against theoretical restrictions on detection and measurement processes, and must allow for all fundamental physical limitations as well as for the failure of practical equipment to operate in an optimum way.

---

* This table has also been used successfully as an outline for a tracking radar specification, with appropriate emphasis on accuracy and precision.

Table 8.1 PERFORMANCE FACTORS FOR SEARCH RADAR

1. *General Function of Radar*
2. *Target Description*
   a. Cross section: mean, median, and amplitude distribution for discrete targets; reflectivity characteristics ($\sigma°$ or η) for continuous targets.
   b. Shape, size, or spatial extent.
   c. Fluctuation frequency spectrum, or aspect angle stability with time.
   d. Velocity: maximum, mean, and distribution of velocities.
   e. Distribution of target locations within coverage.

3. *Required Coverage*
   a. Maximum and minimum range.
   b. Azimuth and elevation sectors.
   c. Arbitrary limits, or regions of special interest.

4. *Detection Criteria*
   a. Single-scan P_d (blip-scan ratio).
   b. Cumulative probability of detection P_c, after given period of time or penetration depth.

5. *Resolution and Ambiguity*
   a. Dimensions of resolution element in three or four coordinates.
   b. Degree of resolution required for adjacent and widely separated targets.
   c. Allowable ambiguities in any coordinate, and criteria for resolving ambiguities.

6. *Accuracy and Precision*
   a. Over-all system accuracy in spherical or rectangular coordinates.
   b. Effect of signal-to-noise ratio on precision.
   c. Special limitations on type or distribution of error.

7. *Data Rate and Output Characteristics*
   a. Data rate, maximum data interval, or effective data bandwidth of system.
   b. Number and type of operator displays.
   c. Video or i-f outputs required.
   d. Type of angle and range outputs.
   e. Corrections and computations to be performed.

8. *Performance in Clutter*
   a. System cancellation ratio, subclutter visibility, or improvement factor.
   b. Detectability of standard targets in standard clutter.
   c. Circuit specifications.

9. *Other Environmental Factors*
   a. Description of radar platform, size, and weight limits.
   b. Type and amount of input power available to radar.
   c. Stabilization requirements.
   d. Compatibility with related equipment.
   e. Meteorological conditions, operating, and survival.
   f. Electronic environment: interference, jamming, limits on radiated noise, etc.

10. *Standards, Component Specifications, Etc.*

Target Characteristics

Search radar targets may be divided into discrete and continuous types, the latter referring to such targets as land and weather masses when they are to be observed by the radar. The system designer will need information on the statistics of the target in amplitude, frequency, and space, as discussed in Chapter 3. When these are not stated in the specification, they must be derived from whatever information is available, or assumptions must be made in order to arrive at the radar design.

One procedure is to specify a particular type of aircraft as a target. If measured values of cross section are used on this type of target, the amplitude distributions can be found and median or average values determined. The actual distributions or mathematical approximations may then be used in detection calculations. The size and shape of the aircraft, combined with measured or assumed data on the stability of its shape and its aspect angle relative to the radar, will yield good estimates of the fluctuation spectrum or Doppler frequency spread, as well as the presence of glint effects. These will be needed both in calculating detectability and in arriving at estimates of the accuracy and data rate of radar position measurements. The maximum target velocity can be used to estimate the distribution of radial velocities to be seen by the radar,* and to calculate the rate of change of aspect angle and the amount of time available for search prior to penetration of the target to a given range.

In some cases, the targets against which the radar must be effective will consist of a large number of aircraft types, or of mixtures of aircraft and other objects. Three procedures are available for specifying radar performance on such mixed classes of targets. The required performance may be considered to be met when the radar performance meets its goals after averaging over all the different target types. The relative probability of encountering each type of target should be known if this procedure is to be used. Alternatively, a specified minimum performance may be required on every type of target, which will require design around the most difficult combination of target characteristics. Lastly, a separate set of goals may be set for each major type of target, to avoid overdesign based on the most difficult characteristics.

Knowledge of spatial distribution of targets within the region of coverage is not essential, but is desirable if the best compromise in radar design is to be obtained. In many cases, this information is implicit in the coverage specified, which ideally would be matched to the target distribution.

---

* A. G. Emslie and R. A. McConnell, "Moving Target Indication," Chap. 16 in *Radar System Engineering*, L. N. Ridenour, ed. (New York: McGraw Hill Book Company, 1947), p. 654.

  Merrill I. Skolnik, *Introduction to Radar Systems* (New York: McGraw-Hill Book Company, 1962), p. 129.

Unless a particular nonuniform distribution is given, one must scan uniformly over the assigned coverage angle, sacrificing detection probability by giving equal attention to regions which may seldom or never contain targets.

Required Coverage

The performance of a search radar was shown in Chapter 5 to depend upon the maximum range and the solid angle to be searched. The solid angle was given in terms of an azimuth sector A_m and elevation sector limits E_m and E_o. In Chapter 6 we discussed some of the common types of coverage actually used in search radar systems, which require nonuniform distribution of power over the solid angle covered. Limitations caused by earth's curvature and surface reflection phenomena were also discussed. With these as a background, the coverage can be defined in a manner appropriate for the particular radar. We shall assume here that the problem has already been reduced to portions which are to be taken care of by individual radars, whose locations are defined relative to the region of coverage. In a later chapter we shall consider the factors which enter into the use of radar networks and dispersed sites which share the coverage.

Detection Criteria

Search radar coverage can be defined only for some stated detection criterion. The most commonly used criterion is the single-scan detection probability P_d, sometimes known as the "blip-scan ratio." A contour of constant P_d, for a given target cross section and scan rate, can be used to define the coverage volume. This criterion is easy to use in practice, and it can be verified by simple tests if care is taken to obtain enough samples of data to account for the effects of target fluctuations and surface reflections. When the 50 per cent probability level is used, the effects of target fluctuations are relatively slight, and exact knowledge of target amplitude distribution is not needed. If higher levels of detection probability are used, the loss caused by fluctuation becomes important, and the amplitude distribution and frequency spectrum are of critical importance in evaluating the system performance.

Although it is not strictly necessary when visual detection is used, a false-alarm rate should be specified for the automatic detection systems. As was shown in Chapter 6, the equivalent false-alarm rate for the human operator corresponds to a false-alarm probability of about 10^{-6}. The formulas for converting the false-alarm probability to false-alarm time or rate were given in Chapter 1. In track-while-scan systems, the false-alarm rate may be determined by internal considerations of storage or channel capacity rather than by the requirements of the output data.

Another way of specifying detection performance is to require that the cumulative probability of detection P_c reach a certain level before the target has penetrated into the coverage volume more than a certain distance, at a given radial velocity. This criterion is more applicable to early-warning radar than to systems which must maintain tracks over an entire volume of coverage. It permits the beam shape and energy distribution to be optimized for earliest warning without unnecessary overlapping of coverage which may be provided by other (surveillance or tracking) radars. It also permits the system to take maximum advantage of the extended detection range which is achieved with low values of P_d on fluctuating targets.

Resolution and Ambiguity

The definition of radar resolution is of critical importance in determining the design which will meet the performance specification. As in Chapter 2, we shall use the term "resolution" to denote the ability of the radar to make separate measurements on two or more targets whose amplitudes are approximately equal. This does not require that the energy of the unwanted target be completely excluded from the measurement channel, but that it be reduced to the point where its influence on the measurement is small. The point at which this condition is met will depend upon the size of the other errors in the system. Although this is the common usage of the term, confusion has arisen because the same word is sometimes applied to denote the smallest *change* in a measured value which can be recognized at the output of the radar system, with only one target present (a quality we refer to as precision). Our usage follows the definitions of resolving power and resolving time given by the IRE,[*] and the definition of resolution used generally in optics and television.[†] Thus, the dimensions of the resolution element will be given roughly by the pulse width in range delay, the beamwidths in angle, and the reciprocal of the observation time in Doppler frequency. More exactly, the resolution in range delay has been shown to depend upon the reciprocal of the system bandwidth, where both transmission and receiving systems utilize this bandwidth (as in pulse compression). When the transmitter rise time τ_c is much shorter than the pulse width τ, there will be small amounts of energy distributed over a wide spectrum, and it becomes theoretically possible to use a "leading edge" discrimination technique to resolve the earliest target in range. This is not usually done, however, because the usable signal-to-noise ratio is that

---

[*] *IRE Dictionary of Electronic Terms and Symbols* (1961), p. 123.

The International Dictionary of Physics and Electronics, Walter C. Michaels, senior ed. (Princeton, N.J.: D. Van Nostrand Company, Inc., 1956), pp. 775–76.

[†] Nelson M. Cooke and John Markus, *Electronics and Nucleonics Dictionary* (New York: McGraw-Hill Book Company, 1960), p. 401.

determined by the energy in the leading edge of the pulse, and because of the special processing circuits required. To achieve high range resolution, the system designer will either use a short pulse or will spread the energy of the entire pulse over a wide spectrum by using one of the pulse compression techniques.

The frequency resolution of search systems is usually limited to the reciprocal of the time-on-target per scan, with phase coherence being maintained over the several pulses exchanged in this interval. Where no means is provided for maintaining a phase reference between pulses, either by storage within the radar or by noncoherent techniques, the frequency resolution is the reciprocal of the pulse length, and is of little importance except in those systems which use very long pulses. Whatever observation time is used to determine resolution, the stability of the transmitting and receiving oscillators must be adequate over that time, and the target must stay within the velocity spread which corresponds to the Doppler resolution. These factors normally preclude the coherent processing of signals received on successive scans.

Ambiguity is closely related to resolution, but refers to possible confusion between targets which are widely separated, rather than adjacent. Errors in position and velocity measurement caused by ambiguities will fall completely outside the normal error distribution, and must be excluded from calculations of rms error if a reasonable statement of system accuracy is to be made. Examples of ambiguities in pulsed search radar are the "second time around" echoes which are returned from targets whose range delay exceeds the pulse repetition period, and side lobe echoes, which may arrive from strong targets far from the beam position. When linear frequency modulation ("chirp") is used for pulse compression, the ambiguities lie along a diagonal line in the range-Doppler plane (see Appendix D), and in "time side lobes" beside the intended response. All systems of modulation have ambiguities in the range-Doppler plane, and the radar specification should guide the designer in his choice of a wave form which relegates the ambiguities to regions which will not cause serious errors and confusion during normal operation.

Further ambiguities can arise in some systems when two or more real targets combine to produce two or more additional false target indications. As an example of this, consider the operation of a height-finder in conjunction with a fan-beam search radar. Two targets at the same range may be resolved by the search radar in azimuth and by the height-finder in elevation, but there exist two possible pairs of targets which could produce the same indications (see Fig. 8.1). Only when the two targets become resolvable in range, or when they separate far enough in one of the angular coordinates to become recognizable by both radars, will the two "ghost" targets be ruled out. Similar ambiguities appeared in the "V-

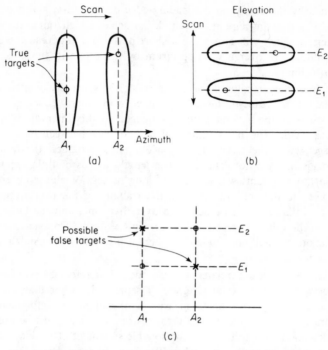

Figure 8.1 Ambiguity in system using search radar and height finder. (a) Search radar view. (b) Height finder view. (c) Results of combined radar data.

beam" radars used during and after the second World War.* In these two examples, it should be noted that the ability of each radar (or beam) to resolve the signal energy permits the "ghosts" to be eliminated when the measurement accuracy on one target discloses the proper pairing of the indications. It is not necessary that the target be completely resolved in the second coordinate. Most types of ambiguity can be resolved by observing targets over a sufficient period of time. The specification should make clear whether the readings taken over several scans may be combined to eliminate ambiguity, whether some external source of information can be relied upon to limit the possible confusion, or whether the radar itself must eliminate all ambiguities on each scan.

---

* L. N. Ridenour, ed., *Radar System Engineering* (New York: McGraw-Hill Book Company, 1947), pp. 192-96.

Merrill I. Skolnik, *Introduction to Radar Systems* (New York: McGraw-Hill Book Company, 1962), p. 459.

Accuracy and Precision

The process of measurement consists of identifying the location of the target in the four-dimensional space which was used to specify the resolution element. Mere detection of the target will serve to identify the resolution element in which it is located, and a simple readout of this information may meet the requirements for accuracy in a simple system. In general, a higher accuracy is needed, and in some cases it may be necessary to approach the limit set by thermal noise, as discussed in Chapter 2. When refinements are introduced to interpolate within the resolution element, it becomes important to distinguish between different types of error and properly to define the requirements. The term *accuracy* is defined as "the quality of correctness or freedom from error," and applies both to random and to bias or systematic components of error. Where the error varies with time, a given reading may be characterized in terms of both accuracy and *precision,* the latter denoting "the quality of being exactly or sharply defined or stated."* The distinction is thus a matter of whether the error is consistent or random in nature, and in the radar application the random error will generally vary with the passage of time. An accuracy specification governs all errors in the system, and must allow a greater tolerance than the precision specification, which covers only random noise. For a more thorough discussion of errors and error specifications, reference may be made to Chapter 10 in the section on tracking radars.

Data Rate and Output Characteristics

The system which uses the radar data will usually impose particular requirements on the rate at which it must receive data, and on the form in which it must arrive. The basic data rate for a search radar is the scan rate of the antenna, or the reciprocal of the time required to perform a complete scan over the assigned volume of coverage. The time between successive readings is referred to as the data interval or "frame time" (by analogy to television picture scanning). As with television, alternate frames may be interlaced or otherwise varied to obtain better coverage of the volume. Many search radars are intended to operate in conjunction with existing standard types of signal-processing equipment, which will determine the form of the radar output signals. Thus, when a standard PPI or RHI display is used, the radar will provide a trigger pulse, video signals, and an angle output to operate the deflection servo. Common units have been developed to provide MTI signal processing, video integration, and interference rejection for many different types of search and height-finding

---

* *The International Dictionary of Physics and Electronics,* Walter C. Michaels, senior ed. (Princeton, N. J.: D. Van Nostrand Company, Inc., 1956), p. 7.

radars.* These accept radar i-f signals from linear, limited, and logarithmic receiver channels, plus the usual triggers and angle data.

Often, a radar specification is expanded to include both signal-processing and computing functions on the output data for multiple targets. Specifications must then cover the detection and measurement processes carried out on the i-f and video signals, plus the other operations such as coordinate conversion, vector velocity estimation, and storage of track-while-scan data, which are performed in analog or digital computing elements. In such cases, the output of the system may take the form of digital data for direct transmission on telephone, teletype, or radio circuits.

Performance in Clutter

The ability of the radar to function on targets embedded in a background of clutter is dependent upon many factors, some of which were enumerated in the preceding chapter. Once the size of the radar resolution element in range, azimuth, and elevation has been established and the required data rate is set, the important considerations relating to clutter may be reduced to the cancellation ratio or subclutter visibility under various conditions of operation. The difficulties in defining, testing, and evaluating these factors have been discussed. Some method of specifying the MTI performance should be adopted which is consistent with the other provisions of the performance specification, and which is susceptible to quantitative testing when the radar becomes available. Common methods include the requirement for cancellation ratio on fixed targets with the antenna fixed, subclutter visibility on simulated signals superimposed on actual ground or sea clutter, and combinations of performance specifications with design specifications, the latter calling for particular characteristics such as rejection-notch width and shape.

The amount of clutter cancellation required in the radar is also dependent upon the amount which enters the receiver in the first place. It was shown in Chapter 3 that the clutter power is a sensitive function of the operating frequency and of the size of the resolution element, as well as of the environmental conditions. Radar polarization may also enter into the clutter power computation, especially for sea clutter. For these reasons, if the major radar parameters are left open for optimization by the radar designer, the amount of clutter rejection should be specified only in terms of final target detectability under given environmental conditions. Otherwise, a superior design may be ruled out because its clutter rejection is based on range or angle resolution, whereas an inferior design appears to

---

* C. A. Fowler, A. P. Uzzo, Jr., and A. E. Ruvin, "Signal Processing Techniques for Surveillance Radar Sets," *Trans IRE*, Vol. **MIL-5**, No. 2 (April 1961), pp. 103–108.

meet the goals by partially cancelling a large amount of clutter which is permitted to enter the receiver.

Other Environmental Factors

We shall not be concerned here with the settings of the environmental test chambers in which the radar components will be forced to operate, but with the broader description of the environment which will determine system parameters. Most basic, of course, is the description of the platform on which the radar will be mounted: land, ship, aircraft, or other. The type of platform, and the means of transportation provided to move the radar to its operating site, will set limits on its over-all size and weight. These limits may be stated in the specification, or they may be set indirectly by specifications relating to compatibility with the platform and other equipment. The available input power for operation of the radar may also be specified if this is critical. Knowing the type of platform, the designer may establish the need for antenna stabilization or conversion of output data to a stable set of coordinates.

Beyond the platform for the radar, the environment consists of the surrounding land or sea surface, the atmosphere, possible sources of interfering signals, and the system into which the radar is to supply its outputs. The presence or absence of human operators and maintenance personnel is also of great concern. When locations for the radar are given, the system designer may consult his geography and meteorology references for much of the needed information. Otherwise, he will need to know the possible altitudes of the sites above sea level and above surrounding reflecting surfaces, the probabilities of various conditions of wind and precipitation, and the range of conditions over which the radar must remain in operation, as opposed to mere survival. Finally, he must have some idea of the electronic environment and the amount of interference, both intentional and "friendly," which his radar must overcome. These last factors pose some of the most difficult problems in preparing realistic specifications, both for reasons of security and because the range of possible conditions is so variable with location, time, and tactics. However, it is better to state some set of conditions in the specification than to permit each successive level in the design procedure to establish its own set of assumptions and contingency factors.

Standards, Component Specifications, Etc.

There are many matters beyond those discussed above which must be covered by the search-radar specification, including reliability, quality control, construction practices, and standardization of parts, circuits, and techniques. Since these are not related directly to the system design or analysis techniques covered by this book, they will not be discussed further. Their omission does not indicate lack of appreciation for their importance or their impact on successful radar design.

8.2 ANALYSIS OF PRACTICAL SEARCH RADARS

To illustrate the application of search radar theory to a practical case, we shall analyze the performance of one type of search radar which has received considerable attention over the period since World War II. This is the lightweight, transportable air-search radar, represented by the AN/TPS-1D. Its major characteristics are given in Table 8.2. We shall

Table 8.2 CHARACTERISTICS OF SEARCH RADAR AN/TPS-1D

1. Frequency	1220–1350 mc
2. Power output	500 kw peak
3. Pulse width	2.0 μsec
4. Repetition rate	400 pps
5. Antenna size	15 ft wide × 5 ft high
6. Antenna gain	28 db
7. Beamwidths	Azimuth 4°; Elevation 10°
8. Scan speed	0–15 rpm
9. Receiver noise factor	10 db
10. Receiver bandwidth	1 mc
11. Coverage	Azimuth 360°; Elevation 0–8°
12. Indicator range	20, 40, 80, 160 nmi
13. Unambiguous range	200 nmi
14. Indicator types	7 in. PPI; 5 in. A-scope
15. Rated accuracy	1 mi + 3 per cent of range; ±1° azimuth
16. Detection range (20 m²)	90 mi (free space)
17. Input power	7500 w; 115 vac; 400 cps
18. Weight (less shelter)	2000 lb nominal

(SOURCE: Air Force T. O. 31P-1-22, 1 Dec. 1960, Raytheon Mfg. Co.)

apply to this radar the analysis procedure developed in Chapter 5, and check this against the results obtained by use of the procedure described by Blake. A coverage chart will be prepared, and the compromises involved in adjusting the coverage and the scan rate will be discussed. The performance of the MTI system will be analyzed for various types of clutter. In the following section, we shall describe the evolutionary process by which this type of radar can be improved and its coverage extended to greater ranges on smaller targets.

Ideal Detection Range

Before applying the search radar equation, we must derive from Table 8.2 the basic search parameters and coverage sector for this radar.

1. *Average power.* From Eq. (5.3), we find $P_{av} = P_t \tau f_r = 400$ w or $+26$ dbw.
2. *Receiving aperture.* $15 \times 5 = 75$ ft². We shall reduce this to 70 ft² to account for the rounded ends. Hence, $A = +18.5$ dbft².
3. *Search time.* Assume initially $t_s = 10$ sec, or $+10$ db, corresponding to a single scan at 6 rpm.
4. *Coverage.* Assume a coverage matched to the beamwidths: $A_m = 360° = 2\pi$ rad $(+8$ dbr$)$; $E_m = 10° = 0.174$ rad $(-7.5$ dbr$)$. See discussion below on coverage charts.
5. *Search loss.* Assume initially the 13 db minimum combined loss given in Chapter 5.
6. *Required* $(S/N)_i$. Assume $P_d = 50$ per cent, $P_n = 10^{-6}$ for the 10 sec period, giving $(S/N)_i = 11.2$ db. We shall later discuss the effect of longer search times and higher probabilities of detection.
7. *Target cross section.* Assume $\bar{\sigma} = 20$ m² $(+13$ dbm²$)$, corresponding to a large bomber or transport aircraft.
8. *Operating noise factor.* The 10 db receiver noise factor given in Table 8.2 may be taken as a first approximation of the system operating noise factor, if it is assumed that the combined input line and antenna temperatures total 290°K.

The above parameters may now be applied to calculate the maximum theoretical range for a radar of this type; Eq. (5.10) is used, with $(S/N)_i$ known and R_m unknown.

$$4(R_m)_{dbnmi} = +26 + 18.5 + 10 + 13 - 11.2 - 8 + 7.5 - 10 - 13 + 52$$
$$= +84.8$$

$$(R_m)_{dbnmi} = +21.2, \quad or \quad R_m = 132 \text{ nmi}$$

Practical Losses

The minimum loss allowance of 13 db was based upon the following

Antenna efficiency loss	3.0 db
Receiver matching loss	1.0 db
Combined loss from integration, scan distribution, beam shape, scanning, and target fluctuation	9.0 db

A more detailed evaluation of these and other losses shows that the AN/

TPS-1D, operated for a single scan with a required detection probability of 50 per cent, has a total search loss of about 21.1 db, distributed as follows

Receiver matching loss ($B_\tau = 2$)	2.9 db
Antenna efficiency loss	4.5 db
Integration-detection loss	3.4 db
Scan distribution loss	0 db
Beam shape loss (two-dimensional)	3.2 db
Scanning loss	0 db
Fluctuation loss	1.4 db
Transmission line loss	2.0 db
Collapsing loss	1.5 db
Integrator weighting loss	1.0 db
Propagation attenuation (90 mi)	1.2 db
Total search loss from ideal case	21.1 db

When the additional 8.1 db loss is inserted in the search radar equation, the maximum range is reduced to

$$4(R_m) = 76.7 \text{ dbnmi}, \qquad (R_m) = 19.2 \text{ dbnmi}, \qquad R_m = 83 \text{ nmi}$$

Optimum Operating Conditions

The apparent large excess loss factor suggests that it might be possible to improve the search performance by varying some of the operating conditions of the radar. Before we investigate this, however, we should consider that the single-scan goal of $P_d = 50$ per cent does not represent the actual criterion of radar performance in most applications. A more accurate statement of the radar problem would be to achieve a cumulative probability of detection of 90 per cent over a longer period, perhaps 40 sec. This is almost exactly what would be found by applying the formula for cumulative detection probability [Eq. (5.12)] to four scans of the radar at 10 sec per scan, if it is assumed that independent results are obtained on each scan.

$$P_c = 1 - (1 - P_d)^j = 1 - 0.5^4 = 0.9375$$

If we look at this 40 sec period of operation and revise the loss figures to reflect the increased value of P_d required for this period, we find that the losses have actually increased relative to the ideal case. By holding the scan rate constant at six revolutions per minute, for instance, we find that all losses remain the same as previously calculated except the following:

Scan distribution loss. Increased from 0 to 4.4 db, if four scans with no scan-to-scan correlation are assumed (see Fig. 5.1).

Beam shape loss. Increased from 3.2 to 4.8 db, unless the target is known to move over a major portion of the elevation beamwidth during the 40 sec period (see Fig. 5.5, adding 1.6 db for the second, or scanning, dimension).

This additional 6 db loss offsets the increased search time, which provided a 6 db increase in the t_s term. The requirement for $(S/N) = 13.2$ db, compared to 11.2 in the previous case, leads to a reduction in range from 83 to 74 mi for this case. This reflects the fact that the original 50 per cent detection probability on one scan was averaged over the entire $10°$ elevation sector, with lower values applying near the edges of the beam. Targets which remain near the edge of the coverage cannot, therefore, be detected with 90 per cent cumulative probability unless the single-scan probability is increased above 50 per cent.

Given the above situation, we may investigate the possible variation in operating conditions which would reduce losses and lead to better performance, when this performance is measured over the 40 sec search period. The total number of pulses n_t exchanged during this period is 176, with 44 pulses per scan at the 6 rpm rate. When we compare this with the case shown in Fig. 5.9(a), where $n_t = 40$ and $P_d = 90$ per cent, we see that the only change is the somewhat increased integration loss caused by the larger number of pulses. The optimum number of scans would, therefore, be slightly higher than for $n_t = 40$. However, the curve for the fluctuating-target case is so broad in this region that there would be little, if any, practical gain in increasing the scan rate. To verify this, let us calculate the losses for the maximum available scan rate of 15 rpm, which would provide ten scans with 17.5 pulses per scan. From Table 5.1 or Fig. 5.1, we see that the scan distribution loss would increase from 4.4 to 7.6 db, while the single-scan detection probability would drop from 44 per cent to 20.6 per cent. The fluctuation loss would, therefore, drop from 1.0 to -1.0 db, whereas the integration loss would drop from 3.4 to 2.4 db (see Fig. 1.15). All other losses remain constant, leading to a net increase of 0.2 db in total loss relative to the case for 6 rpm. When the scan rate is reduced below 6 rpm, the curve of Fig. 5.9(a) gives an accurate indication of increase in loss, although there would be a somewhat more rapid increase owing to the higher integration loss for $n_t = 176$ pulses.

Inspection of the list of losses for the AN/TPS-1D shows very little opportunity for improvement, except possibly in the area of antenna efficiency. The use of an optimum video-integration and detection device, without collapsing loss or weighting loss, could recover as much as 2.5 db in performance, but it is questionable whether this type of equipment could be made small, light, and reliable enough to justify its replacing the PPI display in its present form.

Applying Blake's Range Equation

To show that the search radar analysis method described above is consistent with other commonly used methods, let us apply the procedure described by Blake, which is accepted as a standard of evaluation.* The equation for search range giving a 50 per cent blip-scan ratio is

$$R_{50} = 129.2F\left[\frac{P_{t(\text{kw})}\tau_{\mu\text{sec}}G_t G_r \sigma_{50(\text{sq m})}}{f_{\text{mc}}^2 T_n V_{o(50)} C_b L}\right]^{1/4} \tag{8.1}$$

This expression is similar to Eq. (4.15), and yields the same results with the following substitutions

1. The factor C_b/τ appears in place of the receiver bandwidth B. The bandwidth correction factor C_b, which was plotted in Fig. 6.6b, is given by the empirical equation

$$C_b = \frac{B\tau}{4.8}\left(1 + \frac{1.2}{B\tau}\right)^2 \tag{8.2}$$

   The use of the bandwidth correction factor accounts for that portion of the receiver matching loss which is above the minimum value of 2.3 db shown in Fig. 6.6b.
2. The radar frequency f appears in the denominator of Eq. (8.1) in place of the wave length λ in the numerator of Eq (4.15), with appropriate change in the constants.
3. Receiving system noise temperature T_n is used instead of the equivalent term $T_o\overline{NF}_o$.
4. In place of unity S/N in the denominator, the visibility factor $V_{o(50)}$ is used (see Fig. 6.10). This accounts for the integration gain of the operator and display, leading to the range for $P_d = 50$ per cent. To see the relationship between the visibility factor and these other parameters, we note that $V_{o(50)}$ represents the ratio $S\tau/N_o$ required to achieve $P_d = 50$ per cent for a particular display and number of pulses integrated. For a PPI display with one pulse, $V_{o(50)} = 13.4$ db implies a required S/N ratio of 11.1 db, when averaged over the

* Lamont V. Blake, "Interim Report on Basic Pulse-Radar Maximum Range Calculation," Naval Res. Lab. Report 1106, Nov. 1960.

———, "A Guide to Basic Pulse-Radar Maximum-Range Calculation (Part 1)," Naval Res. Lab. Report 5868, Dec. 28, 1962.

pulse output of an optimum, quasi-rectangular i-f amplifier ($B\tau = 1.2$) with a matching loss of 2.3 db. This would indicate that the performance of the human operator is equivalent to an automatic detector operating on the average pulse output and adjusted for a false-alarm probability $P_n = 10^{-6}$. The factor $V_{o(50)}$ is reduced by the integration gain for n pulses. Hence, the calculation using C_b and V as in Eq. (8.1) is equivalent to applying the matching loss L_m and integration gain G_i to Eq. (4.18) to compute the range for $(S/N)_i = 11.2$ db.

5. The pattern-propagation factor F is included to account for vertical coverage relative to the free-space gain on the axis of the antenna.

6. The median target cross section σ_{50} is used, this being 1.4 db below the average $\bar{\sigma}$ for the Rayleigh target.

7. The losses are evaluated as described in Chapter 4 for the on-axis case and in Chapter 6 for the scanning case, except that the matching and integration losses are excluded (being considered in C_b and V_o), and the fluctuation and scan distribution losses are eliminated by considering only one scan with $P_d = 50$ per cent.

Blake's work sheet is reproduced as Table 8.3, with entries for the AN/TPS-1D. His tables for power and range ratios are reproduced in Appendix E. If a propagation factor of unity is assumed to find free-space range at the center of the beam, a value of 90.5 mi is found. This is not inconsistent with the 83 mi found in the earlier calculation for this radar, which considered the average detection performance at all elevation angles within the beam. The difference is exactly the 1.6 db loss factor inserted in the previous loss calculation to account for the beam shape loss in the second (nonscanning) coordinate. Identical results may also be obtained for targets on the elevation axis of the beam by applying the radar equation in one of the forms from Chapter 4, along with the integration gain from Eq. (1.10), the required $(S/N)_i = 11.1$ db, and the one coordinate beam shape loss of 1.6 db.

The relative merits of the different equations are largely a matter of convenience in use and clear understanding of the phenomena involved. Blake's procedure is particularly attractive when an existing set of radar parameters is to be checked against a required detection range, if the conventional PPI display or A-scope detection with the human operator is assumed. The method given in Chapter 5 is of value when the radar parameters have not been firmly established, or where a range of control settings is available and the optimum set of conditions is to be established. The forms using single-pulse S/N ratio as an intermediate step are of value when measurements of system performance are to be made under controlled conditions, and when electronic integrators are used.

Table 8.3 PULSE-RADAR RANGE-CALCULATION WORK SHEET

AN/TPS-1D Radar, ω = 6 rpm, n = 44

1. Compute system noise temperature, T_n, following outline in section (1) below.
2. Enter range factors known in other than decibel form in section (2) below, for reference.
3. Enter logarithmic and decibel values in section (3) below, positive values in plus column, negative in minus. (Example: If $V_{o\,(50)(db)}$ as given by Figs. 6.9 or 6.10 is negative, then - $V_{o\,(50)(db)}$ is positive, goes in plus column.) To convert range factors to decibel values, use Table E.1. For $C_{B\,(db)}$ use Fig. 6.6 (b).

Radar antenna height: h = ____ ft.　　Target elevation angle: θ = 1° (see Fig. 6. 2).

(1) Computation of T_n:	(2) Range factors		(3) Decibel values	Plus (+)	Minus (-)
$T_n = T_a / L_r + T_r + T_e$	$P_{t\,(kw)}$	500	10 log $P_{t\,(kw)}$	27.	
	$\tau\,\mu sec$	2	10 log $\tau\,\mu sec$	3.	
(a) For general range computation, use Fig. 15.6 for T_a.	G_t	630	$G_{t\,(db)}$	28.	
	G_r	630	$G_{r\,(db)}$	28.	
(b) To find $1/L_r$, given $L_r\,(db)$, use first and third columns of Table E.1.	$\sigma_{50(sq\,m)}$	20	10 log σ_{50}	13.	
	$f\,mc$	1300	-20 log $f\,mc$		62.4
	T_n, °K	2755	-10 log T_n		34.4
(c) Also in Table E.1, opposite $L_r\,(db)$ in first column, read T_r in fourth column. Note: If thermal temperature (T_t) of transmission line is appreciably different from 290° K, multiply Table E.1 values of T_r by $T_t/290$.	$V_{o\,(50)}$	1.12	$-V_{o\,(50)(db)}$		0.5
	C_B	1.12	$-C_{B\,(db)}$		0.5
	L_t	1.26	$-L_{t\,(db)}$		1.
	L_r	1.26	$-L_{r\,(db)}$		1.
	L_p	1.45	$-L_{p\,(db)}$		1.6
	L_o	1.78	$-L_{o\,(db)}$		2.5
(d) Opposite $\overline{NF}_{(db)}$ in first column, read T_e in fifth column.	Range-equation constant (40 log 1.292)			4.45	
	4. Obtain column totals			103.45	103.9
	5. Enter smaller total below larger			.	103.45
	6. Subtract to obtain net decibels			+ .	0.45

T_a	T_a / L_r
$1/L_r$	T_r
$\overline{NF}_{db}$	T_e
T_t	T_n

7. In Table E. 2, find range ratio corresponding to this net decibel value, taking its sign (±) into account. Multiply this ratio by 100. This is R_o 　　**97**

8. Multiply R_o by the pattern-propagation factor. $\boxed{F = 1.0}$ $R_o \times F = R'$ 　　**97**
9. On the appropriate curve of Fig. 15.3, determine the atmospheric-absorption loss factor, $L_{a\,(db)}$, corresponding to R'. This is $L_{a\,(db)\,(1)}$ 　　**1.25**
10. In Table E. 2, find the range-*decrease* factor corresponding to $L_{a\,(db)\,(1)}$, δ_1 　　**0.93**
11. Multiply R' by δ_1. This is a first approximation of the range, R_1 　　**90.5**
12. If R_1 differs appreciably from R', on the appropriate curve of Fig. 15.3 find the new value of $L_{a\,(db)}$ corresponding to R_1. This is $L_{a\,(db)\,(2)}$ 　　
13. In Table E. 2, find the range-*increase* factor corresponding to the *difference* between $L_{a\,(db)\,(1)}$ and $L_{a\,(db)\,(2)}$. This is δ_2.
14. Multiply R_1 by δ_2. This is the radar range in nautical miles. R_{50} 　　**90.5**

Note: If the difference between $L_{a\,(db)(1)}$ and $L_{a\,(db)(2)}$ is less than 0.1 db, R_1 may be taken as the final range value, and steps 12-14 may be omitted. If $L_{a\,(db)(1)}$ is less than 0.1 db, R' may be taken as the final range value, and steps 9-14 may be omitted. (For radar frequencies up to 10,000 mc, correction of the atmospheric attenuation beyond the $L_{a\,(db)(2)}$ value would amount to less than 0.1 db.)

Preparation of Vertical-Coverage Charts

The vertical-coverage chart of Fig. 6.2 may be used to determine the limits of the AN/TPS-1D for aircraft detection by plotting the antenna gain (voltage) pattern as a function of elevation, the maximum range being used as a reference. Free-space patterns for a 10° beam centered at 5 deg elevation and at 0° are plotted in Fig. 8.2, with maximum range corresponding to the 20 m² target. The low-altitude gap which is present for the upper pattern explains why it is necessary to use a 10° beam centered near zero to cover an elevation sector of 6° or 8°. A portion of the extra energy which is radiated at negative elevation angles is reflected to produce the lobe pattern shown in Fig. 8.3. The maximum value of the pattern-propagation factor F has been taken as 1.2 in this figure, corresponding to operation over a rough land surface whose reflection coefficient is about 0.2. An antenna height of 22 ft above the surface is assumed, giving a basic lobing interval of 1°. When the radar is operating over water, the lobing effect would be exaggerated, approaching twice the normal 90 mi range on the lowest lobe and punctuated by nulls of near-zero range. The

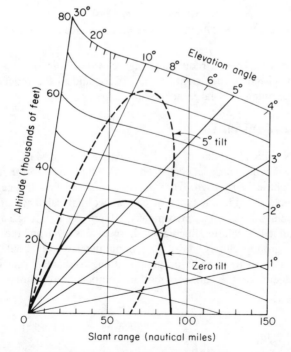

Figure 8.2 Free-space patterns for 10° vertical beamwidth (normalized to 90 mi range).

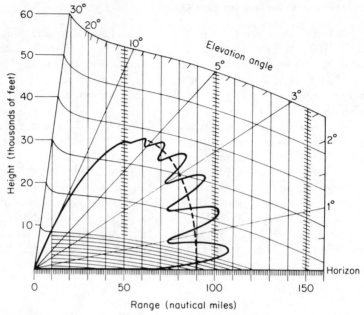

Figure 8.3 AN/TPS-1D coverage including ground reflection.

amplitude of the lobing is reduced at high elevation angles, owing both to the reduced reflection coefficient of the surface and to the reduced illumination by the antenna at large depression angles.

Performance with Clutter

The AN/TPS-1D radar is equipped with coherent MTI, a single recirculating delay line in the canceller and a COHO locked to the magnetron transmitter phase being used. To enter Fig. 7.7 and find the cancellation ratio attainable on different types of clutter, we first calculate the product $\lambda f_r = 9200$ cm per sec, or approximately 10^4 cm per sec. The cancellation ratio determined by clutter fluctuation is seen to vary from the 30 or 50 db level for ground clutter, down to a minimum level of about 20 db for airborne clutter such as chaff or precipitation. At a scan rate of 6 rpm, the rms velocity of the antenna motion component of clutter [see Eq. (7.56)] is

$$\sigma_{va} = \frac{\lambda \omega}{7.3 \theta_a} = \frac{23 \times 0.63}{7.3 \times 0.07} = 28 \text{ cm per sec or } 0.95 \text{ ft per sec}$$

This sets a limit of about 27 db on cancellation of stable clutter. Operation at reduced scan rate would make possible an increase in cancellation of ground clutter, but would not substantially improve the performance against airborne clutter. In addition, the residue from imperfect radar stability would probably limit cancellation near the 30 db level.

The amount of weather clutter which would be encountered with this radar can be estimated from Fig. 3.16. At a wave length of 23 cm, with a rain rate of 100 mm per hr (very heavy rain), the radar reflectivity would be about 3×10^{-7} m²/m³. If it is assumed that this rainfall could fill the atmosphere below 20,000 ft, the volume in the resolution cell at a range of 50 mi would be [Eq. (3.23)]

$$V_c = \frac{\pi}{16}\theta_a\theta_e R^2 \tau c \times \frac{20,000}{40,000} = 5.2 \times 10^9 \text{ m}^3$$

The corresponding maximum rain cross section would be about 1500 m², which would obscure almost any target. With a cancellation of 20 db, the rain return would be brought below the level of the largest targets, and detection would be possible. The uncancelled residue would have characteristics similar to random noise, and the integration gain of the display would be about the same as for detection in thermal noise.

Application of Eq. (7.3) shows the blind speeds of the AN/TPS-1D to occur at intervals of 150 ft per sec or 90 knots. Aircraft targets would be distributed fairly uniformly over the responses of the canceller, which would provide an average gain of 3 db relative to the reference level of the uncancelled signal. This gain would not improve the detectability in thermal noise, since it applies equally to noise power, but it would have the effect of doubling the average signal-to-clutter power ratio after cancellation.

8.3 SYNTHESIS OF A RADAR TO MEET PERFORMANCE GOALS

Just as there is no "correct" design for a radar, there is no single path by which an optimum set of characteristics may be determined. The synthesis of a search radar, starting from performance goals, is still an art rather than a science, and the methods described here are intended only as a guide to those engineers who have not developed their own approach through experience. Differences in opinion as to the relative importance of the intangible factors (such as reliability and adaptability to various operating environments) often lead to quite different approaches to radar design.

Extrapolation from Existing Systems

The most common of all approaches to radar system synthesis is the use of extrapolation from an existing radar. This method is also one of the most satisfactory, provided that the radar chosen for growth is one which performs its present task satisfactorily and provided that this task is not too different from that required in the new specification. Since most "new" requirements are actually based on a moderate extension of performance of existing systems, in order to meet some new step in military or commercial technology, it is entirely reasonable that the existing equipment should be modified or scaled up to meet the new requirements. In some cases, the performance required is the same as that provided by existing equipment, and it is the size, weight, power consumption, or cost which is to be scaled down.

The first step in the extrapolation process consists of an accurate analysis of the present equipment performance, as decribed in the previous section. In some cases, this analysis will show one or more obvious means of obtaining the new level of performance. For example, consider the case of the AN/TPS-1D radar. The development of higher-power magnetrons during the early 1950's made it possible to increase both peak and average power in this frequency band, and the availability of improved crystal mixers provided lower receiver noise factors. An improved version of the radar was desired to give a longer detection range, improved MTI performance, and decreased assembly time. The AN/UPS-1 radar was developed to use the same frequency band, the same type of antenna, and the same portable-generator capacity for prime power. Table 8.4 gives the major characteristics of the new radar.

Table 8.4   CHARACTERISTICS OF SEARCH RADAR AN/UPS-1

Frequency	Tunable 1250–1350 mc
Power output	1.0 Mw peak, 1120 w average
Pulse widths	4.2 and 1.4 μsec
Repetition rates	267 and 800 pps
Antenna size	16 ft wide × 4.8 ft high
Antenna gain	28.5 db
Beamwidths	Azimuth 3.8°; Elevation 10°
Scan speeds	0–15 rpm
Receiver noise factor	9.0 db
Receiver bandwidth	0.45 and 1.5 mc
Coverage	Azimuth 360°; Elevation 0–10°
Indicator range	20, 40, 80, and 275 nmi
Unambiguous range	300 and 100 nmi
Indicator types	10 in. PPI, 3 in. A-scope
Rated accuracy	1 mi + 3 per cent of range; ±1° azimuth
Detection range (20 m²)	151 mi (free-space)
Input power	7500 w; 115 v; 400 cps
Weight	2280 lb

(SOURCE: Technical Manual NAVSHIPS 94122, 12 June 1961)

Table 8.5 PULSE-RADAR RANGE-CALCULATION WORK SHEET

AN/UPS-1 Radar, $\omega = 6$ rpm, $n = 27.5$

1. Compute system noise temperature, T_n, following outline in section (1) below.
2. Enter range factors known in other than decibel form in section (2) below, for reference.
3. Enter logarithmic and decibel values in section (3) below, positive values in plus column, negative in minus. (Example: If $V_{o\,(50)(db)}$ as given by Figs. 6.9 or 6.10 is negative, then - $V_{o\,(50)(db)}$ is positive, goes in plus column.) To convert range factors to decibel values, use Table E.1. For $C_{B\,(db)}$ use Fig. 6.6(b).

Radar antenna height: $h =$ ____ ft. Target elevation angle: $\theta = 1°$ (see Fig. 6.2).

(1) Computation of T_n:	(2) Range factors		(3) Decibel values	Plus (+)	Minus (-)
$T_n = T_a/L_r + T_r + T_e$	$P_{t\,(kw)}$	1000	$10 \log P_{t\,(kw)}$	30.	
	$\tau_{\mu sec}$	4.2	$10 \log \tau_{\mu sec}$	6.25	
(a) For general range computation, use Fig. 15.6 for T_a.	G_t	700	$G_{t\,(db)}$	28.5	
	G_r	700	$G_{r\,(db)}$	28.5	
(b) To find $1/L_r$, given $L_{r\,(db)}$, use first and third columns of Table E.1.	$\sigma_{50(sq\,m)}$	20	$10 \log \sigma_{50}$	13.	
	f_{mc}	1300	$-20 \log f_{mc}$		64.4
	$T_n, °K$	2438	$-10 \log T_n$		33.85
(c) Also in Table E.1, opposite $L_{r\,(db)}$ in first column, read T_r in fourth column.	$V_{o\,(50)}$	1.41	$-V_{o\,(50)(db)}$		0.5
	C_B	1.15	$-C_{B\,(db)}$		0.6
	L_t	1.12	$-L_{t\,(db)}$		0.5
Note: If thermal temperature (T_t) of transmission line is appreciably different from $290°K$, multiply Table E.1 values of T_r by $T_t/290$.	L_r	1.12	$-L_{r\,(db)}$		0.5
	L_p	1.45	$-L_{p\,(db)}$		1.6
	L_o	1.26	$-L_{o\,(db)}$		1.0

(d) Opposite $\overline{NF}_{(db)}$ in first column, read T_e in fifth column.	Range-equation constant $(40 \log 1.292)$	4.45	
	4. Obtain column totals	110.7	101.95
	5. Enter smaller total below larger	101.95	
	6. Subtract to obtain net decibels	+ 8.75	

T_a	T_a/L_r	
$1/L_r$	T_r	
$\overline{NF}_{db}$	T_e	
T_t	T_n	

7. In Table E.2, find range ratio corresponding to this net decibel value, taking its sign (±) into account. Multiply this ratio by 100. This is R_o $\boxed{165}$

8. Multiply R_o by the pattern-propagation factor. $\boxed{F = 1.0}$ $R_o \times F = R'$ $\boxed{165}$
9. On the appropriate curve of Fig. 15.3, determine the atmospheric-absorption loss factor, $L_{a\,(db)}$, corresponding to R'. This is $L_{a\,(db)\,(1)}$ $\boxed{1.6}$
10. In Table E.2, find the range-*decrease* factor corresponding to $L_{a\,(db)\,(1)}$, δ_1 $\boxed{0.912}$
11. Multiply R' by δ_1. This is a first approximation of the range, R_1 $\boxed{151}$
12. If R_1 differs appreciably from R', on the appropriate curve of Fig. 15.3 find the new value of $L_{a\,(db)}$ corresponding to R_1. This is $L_{a\,(db)\,(2)}$ $\boxed{1.6}$
13. In Table E.2, find the range-*increase* factor corresponding to the *difference* between $L_{a\,(db)\,(1)}$ and $L_{a\,(db)\,(2)}$. This is δ_2. $\boxed{}$
14. Multiply R_1 by δ_2. This is the radar range in nautical miles, R_{50} $\boxed{151}$

Note: If the difference between $L_{a\,(db)(1)}$ and $L_{a\,(db)(2)}$ is less than 0.1 db, R_1 may be taken as the final range value, and steps 12-14 may be omitted. If $L_{a\,(db)(1)}$ is less than 0.1 db, R' may be taken as the final range value, and steps 9-14 may be omitted. (For radar frequencies up to 10,000 mc, correction of the atmospheric attenuation beyond the $L_{a\,(db)(2)}$ value would amount to less than 0.1 db.)

When the search radar equation is used, the improvement in performance should be 5.3 db over the AN/TPS-1D, owing to average power (4.4 db increase), aperture area (0.1 db increase), and noise factor (1.0 db improvement). The actual improvement under free-space conditions is a factor of 1.56 in range, or 7.6 db. The additional 2.1 db comes about through reduction of losses, both in transmission line and in collapsing losses. A range calculation using Blake's method is shown in Table 8.5. The vertical-coverage pattern for this radar with a target cross section of 3.0 sq m is basically the same as that shown for the AN/TPS-1D in Figs. 8.2 and 8.3, where a 20 sq m target was assumed. When the radar is used over water, experimental results have shown that the lobes are extended as shown in Fig. 8.4. A variable tilt-angle control on the antenna makes it possible to control the lobe intensity and vertical coverage to take maximum advantage of surface reflections or to avoid holes in the coverage at shorter range.

Improved MTI performance could not be combined with the lower repetition rate used for extended range. A second selection of repetition rate and pulse width was therefore provided, permitting operation at ranges

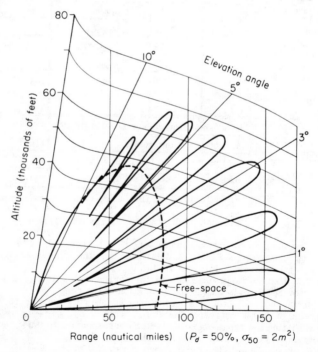

Figure 8.4   AN/UPS-1 coverage over water (10° vertical beam with 2° upward tilt angle).

out to 80 mi with about six decibels of additional clutter cancellation when compared with the AN/TPS-1D. As in the previous case, a single-delay recirculating delay canceller was used.

Further evolution of this type of radar could follow several courses. Since the number of pulses per beamwidth is still quite high, the antenna could be extended to perhaps twice the width without encountering appreciable scanning loss. Additional height could be added, but at the expense of more complexity in preserving the vertical coverage angle (e.g., dual or stacked-beam configurations). With the wider antenna, the performance of the MTI would be severely impaired unless a two-stage canceller were used. Another fruitful area of improvement is in the receiver, where reduction of noise temperature by a factor of eight to ten is possible. Improvement in this direction, of course, would be of no value if clutter and external noise were the limiting factors. Increase in transmitter power is also possible, but any significant improvement along this line would necessarily add to the size and weight of both the radar and the prime power source. The AN/UPS-1 in its original form converts input power to r-f output with an efficiency of about 15 per cent, including all tube heaters and operation of all parts of the radar. Only the highest-power transmitters can provide much improvement over this efficiency in the microwave region, when filament heating and equipment cooling power is considered.

The Systems Approach

In many cases the use of the extrapolation process is inconvenient, either because the problem is different from those solved by existing radars or because some new component or technique makes possible a different approach to an old problem. Direct synthesis of a new radar design from performance requirements is then the most practical procedure. The following methods are presented as a logical approach to this sort of problem, although they do not represent the only approach and they may not lead to a unique solution. As an initial step in this process, it may be necessary to review the stated requirements, and to fill in gaps with assumed needs or background information from other system studies. However, if all the information covered by Table 8.1 is available, one should be able to proceed at once, using the method described in Chapter 5, to estimate the scale of the radar required for the problem.

An example of this procedure was given in Section 5.2, in which the requirements called for search out to a range of 1000 mi, over a large solid angle. The resulting power-aperture product provides a basis for rechecking the reasonableness of the original requirements, and determining whether the need justifies the expenditure for such a piece of radar equipment. Let us asume that no substitute is found for complete coverage as originally specified: 90° azimuth sector, zero to 45° elevation, 1000 mi range. The power-aperture product for search of this volume in 10 sec with 90 per cent detection probability on 1 m^2 targets was found to be $+88.5$ db relative to one w-ft². This value assumed that the total search loss could be held within 16 db, and the resulting radar configurations were in the following size class.

$$P_{av} = 50 \text{ kw}, \qquad A = 14{,}000 \text{ ft}^2 \quad (120 \times 120 \text{ ft}),$$

or

$$P_{av} = 500 \text{ kw}, \qquad A = 1400 \text{ ft}^2 \quad (38 \times 38 \text{ ft})$$

Choice of Frequency

Although the theoretical performance of the ideal search radar is independent of frequency, the search losses and the ability to approach the ideal procedures assumed in the analysis are heavily dependent upon frequency and beamwidth. The next logical step is to determine what limits, if any, are imposed on the frequency to be used for the radar. The "radar window" through the atmosphere extends from about 30 mc to the vicinity of 10,000 mc. Below this band, the attenuation and refraction of the ionosphere would prevent reliable operation out to 1000 mi range, and above it the attenuation of the air introduces losses which are intolerable for high-power systems. More stringent limitations are present if the system must operate in all types of weather (limiting the upper frequency to perhaps 6000 mc) or if high accuracy or resolution is required (raising the lower limit toward 1000 or 2000 mc).

In our example, let us assume that all-weather operation is a requirement, and that the output data is to be accurate to about 1 mi rms. The data given in Chapter 15 show that the frequency should be above 200 mc because of ionospheric effects, and that the temperature of the sky background will be lowest in the region from 1000 to 6000 mc. Considerations of power tend to favor the lower frequencies, where the required average powers for detection are more easily generated and carried by wave guides and transmission lines. The need for a large aperture area and extended search angle also favors the low frequencies, as the number of beam positions to be searched varies directly with the product of aperture times the square of frequency. On the other hand, the accuracy specification makes it necessary to use very large apertures at the low-frequency end of the band in order to reduce the size of the resolution element and make possible the interpolation of target position to 1 mi. Table 8.6 compares the features of several possible frequency choices, assuming an intermediate aperture area of 4500 sq ft. The number of receiving beams required for coverage is calculated on the assumption that each beam must remain on

Table 8.6 FACTORS DETERMINING CHOICE OF FREQUENCY

Frequency (mc)	100	300	600	1200	2500	5000	10,000
Wave length (ft)	10	3.3	1.63	0.83	0.4	0.2	0.1
Sky temp. (max) (°K)	1200	200	60	25	40	70	145
Two-way loss L_a (db)	0.3	0.3	0.5	0.6	0.8	1.5	3.8
Relative noise level $(L_a T_s/T_o)$ (db)	+6	−1.3	−4.5	−10	−8	−4	+0.9
Width of circular beam (dg)	9.6	3.2	1.6	0.8	0.38	0.19	0.096
Beam angle ψ_b (dg²)	72	8	2	0.5	0.12	0.029	0.0072
Beam positions n_s	50	450	1800	7500	31,000	125,000	500,000
Minimum number of beams	1	2	9	37	150	600	2,500
Beamwidth (mils) Resolution (mi)	170	55	28	14	6.5	3.3	0.17
Error σ_θ (mils) Error $R_{max}\sigma_\theta$ (mi)	17	5.5	2.8	1.4	0.65	0.33	0.17

CONDITIONS: $A = 4500$ ft², $P_{av} = 160$ kw, $A_m = 90°$, $E_m = 45°$, $\psi_s = 3600$ deg² = 1.1 steradians target for at least four times the round-trip signal delay time, in order to avoid excessive scanning loss. Since the delay time for maximum-range targets in our example is 0.0125 sec, each beam position must be occupied for at least 0.05 sec, and a single beam may not scan more than 200 positions in the 10 sec allowed for search.

When the requirement for accuracy is considered, the use of frequencies below 1000 mc is ruled out, in spite of the benefits of simplicity and power-generation capability. If a frequency just above 1000 mc is used, the sky temperature will be only about $30°$ K, and an over-all operating noise factor of -7 db appears possible for a system with low-noise antenna and receiving system. This permits the average power to be reduced by a factor of five from that given earlier, which was based on a 0 db noise factor. Reduction in aperture is not permitted unless the power-aperture product is to be governed by accuracy instead of detectability. At this frequency, about 40 separate search channels must be used, each derived from an aperture 70×70 ft and each transmitting about 1 kw of average power. The beams would be about $0.75°$ in width, and would scan at $15°$ per sec. Configurations using elliptical or fan beams with different scan rates would also meet the detection requirements, but would produce larger errors in the coordinate having the broader beam.

If we consider detection only and use a frequency near 300 mc, the operating noise factor would remain near unity, as assumed in the original power-aperture computation. By using the largest average power shown, it would then be possible to reduce the aperture sufficiently to scan the angle with a single beam. Alternatively, at this frequency, the aperture could be permitted to approach the 120×120 ft figure, and five or six separate channels could be used with about 10 kw average power per channel. The accuracy requirement could not be met in this frequency band without further increase in aperture size or power. At frequencies below 300 mc, the increased sky temperature would call for an increase in power-aperture product even for detection, and the accuracy attained would be so poor that it might not be possible to designate targets successfully to a second radar used for measurement, as described below.

Multiple-Stage Search and Measurement

The conflict in frequency choice imposed by combining search of a large solid angle with accurate measurement requirements can sometimes be avoided by separating the radar functions into two or more parts, as when a search radar is used in conjunction with height-finding or tracking radar. In our example, a low-frequency search radar using a single broad beam might well be combined with a second radar which would investigate the regions where potential targets have been detected, verifying the presence of actual targets and performing the range and angle position measurements. The advantages of such a procedure are twofold. First, the false-alarm rate of the search system may be increased to a much larger value, since the system will send out an alarm only after the tracker has verified the presence of a target. The required S/N ratio for the search radar may be reduced as much as 4 db when the second stage of detection is provided by another radar. Second, there is no stringent requirement for accuracy placed on the search radar, which may use a wide beam at low frequency to cover the large search angle without excessive amounts of equipment.

In the example above, the problems may be solved either by a single radar with many channels or by a combination of two radars, one for search and coarse designation of targets and another for verification and accurate measurement. If a single radar is to be used, the choice of frequency must represent a compromise between the conflicting requirements for search and measurement, and the region near 1000 mc would probably by used. In the two-stage system, a frequency near the limiting value for the low end of the radar band would be preferred for search, possibly around 300 mc. Depending upon the requirement for traffic-handling capacity, one or more smaller microwave radars would be used for the second function.

Matching Beam Shape to Coverage

Knowledge of the required aperture, average power, and frequency does not necessarily determine the number and shape of the search beams. The aperture may vary from circular or square in shape to an elongated form, approaching as a limit the line-source which is directional only in one coordinate. In any situation where the search angle is larger than the solid angle of the beam, the system designer has the choice of covering the assigned search angle sequentially, with scanning beams, or simultaneously, with many fixed beams. A few examples will be given to show different means which may be used to match beam shape to coverage. Many possibilities exist which cannot be discussed in detail here.

Cosecant-squared coverage for air surveillance. The simple fan-beam coverage shown in Figs. 8.2 through 8.4 is adequate for warning of the approach of aircraft and for maintaining track-while-scan operations at long and intermediate ranges. Targets at high altitudes are lost, however, when they approach within 50 mi of the radar. A slight modification of the reflector can provide the cosecant-squared pattern shown in Fig. 8.5. The height of the reflector is increased from 4.8 to 6 ft, and the gain is reduced from 28.5 to 27.5 db, but otherwise the system remains the same. There is no increase in complexity of the equipment, as a single feed-horn and receiver channel is still used.

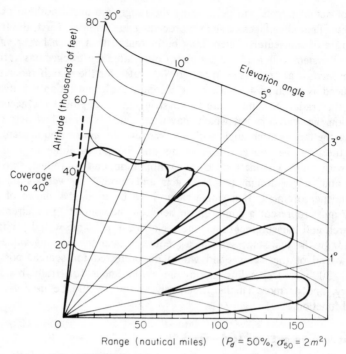

Figure 8.5 AN/UPS-1 with extended vertical coverage using cosecant-squared beam.

A second approach to this coverage problem uses a series of separate beams in elevation, each produced by a separate feed horn and connected to a separate receiver channel. A single transmitter may feed all horns through separate duplexers. By matching the distribution of transmitter power to the coverage shown in Fig. 8.5, the effective transmitting gain in the main lobe will be the same as provided by the shaped reflector (27.5 db for the case discussed). However, the separate receiving beams may approach the gain of the full aperture, which may be increased in height for gain levels of 30 db or more. In addition, the elevation resolution may be used to estimate the height of the targets, and clutter will be reduced by virtue of the smaller beams. Ground clutter will appear at full amplitude only in the lowest beam, and noncoherent MTI may be used in the upper beams to reduce moving clutter such as rain return. The complexity of such a system is its major drawback, especially when the number of stacked beams exceeds five or six. A major advantage is the ability to produce intense beams near the horizon without illuminating the ground, as was done with the fan-beam antenna.

A third means of obtaining cosecant-squared coverage has been applied to radars which have electronically controlled scan in elevation. By increasing the scan rate at high angles, the number of pulses per scan is reduced and the integrated signal-to-noise ratio may be programmed to follow the desired coverage. In such a case, although the peak power and antenna gain remain constant at the different elevation angles, the energy distribution is varied. Unless the repetition rate is high, this method will be limited by the rapid increase in scanning loss which is incurred when the beam moves more than one-third its width in the maximum range-delay time. The increase in this loss would reduce the received energy at the higher angles to conform with the desired pattern, but without making the unneeded energy available for use at the longer ranges.

Broad sector coverage for missile surveillance. The problem given in the earlier search-radar example will serve as a second type of coverage for discussion. The required coverage consisted of an azimuth sector of 90°, with elevation coverage from 0° to 45°. Consider first how this coverage would be obtained if the 70 × 70 ft aperture at 300 mc is used. Two beams of approximately 3° width could scan the volume in 10 sec. Mechanical motion of the aperture at this rate would present difficulties, but one of the electronic scan techniques could be used in the rapid-scan coordinate to avoid excessive mechanical rates. The 45° elevation sector would lend itself to electronic scanning without any need of increasing the aperture, since the entire antenna could be tilted to about 22° elevation to center the scan at a point normal to the antenna axis. Either mechanical or electronic scan could be used in the slow (azimuth) coordinate. The coverage would be interlaced to get as close as desired to uniform energy distribution, with only the extreme edges of coverage suffering from the absence of overlapping scans.

To illustrate the variety of possible solutions in the same frequency band, consider the use of a stacked-beam system, with relatively slow azimuth sector scan accomplished mechanically by moving the entire antenna (10 sec to scan the 90°, 10 sec to return). The coverage could be provided with as few as two beams, each 22° in elevation and 0.45° in azimuth. The antenna would now be 500 ft wide by 10 ft high. The gain would be the same as in the previous case, and the antenna would still scan 20 beamwidths per second, or one-fourth beamwidth in the range-delay time. Coverage would be nonuniform in elevation, dropping off near the edges of the broad beams. Alternatively, more beams could be used to cover the 45° elevation sector, the receiving gain being held constant by reducing the antenna width as the height is increased. In the limit, the original 70 × 70 ft aperture would be used with 14 circular beams stacked one above the other. Each beam would be fed with one-fourteenth of the transmitter power, and would provide a time-on-target of about $\frac{1}{3}$ sec. In a pulsed system, 28 pulses would be received per beamwidth when the maximum unambiguous repetition rate is used.

In these examples, the fluctuation loss for Rayleigh-distributed targets would probably require more rapid azimuth scanning in order to achieve 90 per cent detection probability during a 10 sec period. Two or more scans during this period would normally be used, doubling the rates and increasing the mechanical problems of moving the entire antenna. All the above systems would provide comparable performance from the standpoint of detection, and the choice between them must be based on considerations other than detection: accuracy, resolution, design feasibility, etc. No attempt will be made here to arrive at the criteria for the "best" system, since we are concerned with exploring the theoretical limitations imposed by detection, resolution, and measurement, and with methods for synthesizing systems which do not violate any of the fundamental limitations.

Choice of Transmitted Wave Form

It has been established earlier that all types of modulation are equally good in providing detection against a background of thermal noise, so long as the receiving system is matched to the transmitted spectrum. The high incidence of pulsed-radar systems in actual use testifies to their advantages in a number of practical matters, however. First is the available isolation between transmitter and receiver, which is accomplished by time switching in the duplexer of the pulsed radar. Systems using modulated c-w transmissions have been built for a number of applications, but there are always serious problems encountered in isolation of the receiver and transmitter, which often raise the receiver noise level above that of thermal noise. Second, the time resolution between adjacent targets and between targets and short-range clutter is usually better in the pulsed systems than is the frequency resolution of c-w systems. The receiver filtering and processing circuits are more simple for the conventional pulsed system than for the various forms of modulated c-w transmission. For all these reasons, the c-w transmission techniques have been limited largely to applications where extreme sensitivity and resolution between multiple targets are not important, as with radar altimeters, Doppler navigators, and automobile speed-control radar.

Assuming that a pulsed transmission has been selected, the choice of pulse width, repetition rate, and intrapulse modulation can be made to optimize performance of the system and to balance performance against complexity. In a conventional radar, the repetition rate is first set to provide unambiguous ranging over the entire coverage volume. The pulse width is then established to provide the required range resolution or minimum range (determined by the dead time during and immediately following the transmission). Peak power is then set to achieve the average power level needed for detection or measurement at maximum range. For example, in the previously discussed search radar, at 1000 mc, it was found that approximately 40 search channels were required, with average powers of about 1 kw per channel. Unambiguous search to a range of 1000 mi would require that the repetition rate be below 80 pulses per second. In the absence of more stringent specifications for range resolution, the pulse width can be established on the basis of range accuracy, which was to be within the over-all 1 mi tolerance. Since detection of targets is possible with an integrated signal-to-noise ratio of about 25 (14 db), the range precision limit set by thermal noise will be in the order of one-tenth the pulse width [see (Eq. 2.3)]. A pulse width of about 100 μsec could, therefore, be used without any special type of intrapulse modulation, to achieve the accuracy tolerance at maximum range. Since some targets would be detected with lower signal-to-noise ratios, the pulse width could be reduced by a factor of two to permit the accuracy to be obtained on all detected targets. The peak power per beam for a 50 μsec pulse at 80 pulses per second would be about 250 kw, a value well within the capabilities of r-f tubes and wave guide at this frequency.

When the peak power level is difficult or impossible to obtain, there are a number of alternative approaches to the system design problem. The coverage may be divided into more separate search channels with reduced power per channel. Higher repetition rates may be used, with some coding technique to resolve the range ambiguities during the normal search cycle. More flexibility is provided by the use of some form of pulse compression, which permits the needed range resolution or accuracy to be achieved with much longer pulses and reduced peak power levels. In principle, the pulse may be widened to approach a modulated c-w type of transmission, the limit being set by the factors which discourage the use of continuous transmissions. The advantage of most pulse-compression schemes is that they combine the merits of wide-bandwidth phase or frequency modulation with the isolation of pulsed systems. The pulse width is set to permit recovery of the receiver within the minimum search range (perhaps 10 to 50 mi in the type of radar used in the example), and the bandwidth or compression ratio is then set to obtain the range accuracy or resolution specified. In our example, the pulse width might be set at 100 or 200 μsec, and a compression ratio of 20 or 40 to 1 used to achieve a comfortable safety factor in range accuracy along with reasonable "time side-lobe" levels. The peak power per channel could thereby be reduced to 60 or 120 kw, although an additional matching loss of about 1 db might be introduced by the characteristics of the compression filters.

The exact form of modulation to be used for pulse compression is a subject of considerable difference of opinion between groups and individuals working in the field, and will not be discussed here. Many types of modulation have been developed for special applications, starting with the linear f-m or "chirp" system described by Darlington and by Cook. From the viewpoint of the systems engineer, the primary consideration which distinguishes the different modulations is the type of ambiguities which are generated in the time and frequency domains. These ambiguities are described briefly in Appendix D for the most common types of signal. The "regular" forms of modulation such as linear f-m are characterized by generation of well-defined regions of ambiguity, often distributed in resolution cells near the main response. For example, when linear f-m is used, the ambiguity consists of a diagonal ellipse, such that a target with Doppler shift appears at a slightly different range delay than would be produced by a fixed target at the same range. The "pseudonoise" modulations, on the other hand, have widely distributed ambiguities in the time-frequency plane. Both types of ambiguity function repeat at intervals equal to the repetition period in the time domain and the repetition rate in the frequency domain. These intervals may be extended by introducing variable coding into successive pulses, forming longer trains of repetitive pulse groups. The ultimate in extension of the ambiguity intervals is obtained when each pulse is coded differently, according to some random input which is stored and used to process the received signals.

Signal Processing

The field of radar signal processing has undergone a radical transformation since the end of World War II. This has been due, in large measure, to the introduction of digital computer techniques into both video and i-f areas of the radar receiving system. Allied to this has been the improvement in means of storing and integrating signals in both digital and analog form. Some of these techniques have been discussed in connection with signal detection and integration, while others are more closely related to position and velocity measurement. Without going into a discussion of specific circuits and components, we may characterize the various processing schemes by their transfer functions, weighting functions, bandwidths, sampling rates, and accuracies. The search radar PPI display, for example, may be described in terms of a nonlinear input-output function which relates trace intensity to i-f signal power, an exponential weighting function describing the decay of luminescence, a video bandwidth restricted by the input amplifiers or by sweep speed and spot size, a sample rate dependent upon the scan rate, and an accuracy measured in percentage of the range sweep or angular sector displayed. More refined processing schemes are capable of greater dynamic range before reaching saturation, and may be adjusted to give weighting functions other than exponential functions when required. They may operate at a basic sample rate limited only by the radar repetition rate, and with accuracies limited only by thermal and target noise and considerations of economics and circuit complexity. The theory presented in the foregoing discussions is intended to provide a basis for specifying and estimating the performance of any processing system, in terms of the fundamental informational content of the received signals. No attempt will be made here to cover the many possible applications, or the analog and digital techniques which have been devised to perform the different operations on the received signals.

The function of a tracking radar is to select a particular target and follow its course in range, angle, and sometimes in frequency (or velocity) coordinates. This chapter will deal with the techniques used in angle tracking and in extracting the angular data from the received signals. These techniques are closely related to those of radio direction finding, and with the development of communications and telemetry systems in the UHF and microwave bands the distinction between radar angle tracking and radio D/F has been almost eliminated. For this reason, much of the following discussion, although directed primarily at pulsed radar systems, may be applied directly to angle-tracking devices operating on continuous, one-way transmissions. The essential element in both cases is measurement of the angle of arrival of the received signal, a directive antenna beam which is pointed towards the target or transmitter being used. The characteristics of the antenna must be such that errors in pointing are measured and made

ANGLE MEASUREMENT AND TRACKING

9

available as signals to control the position of the antenna.

9.1 METHODS OF ANGLE TRACKING

The trackers discussed here consist of high-gain, pencil-beam antennas, mounted on two- or three-axis pedestals which point the beam under the control of tracking servomechanisms (or servos). The antennas may, in principle, consist of arrays, reflectors, or lenses, or of some combination of these types. Although the error data is derived initially in two coordinates normal to the beam axis (elevation and traverse angles), the pedestal usually converts the output data to angles which are referred to a fixed coordinate system, such as azimuth and elevation. Data takeoff devices such as synchros, potentiometers, resolvers, or digital encoders are mounted on the pedestal to translate the mechanical rotations of the shafts into output data for further electrical processing.

All tracking antennas use some arrangement of offset feeds to sense the angle error of the target with respect to the tracking axis. A major distinction may be drawn between systems in which the several feed positions are sensed sequentially, and those in which simultaneous samples are taken from each of the positions. The earliest of the tracking radars employed "lobe switching" to develop two pairs of signals, using beams which were displaced alternately by small amounts each side of the center axis. These signal pairs were presented on oscilloscopes to the azimuth and elevation operators, who acted as human servo actuators to direct the antenna and equalize the signal pairs. More advanced "sequential-lobing" radars sense the tracking errors automatically and correct the beam position with electromechanical servos of higher performance than could be achieved with human operators.

The conical-scan radar uses a single offset beam to sense the error, rotating it rapidly about the tracking axis to generate a narrow cone in which the target is centered to produce a steady signal return. This is also a sequential technique, and it shares with all such systems the requirement that at least three (and in practice four or more) pulses must be received during a complete scan cycle, in order to sense the direction and magnitude of errors in both tracking coordinates. The best-known example of the conical-scan radar is the SCR-584, developed by the Massachusetts Institute of Technology Radiation Laboratory during World War II for anti-aircraft fire control.* This radar has remained in use in this country and abroad for almost 20 years, largely as a result of its reliability and flexibility. Conical-scan radars are still important in the field of airborne fire control, owing to their simplicity and their ability to achieve the moderate accuracy required in air-to-air combat systems.

At the end of World War II, work was proceeding in several laboratories on a new type of tracking radar using the principle of "simultaneous lobe comparison." This technique is often referred to as "monopulse," since it permits the extraction of complete error information from each

---

* L. N. Ridenour, ed., *Radar System Engineering* (New York: McGraw-Hill Book Company, 1947), pp. 284-86. Also see the anonymous article, "The SCR-584 Radar," *Electronics* (Feb. 1946), p. 110.

received pulse. A discussion of its early development and an analysis of various arrangements of antenna and receiver components has been published.* Three major types of monopulse system are recognized, with various subtypes combining features of these types: the amplitude comparison type, the phase comparison type, and the mixed or phase-amplitude comparison type. A common feature of all three is the use of at least three (and usually four) receiving antennas or feeds, which are displaced from the center axis, operating simultaneously in the reception of target signals during each repetition period. Pairs of signals are compared as to amplitude or phase, to extract error data in both tracking coordinates. The transmitted pulse is applied to the feed system in such a way as to produce a single beam on the central axis of the antenna, and a receiving channel is similarly connected to provide a signal for detection, ranging, and display. In return for the additional number of channels in its antenna and receiving system, the monopulse technique offers reduction in tracking errors, as compared to a conical-scan system of similar power and size. The data rate or servo bandwidth may be higher in a monopulse radar, for the same repetition rate, leading to further reduction in tracking error on maneuvering targets. The agc bandwidth may also be increased to minimize signal fading. For these reasons, the monopulse technique is now used extensively in surface-based radars, especially where great accuracy and precision are required. Communications and telemetering antennas are also making increasing use of the monopulse technique, to obtain improved efficiency in combination with a self-tracking capability.

A complete description of the many different systems which may be used in tracking radar is beyond the scope of this book. Instead, general descriptions will be given of the two systems which have been found the most useful in actual equipment: the conical-scan type of sequential system, and the four-horn, amplitude comparison type of monopulse system. The performance of these two systems will be analyzed and compared quantitatively, and related qualitatively to that of other systems.

Conical-Scan Radar

A typical conical-scan radar is shown in the block diagram of Fig. 9.1. The provisions for display, synchronization, transmitter, receiver, and duplexing equipment are those which characterize most pulsed radars, regardless of application. The tracking features consist of the mechanically driven scanner, the range-gated automatic gain control (agc), the error

---

* Donald R. Rhodes, *Introduction to Monopulse* (New York: McGraw-Hill Book Company, 1959). He classifies systems as phase or amplitude comparison, with a sum-and-difference variation which may be used with either system.

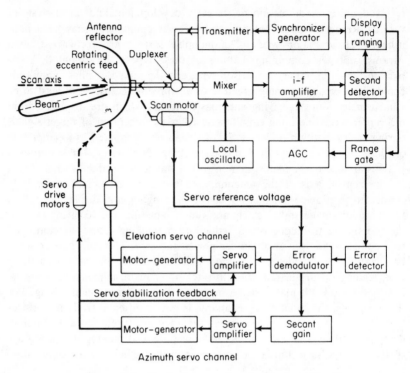

Figure 9.1 Block diagram of conical-scan radar.

detector, and the two servo channels. The scanner may take the form of a rotating dipole feed, electrically unbalanced to cause an offset in the beam relative to the mechanical axis; a nutating feed device which maintains constant polarization; or of a lens or metal plate which rotates in front of a fixed feed to provide displacement of the beam. The maximum rate of rotation is limited by the radar repetition rate and the requirement for at least four received pulses per scan cycle. Scan rates from one-tenth to one-hundredth of the repetition rate are commonly used, although higher rates may be used if the scan motor is synchronized with the received pulses. Where c-w signals are received, as in telemetry trackers, the only limit on scan rate is mechanical.

The tracking-radar display unit provides a full-range presentation of received signals, but the automatic-gain-control and error-detector channel is controlled by the range gate to operate only during reception of the signal from the selected target. The agc then maintains constant amplitude of the selected signal, averaged over a period which exceeds the scan cycle, providing consistent performance of the tracking loops regardless of target size and range. The error detector consists of an envelope detector, sensitive to the scan-rate components of signal modulation and rejecting the d-c level and components at the repetition rate and above. The a-c output at the scan frequency can be calibrated in terms of off-axis error angle, and its phase relative to the reference-generator voltage indicates the direction of the error. The error demodulator resolves the error voltage into elevation and traverse components, which control the two servos. In an azimuth-elevation pedestal, the traverse error voltage is multiplied by a gain function approximating the secant of elevation angle, in order to drive the azimuth servo with constant loop gain as the target nears zenith.

The relationships between the scanning beam, the target, and the error voltage in a conical-scan system are shown in Fig. 9.2. The received pulse train is held at an average amplitude E_o by the agc loop, operating with a

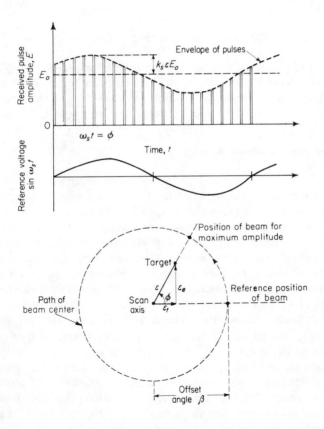

Figure 9.2 Error detection in conical-scan radar.

long time constant. The instantaneous pulse amplitude, on a steady target near the axis, is then

$$E(t) = E_o[1 + k_s\epsilon \cos{(\omega_s t + \phi)}] \tag{9.1}$$

where ϵ is the magnitude of the error angle, ϕ is the direction of the error with respect to the horizontal, ω_s is the radian scan rate ($\omega_s = 2\pi f_s$), and k_s is the error-slope factor of the antenna. The error slope is expressed in units of fractional modulation of the signal per unit off-axis error (e.g., $k_s = 1$ indicates that the modulation would be 1 per cent for an error equal to 1 per cent of the beamwidth). In normal operation, the error will not exceed a small fraction of the beamwidth, and the error slope will be a constant which can be evaluated on the tracking axis. If the error is considered to be composed of two separate components, Eq. (9.1) may be written as

$$E(t) = E_o(1 + k_s\epsilon_t \cos{\omega_s t} + k_s\epsilon_e \sin{\omega_s t}) \tag{9.2}$$

where ϵ_t and ϵ_e are the traverse and elevation errors, respectively. The a-c portion of this function is reproduced at the output of the error detector and passed to the error demodulator, which separates the two components for use by the servo amplifiers, and transforms the error voltage into d-c form.

Error Slope in Conical Scan

The error-slope factor k_s is a function of the beam offset angle β, and may be calculated if the shape of the beam is known in the region near the scan axis. The curves shown in Fig. 9.3 apply to a beam whose voltage pattern follows the $(\sin x)/x$ curve, and are given for both one-way and two-way tracking cases (for beacon or reflection targets). Although k_s is seen to increase rapidly for offset angles aproaching one beamwidth, the "crossover loss" L_k, which shows loss in received signal power for a target on the tracking axis, also increases. This places a practical limit on k_s in the order of 1.0 to 1.5, or 100 to 150 per cent modulation of the received signal per beamwidth of error. The dashed curves in Fig. 9.3 show the ratio $k_s/\sqrt{L_k}$, which will be shown to represent the measure of tracking sensitivity of the system. The offset angle is usually set at about one-half beamwidth for one-way operation, giving a power gain of 50 per cent of maximum for targets on the axis. For reflection tracking, the offset angle should be nearer one-fourth beamwidth, corresponding to 80 per cent power gain at crossover. Although this does not quite provide maximum angle-tracking sensitivity, it represents a good balance with respect to system loss and range-tracking performance.

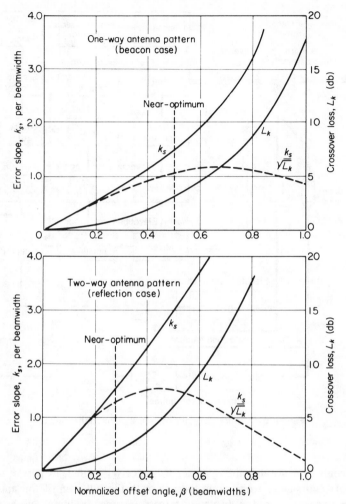

Figure 9.3　Error slope and crossover loss.

Amplitude Comparison Monopulse

The four-horn, amplitude comparison monopulse system shown in Fig. 9.4 uses the sum-and-difference technique, and operates as follows: the four horns are located symmetrically around the axis, and are fed in phase by the transmitter to produce a single pencil beam along the mechanical axis of the reflector. The in-phase or reference channel of the antenna is

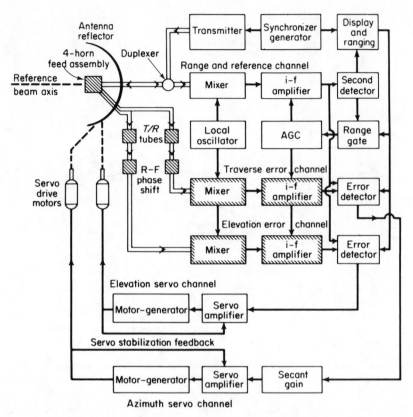

Figure 9.4 Block diagram of typical monopulse radar.

also connected through the duplexer to a receiver chanel, to permit reception of echo signals from along the axis with the full gain of the antenna. These signals are used in display of targets, in range tracking, and as reference signals for the angle error detectors. Up to this point, the monopulse and conical-scan radars are equivalent, as are the servos required to drive the antenna. The additional receiving channels required in the monopulse radar, which have no counterparts in the conical-scan system, are indicated by the shaded blocks in Fig. 9.4. These start with the four-horn feed assembly, which picks up the off-axis components of received signal and transforms them into elevation and traverse error signals. These are brought through T/R switch tubes and adjustable r-f phase-shifters to the error channel mixers. The electrical length of the paths between feed assembly and mixers must be held approximately equal to preserve the sensitivity of the error detectors and minimize errors in location of the tracking axis.*

After conversion to i-f, the reference and error signals are amplified in separate i-f channels, which deliver all signals to the error detectors in the same phase relationships as applied at the feed, and preserve the amplitude ratios. The agc loop is based upon the output of the reference channel, and controls all three i-f amplifiers in the same way to preserve these ratios, while achieving a constant signal level out of the reference channel. The gain of the tracking loop is thus held constant for varying target size and range. The phase-sensitive error detectors now operate on the error signals, producing bipolar video pulses proportional to the components of tracking error and indicative of its direction. A simple stretching or "boxcar" operation, followed by low-pass filtering, now supplies d-c inputs to the servo amplifiers. The signals supplied to the error detectors are range-gated, to select the desired target, as in the conical-scan system described previously.

Monopulse Feed Principles and Antenna Patterns

The operation of an early type of monopulse feed assembly is shown in Fig. 9.5, which indicates diagramatically the coupling between the four

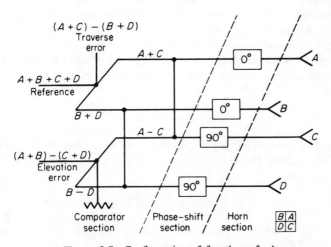

Figure 9.5 Configuration of four-horn feed.

* John H. Dunn and Dean D. Howard, "Precision Tracking with Monopulse Radar," *Electronics*, Vol. 35, No. 17 (April 22, 1960), pp. 51–56.

William Cohen and C. Martin Steinmetz, "Amplitude- and Phase-Sensing Monopulse System Parameters," *Microwave Journal*, Vol. 2, No. 10 (Oct. 1959), pp. 27–33; and No. 11 (Nov. 1959), pp. 33–38.

horns and the three outputs. A target signal received from along the axis
will pass entirely into the sum or reference channel, with essentially perfect
cancellation at the two error-channel terminals and at the unused termination. However, a signal received from a target slightly above the axis will
be reflected from the antenna surface more strongly into the lower pair of
horns, marked D and C in the illustration, and cause an unbalance signal
to appear at the elevation terminal. Similarly, any traverse error component
will cause an unbalance signal to appear at the traverse terminal, with the
signal amplitude in each case proportional to the magnitude of the corresponding error component. In an amplitude comparison system, the unbalance signal will be either in phase or 180° out of phase with the reference signal, the phase relationship serving to establish the sense of the
error (up or down, right or left). The phase shifts through the r-f and i-f
portions of the system must be carefully equalized to maintain this ability
to sense the direction of the error, but need not be held to the tolerance
required between the horns and the comparator section of the feed assembly.
It has been shown that the energy appearing in the two error channels
represents only that part of the signal which would, in a single-feed system,
be lost from the antenna as a result of the off-axis position of the target.
At the tracking axis, or with small values of error, this energy will represent
a negligible portion of the total received signal.

Reference and error patterns of a four-horn antenna system must be
determined experimentally in order to account for the effects of coupling
between horns, and of introduction of high-order transmission modes at the
mouth of the horn assembly.* A study of actual antenna patterns† showed
that the reference pattern could be represented very closely by the following
equation:

$$E_r = \cos^2(1.14\Delta)$$

where Δ in radians is the angle from the beam axis, normalized to the half-
power beamwidth. The measured error pattern for this same antenna was
represented by

$$E_e = 0.707 \sin(2.28\Delta)$$

These patterns are plotted in Fig. 9.6, along with the $(\sin x)/x$ pattern,
where $x = 2.78\ \Delta$. The only difference appears at the edges of the main
lobe, where the actual pattern is modified by side lobes and neither equation can be considered reliable. The error pattern in the case shown here is

---

* Peter W. Hannan, "Optimum Feeds for All Three Modes of a Monopulse Antenna," *Trans. IRE*, Vol. **AP-9**, No. 5 (Sept. 1961), pp. 444–61.

† Samuel F. George and Arthur S. Zamanakos, "Multiple Target Resolution of
Monopulse vs. Scanning Radars," *Proc. NEC*, **15** (1959), pp. 814–23.

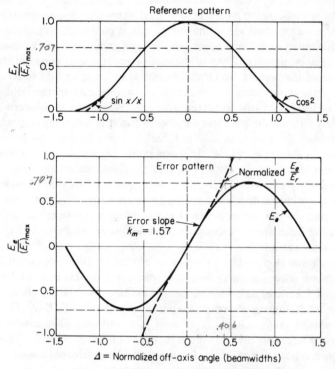

Figure 9.6 Monopulse antenna patterns.

characterized by a slope, near the axis, equal to 1.57 (in units of reference pattern voltage per beamwidth error). The error pattern is approximately linear to one-third beamwidth off the axis. The normalized error pattern E_e/E_r, which results from the agc action in the receiver system, is also plotted in Fig. 9.6, and shows linearity over a greater distance from the axis. In this case, the slope increases as the target approaches the half-power point of the reference beam. It will be shown that these patterns represent a close approach to the ideal slope attainable with tapered illumination of the reflector, as is required for reasonable side lobe levels. It may also be noted that the error and sum patterns are not simply the sum and difference of a pair of offset lobes, as is often assumed in analysis of monopulse radar.

The operation of monopulse tracking antennas has been discussed by several writers,* using three-dimensional plots of field intensity across the

---

* Dunn and Howard (1960); Hannan (1961); *op. cit.*

image plane at the focus of the antenna. Using this concept, we may see how the in-phase combination of four feed horns can pass to the reference receiver a high percentage of the energy intercepted by the aperture. Properly designed systems may approach the high gain and narrow beamwidth of a single, on-axis feed horn of optimum size. It is also apparent that the presence of the septum which divides the horns will lead to some loss in signal, since it introduces a mismatch or short circuit in the field at the center of the focal plane, reflecting energy back into space. Modern monopulse feeds for linear polarization tend to avoid this loss by removing the septum from the mouth of the feed, matching the feed to the energy distribution in the image plane. The generation of new transmission modes within the feed horn permits the septum to be introduced at a point somewhat behind the mouth of the horn with less loss. In other designs, the septum is omitted entirely and error signals are extracted from ports which are coupled only to modes excited by the off-axis components of the signal. For further discussion of monopulse feed design, reference should be made to Hannan's papers.

Our "error slope factor" k_m may be related to the fundamental limitations imposed by antenna theory, as discussed by Hannan and others. Hannan's "relative difference slope" is defined as "the difference slope relative to the maximum possible sum voltage": $K \equiv E_e/E_{ro}\epsilon$. The maximum possible sum voltage E_{ro} results from uniform illumination of the aperture, which also provides the minimum sum beamwidth θ_o. The maximum theoretical value* of K, for a rectangular aperture of width w, is: $K_o = \pi w/\sqrt{3}\,\lambda$. Recalling that the beamwidth for uniform illumination is $\theta_o = 0.887\,\lambda/w$, we may express the difference slope ratio as

$$K_r \equiv \frac{K}{K_o} = \frac{\sqrt{3}\,E_e\lambda}{\pi E_{ro}\epsilon w} = \frac{E_e\theta_o}{1.61 E_{ro}\epsilon}$$

It has also been shown that the final expression for K_r, with the same constant 1.61, will apply to the circular aperture.†

We may now express our error slope factor in terms of the above parameters and two related antenna design constants

$$k_m \equiv \frac{E_r\theta}{E_r\epsilon} = \frac{1.61 K_r(\theta/\theta_0)}{(E_r/E_{ro})} = \frac{1.61 K_r K_\theta}{\sqrt{K_g}}$$

Here, $K_\theta \equiv \theta/\theta_o$ is the ratio of actual beamwidth to the optimum (or minimum) value (Hannan's "beamwidth ratio"), and

---

* Merrill I. Skolnik, *Introduction to Radar Systems* (New York: McGraw-Hill Book Company, 1962), p. 477.

† Richard P. Kinsey, "Monopulse Difference Slope and Gain Standards," *Trans. IRE*, Vol. AP-10, No. 3 (May 1962), pp. 343–44.

$$K_g \equiv \left(\frac{E_r}{E_{ro}}\right)^2 = \eta_a$$

is the ratio of actual one-way antenna gain to its optimum value (antenna efficiency, or Hannan's "reference channel gain ratio"). The true measure of angle-tracking efficiency, for a given aperture size, is $K_r\sqrt{K_g}$ or $k_m K_g/K_\theta$, for two-way radar operation. For one-way (beacon) tracking, the measure is simply K_r or $k_m\sqrt{K_g}/K_\theta$. The triangular form of aperture illumination provides the optimum error pattern, as originally stated by Kirkpatrick in his 1953 paper. (It is interesting to note that this is also the shape of the weighting function for optimum differentiation of data which is contaminated by thermal noise. (See Section 13.3.)

Referring to Hannan's charts for monopulse feed design, we find that practical values of k_m will fall between the following limits.

1. For a four-horn feed that uses the same horn dimensions in reference and error channels: $k_m \cong 1.2$, when the total width of the feed is $1.5\ \lambda F/w$. (F is the focal length of the reflector or lens.)
2. For a feed in which the effective horn sizes are optimized independently for reference ($1.3\ \lambda F/w$) and error ($2.5\ \lambda F/w$) channels: $k_m \cong 2$.

Thus, the measured $k_m = 1.57$ shown in Fig. 9.6 may be taken as representing a good value for a four-horn feed in which multimode operation gives some improvement over single-mode horns. In the case of an antenna with low spillover loss in the reference channel, we will find that $K_\theta \cong 1/\sqrt{K_g}$, leading to $k_m \cong 1.61\ K_r/K_g$. On this basis, the twelve-horn and the four-horn, triple-mode feeds described by Hannan will have values of k_m equal to 1.9 and 1.7, respectively, with efficiencies of 0.58 and 0.75. Both designs represent considerable improvements over the basic four-horn systems. It should be noted that feeds using square horns are required for operation on signals of varying polarization. The systems using four or twelve square horns are thus suitable for omnipolarization reception, but exhibit lower efficiency than the polarization-sensitive types that use multimode horns.

9.2 TRACKER RESPONSE TO RECEIVER NOISE

The effects of thermal noise, originating in the radar receiver, on the accuracy of angle tracking can be evaluated easily for the case where the single-pulse signal-to-noise ratio S/N is high. By considering the effects of noise on the agc loop and the error detectors, the results of the analysis can be extended to values of S/N near and below unity.

Conical-Scan Signal and Noise Spectra

The i-f spectrum of the conical-scan radar is shown in Fig. 9.7(a). The signal power is divided into a number of discrete lines separated by the repetition rate f_r and extending out to the limits of the transmission bandwidth $B_t = 1/\tau$. The noise power is distributed across the entire receiver bandwidth B with an average spectral density $N_o = N/B$. The detail of the spectrum near one of the signal lines is shown in Fig. 9.7(b). The sidebands

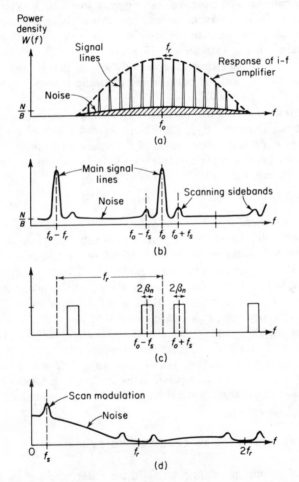

Figure 9.7 Signal and noise spectra for conical scan. (a) Conical scan i-f spectrum. (b) Detail of conical scan i-f spectrum. (c) Passbands of tracking system. (d) Audio spectrum after detection and boxcar stretching.

owing to scanning, for a target slightly off the axis, are shown at frequencies $\pm f_s$ from the main signal line. If we assume that this particular region of the spectrum is near the central line, then the signal power S is proportional to the area under the main spectral line, and the noise power is proportional to the area under the noise curve, in the interval from $f_o - f_r/2$ to $f_o + f_r/2$. The area under each sideband is proportional to $m^2 S/4$, where m is the fractional modulation resulting from a small error in tracking. The constant of proportionality will be the same for the signal and its sidebands, and for the case where $B = B_t = 1/\tau$ the same constant will also hold for noise.

The effective passband of the tracking system consists of a series of responses centered on each scanning sideband and extending over an interval β_n each side of the sideband frequency, where β_n represents the equivalent noise bandwidth of the servo system [Fig. 9.7(c)]. There are two such response bands for each main signal line in the spectrum, resulting in a servo noise output equal to $4\beta_n/f_r$ times the noise power in the i-f system. The effect of the second detector, range gate, and error detector on the spectrum is shown in Fig. 9.7(d). The a-c portion of this spectrum below $f_r/2$ becomes the input to the error demodulator. The demodulator separates both scan modulation and noise into traverse and elevation components to drive the two servo amplifiers, and as a result only half the noise power appears in each channel, along with whatever error component is appropriate. The noise power per servo channel can thus be given as $2\beta_n/f_r$ times the noise power in the i-f system.* The servo system will drive the antenna off axis far enough to balance each component of scan modulation against the noise power in each channel.

Balance of Modulation and Noise

The degree of modulation required to produce detected audio power equal to noise power has been derived for target detection applications.† The results are equally applicable to analysis of servo noise, so long as the separation of noise into the two servo channels is taken into account. The equation for modulation required to balance noise in one servo channel is

$$m_{tr} = m_c = \frac{1}{\sqrt{(S/N)(f_r/2\beta_n)}} \qquad \textbf{(9.3)}$$

---

* There are B/f_r main spectral lines within the receiver passband, and with each of these is associated a servo noise response bandwidth of $4\beta_n$. For a noise spectral density $N_o = N/B$, the total noise accepted by the servos is $(4\beta_n B/f_r)N_o = (4\beta_n/f_r)N$, where N is the total i-f noise power. This servo noise is divided equally between the two channels, as noted above, leading to a noise power of $(2\beta_n/f_r)N$ in each channel.

† James L. Lawson and George E. Uhlenbeck, *Threshold Signals* (New York: McGraw-Hill Book Company, 1950), pp. 278-88.

Since Eq. (9.2) gives the modulation in terms of error slope k_s and tracking error ϵ, the expression for rms tracking noise owing to thermal noise can be given for each tracking channel as $\sigma_t/\theta = m/k_s$, resulting in

$$\sigma_t = \frac{1.4\theta}{k_s\sqrt{(S/N)(f_r/\beta_n)}} \qquad \text{(matched system, } S/N > 4) \qquad \textbf{(9.4)}$$

The above equation is accurate to within 10 per cent for the conical-scan radar with S/N greater than 6 db (a factor of four), and for $B\tau = $ unity. When a wider bandwidth is used, with S/N still well above unity, the increased noise will appear in regions of the i-f spectrum beyond the signal lines, and will not affect the noise density N_o which is applied to the error detector. As a result, the effective value of S/N to be used in Eq. (9.4) will be $B\tau$ times the actual value of i-f S/N, leading to

$$\sigma_t = \frac{1.4\theta}{k_s\sqrt{B\tau(S/N)(f_r/\beta_n)}} \qquad (S/N > 4) \qquad \textbf{(9.5)}$$

In all the above equations, the signal-to-noise ratio is that which actually applies to the target on the tracking axis, or $1/L_k$ times the maximum attainable at the center of the beam. For a given target and antenna size, it is obvious that the "absolute" error slope given by $k_s/\sqrt{L_k}$, which represents fractional modulation of the potential maximum signal (at the center of the beam) is the proper measure of tracking efficiency on strong signals.

Tracking at Low S/N Ratio

As the S/N ratio drops to unity and below, three factors operate to modify Eq. (9.4). First, the agc action tends to hold constant the total receiver output power $S+N$, rather than S alone. This reduces the loop power gain of the servo by a factor $C_a = (S + N)/S$ relative to its normal value. In addition, the "small-signal suppression effect"* in the second detector causes a reduction in loop gain, as well as an increase in the ratio of detected-noise to signal power. This detector factor is approximately $C_d = (2S + N)/2S$, although it differs slightly depending upon the type of second detector used.† Lastly, the reduced servo loop gain causes a reduction in the bandwidth β_n and in the torque available for overcoming friction in the pedestal. The bandwidth and torque constants will both vary approximately as $1/\sqrt{C_a C_d}$. When these three effects are taken into account (for the moment, the nonlinear effect of friction being neglected), Eq. (9.5) can be rewritten in the following form:

---

* Wilbur B. Davenport, Jr. and William L. Root, *An Introduction to the Theory of Random Signals and Noise* (New York: McGraw-Hill Book Company, 1958), p. 267.
† Lawson and Uhlenbeck (1950), *op. cit.*, p. 283.

$$\sigma_t = \frac{1.4\theta}{k_s\sqrt{B_\tau(S/C_dN)(f_r/\beta_s)}}$$ (9.6)

where β_s is the actual noise bandwidth of the servo for the operating conditions encountered.* Assuming the usual type of servo transfer function (see Fig. 9.15), with low friction, we shall find that $\beta_s = \beta_n/\sqrt{C_aC_d}$, where β_n is the nominal bandwidth, measured at high S/N ratio. The result for tracking noise, in terms of S/N and nominal bandwidth, is

$$\sigma_t = \frac{1.4\theta \sqrt[4]{C_d/C_a}}{k_s\sqrt{B_\tau(S/N)(f_r/\beta_n)}}$$ (9.7)

The factors C_d and C_a are plotted in Fig. 9.8, along with the ratio C_d/C_a

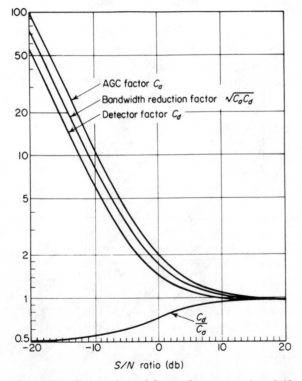

Figure 9.8 Factors determining performance at low S/N.

---

* David K. Barton, "Tracking Radars," Lecture 25 of special summer course, University of Pennsylvania (June 1961), to be published.

Jean A. Develet, Jr., "Thermal-Noise Errors in Simultaneous-Lobing and Conical-Scan Angle-Tracking Systems," *Trans. IRE*, Vol. **SET-7**, No. 2 (June 1961), Eq. 39.

and the bandwidth reduction factor $\sqrt{C_a C_d}$. It should be noted that these factors are dependent upon the actual S/N ratio, and that the performance of the system is degraded when the receiver bandwidth B is greater than the optimum value. The additional noise energy which appears at frequencies beyond the signal spectrum does cause an increase in the spectral density of noise following the envelope detector, and hence increases the noise power applied to the error detector. Although Eq. (9.7) shows little change in σ_t relative to the strong-signal case [Eq. (9.5)], the actual effect on tracking performance is pronounced as S/N goes below unity. The rapid reduction in loop gain causes the system to develop errors on moving targets, and the steady increase in σ_t with reduced bandwidth implies a very rapid increase in the spectral density of tracking noise. This spectral density, which is important in determining the amount of error following smoothing or processing of the radar data, is expressed in (units of angle)^2 per cps, and can be calculated from

$$W(f) = \frac{2\theta^2 C_d}{k_s^2(S/N)f_r} \qquad 0 < f < \beta_s \qquad \textbf{(9.8)}$$

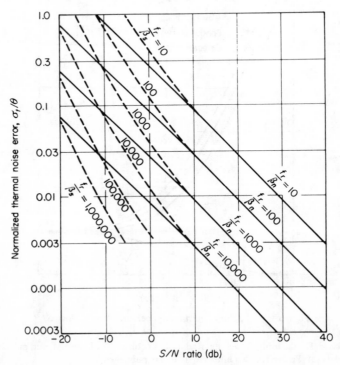

Figure 9.9 Normalized tracking error vs. S/N ratio for conical-scan radar ($k_s = 1.5$).

For low values of S/N, the detector factor C_d approaches the ratio $N/2S$, so the tracking noise power density will vary inversely as the square of S/N.

Figure 9.9 shows the normalized tracking error σ_t/θ for various values of f_r/β_n, plotted as a function of S/N between $+40$ and -20 db for a matched system ($B\tau = 1$). The actual servo bandwidth β_s can also be determined from this figure by noting the position of the error curve with respect to the dashed lines of constant f_r/β_s. In practice, the radar will usually lose track as a result of target motion, servo unbalance, or low-frequency oscillations (induced by the reduced gain of the servo) before the S/N ratio reaches -10 db.

Thermal Noise in Monopulse Tracking

The analysis of the amplitude comparison monopulse system is similar to that of conical scan, but with three differences. First, although the error slope for the patterns of Fig. 9.6 is only slightly higher than that of the conical-scan system (1.57 as compared to 1.5), there is no crossover loss in the monopulse case. Thus, the value of S/N to be used will be that which applies to the axis of the beam, with full antenna gain. Second, the tracking system passband consists of a single response, of width $2\beta_n$, centered at each line of the signal spectrum, rather than the pair of responses at $\pm f_s$ (see Fig. 9.10, as compared to Fig. 9.7). As a result of this, the noise power input to each servo is only half as great as in conical scan, for equal i-f S/N ratio and servo bandwidth. The phase-sensitive error detector in the monopulse system performs in much the same manner as the error demodulator of the conical-scan system, passing only the in-phase component of noise and reducing the noise power by a factor of two. Lastly, in the low-signal case for monopulse (reference-channel S/N near or below unity), the suppression of signal by noise will be governed by the characteristics of the phase-sensitive detector with noisy reference input. This will reduce the gain and sensitivity somewhat more rapidly, for the same S/N ratio, than they were reduced in the conical-scan case.

In the calculation of thermal-noise error for a monopulse tracker, it is convenient to start with the analysis of signal and noise for a single received pulse. This will then be extended to the average of many pulses, as formed at the output of the low-pass filter provided by the servo. Initially, it will be assumed that the S/N ratio is high, that the system is matched ($B\tau = 1$), and that the target is on the antenna axis. The output of the error i-f channel will then consist of noise alone, and the action of the agc system will hold this noise level equal to that of the reference channel. The noise voltage can be expressed in terms of the signal voltage in the reference channel and the i-f S/N ratio: $E_n = E_r/\sqrt{S/N}$.* If it is assumed that the error detec-

---

* All voltages are given as rms values here.

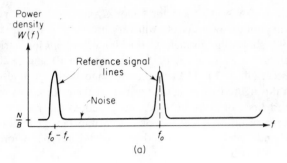

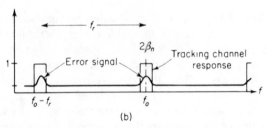

Figure 9.10 Signal and noise spectra for monopulse. (a) Reference channel i-f spectrum. (b) Error-channel and tracking passband.

tor is designed to have a gain of unity for in-phase components of signal and noise, the rms output to the servo amplifier will be $E_{nr} = E_r/\sqrt{2S/N}$. The motion required of the antenna to balance this noise voltage with signal output will be a function of the error slope k_m (Fig. 9.6) and can be given as

$$\sigma_1 = \frac{\theta}{k_m \sqrt{2S/N}} \quad \text{(unfiltered, } S/N > 4\text{)} \quad \textbf{(9.9a)}$$

where σ_1 refers to the rms error on a one-pulse basis. This expression is valid for all values of $B\tau$, if the peak value of the error detector output is used as the input to the servo amplifier. However, if the error detector output is averaged over the pulsewidth, after removal of adjacent range resolution elements in a range gate, the error can be reduced to that which would have been observed in the matched system. Thus, so long as the actual S/N ratio remains well above unity, the filtered error output of a monopulse error detector will be independent of i-f bandwidth.

$$\sigma_1 = \frac{\theta}{k_m \sqrt{2B\tau(S/N)}} \quad \text{(filtered, } B\tau > 1\text{)} \quad \textbf{(9.9b)}$$

When the effect of the servo is considered, the noise output power will be reduced by the factor $n = f_r/2\beta_n$, which represents the effective number of pulses integrated within the servo bandpass. The rms error at the antenna will then be

$$\sigma_t = \frac{\theta}{k_m\sqrt{(S/N)(f_r/\beta_n)}} \qquad \text{(unfiltered)} \qquad \textbf{(9.10a)}$$

Again, if an average is taken over the pulsewidth, an additional improvement can be had for $B\tau > 1$, given by

$$\sigma_t = \frac{\theta}{k_m\sqrt{B\tau(S/N)(f_r/\beta_n)}} \qquad \text{(filtered)} \qquad \textbf{(9.10b)}$$

Comparing Eq. (9.10) with Eq. (9.5), we can express the improvement achieved by monopulse over conical scan in terms of the decrease in required power (or receiver sensitivity) to obtain equivalent tracking accuracy. Using the error slope values $k_m = 1.57$ and $k_s = 1.5$, and considering the crossover loss for conical scan, we find that the advantage of the monopulse system over conical scan is

> 5.2 db, for echo tracking
> 6.3 db, for beacon tracking

The results of Eq. (9.9) may also be compared with those derived in Chapter 2 for the optimum angular accuracy of a measuring system. Manasse's result, applied to the circular aperture, was

$$\sigma_\theta = \frac{2\lambda}{\pi D\sqrt{\mathcal{R}}} = \frac{0.62\theta}{\sqrt{\mathcal{R}}} \qquad [\text{see Eq. (2.14)}]$$

The factor λ/D corresponds to 0.98 times the beamwidth of a circular aperture with uniform illumination.* The energy ratio $\mathcal{R}$ for a one-pulse measurement is equal to $2B\tau(S/N)$, leading to an optimum accuracy.

$$\sigma_\theta = \frac{0.44\theta}{\sqrt{B\tau(S/N)}}$$

Assuming $k_m = 1.57$, we see that Eq. (9.9b) gives essentially the same result, except that the factor in the numerator is 0.45 instead of 0.44. Of course, the beamwidth for a practical antenna will be broader than 1.02 (λ/D), and the error will be correspondingly higher. However, the mono-

---

* Samuel Silver, *Microwave Antenna Theory and Design* (New York: McGraw-Hill Book Company, 1949), p. 195.

pulse radar is seen to represent a close approach to the optimum angle measurement device for a given aperture.

Effect of Low S/N on Monopulse

The degradation in performance at low values of reference channel S/N is much the same for monopulse radar as for conical scan. The agc system reduces the power gain of the servo loop to signals by the factor $C_a = (S + N)/S$, shown in Fig. 9.8. The monopulse error detector exhibits a small-signal suppression effect similar to that in the envelope detector, causing further reduction in signal loop gain and additional degradation in the signal-to-noise ratio of the servo channel. An analysis of this effect in the coherent or product detector, such as used as the error detector in this system, has been carried out.* For input signal-to-noise ratios represented by x_1 and x_2, with independent noise sources, the output signal-to-noise ratio x_o is

$$x_o = \frac{2x_1 x_2}{1 + x_1 + x_2} \tag{9.11}$$

If we let x_2 be the value of S/N for the the reference channel, and consider targets near the tracking axis ($x_2 \gg x_1$), Eq. (9.11) reduces to

$$x_o = \frac{2x_1 x_2}{1 + x_2} = \frac{2x_1}{C_a} \tag{9.12}$$

Thus the error detector, which doubles the error channel signal-to-noise ratio for high values of reference S/N, is degraded by the factor C_a for low values of S/N.

The combined effects of agc and detector-loss factors in monopulse result in reduction of servo bandwidth to a value $\beta_s = \beta_n/C_a$, where again the bandwidth β_n represents the nominal value measured at high S/N. Equation (9.10a) or (9.10b) can, therefore, be used for all values of S/N, if it is kept in mind that β_n will not represent the actual tracking bandwidth for S/N ratios below about 6 db. If it is desired to express the tracking noise in terms of the actual bandwidth β_s, the equation becomes

$$\sigma_t = \frac{\theta}{k_m \sqrt{B\tau (S/C_a N)(f_r/\beta_s)}} \tag{9.13}$$

---

* T. W. R. East, "The Coherent Detector with Noisy Reference Input," McGill University Physics Dept. Report No. 563, Contract P 69-8-442 DRB (1955), Astia Document AD 90, 783.

Jean A. Develet, Jr. (June 1961), *op. cit.*, Eq. 24.

As in the strong-signal case, use of a filter following the error detector can reduce the noise power by the factor $B\tau$. However, the detector loss must first be calculated by using the actual i-f S/N, so the potential performance is not recovered by such filtering. Figure 9.11 shows the results in terms of both nominal bandwidth β_n and actual bandwidth β_s for a matched system ($B\tau = 1$). In a given system, the bandwidth β_s may be held constant at low S/N only by increasing the servo loop gain in accordance with signal strength, some means other than agc being used. This is not generally possible, unless the target is known to follow a predictable course with constant cross section or power output.

In using these equations and making comparisons with conical-scan radar, two points should be kept in mind. First, the value of S/N to be used is that obtained in the reference channel during tracking. It is to be expected that the error channel will have a signal-to-noise ratio below unity at all times, just as the modulation sidebands in the conical-scan system are normally well below the noise level of the i-f amplifier. Second, the equations show an advantage for monopulse systems which exceeds the crossover loss of the conical-scan system, if equal antenna gain is assumed at the center of the beam. This is because Eq. (9.5) uses the actual value of S/N obtained with the target on the tracking axis, rather than the potential maximum value which would be obtained at the center of the beam. The values of C_a and C_d must be calculated for the actual S/N in each case, and as a result the conical-scan system will lose performance more rapidly than the monopulse system using the same antenna size and power. If, however, the monopulse system is operated at a lower power output to obtain equal performance on strong signals, its detector factor will cause a more rapid deterioration for the weak-signal case.

Extension to C-W Systems

For the pulsed radar case, the bandwidth-reduction factor used to account for the integration of pulses in the servo system is given by $n = f_r/2\beta_n$. The S/N ratio in Eqs. (9.3) through (9.13) is calculated on the basis of peak power P_t and i-f bandwidth B. Peak power is multiplied by the duty cycle of the transmission ($D_u = \tau f_r$) to find average power.

$$P_{\mathrm{av}} = P_t \tau f_r = \frac{P_t f_r}{B_t}$$

In either pulsed or c-w tracking systems, the thermal-noise error may be expressed in terms of the average signal-to-noise ratio, and the factor B_t/β_n may be substituted in Eqs. (9.3) through (9.13) in place of f_r/β_n. Thus, for the monopulse tracker, Eq. (9.10b) could be written

$$\sigma_t = \frac{\theta}{k_m \sqrt{(S/N)_{\mathrm{av}}(B/\beta_n)}} \qquad (9.14)$$

where $(S/N)_{\mathrm{av}}$ represents the average signal-to-noise ratio measured over the entire repetition period or over any period long enough to average the effects of modulation in a c-w system. The result is equivalent to using, in the denominator, the error slope times the square root of an "integrated signal-to-noise ratio" based on the signal and noise power within the servo bandwidth β_n.* It is also equivalent to using the energy ratio $\mathcal{R}$ evaluated over an observation time $t_o = 1/2\beta_n$. The tracking performance is thus shown to be independent of the type of signal modulation and the pre-detection bandwidth, except for the loss factors introduced by the detector at low signal-to-noise ratio. These losses, however, can be calculated only on the basis of the actual signal-to-noise ratio at the second detector or error detector, so the use of the equations containing i-f S/N is to be recommended.

Figures 9.9 and 9.11 show the potential advantage of using a coherent system for angle tracking (see Chapter 12). With a noise-free local reference signal, locked to the received signal in frequency or phase, the system bandwidth may be reduced from B to a figure near $2\beta_n$ prior to the introduction of nonlinear detectors. Small-signal suppression will thereby be avoided at values of i-f S/N well below unity. The noise will increase the tracking error along the solid curves of these figures, but the nominal servo bandwidth will be maintained to lower S/N values. Only when the signal at the second detector nears the level of noise within $2\beta_n$ will there be small-signal suppression.

9.3 RESPONSE TO TARGET NOISE

Target noise in angle tracking is the result of two different but closely related effects: angle noise, or "glint," and amplitude noise, or "scintillation" (see Section 3.3). The existence of these two effects was recognized during World War II, and led to the development of monopulse techniques

---

* A more direct derivation for the c-w case may be based upon the assumption that the ratio $(S/N)_{\mathrm{av}}$ is high at the detector, so that cross products of noise components are not formed. For an i-f noise spectral density $N_o = N/B$, the servo system will then accept noise power equal to $2\beta_n N_o$. As before, this noise is shared by the two tracking channels, so that each channel has a noise power $\beta_n N_o$, as compared to $B N_o$ for total i-f noise. The improvement owing to bandwidth reduction and phase-sensitive error detection is, therefore, equal to the ratio B/β_n, and the average i-f signal-to-noise ratio may be multiplied by this ratio, as in Eq. (9.14), when the thermal-noise error of a single tracking channel (elevation or traverse) is to be found.

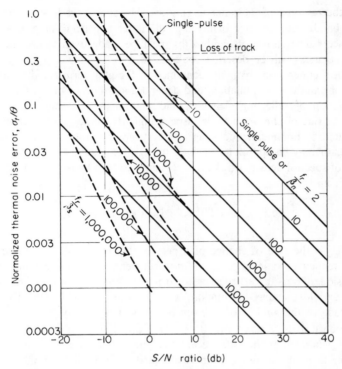

Figure 9.11 Normalized tracking error vs. S/N ratio for monopulse radar ($k_m = 1.57$).

as a means of eliminating the scintillation component. The material discussed in Chapter 3 will be applied here to the two types of angle tracker: conical scan and monopulse. Much of this discussion is based on the work of Dunn and Howard and their associates at Naval Research Laboratory.

Glint or Angle Noise

It was shown in Chapter 3 that the glint error may have peak values greater than the physical extent of the target itself, and that a tracking system with unlimited bandwidth and dynamic range would have an infinite rms error when tracking certain types of target. The special case of the two-element (dumbbell) target was used as an illustration. When observed with microwave radars of conventional design, most complex targets are found to have an rms glint error in the order of one-sixth to one-third of the total target span, measured either in terms of linear distance at the target or of angle subtended at the radar. The factors which determine the response of the radar to glint are the radar wave length, aperture, servo bandwidth, and agc bandwidth, the latter two being of primary importance.* The choice of bandwidths must be made to minimize the total radar error, as will be described below.

The frequency components contained in the glint term are the same as those of amplitude scintillation, and depend upon the size of the target, the distribution of reflecting elements, and the radar frequency, as well as upon the rate of change of target aspect angle. If we neglect the harmonics generated by nonlinear processing in the radar, the power spectrum of the glint error may be approximated by a triangle, extending from a maximum value of spectral density W_o at zero frequency to zero density at a frequency f_{max} given by

$$f_{max} = \frac{2\omega_a L}{\lambda} \quad \text{[see Eq. (3.8)]}$$

Here, ω_a is the rate of change of aspect angle, in rad per sec, caused by target maneuver or change in radar-target geometry; L is the maximum dimension of the target, measured normal to the direction of the radar beam and to the axis of rotation; λ is the radar wave length. The rms amplitude of the glint error is given by the square root of the area under the spectral density curve, or $\sqrt{W_o f_{max}/2}$. Considering the presence of harmonics generated by the radar receiver response to sharp nulls in the signal, a better fit to the glint spectrum is obtained by using a Markoffian curve with the same value of W_o and with a half-power frequency f_g equal to f_{max} of the previous equation.

$$W(f) = W_o \frac{f_g^2}{f_g^2 + f^2} \tag{9.15}$$

In this case, the amplitude of the glint error is given by $\sigma_g = \sqrt{\pi W_o f_g/2}$. The density W may be expressed in such units as mils^2/cps or ft^2 per cps, corresponding to glint errors σ_g in mils angle subtended at the radar or feet measured across the target. The latter units are generally preferred in descriptions of target characteristics, since they lead to values which are largely independent of radar range and antenna parameters, over a broad set of operating conditions. When aircraft are tracked in stable flight, values of f_g in the order of a few cps are generally measured by microwave radars. The greater stability and lower values of ω_a for large aircraft tend to cancel the effect of their larger dimensions, leading to glint spectra which have

* Richard H. Delano and Irwin Pfeffer, "The Effect of AGC on Radar Tracking Noise," *Proc. IRE*, Vol. **44**, No. 6 (June 1956), pp. 801–10.

John H. Dunn and Dean D. Howard, "The Effects of Automatic Gain Control Performance on the Tracking Accuracy of Monopulse Radar Systems," *Proc. IRE*, Vol. **47**, No. 3 (March 1959), pp. 430–35.

the same form for large and small aircraft. The difference between the two lies principally in the value of W_o.

Scintillation or Amplitude Noise

The amplitude modulation of the returned echo signal produces two separate effects on tracking radars. First, in sequential lobing systems, it produces a direct noise input to the servo, within the passband which is centered at the scan frequency f_s. This causes an error in tracking, as the servo will force the antenna off target far enough to balance the input to each servo channel. This effect is entirely absent from monopulse radar, which has no carrier in the error channel to be modulated in this way. The second effect is common to all tracking systems, and is caused by modulation of the signal in the low-frequency region (within the servo bandwidth β_n), which brings about a modulation in loop gain of the servo. If there is any component of lag error owing to target velocity or acceleration (see Section 9.4), this component will be modulated by the amplitude fluctuation of the target, adding noise to the radar output data.* Virtually all tracking systems use agc to reduce this type of error, and the monopulse systems are able to use agc loops with high gain and wide bandwidth to suppress the error almost entirely. The response of the agc loop used in conical scan is limited by the necessity of preserving modulation components at the scan frequency, and as a result most conical-scan systems are more susceptible to the second type of error than are monopulse systems.

The analysis of direct scintillation error for conical scan follows closely the method used earlier for determining the response to thermal noise. The fractional modulation due to scintillation within the servo passband at f_s is given by

$$m_s = \sqrt{2W(f_s)\beta_n}$$

where $W(f_s)$ is the power spectral density of video modulation near f_s, expressed in units of (fractional modulation)^2/cps. If it is assumed, as before, that the power is divided equally between the elevation and traverse channels, the resulting scintillation error σ_s required to balance each servo channel is[†]

---

* John H. Dunn, Dean D. Howard, and A. M. King, "The Phenomena of Scintillation Noise in Radar Tracking Systems," *Proc. IRE*, Vol. 47, No. 5 (May 1959), p. 858.

† Dunn, Howard, and King refer to earlier work by Meade, Hastings, and Gerwin as the source of an error equation for this term: $\sigma_{amp} = 0.85BA_{amp}\sqrt{\beta}$. Here, B is the two-way beamwidth of the antenna ($= 0.72\theta$), A is the amplitude spectral density $[= \sqrt{W(f_s)}]$, and β is servo bandwidth as used above. With these substitutions, their equation reduces to $\sigma = 0.61\theta\sqrt{W(f_s)\beta_n}$. The difference in constant is due to a higher assumed value for k_s in the NRL work, obtained at the expense of greater loss L_k.

$$\sigma_s = \frac{m_s \theta}{k_s \sqrt{2}} = \frac{\theta \sqrt{W(f_s) \beta_n}}{k_s} \cong 0.67 \theta \sqrt{W(f_s) \beta_n} \qquad \text{(9.16)}$$

The scintillation spectrum may be assumed to be identical with the glint spectrum [Eq. (9.15)], since they both result from the same interference phenomena at the target. The total fractional modulation at all frequencies is typically around 0.5 rms (a value of 0.52 rms would apply to the Rayleigh target). Using the Markoffian model of the spectrum, and setting the total fractional modulation equal to the square root of the area under this spectrum, or $\sqrt{\pi f_g W_o / 2}$, the value of W_o is found to be given by $1/(2\pi f_g)$. If the scan frequency f_s is well above f_g, the value of $W(f_s)$ will be approximately $W_o f_g^2 / f_s^2$, or $f_g/(2\pi f_s^2)$. The resulting tracking error caused by scintillation is

$$\sigma_s = 0.67 \theta \sqrt{\frac{f_g \beta_n}{2\pi f_s^2}} = \frac{0.27 \theta}{f_s} \sqrt{f_g \beta_n} \qquad \text{(9.17)}$$

For example, if $f_g = 5$ cps, $f_s = 30$ cps, $\beta_n = 2$ cps, and $\theta = 40$ mils, the error would be 1.1 mils rms. This is consistent with the analysis of tracking accuracy of the SCR-584,* and with many practical observations on similar systems. The possibility of reduction in scintillation error by using higher scan rates was recognized during World War II, and has been exploited in postwar radar designs.† The benefits of high lobing rate are limited, however, by the presence of modulation components at frequencies as high as several hundred cps, owing to propeller modulation, aircraft vibration, and similar causes.‡ These spectral components invalidate the Markoffian model used in the above analysis, for cases where the scan rate is more than a few octaves above f_g. It should be noted that the agc characteristics of the radar will not have any appreciable effect on this component of scintillation error, even if a wide-band loop is used, since both the scintillation and the scan modulation will be reduced by the same factor.

Effect of agc on Target Noise

The second component of scintillation error will be governed largely by the characteristics of the agc system, and by the magnitude of dynamic lags or other bias errors present in the tracking system. Let it be assumed

---

* Hubert M. James, Nathanial B. Nichols, and Ralph S. Phillips, *Theory of Servomechanisms* (New York: McGraw-Hill Book Company, 1947), p. 296.

† John H. Dunn, Dean D. Howard, A. M. King, *op. cit.*, p. 856.

‡ James L. Lawson and George E. Uhlenbeck, *Threshold Signals* (New York: McGraw-Hill Book Company, 1950), p. 288.

that the servo is balanced, at a given time, with a fixed or bias component of tracking error given by ϵ_o. This could be the result of the requirement for driving the antenna at a constant rate, of the need for torque to overcome wind or friction, or of the presence of electrical or mechanical unbalance in the servo loop. The presence of signal modulation within the servo bandwidth will cause this bias to be modulated inversely as the signal amplitude, so that the instantaneous value of error, for small modulation m, will be

$$\epsilon(t) = \frac{\epsilon_o}{1 + m \sin \omega_m t} \simeq \epsilon_o(1 - m \sin \omega_m t) \qquad \textbf{(9.18)}$$

The noise component of this error is thus given by the sinusoid $-\epsilon_o m \sin \omega_m t$, with an rms value

$$\sigma_{s_2} = \frac{\epsilon_o m}{\sqrt{2}}$$

Modulation on the returned signal will be suppressed by the action of agc. The factor by which the modulation is reduced may be written

$$\frac{m}{m'} = |1 + Y_a(f_m)|$$

where m' is the remaining modulation, and $Y_a(f_m)$ is the open-loop voltage gain transfer function of the agc loop at the frequency $f_m = \omega_m/2\pi$. Expressing the gain of the servo by its closed-loop transfer function Y_c, we arrive at a "scintillation error factor" Y_s, which describes the response of the system to scintillation at any frequency other than those within the scan-rate region.

$$Y_s = \frac{Y_c}{|1 + Y_a|}$$

Curves for this function, in a typical case, are shown in Figs. 9.12 and 9.13, which apply to a monopulse radar with fast agc and to a system with slow agc. The amplitude of scintillation error at any low frequency may be found as

$$\epsilon_n(f_m) = Y_s(f_m)\epsilon_o m \sin(\omega_m t + \phi) \qquad \textbf{(9.19)}$$

If the power spectrum of scintillation $W(f)$ is known, the spectrum of the scintillation error $W_n(f)$ will be

$$W_n(f) = W(f)Y_s^2(f)\epsilon_o^2 \qquad (9.20)$$

The total error σ_{st} will then be found by integrating $W_n(f)$ from zero frequency to infinity, adding the square of the direct component of scintillation error [from Eq. (9.16) or Eq. (9.17)], and taking the square root of the sum. The integration may be carried out simply when f_g is considerably greater or less than the servo bandwidth β_n, and graphical procedures are usually adequate for any intermediate case.

As an example of application of this analysis, the scintillation error will be calculated for the Markoffian model of spectrum used earlier, a Rayleigh distribution of echo signal amplitudes, a servo bandwidth of five cps, and three possible agc characteristics being assumed. The normalized scintillation spectrum, the error factor Y_s^2 for each agc model, and the resulting tracking error spectra are shown in Fig. 9.14. The zero-frequency density W_o for the Rayleigh distribution is $1/(2\pi f_g)$. The square of the tracking error is found by multiplying the area under the normalized error spectrum by this quantity and then by the square of the bias error ϵ_o. For a bias error equal to the target span L, the rms errors will be $0.006L$ for the fast-agc case, $0.14L$ for the slow-agc case, and $0.77L$ without agc (plus

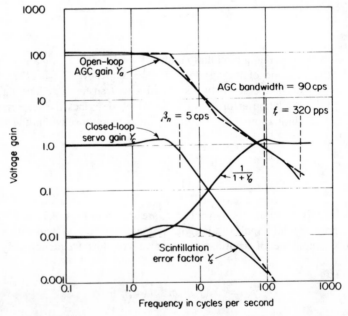

Figure 9.12 Factors determining extent of noise modulation of servo lag error (for monopulse).

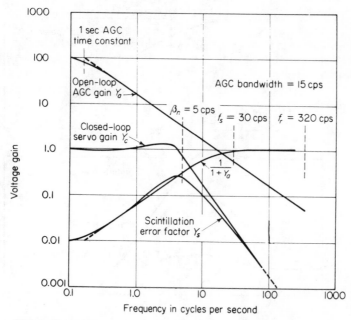

Figure 9.13 Factors determining extent of noise modulation of servo lag errors (for conical scan).

any direct scintillation error if sequential lobing is used). These errors will increase linearly with ϵ_o, and will greatly exceed the glint error when ϵ_o is greater than L, unless fast agc is used. In a narrow-beam, conical-scan system, they may also exceed the direct scintillation error σ_s. For this reason, the performance of some conical-scan systems may be improved by using a properly equalized, fast-agc loop, similar to that shown in Fig. 9.12 for the monopulse system, with a high degree of modulation suppression over the entire servo bandwidth. The reduction in sensitivity to scan modulation may be offset by more gain in the servo amplifier, without changing the thermal noise response or other characteristics. Actual simulations of tracking loops with varying agc bandwidths* give results consistent with this analysis.

9.4 RESPONSE TO TARGET MOTION

The preceding sections have been concerned with the response of tracking systems to receiver noise and to noise generated by wave interference pheno-

---

* John H. Dunn and Dean D. Howard, "The Effects of Automatic Gain Control Performance on the Tracking Accuracy of Monopulse Radar Systems," *Proc. IRE*, Vol. 47, No. 3 (March 1959), pp. 430–35.

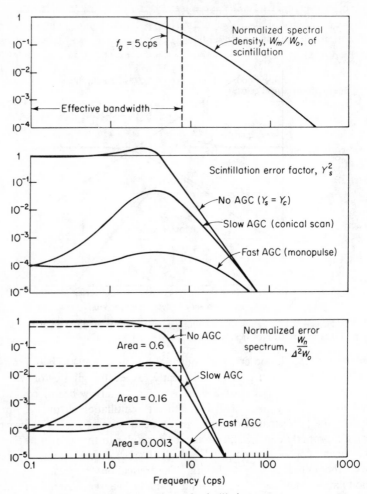

Figure 9.14 Comparison of scintillation errors.

mena at the target. In both cases, the noise was found to be widely distributed in the frequency spectrum, and its effect on the tracking data was
increased by the use of wide servo bandwidth. In order to evaluate fully
the response of the tracker to the noise terms, it is necessary to consider
also those factors which dictate the choice of servo bandwidth in the practical tracking system. Without going into the details of servo design, this
section will establish the relationships between servo bandwidth and dynamic lag errors for various types of target motion, and will provide simple
procedures for choosing an appropriate value of bandwidth for any given

situation. For the present, we shall consider that the error in radar data output is the criterion of system performance. In a later chapter, the effect of smoothing on the output data will be considered, and these results will be modified.

Servo Error Coefficients

In a stable servo system, the steady-state response to an arbitrary input function $\theta_i(t)$ may be expressed in terms of a power series, which gives the error $\epsilon(t) = \theta_i(t) - \theta_o(t)$ in terms of the input function, its several derivatives, and the servo "error coefficients" or "error constants."* If initial transients are neglected,† the error is

$$\epsilon(t) = \frac{\theta_i}{K_o} + \frac{\dot{\theta}_i}{K_v} + \frac{\ddot{\theta}_i}{K_a} + \frac{\dddot{\theta}_i}{K_3} + \cdots \tag{9.21}$$

The time derivatives of the input function are represented by the dotted symbols: $\dot{\theta}_i = d\theta_i/dt$, etc. The constants K_o, K_v, etc. represent the error constants for position, velocity, acceleration, and higher derivatives. Methods of calculating the several significant error constants for systems of different sorts are discussed in the literature.‡ One convenient approach is based upon the use of the asymptotic open-loop response curve, or Bode diagram, of which a typical example is shown in Fig. 9.15. This response represents the magnitude of the open-loop gain function $Y_{11} = \theta_o/\epsilon$ plotted as a function of radian frequency ω. Two common types of servo equalization are shown, both characterized by four response regions of different slopes, separated by three "break frequencies" ω_1, ω_2, and ω_3. In the first case, the slopes of the four regions are -20, -40, -20, and -40 db per decade of frequency, starting from the low-frequency end of the plot. This type of response is used to assure servo stability by placing the "crossover frequency" ω_c (at which $|Y_{11}|$ is unity) near the center of the four-octave region which extends from ω_2 to ω_3 with a slope of -20 db/decade. The second example of response shows a steeper slope between ω_1 and ω_2 to

* W. P. Manger, "General Design Principles for Servomechanisms," Sects. 4.2–4.9 of *Theory of Servomechanisms*, Hubert M. James, Nathaniel B. Nichols, and Ralph S. Phillips, eds. (New York: McGraw-Hill Book Company, 1947), p. 147.

† John G. Truxal, *Automatic Feedback Control System Synthesis* (New York: McGraw-Hill Book Company, 1955). See pp. 80-84 for a discussion of the limitations in applying error coefficients.

‡ W. P. Manger, *op. cit.*

Leonard H. King, "Reduction of Forced Error in Closed-Loop Systems," *Proc. IRE*, Vol. 41, No. 8 (Aug. 1953), p. 1037.

Sidney Shucker, "Error Coefficients Ease Servo Response Analysis," *Control Engineering*, Vol. 10, No. 5 (May 1963), pp. 119-23.

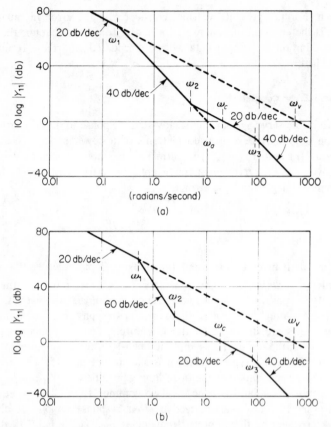

Figure 9.15 Open-loop frequency response of typical servos. (a) First-order servo with single integration. (b) First-order servo with double integration.

achieve a more rapid buildup of gain in the low-frequency region, with corresponding reduction in lag error. In both cases, ω_3 is limited in value by the mechanical response of the pedestal and servo motor, and this places an upper limit on the closed-loop bandwidth β_n which can be used with a given type of antenna. The object of servo equalization is to maintain stability and at the same time to achieve the necessary high values of error constants, to minimize lag errors.

Manger shows that the value of K_v may be determined graphically from the Bode diagram by extending the low-frequency segment of the plot at −20 db/decade until it intersects the 0 db gain axis at $\omega_v = K_v$ rad per sec. Also, for the type of response shown in Fig. 9.15(a), an approximate value of K_a is found by extending the −40 db/decade slope beyond ω_2

until it intersects the axis at $\omega_a = \sqrt{K_a}$ rad per sec. Shucker has prepared tables from which the first three finite error coefficients may be calculated in terms of the equalizer time constants $t_1 = 1/\omega_1$, etc., and the gain setting which establishes K_v. His formulas for first-order servos (in which the slope at zero frequency is -20 db per decade) are shown in Table 9.1. The results are in agreement with those of Manger for K_v, and show that the approximation $K_a = \omega_a^2$ is quite good for the servo whose slope in the second region is -40 db/decade. In Shucker's formula for the response with slopes of -20, -40, -20, and -40, the same approximation can be derived if $t_1 \gg t_2 \gg t_3$ and $K_v \gg 1$.

$$K_a = \frac{K^2}{K(t_1 + t_3 - t_2)} \cong \frac{K}{t_1} = \omega_1 \omega_v = \omega_a^2$$

For the same case, the third error constant may be found approximately as $-K/t_1 t_2 = -K_a \omega_2$. This result differs from that derived by Manger, as does the exact expression in Table 9.1, owing to an error in Eq. (66) on p. 149 of Manger.

For the system response of Fig. 9.15(a), the significant time constants, frequencies, and error constants are

$$
\begin{array}{ll}
\omega_1 = 0.2 \text{ rad per sec} & t_1 = 5 \text{ sec} \\
\omega_2 = 5 \text{ rad per sec} & t_2 = 0.2 \text{ sec} \\
\omega_3 = 80 \text{ rad per sec} & t_3 = 0.0125 \text{ sec} \\
\omega_v = 500 \text{ rad per sec} & \omega_a = 10 \text{ rad per sec} \\
K_v = 500 \text{ per sec} & K_a = 100 \text{ per sec}^2 \\
K_3 = -500 \text{ per sec}^3 & \beta_n = 6 \text{ cps}
\end{array}
$$

For the response of Fig. 9.15(b), time constants were chosen to obtain the same K_v and β_n as above, but the slope is steeper between ω_1 and ω_2. The following values apply to this case

$$
\begin{array}{ll}
\omega_1 = 0.5 \text{ rad per sec} & t_1 = 2 \text{ sec} \\
\omega_2 = 2.5 \text{ rad per sec} & t_2 = 0.4 \text{ sec} \\
\omega_3 = 80 \text{ rad per sec} & t_3 = 0.0125 \text{ sec} \\
K_v = 500 \text{ per sec} & K_a = 156 \text{ per sec}^2 \\
K_3 = 385 \text{ per sec}^3 & \beta_n = 6 \text{ cps}
\end{array}
$$

The value of K_a has been increased at some expense in K_3, as compared to the first case. It should be noted that the region of -20 db per decade slope between ω_2 and ω_3 was made somewhat longer in the second case, to make the phase margins approximately equal and to provide a true comparison, with equal bandwidths and stabilities.

Table 9.1 ERROR CONSTANTS FOR FIRST-ORDER SERVO

System slopes	Open-loop transfer function Y_{11}	K_v	K_a	K_3
$-20, -20,$ $-20, -20$	$\dfrac{K}{p}$	K	$-K^2$	K^3
$-20, -20,$ $-20, -40$	$\dfrac{K}{p(pt_3+1)}$	K	$\dfrac{K^2}{Kt_3-1}$	$\dfrac{-K^3}{2Kt_3-1}$
$-20, -20,$ $-20, -60$	$\dfrac{K}{p(pt_3+1)^2}$	K	$\dfrac{K^2}{2Kt_3-1}$	$\dfrac{K^3}{K^2t_3^2-4Kt_3+1}$
$-20, -40,$ $-20, -20$	$\dfrac{K(pt_2+1)}{p(pt_1+1)}$	K	$\dfrac{K^2}{K(t_1-t_2)-1}$	$\dfrac{-K^3}{K^2t_2(t_1-t_2)+2K(t_1-t_2)-1}$
$-20, -40,$ $-20, -40$	$\dfrac{K(pt_2+1)}{p(pt_1+1)(pt_3+1)}$	K	$\dfrac{K^2}{K(t_1+t_3-t_2)-1}$	$\dfrac{K^3}{K^2(t_2^2-t_1t_2-t_2t_3+t_1t_3)-2K(t_1+t_3-t_2)+1}$
$-20, -40,$ $-20, -60$	$\dfrac{K(pt_2+1)}{p(pt_1+1)(pt_3+1)^2}$	K	$\dfrac{K^2}{K(t_1+2t_3-t_2)-1}$	$\dfrac{K^3}{K^2(t_2^2-t_1t_2-2t_2t_3+2t_1t_3+t_3^2)-2K(t_1+2t_3-t_2)+1}$
$-20, -60,$ $-20, -20$	$\dfrac{K(pt_2+1)^2}{p(pt_1+1)^2}$	K	$\dfrac{K^2}{2K(t_1-t_2)-1}$	$\dfrac{K^3}{K^2(3t_2^2-4t_1t_2+t_1^2)-4K(t_1-t_2)+1}$
$-20, -60,$ $-20, -40$	$\dfrac{K(pt_2+1)^2}{p(pt_1+1)^2(pt_3+1)}$	K	$\dfrac{K^3}{K(2t_1-2t_2+t_3)-1}$	$\dfrac{K^3}{K^2(3t_2^2+t_1^2-4t_1t_2-2t_2t_3+2t_1t_3)-2K(2t_1-2t_2+t_3)+1}$
$-20, -60,$ $-20, -60$	$\dfrac{K(pt_2+1)^2}{p(pt_1+1)^2(pt_3+1)^2}$	K	$\dfrac{K^2}{2K(t_1+t_3-t_2)-1}$	$\dfrac{K^3}{K^2(3t_2^2+t_1^2+t_3^2+4t_1t_3-4t_1t_2-4t_2t_3)-4K(t_1+t_3-t_2)+1}$

NOTE: K_0 is infinite for all cases.

Table 9.2 ERROR CONSTANTS FOR ZERO-ORDER SERVO

System slopes	Open-loop transfer function Y_{11}	K_o	K_v	K_a
$-0, -20, -20$	$\dfrac{K}{pt_1 + 1}$	$K+1$	$\dfrac{(K+1)^2}{Kt_1}$	$\dfrac{-(K+1)^3}{Kt_1^2}$
$0, -20, -40$	$\dfrac{K}{(pt_1+1)(pt_3+1)}$	$K+1$	$\dfrac{(K+1)^2}{K(t_1+t_3)}$	$\dfrac{(K+1)^3}{K^2 t_1 t_3 - K(t_1^2 + t_1 t_3 + t_3^2)}$
$0, -20, -60$	$\dfrac{K}{(pt_1+1)(pt_3+1)^2}$	$K+1$	$\dfrac{(K+1)^2}{K(2t_3+t_1)}$	$\dfrac{(K+1)^3}{K^2(2t_1 t_3 + t_3^2) - K(t_1^2 + t_1 t_3 + 3t_3^2)}$
$0, -40, -20$	$\dfrac{K(pt_2+1)}{(pt_1+1)^2}$	$K+1$	$\dfrac{(K+1)^2}{K(2t_1 - t_2)}$	$\dfrac{(K+1)^3}{K^2(t_1 - t_2)^2 - Kt_1(3t_1 - 2t_2)}$
$0, -40, -40$	$\dfrac{K(pt_2+1)}{(pt_1+1)^2(pt_3+1)}$	$K+1$	$\dfrac{(K+1)^2}{K(2t_1 + t_3 - t_2)}$	$\dfrac{(K+1)^3}{K^2(t_1^2 + 2t_1 t_3 - t_2 t_3 + t_2^2 - 2t_1 t_2) - K(3t_1^2 + 2t_1 t_3 + t_3^2 - 2t_1 t_2 - t_2 t_3)}$
$0, -40, -60$	$\dfrac{K(pt_2+1)}{(pt_1+1)^2(pt_3+1)^2}$	$K+1$	$\dfrac{(K+1)^2}{K(2t_1 + 2t_3 - t_2)}$	$\dfrac{(K+1)^3}{K^2(t_1^2 + 4t_1 t_3 + t_3^2 + t_2^2 - 2t_1 t_2 - 2t_2 t_3) - K(3t_1^2 + 4t_1 t_3 + 3t_3^2 - 2t_1 t_2 - 2t_2 t_3)}$

Table 9.3 ERROR CONSTANTS FOR SECOND-ORDER SERVO

System slopes	Open-loop transfer function Y_{11}	K_a	K_3	K_4
-40, -40, -20, -20	$\dfrac{K(pt_2 + 1)}{p^2}$	K	$\dfrac{-K}{t_2}$	$\dfrac{K^2}{Kt_2^2 - 1}$
-40, -40, -20, -40	$\dfrac{K(pt_2 + 1)}{p^2(pt_3 + 1)}$	K	$\dfrac{-K^2}{t_2 - t_3}$	$\dfrac{K^2}{Kt_2(t_2 - t_3) - 1}$
-40, -40, -20, -60	$\dfrac{K(pt_2 + 1)}{p^2(pt_3 + 1)^2}$	K	$\dfrac{K}{2t_3 - t_2}$	$\dfrac{K^2}{K(t_3^2 + t_2^2 - 2t_2t_3) - 1}$
-40, -60, -20, -20	$\dfrac{K(pt_2 + 1)^2}{p^2(pt_1 + 1)}$	K	$\dfrac{K}{t_1 - 2t_2}$	$\dfrac{K^2}{K(3t_2^2 - 2t_1t_2) - 1}$
-40, -60, -20, -40	$\dfrac{K(pt_2 + 1)^2}{p^2(pt_1 + 1)(pt_3 + 1)}$	K	$\dfrac{K}{t_1 + t_3 - 2t_2}$	$\dfrac{K^2}{K(3t_2^2 + t_1t_3 - 2t_1t_2 - 2t_2t_3) - 1}$
-40, -60, -20, -60	$\dfrac{K(pt_2 + 1)^2}{p^2(pt_1 + 1)(pt_3 + 1)^2}$	K	$\dfrac{K}{t_1 + 2t_3 - 2t_2}$	$\dfrac{K^2}{K(3t_2^2 + 2t_1t_3 + t_3^2 - 2t_1t_2 - 4t_2t_3) - 1}$

NOTE: K_o and K_v are infinite for all cases.

Shucker's results for the zero-order and second-order servo systems are given in Tables 9.2 and 9.3. Whereas the first-order system had an infinite position error constant K_o, the second-order system has infinite K_o and K_v. The zero-order system has finite values for all error constants. The value of K_a for the second-order system may be calculated graphically in the same way as for the first-order system: by extending the -40 db per decade slope to intersect the axis at $\omega_a = \sqrt{K_a}$.

Derivatives of Target Angle

The servo lag error in tracking a target is determined by the values of the error constants and by the real and apparent velocity, acceleration, and higher derivatives of target angle, viewed by the radar. If the target has a linear velocity v_t and acceleration a_t, the radar will see angular components given by

$$\omega_t = \frac{v_t}{R} \sin \alpha \quad \text{rad per sec}$$

and

$$\dot{\omega}_t = \frac{a_t}{R} \sin \beta \quad \text{rad per sec}^2$$

where ω_t is the real angular rate of the target, R is the target range, and α and β are the angles between the v_t and a_t vectors, respectively, and the radar beam. In addition to these rates, and higher derivatives due to $\dot{a}_t$, etc., which are limited by the maneuverability of the target, there are apparent or "geometrical" accelerations and higher derivatives, owing to the fact that the radar observes the target in a spherical coordinate system. Thus, even a target moving at a constant rate in a straight line will cause the angle servos to operate with varying rates, depending upon the location of the target with respect to the radar, and the direction of the velocity vector. These geometrical components have been discussed by several authors under the general heading of the "pass-course" problem.* Because the problem is of great importance in establishing the bandwidth of the tracker and in evaluating errors, the results of these earlier discussions will

---

* Nathaniel B. Nichols, "General Design Principles for Servomechanisms: Applications," Sects. 4.18 and 4.19 of *Theory of Servomechanisms*, Hubert M. James, Nathaniel B. Nichols, and Ralph S. Phillips, eds. (New York: McGraw-Hill Book Company, 1947), p. 229.

Charles F. White, "Servo System Theory," Chap. 7 of *Guidance*, A. S. Locke, ed. (Princeton, N.J.: D. Van Nostrand Company, Inc., 1955), p. 246.

Sidney Shucker, "Error Coefficients Ease Servo Response Analysis," *Control Engineering*, Vol. 10, No. 5 (May 1963), pp. 119-23.

be summarized and extended here to elevation tracking as well as to the azimuth tracking problem which they discuss.

The basic geometry of the pass-course problem is shown in Fig. 9.16. The analysis will first be carried out for the ground triangle, involving azimuth angle and ground range, for a target in level flight. The projected component of target velocity in the horizontal plane will be taken as v_t, and the minimum ground range (at "crossover") is R_c. The azimuth angle A will be taken at 90° at crossover, although the results can be applied to a target approaching from any azimuth. The target will be assumed to travel at constant velocity, and any real acceleration component may be added directly to the results of the geometrical calculations. Table 9.4 shows the equations for the first four derivatives of A, with their peak values and the

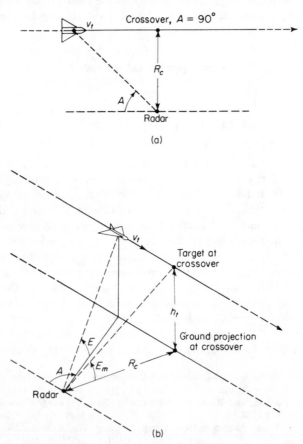

(a)

(b)

Figure 9.16 Geometry of pass-course problem. (a) Ground projection of flight path. (b) Pass course with target in level flight.

Table 9.4 DERIVATIVES OF AZIMUTH ANGLE FOR PASS COURSE

Derivative	Equation	Peak value	Peak azimuth
$\dot{A} = \omega_a$	$\omega_m \sin^2 A$	$\omega_m = \dfrac{v_t}{R_c}$	$90°$
$\dot{\omega}_a$	$2\omega_m^2 \sin^3 A \cos A$	$\pm \dfrac{3\sqrt{3}\,\omega_m^2}{8} = \pm 0.65\omega_m^2$	$60°, 120°$
$\ddot{\omega}_a$	$2\omega_m^3 \sin^4 A(4\cos^2 A - 1)$	$-2\omega_m^3$	$90°$
$\dddot{\omega}_a$	$24\omega_m^4 \sin^5 A \cos A \cos 2A$	$\mp 4.5\omega_m^4$	$72°, 108°$

azimuth angles at which the peak values occur. Figure 9.17 is a plot of the four derivatives, normalized with respect to $\dot{A}_{\max} = \omega_m$.

In elevation, the same equations may be used for the special case of a target which passes directly overhead, with the elevation angle E replacing A, the altitude h_t replacing R_c, and elevation tracking assumed to be continuous through $180°$. If the elevation coordinate is limited to $90°$ of travel by mechanical or electrical features, there will be a discontinuity at zenith, and tracking will be lost as the antenna attempts to slew through $180°$ in azimuth in zero time. For the more general case shown in Fig. 9.16(b), the elevation angle will increase to a maximum value at crossover, given by $E_m = \tan^{-1}(h_t/R_c)$, and then will decrease as the target continues beyond crossover. Table 9.5 and Fig. 9.18 give the resulting first three derivatives of elevation angle, as functions of the azimuth A, the peak azimuth rate ω_m, and the parameter $X = h_t/R_c$. The peak values of all derivatives of E

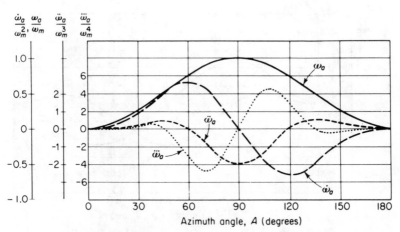

Figure 9.17 Derivatives of azimuth angle for pass course.

Table 9.5   DERIVATIVES OF ELEVATION FOR PASS COURSE
(CONSTANT VELOCITY TARGET IN LEVEL FLIGHT BEING ASSUMED)

First derivative	*Peak values*	*Peak A*
	$\pm 0.38 X\omega_m$	$55°$, $125°$ $(X \ll 1)$
$\dot{E} = \omega_e = X\omega_m \dfrac{\sin^2 A \cos A}{1 + X^2 \sin^2 A}$	$\pm 0.24\omega_m$	$48°$, $132°$ $(X = 1)$
	$\pm\omega_m/X$	$0°$, $180°$ $(X \gg 1)$

Second derivative

$$\dot{\omega}_e = X\omega_m^2 \sin^3 A \frac{(2 - 3\sin^2 A - X^2 \sin^4 A)}{(1 + X^2 \sin^2 A)^2}$$

	Peak values	*Peak A*	
	$-X\omega_m^2$	$90°$	$(X \ll 1)$
	$-0.5\omega_m^2$	$90°$	$(X = 1)$
	$-\omega_m^2/X$	$90°$	$(X \gg 1)$

Third derivative

$$\ddot{\omega}_e = X\omega_m^3 \sin^4 A \cos A \frac{6 - 15\sin^2 A - X^2 \sin^2 A(2 + 10\sin^2 A + 3X^2 \sin^4 A)}{(1 + X^2 \sin^2 A)^3}$$

	Peak values	*Peak A*	
	$\mp 1.96 X\omega_m^3$	$70°$, $110°$	$(X \ll 1)$
	$\mp 0.77\omega_m^3$	$67°$, $113°$	$(X = 1)$
	$\mp 0.85\omega_m^3/X$	$63°$, $117°$	$(X \gg 1)$

NOTE: $A =$ azimuth angle, $90°$ at crossover.
$X = h_t/R_c = \tan E_m$.
$X \cong 1$ results in maximum values of derivatives for given value of $\omega_m = v_t/R_c$.

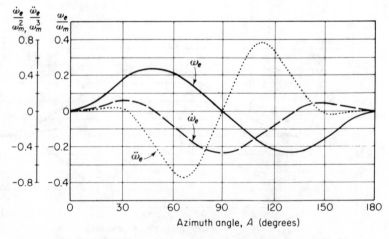

Figure 9.18   Derivatives of elevation angle for pass course $(X = 1)$.

are lower than the corresponding values for derivatives of A. This justifies the use of lower tracking rates and bandwidth in the elevation servo, as compared to the azimuth servo, in most radars.

The azimuth and elevation angle derivatives calculated above may be used to evaluate the dynamic lag errors in tracking a given target, by using the servo error constants and Eq. (9.21). For a servo with infinite position error constant K_o, the azimuth error ϵ_a will be

$$\epsilon_a = \frac{\omega_a}{K_v} + \frac{\dot{\omega}_a}{K_a} + \frac{\ddot{\omega}_a}{K_3} + \frac{\dddot{\omega}_a}{K_4} + \cdots \qquad (9.22)$$

The calculation is carried beyond the second term here to show the decreasing importance of the higher derivatives. In most cases, terms beyond the acceleration error are of no importance. For example, if a high-speed aircraft ($v_t = 2000$ ft per sec) passes the tracker at a ground range of 10,000 ft, the peak values of the derivatives are

$$\begin{aligned}
&0.2 \text{ rad per sec} &&\text{for } \omega_a \\
&0.026 \text{ rad per sec}^2 &&\text{for } \dot{\omega}_a \\
&0.016 \text{ rad per sec}^3 &&\text{for } \ddot{\omega}_a \\
&0.0072 \text{ rad per sec}^4 &&\text{for } \dddot{\omega}_a
\end{aligned}$$

A complete calculation of the higher-order constants for the servo response of Fig. 9.15(a) shows that K_4 is about 10,000 per sec⁴, and K_5 is near 25,000 per sec⁵. The magnitudes of the resulting peak error components, which occur at different times during the track, would be

$$\begin{aligned}
&0.2/500 = 0.0004 \text{ rad, or } 0.4 \text{ mils for velocity} \\
&0.026/100 = 2.6 \text{ mils for acceleration} \\
&0.016/500 = 0.033 \text{ mil for third derivative} \\
&0.0072/10,000 = 0.0007 \text{ mil for fourth derivative}
\end{aligned}$$

In a majority of pass-course problems, the acceleration component proves to be the chief source of error. Velocity may also contribute a significant component, and the two components will be additive just prior to crossover.

If the angle acceleration owing to real target maneuver is compared to the geometrical component, it will be seen that the importance of the real component grows with increasing target range. There will be a critical range R_b at which the two have equal potential peak values.

$$R_b = 0.65 \frac{v_t^2}{a_t} \qquad (9.23)$$

Aircraft, with velocities of about 1000 ft per sec and accelerations of a few g's, will be characterized by values of R_b near 1 mi. Certain missiles, which are capable of velocities up to 25,000 ft per sec and accelerations to 100 g, will have an R_b of several miles, whereas objects in free flight at orbital velocities may have R_b of thousands of miles. The last figure shows that satellite trackers will be influenced almost entirely by geometrical considerations in choice of servo bandwidth.

Relationship of Bandwidth to Tracker Design

The preceding discussion has established the methods of calculating dynamic lag error as a function of target motion and servo error constants. The acceleration component is seen to be of critical importance, first, because of the possibility of large geometrical components, and second, because of the fact that K_a is closely tied to servo bandwidth. From Fig. 9.15, it is evident that K_v can be increased without limit by reducing ω_1 and increasing gain, without affecting that portion of the response curve near ω_c which controls the bandwidth and stability of the servo. Indeed, in a second-order servo, K_v is made infinite by inclusion of a second integrator in the servo loop. The acceleration error constant, however, can be increased only by increasing ω_2. This, in turn, forces the servo crossover frequency ω_c to a higher value, unless the length of the -20 db per decade region is reduced at the expense of stability. The crossover frequency is approximately proportional to servo bandwidth β_n, and is limited both by mechanical resonances in the tracking pedestal and by considerations of thermal and target noise. The basic compromise which must be made in any tracker is to choose β_n in such a way that a proper balance is found between noise and dynamic lags, and in many cases the desired high value of β_n requires special mechanical design in the pedestal.

A numerical relationship between β_n and K_a may be estimated from the response curves of Fig. 9.15. If it is assumed that there is to be a four-octave region between ω_2 and ω_3, in the system with a single integration for equalization, then ω_a is one octave above ω_2 and one octave below ω_c. The noise bandwidth β_n of such a servo is approximately twice the crossover frequency in cps ($\omega_c/2\pi$), or $4/2\pi$ times ω_a. The value of K_a is thus

$$K_a = \omega_a^2 = \left(\frac{\omega_n}{4}\right)^2 = \left(\frac{2\pi}{4}\right)^2 \beta_n^2 = 2.5\beta_n^2 \qquad (9.24)$$

When a double integration is used in the equalizer, as in Fig. 9.15(b), a slightly higher value of K_a may be achieved with the same bandwidth, resulting in a factor of about 3.5 instead of 2.5 in Eq. (9.24). These values will hold approximately for both first- and second-order servo systems.

The only limit on K_v is a practical one set by the period of the transient response which occurs when the tracker initially locks onto a target. In a first-order servo, if $t_1 = 1/\omega_1$ is made very large, the settling time of the servo may be excessive. Any initial error in charge on the equalization capacitors (or in setting of the first integrator in a second-order system) will result in a tracking error which will decay with the time constant t_1. Such an error would result if the tracker received designation from an external source which differed from the true rate of the target, and if the capacitor or integrator were not set to the proper value before initiation of automatic tracking. For an initial rate error $\Delta\omega$, the initial position error at lockon would be $\epsilon_1 = \Delta\omega/K_v$. If ϵ_1 exceeds the allowable error of the tracker by a factor of, say, three, then a period equal to $3t_1$ must elapse before the output data would be usable. Even for the wide-bandwidth system in Fig. 9.15(a), this could require about 15 sec, a time far greater than the settling time for an initial position error. The problem may be overcome by proper precharging of the equalizer or presetting of the first integrator, during the period preceding transition to automatic tracking. It is also desirable to switch through a wide-bandwidth tracking mode prior to reducing bandwidth for best tracking conditions, in order to permit the initial transients to settle before the long time-constant circuit is used.

If it is assumed that the velocity lag error may be reduced to negligible proportions, the choice of bandwidth may be based almost entirely upon a balance of acceleration lag error and total noise error. For a monopulse system at medium or long range, the target noise may be neglected, and the total noise may be estimated by using the equation for thermal noise [Eq. (9.10) or a related form]. The total tracking error is then expressed as the rms sum of acceleration lag and thermal noise.

$$\sigma^2 = \sigma_t^2 + \left(\frac{\dot{\omega}_t}{K_a}\right)^2$$

The optimum bandwidth will minimize the total tracking error given by this expression. Considering, for the present, the results of a real target acceleration component $\dot{\omega}_t = a_t/R$, and using the relationship $K_a = 2.5\beta_n^2$, we find that the total mean square error will be

$$\sigma^2 = \frac{\theta^2\beta_n}{k_m^2 f_r B_\tau (S/N)} + \frac{a_t^2}{6.25R^2\beta_n^2} \qquad (S/N \gg 1)$$

Differentiating this with respect to β_n and setting the result equal to zero gives the following expression for optimum bandwidth β_o:

$$\beta_o = \left[\frac{a_t^2 k_m^2 f_r B_T (S/N)}{1.57 \theta^2 R^2} \right]^{1/5} \tag{9.25}$$

In a pass-course problem, the real acceleration component a_t/R may be replaced by the geometrical term $\ddot{\omega}_t$ obtained from Tables 9.4 or 9.5. Since the peak value of $\ddot{\omega}_t$ will lie between $0.5 v_t^2/R_c^2$ and $0.65 v_t^2/R_c^2$, depending upon whether azimuth or elevation is considered, the optimum bandwidth for tracking near crossover will vary as the fifth root of the term $B_T(S/N)(v_t/R_c)^2$. The signal-to-noise ratio is itself a function of range, so the optimum bandwidth will be proportional to $(1/R_c)^{6/5}$ for echo tracking or $(1/R_c)^{4/5}$ for beacon tracking. It may be concluded, as a close approximation, that the optimum bandwidth is inversely proportional to the ground range at crossover for both cases.

As an example of bandwidth calculation for the monopulse case, assume that a target is traveling at 2000 ft/sec in straight flight, with $h_t = R_c = 10,000$ ft, and with the tracker beamwidth $\theta = 20$ mils, $f_r = 300$ pps, $B_T = 1$, and an initial value of S/N equal to ten at a range of 20 mi. This S/N will increase steadily, reaching 48 db at crossover. The values of $\ddot{\omega}_a$, $\ddot{\omega}_e$, and optimum bandwidths are shown in Fig. 9.19, as a function of time from crossover. The same values of β_n would apply if the target velocity were 20,000 ft per sec and $h_t = R_c = 100,000$ ft, as in the case of a high-speed missile near re-entry into the atmosphere. The bandwidths indicated as optimum are easily attainable in tracking pedestals, except near crossover, where they increase beyond the mechanical limit of most power servos.

The situation for conical-scan trackers is similar, but an additional error term is present at all ranges, owing to target amplitude noise at the scan frequency f_s. This noise term is also a function of beamwidth and bandwidth, as shown by Eq. (9.16), and it may be included in the analysis as a term similar to thermal noise. The following expression for optimum bandwidth of a conical-scan system results

$$\beta_o = \left[\frac{a_t^2 k_s^2 f_r B_T (S/N)}{1.57 \theta^2 R^2 [W f_r (S/N) + 2]} \right]^{1/5} \tag{9.26}$$

Here, W is the power density of video signal modulation near the scan frequency f_s [see Eq. (9.16)]. If this modulation is absent, the only difference between the expressions for monopulse and conical scan is the factor of two in the denominator, which indicates the increased tracking efficiency of the monopulse system in thermal noise. This makes a difference of only $(\frac{1}{2})^{1/5}$ or 0.87 in optimum bandwidth. For the target with Rayleigh scintillation and Markoffian spectrum, $W = f_g/2\pi f_s^2$ [Eq. (9.17)], and the term $W f_r(S/N)$ in the denominator becomes important, especially at high values

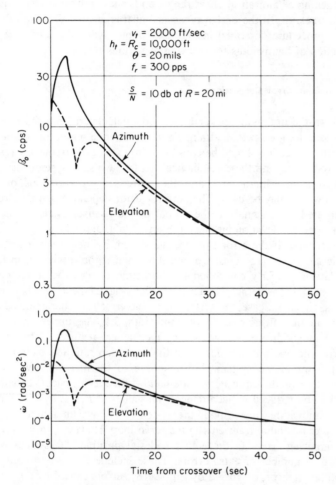

Figure 9.19 Optimum bandwidths for pass course.

of S/N. For example, with $f_g = 5$ cps, $f_s = 30$ cps, and $f_r = 300$ pps, this term becomes $0.26\ S/N$, and will limit tracker performance for $S/N > 10$. If we apply this to the previous example of the pass course, the optimum bandwidth would remain very nearly constant at 0.25 cps throughout the track, with resulting serious lag and scintillation errors. When the target is of appreciable length, a term representing glint would also have to be added to the above analysis, modifying the optimum bandwidth for both monopulse and conical-scan radars. This problem is almost certain to arise in echo tracking of aircraft at short range, and is less probable in missile and satellite tracking. In the design of an actual tracking system, the full error analysis procedure described in the next chapter would be used as a guide in selection of bandwidths for tracking of various targets.

Open-Loop Error Correction

Where the optimum tracker bandwidths calculated from the preceding expressions are greater than attainable in the antenna servo, the use of open-loop error correction should be considered to extend the frequency response of the system beyond these mechanical limits. Two such systems have been described in the literature,* and there are several variations on these schemes which may be used. The data corrector depends upon the presence of calibrated error signals at the output of the receiver system, which can be converted to elevation and azimuth error data, and added to the outputs of the pedestal. If the error slope of the system is linear over the range of possible lag errors (see Figs. 9.3 and 9.6), and if the receiver gain for the error data is properly controlled by the action of agc, the corrected data can be made considerably better than the original pedestal shaft output readings. In such a case, the system as a whole may be characterized by an increase in the effective value of bandwidth β_n. The upper limit of this system bandwidth is no longer set by mechanical features of the pedestal and drive motors, but depends upon the repetition rate and the amount of noise present in the system. The response to thermal noise will be the same as that of the equivalent wide-bandwidth servo system, and the lag errors will be reduced in accordance with values of K_a set by the increased system bandwidth. In addition, the system will have an infinite K_v, and will correct for initial transients in the servo loop. However, in the presence of dynamic lag errors in the actual pedestal position, the system will be subject to a number of extra errors. These include the second component of scintillation error, expressed by Eq. (9.18), errors caused by nonlinearity of the error slope, errors caused by failure to normalize the error signal with respect to the reference signal amplitude, and errors caused by different time lags in the correction system, when compared to the pedestal data output. When the data bandwidth approaches the repetition rate, as is possible in these systems, some form of instantaneous agc must be used for normalization.

* Roger P. Cheetham and Warren A. Mulle, "Enhanced Real-Time Data Accuracy for Instrumentation Radars by Use of Digital-Hydraulic Servos," *IRE Wescon Record* (1958), Part 4.

M. Korff, C. M. Brindley, and M. H. Lowe, "Multiple-Target Data Handling with a Monopulse Radar," *Trans. IRE*, Vol. **MIL-6**, No. 4 (Oct. 1962), pp. 359-66.

9.5 SUMMARY OF TRACKER CHARACTERISTICS

The above analysis covers the major features of the amplitude comparison monopulse and the conical-scan systems of angle tracking, with respect to their response to thermal and target noise and signals. The fundamental differences between these two types of tracker have been shown as follows:

1. The monopulse system requires a fixed, multihorn feed assembly with three receiver channels approximately matched in gain and phase, as compared to the single, scanning feed and receiver channel used in conical scan.

2. The response of a typical monopulse system in the presence of thermal noise is some 6 db better than that of the best comparable conical-scan system, for equal bandwidth and target characteristics. This extends the range of accurate tracking for monopulse by about 40 per cent.

3. The monopulse system has virtually no response to target amplitude noise, which limits the accuracy of conical-scan systems at intermediate ranges.

4. The monopulse system forms a complete estimate of tracking error on each received pulse (or within a time given by $1/B$ for a c-w system), and hence may provide output data at a higher rate than a conical-scan system, which requires at least one scan period to form such an estimate. At low repetition rates, this may permit the monopulse system to operate with a wider servo bandwidth, with consequent reduction in dynamic lag errors.

Both systems respond alike to "glint" or target angle noise, and both suffer a loss in tracking gain and efficiency from "small signal suppression" in the second detector, when the i-f S/N ratio is near or below unity. The relative merits of the two systems for a given tracking problem depend upon the relative importance of accuracy and bandwidth, on the one hand, as compared with simplicity, low cost, and easy maintenance, on the other.

Alternate Forms of Tracking Radar

Some of the limitations of conical scan may be overcome by use of sequential-lobing techniques, in which the receiver is switched rapidly between four antenna feed points, sampling these outputs one at a time or in pairs. The relative performance of two such systems has been compared with monopulse and conical scan, and has been found to lie between these with respect to error sensitivity and resolution.* Both the "conventional"

---

* Samuel F. George and Arthur S. Zamanakos, "Multiple Target Resolution of Monopulse vs. Scanning Radars," *Proc. NEC*, **15** (1959), pp. 814–23.

and the "paired-lobing" methods were analyzed on the assumption that the transmitting pattern consisted of a single lobe on the antenna axis, providing stronger illumination of the target than the offset beam used in conical scan. In the conventional sequential-lobing system, the receiver is switched alternately between the four feed points to compare signal outputs, whereas in paired lobing the horns are added in pairs prior to sampling by the single receiver. Both methods operate with scan rates up to one-fourth the repetition rate. The spectral density of the target scintillation noise is usually lower than in conical scan, and the available data bandwidth is higher. Otherwise, these two sequential systems are very similar to conical scan, in that the tracking error is superimposed on a subcarrier at the scan frequency. The noise bandwidth is, therefore, twice that of the monopulse tracker having the same servo bandwidth, and the system is sensitive to whatever scintillation component exists at the scan rate.

Other Monopulse Configurations

Several other types of monopulse radar are discussed in the literature.* In the phase comparison monopulse system, three or four receiving antennas are separated by many wavelengths, and the off-axis position of the target is indicated by the difference in phase of the signals received at these points. Since the antenna spacing is at least equal to the diameter of the individual aperture, the phase difference builds up rapidly as compared to changes in amplitude of the signals, and the amplitudes may be considered constant over the normal region of operation. This system may be considered as a tracking adaptation of the short-baseline interferometer widely used in radio astronomy. Except for the difference in antenna design, it has been established that the phase comparison scheme has performance equivalent to that of the amplitude comparison monopulse with equal aperture area, both with respect to thermal and target noise components.†

A "phase-amplitude comparison" monopulse system was developed to reduce from three to two the number of receivers required to process the reference and error signals. This system uses amplitude comparison in one tracking coordinate and phase comparison in the other. The reference and error signals are combined in the feed assembly, and the two receivers must maintain both amplitude and phase equality to avoid tracking error. Other

---

* Donald R. Rhodes, *Introduction to Monopulse* (New York: McGraw-Hill Book Company, 1959).

† Ibid., p. 15.

Leon Peters, "Accuracy of Tracking Radar Systems," Ohio State University Research Foundation, Report 601-29 (Dec. 31, 1957), Astia Document AD 200, 027, p. 39.

than the presence of a crosstalk term due to target noise, and the possible occurrence of ambiguous tracking points, the system is equivalent to the amplitude or phase-comparison types of monopulse.*

When one is using amplitude comparison monopulse, there are a number of ways of normalizing the error signals and of amplifying them without introducing tracking errors. In one such system, the sum and difference signals are added with phase shifts of $+90°$ and $-90°$, and then passed through limiting amplifiers. The limited outputs are compared in a phase-sensitive detector, whose output provides an error signal proportional to the off-axis angle of the target. A variation of this technique has been described† in which the two signals are placed on different i-f carriers and amplified in a single i-f channel whose bandwidth is wide enough to pass both signals. When the combined signal is limited, the difference signal is normalized. The sum and difference signals are then separated in bandpass filters and applied to the error detector. All the monopulse systems may be analyzed by using the relationships of Sections 9.2, 9.3, and 9.4, provided that the actual values of error slope, beamwidth, and signal-to-noise ratio are used. Since the small-signal suppression loss is almost identical for envelope detectors (linear and quadratic) and for product detectors, the performance of the various systems at low S/N ratio will also be comparable to that of the four-horn amplitude comparison system which was analyzed in detail in those sections, so long as the actual i-f S/N ratio at the nonlinear detector is used when one is entering Fig. 9.8. The signal-to-noise ratio is not changed in passing through a limiter and narrow-band filter, if the output is to be processed in a phase-sensitive detector. In any specific case, the circuits preceding the error detectors must be analyzed carefully to determine the noise bandwidth which must be used to evaluate the signal-to-noise ratio. When separate r-f amplifiers or mixers are used for each of the monopulse horns, the sum and difference channels may be formed after amplification or frequency conversion, and the signal-to-noise ratio may be 3 or 6 db lower in the initial stages of the receiver. However, so long as the sum signal is formed before nonlinear detection, there will be no loss in performance. There may, however, be an increase in error owing to phase and amplitude instabilities in the active elements that lie between the horns and the comparators. Conventional amplitude comparison monopulse has high accuracy, because the sum and difference signals are formed at the feed after passage of the individual horn signals through passive elements of short electrical length. Unless some form of automatic phase and amplitude compensation is used in the separate amplifiers, it is impossible

---

* Peters (1957), *Ibid.*, p. 39.

† W. L. Rubin and S. K. Kamen, "SCAMP—A Single-Channel Monopulse Radar Signal Processing Technique," *Trans. IRE*, Vol. **MIL-6**, No. 2 (April 1962), pp. 146–52.

even to approach the accuracy of the properly designed sum-and-difference system.

Factors Determining the Choice of Tracking System

Given the wide variety of possible circuits and antenna designs, there may be some question as to which would prove most advantageous in a given application. A few general considerations will be discussed here, as a guide to choosing an appropriate tracking system.

1. *Antenna gain and efficiency.* If the antenna is very large (and hence expensive), it will generally be found more economical to use one of the monopulse systems, since the on-axis gain will be several decibels higher than in the sequential systems. A smaller monopulse antenna can be used with two or three receivers to get the same signal power and tracking accuracy as a larger antenna with sequential scan.

2. *Tracking accuracy vs. signal recovery.* In some cases, the primary use of the signal is for detection, ranging, or recovery of information contained in its amplitude or frequency characteristics. Angle tracking is then used only to keep the antenna pointed at the signal source, and the accuracy of tracking is of secondary importance. Any system which provides a high-gain, on-axis channel for reception is then suitable for the tracker, and error slope may be compromised to obtain optimum performance in the signal channel.

3. *Use of low-noise receivers.* With the advent of parametric amplifiers and masers, it became possible to use relatively small antennas in some applications which had previously required large antennas and high-power transmitters. In the interests of simplicity and reliability of operation, tracking systems requiring only one receiver channel were preferred in many of these applications, since the design of low-noise r-f amplifiers was much easier if the gain and phase tolerances could be relaxed. Both masers and parametric amplifiers have been successfully applied to monopulse radar, using circuits designed to maintain the tolerances previously applied to i-f amplifiers.

4. *Tuning range and polarization diversity.* The design of optimized feed assemblies for the amplitude comparison monopulse system requires a number of compromises in the area of r-f bandwidth and polarization sensitivity. When the conventional 10 per cent bandwidth is not sufficient, or when polarization diversity is of primary importance, the use of the phase comparison type of system may be preferred. In this system, four separate apertures may be used with

independent feeds, and the feed horns may be designed to accommodate special signal characteristics.

5. *Data bandwidth.* As discussed in the previous section, the requirement for wide bandwidth in the angle data channel may dictate the use of a monopulse system. The ultimate in this respect is the single-pulse, digital tracker, in which the error estimate from each received pulse is formed by the receiver and made available to an associated digital computer. The radar using this type of processing may make accurate angle measurements during search of an extended region, and may track targets which would exceed the dynamic capabilities of the conventional analog servo. This is due to the ability of the computer to operate in rectangular coordinates, and to anticipate the accelerations required to track the target in spherical coordinates. The subject of data processing, smoothing, and coordinate conversion will be covered in more detail in the following chapter.

6. *Long-range tracking.* When radar echo tracking is to be carried out at long range, the use of a monopulse antenna permits the system to maintain a wide tracking bandwidth, through the use of repetition rates above the unambiguous value. Conical-scan radars, or lobing radars in which the transmitting beam is moved, are limited to scan periods considerably longer than the range-delay time (regardless of prf), with correspondingly lower tracking bandwidths.

The performance of an angle-tracking instrument is measured primarily by its accuracy under intended conditions of operation. Accuracy has been defined* as freedom from error, and hence the tracking performance will be discussed here in terms of the magnitudes and other characteristics of the error components which affect typical angle-tracking systems. Before discussing these individual error components in detail, we shall consider the general properties of errors in measurement and the methods by which errors may be described. Errors in angle tracking will then be classified by source, by frequency distribution, and by the point at which they enter the tracking system. This will provide a basis for construction of error models which will describe accurately the performance of any specific system. Later chapters will cover the analysis of errors in range and Doppler trackers, and the methods by which radar data may be smoothed and processed to obtain more accurate average values or to extract velocity and higher derivatives from the position data.

ERROR ANALYSIS OF THE ANGLE TRACKER

10

10.1 MATHEMATICAL MODELS OF ERROR

In the discussion of radar targets (Chapter 3), we used statistical procedures to describe the amplitude and frequency distributions of radar

* See Chap. 8, p. 235.

cross section. The same procedures are applicable to errors in tracking or measurement, which have both systematic and random properties. By choosing mathematical models which are close approximations of the actual error components, we may obtain an accurate analysis of the error in a particular system by using available test data and theory for each major system element, and may extrapolate the results to a wide variety of operating situations.

The mathematics of radar error need not be made so difficult that it requires knowledge of advanced statistical theory. In the following discussion, no special background will be assumed, but a general appreciation of the nature of measurement processes will be needed. The treatment will probably not satisfy those who place a high value on rigorous derivations, but the results of the analysis method given have been proven valuable in many radar system problems. They are offered for the use of those who may lack the time or the background to approach the problem from a rigorous mathematical viewpoint.

Definition of Error

The error in a given measurement may be defined as the difference between the value indicated by the measuring instrument and the "true" value of the measured quantity.

$$x = U_{measured} - U_{true}$$

The purpose of error analysis is to provide a description of the error which will permit its magnitude to be estimated for any set of operating conditions, without the necessity of running calibrations or tests for all possible combinations of conditions which may be encountered. In general, the error will vary with the time at which the measurement is made, the value of the quantity to be measured, and with the environmental conditions.

It is common practice to divide errors into systematic (or bias) and random (accidental or noise) components. The former are characterized by their predictability, and may be corrected, at least partially, by a process of calibration applied to the instrument before and after the measurement. In an extreme case, if the error is constant for all conditions, a single number will describe its magnitude, and subtraction of this number from all values measured by the instrument will provide error-free data. One calibration will determine the value of initial bias error and will permit the reduction of residual bias error to zero. More generally, the error will assume some value within a limited range of values which are centered about a mean (or true bias) error, indicated in Figs. 10.1 and 10.2. For any measured set of n error values, we may determine this mean error from the expression

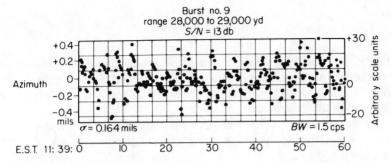

Figure 10.1 Typical radar tracking error plot derived from bore-sight telescope.

$$\bar{x} = \frac{1}{n} \sum_{i=1}^{i=n} x_i \tag{10.1}$$

Other terms used to describe the magnitude of the error include the maximum (or peak) error x_m, the peak-to-peak error $x_{\text{p-p}}$, the rms error x_{rms}, and the probable error x_{50}. The rms error is defined for any distribution by the expression

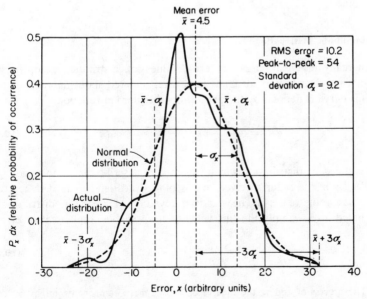

Figure 10.2 Distribution of amplitude of error in typical radar track, compared with normal distribution.

$$(x_{\text{rms}})^2 = \frac{1}{n} \sum_{i=1}^{i=n} x_i^2 \qquad \text{(10.2)}$$

Thus, x_{rms} is exactly what its title states: the square root of the mean value of the square of the individual error values. The probable error is the value which is exceeded in 50 per cent of the readings, and can have any value from zero to the value of the peak error.

In those cases where the mean error is not zero, the rms error may be written as the sum of the bias and the variable components, added in an "rms fashion."*

$$(x_{\text{rms}})^2 = \bar{x}^2 + \frac{1}{n} \sum_{i=1}^{i=n} (x_i - \bar{x})^2 \qquad \text{(10.3)}$$

The above relationships are not limited to any particular types of error, and may be applied to any arbitrary or measured distribution of error values. Figure 10.2, for example, was obtained from the experimental data of Fig. 10.1 by simply counting the errors occurring within each increment of three scale divisions (0–2, 3–5, etc.), and plotting a smooth curve through the resulting points. The rms values shown were computed by using Eq (10.3), and the equivalent normal distribution curve (with equivalent values of $\bar{x}$ and σ_x; see below) was then drawn to show how well the data could be approximated by a simple mathematical distribution function.

Amplitude Distribution

In many practical cases, the error values are distributed according to the "normal distribution" or Gaussian curve, centered about the mean value $\bar{x}$. This curve is defined by the probability distribution function

$$dP_x = \frac{1}{\sqrt{2\pi\sigma_x^2}} \exp\left[-\frac{(x - \bar{x})^2}{2\sigma_x^2}\right] dx \qquad \text{(10.4)}$$

This is the same distribution which was used to describe the i-f noise voltage amplitudes in the radar receiver, and is plotted in Fig. 1.6(a) with respect to a mean value of zero. The standard deviation of x for this distribution, designated σ_x, is simply the rms value of the variable component of error.

$$\sigma_x^2 = \frac{1}{n} \sum_{i=1}^{i=n} (x_i - \bar{x})^2 \qquad \text{(10.5)}$$

---

* The process of adding the squares of two or more components might be described more accurately as "rss addition," since it refers to the root of the sum of the squares. We shall use the more general term, rms addition.

The square of the standard deviation is the "variance" of the error. Although the term *standard deviation* should be limited to cases where the normal distribution is found, the symbol σ_x, is often applied to designate the rms value of any type of variable error, and will be so used here. The probable error x_{50}, for the special case of the normal distribution, is given by

$$x_{50} = 0.6745\sigma_x \qquad\qquad \text{(10.6)}$$

The normal distribution is often assumed to represent errors of unknown characteristics, and it closely approximates the distribution of many actual errors. In most cases, the peak error observed will correspond to a deviation of $3\sigma_x$ from the mean value, and the peak-to-peak error will be about $6\sigma_x$, when the error appears "noisy" in character. The probability of exceeding the 3σ deviation for the normal distribution is 0.3 per cent, which implies that one out of 300 independent observations will deviate from the mean by an amount greater than 3σ. When a set of data points is taken over a period of one minute, using a system with a bandwidth of several cps, we would expect only one or two excursions beyond this value. A deviation of 4σ would be expected only once in 20 or 30 such sets of data, if it followed the normal distribution, and hence we would be likely to measure our peak error near the 3σ point.

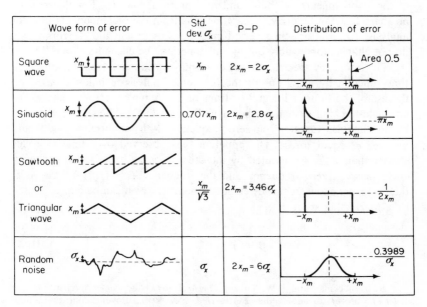

Figure 10.3 Relationships between rms, peak, and peak-to-peak errors.

The error contributed by a single element in the system may follow a regular pattern, such as those shown in Fig. 10.3, rather than the random noise pattern. It will be noted that the wave forms are given in order of increasing ratio of peak to rms error, and that these ratios vary from unity for the square wave to three for random noise. True Gaussian noise, of course, would have an infinite "peak" error, but we would have to wait a long time, as mentioned above, to see the 3σ level exceeded.

Time Functions and Frequency Spectra

The variation of error with time, and the resulting frequency spectrum of the error are important factors in classifying and describing the performance of a tracking system, and must be known if the effect of smoothing or differentiation is to be determined. The bias error defined above, if it is truly constant for periods of time which are long compared to the calibration and operation times of the system, can be removed by the calibration procedure. Hence, it is of little importance in error analysis. The residual bias error observed in most tests and evaluations of equipment is taken as that portion of error which does not change during the period of an individual test or operation, which may be as short as a few seconds or one minute. When errors are introduced as functions of the measured quantity, the speed at which this quantity varies will determine how much of the error appears as "bias" and how much as "noise." For this reason, no sharp line can be drawn separating bias from noise components in error analysis. An arbitrary time period must be chosen, and those errors which do not change appreciably during this time may be classed as "bias."

The frequency spectrum of the error may be obtained from the observed time function by using one of the harmonic analysis techniques based on the use of the Fourier integral or transform.* A typical error spectrum for an angle-tracking system is shown in Fig. 10.4. True bias is represented by an "impulse function" at zero frequency, with infinitesimal width and an area σ_o^2 equal to $\bar{x}^2$. The apparent bias observed over time intervals shorter than t_o is represented by the area σ_b^2 beneath the spectral density curve between zero frequency and the frequency $f_1 = 1/(2t_o)$. Above f_1 there appears a low-frequency error component, which can be approximated by a Markoffian spectrum.

$$W_a(f) = W_o \frac{f_a^2}{f_a^2 + f^2} \tag{10.7}$$

---

* R. B. Blackman and J. W. Tukey, *The Measurement of Power Spectra from the Point of View of Communications Engineering.* New York: Dover Publications, Inc., 1958.

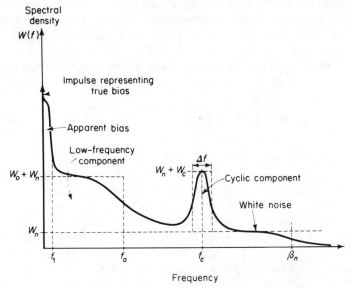

Figure 10.4 Typical angle-tracking error spectrum.

The variance of this component is given by the value of the integral of W_a between zero and infinity.

$$\sigma_a^2 = \frac{\pi}{2} W_o f_a \qquad \textbf{(10.8)}$$

This is the same type of spectrum which we used to describe the fluctuation in target cross section due to scintillation [Eq. (3.14)]. It represents the spectrum which results when broad-band (white) noise is passed through a single-section low-pass filter consisting of a series resistor and parallel capacitor, where $RC = t_a = 1/(2\pi f_a)$.

Above the frequency f_a in Fig. 10.4 we see two more error components. The random or white-noise component extends with uniform spectral density W_n to the limit of the observed spectrum, as set by the bandwidth of the measuring device. The variance of the random noise error is

$$\sigma_n^2 = W_n \beta_n \qquad \textbf{(10.9)}$$

Superimposed upon the white noise is a cyclic component, occupying a narrow band of frequencies centered at f_c. If this component is a pure sinusoid, given by the wave form $x_c = X_c \sin(2\pi f_c t)$, it should appear as an impulse function with area $\sigma_c^2 = X_c^2/2$ at the frequency f_c. In a spectrum obtained from experimental data over a finite period, the same area will be distributed over a narrow band of frequencies, and the error may be approximated by a rectangular spectrum of amplitude W_c and width Δf, chosen so that $\sigma_c^2 = W_c \Delta f$.

The variance of the total error is given by the area under the entire spectrum, or

$$\sigma_x^2 = \int_0^\infty W(f)\,df = \sigma_o^2 + \sigma_b^2 + \sigma_a^2 + \sigma_n^2 + \sigma_c^2 \qquad \text{(10.10)}$$

In the case shown, no error is correlated with any other, and the process of rms addition yields the total variance. It might be expected that two bias components would add directly, rather than in an rms fashion. However, the area σ_b^2 represents a slowly varying error which will sometimes have the same polarity as the true bias and sometimes oppose it. When evaluated over a large number of intervals, each of duration t_o, the sum $(\sigma_o^2 + \sigma_b^2)^{1/2}$ represents the rms value of the bias observed, and the probable bias will be approximately 67 per cent of this value, in accordance with the normal distribution.

Validity of the Error Model

The statistical description of errors permits us to break down a complex tracking system into many elements, to calculate or to measure the corresponding components of error, and then to combine these in an rms addition process to obtain the over-all system error. The underlying assumption is that the several error components are completely uncorrelated with each other, and in practice it has been found safe to ignore possible correlation effects unless there are clear physical links causing the common variation of two or more error components. Two examples taken from analysis of actual radar systems will illustrate the validity of this assumption.

In one case, the radar was investigated to find the extent of the angle-measurement errors resulting from thermal noise in the receiver and from target fluctuation. A lengthy series of digital computer simulations was carried out, using various combinations of mean S/N ratio and modulation, superimposed on the signal to represent target scintillation. After the results were plotted and reviewed, it was found that the entire family of curves resulting from the simulations could be represented accurately by the rms sum of a thermal noise component, varying with mean S/N, and an independent scintillation component. Similar results have been obtained by using the data from Swerling's analysis, as plotted in Fig. 2.4.

In a second case, a tracking pedestal was tested piece by piece with optical procedures, and error curves were determined for each gear mesh, shaft, and major structural member. A detailed analysis showed that the over-all pointing error of the antenna axis, evaluated over many points in the hemisphere, was the same as the rms sum of the individual error components. This was true in spite of the fact that each of the errors was a function of the azimuth and elevation angle of the antenna. The several components were characterized by different periods and phase relationships, as referred to the two tracking angles, and they added as though they were completely uncorrelated.

10.2 CLASSIFICATION AND DESCRIPTION OF ANGLE ERRORS

In addition to separation and classification by their amplitude distributions, wave forms, and spectral properties, errors may be further classified by source and by the degree of their dependence upon target characteristics, propagation conditions, or upon parameters of the tracker itself. This classification will serve to clarify the operating conditions under which a given level of performance can be achieved, and should also simplify the problem of devising tests to verify the results of theoretical analysis. In testing and evaluating angle trackers, an additional source of error is introduced by the reference instrumentation against which the tracker is compared. The apparent errors from this source must be isolated from the actual errors of the tracker under test. Another classification of errors is carried out acording to the point of entry into the tracking system. Those errors which are due to the tracking axis leaving the target may be termed *tracking errors,* whereas the errors in reading the position of the axis and providing numerical values relative to the fixed reference system may be called *translation errors.* This classification is of particular importance when performance tests are to be run by using an optical reference (e.g., boresight telescope), which may yield direct readings of tracking error but no data on translation error.

Using these two classifications and the frequency spectrum classification (bias vs. noise), we may list the error sources in a typical radar angle-tracking system as shown in Table 10.1. We shall discuss each type of error in more detail below, and give the relationships by which the error may be predicted from the known parameters of the radar and target situation. The definitions and descriptions of glint and scintillation, listed under target-dependent errors, have been given in Section 3.3.

Radar-Dependent Tracking Error

The tracking axis of radar may be defined optically or mechanically in terms of a line normal to a machined surface on the antenna pedestal,

Table 10.1 SOURCES OF ANGLE ERROR

Class of error	Bias components	Noise components
Radar-dependent tracking errors	Boresight axis setting and drift Torque caused by wind and gravity Servo unbalance and drift	Thermal noise Multipath Torque caused by wind gusts Servo noise Deflection of antenna caused by acceleration
Radar-dependent translation errors	Pedestal leveling Azimuth alignment Orthogonality of axes Pedestal flexure caused by gravity force Pedestal flexure caused by solar heating	Bearing wobble Data gear nonlinearity and backlash Data takeoff nonlinearity and granularity Pedestal deflection caused by acceleration
Target-dependent tracking errors	Dynamic lag	Glint Dynamic lag variation Scintillation or beacon modulation
Propagation errors	Average refraction of troposphere Average refraction of ionosphere	Irregularities in refraction of troposphere Irregularities in refraction of ionosphere
Apparent or instrumentation errors	Stability of telescope or reference instrument Stability of film base or emulsion Optical parallax	Vibration or jitter in reference instrument Film transport jitter Reading error Granularity error Variation in parallax

passing through the center of the antenna. The data output devices are set to zero when this axis is pointed in an arbitrary reference direction (e.g., horizontally and northward). The electrical axis is then collimated with the mechanically defined axis by using such instruments as a boresight telescope, in conjunction with suitable radar targets which are visible in the telescope. Any errors in the collimation process, as well as subsequent drifts and deviations in the position of the electrical axis appear as tracking errors when the radar is in operation. Those portions of the tracking error which are under control of the radar designer will be considered "radar dependent." Those which depend primarily on the target itself will be discussed later. In the class of radar-dependent errors, we have listed the thermal-noise error described in Chapter 9. The following discussion will cover the other components which are encountered in most tracking radars, and which appear also in instruments such as height finders and nonradar angle-tracking antennas.

Multipath Error

Multipath error is the result of the same interference phenomenon which leads to the generation of the lobing structure in search-radar coverage, as described in Chapter 6. Target signals are reflected from the surface of the earth to the radar antenna, and enter the radar receiver with an amplitude which depends upon the off-axis response of the antenna. The relatively narrow beams used in tracking tend to reduce the reflected signal to low levels as soon as the target rises one beamwidth or more above the horizon, so the lobing effect is seldom observed in the pattern of the antenna. Even very small signals, however, will disturb the null position of the tracking servo, leading to errors in elevation measurement.

The effect may be illustrated most easily for the case of the monopulse radar described in the previous chapter, although the error appears in similar form in all tracking systems. Figure 10.5 shows typical antenna patterns for an amplitude comparison monopulse radar in the elevation plane. When tracking a target at elevation angle E, the antenna axis will be tilted upwards to this angle, and surface reflections will be received at an angle $2E$ below the axis (as in Fig. 6.4). The resulting reference and error signals are shown in the vector diagram, Fig. 10.6. The phase and amplitude of the direct ray in the reference pattrn are used to define a reference vector E_r at zero phase angle. The direct ray in the error channel will appear in phase or $180°$ out of phase with this reference, depending upon whether the target is slightly above or below the axis, and the servo will act to drive this error to a null. The presence of the reflected component is described by the addition of a small vector in each channel, which is rotated relative to the reference phase by an angle

$$\phi = \frac{4\pi h}{\lambda} \sin E + \psi \qquad (10.11)$$

where h is the height of the antenna and ψ is the phase angle introduced by reflection from the surface. Since the ratio of reflected ray amplitude to direct ray is small (for targets one beamwidth or more above the ground), the effect on the reference channel is to cause a small variation in both amplitude and phase, which have no appreciable effect on system operation. The reflected ray in the error channel, however, simulates an error signal and drives the servo until it is balanced by an equal and opposite component of error which is cyclic as a function of varying elevation angle.

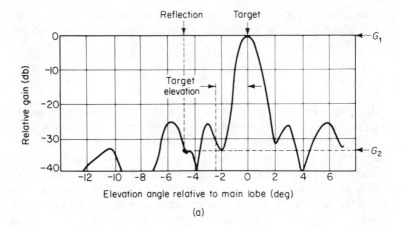

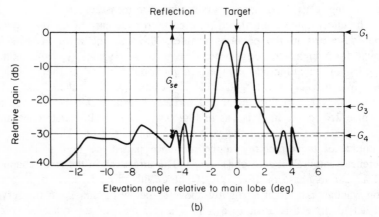

Figure 10.5 Elevation antenna patterns for multipath error computation. (a) Monopulse reference pattern. (b) Monopulse error pattern.

$$x_{\max} = \frac{\theta_e \rho}{k_m \sqrt{G_1/G_4}} \sin\left(\frac{4\pi h}{\lambda}\sin E + \psi\right)$$

$$\cong \frac{\theta_e \rho}{k_m \sqrt{G_{se}}} \sin\left(\frac{4\pi h E}{\lambda} + \psi\right)$$

(10.12)

Here, ρ is the surface reflectivity coefficient (see Chapter 15), k_m is the error slope defined in the previous chapter, and $G_{se} = G_1/G_4$ is the side lobe ratio for the error pattern at an angle $2E$ below the axis. Since this ratio varies in accordance with a gradually decaying lobe structure, and

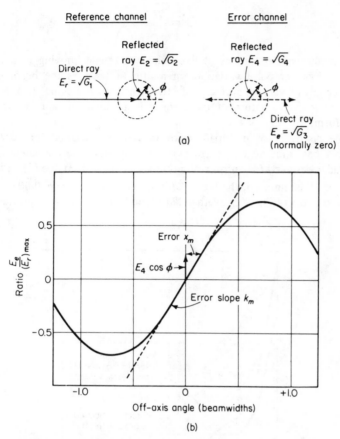

Figure 10.6 Multipath error relationships. (a) Direct and reflected rays. (b) Error generated by reflected ray.

since the surface reflectivity and slope may vary at different points around the radar, it is difficult to predict the exact amplitude of the multipath error or its phase angle. An rms value may be found for G_{se} in the region of interest, and the rms multipath error found from this.

$$\sigma_m = \frac{\theta_e \rho}{k_m \sqrt{2(G_{se})_{rms}}} = \frac{\theta_e \rho}{\sqrt{8(G_{se})_{peak}}} \tag{10.13}$$

Here, it is assumed that $k_m = 1.4$, and that $(G_{se})_{peak}$ is evaluated in the region of interest. If the target is moving in elevation at a rate $\dot{E}$, the frequency of the cyclic error is given by

$$f_m = \frac{2h\dot{E}}{\lambda} \qquad\qquad (10.14)$$

Even in the case of targets at fixed elevation angle, moving in azimuth, there will be temporal variations in multipath error as the beam passes over regions of varying surface reflectivity, height, and slope. In this case, the regular cyclic nature of the error will give way to a more random wave form.

The discovery of multipath error in systems designed for very high accuracy has come as a shock to more than one radar development group. In a radar designed for an over-all accuracy of 0.1 mil (0.0057°), a beam-width of 1° being used, the factor $\rho/\sqrt{8G_s}$ must be less than 0.005. When the radar is operating over average flat ground surfaces ($\rho = 0.3$), the side-

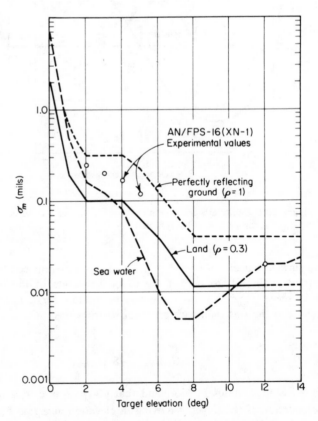

Figure 10.7 Multipath error vs. target elevation.

lobe ratio must be at least 27 db in the error channel, and this level is seldom reached within several beamwidths of the axis. For the antenna patterns shown in Fig. 10.5, the rms multipath error will be as shown in Fig. 10.7. Experimental verification is shown for the region adjacent to the main lobe. In one tracking radar development, this error first became evident in field tests when the radar was used as part of a missile guidance system, and was attributed to some subtle mechanical problem (a common occurrence in a program conducted by electrical experts). It was only after repeated testing and analysis that the source of the difficulty was traced to the high level of side lobes in the error pattern of the lens antenna. When the lens was replaced by a well-designed reflector, these side lobes were reduced and satisfactory error levels were obtained.

In tracking at very low angles, where the main lobe illuminates the surface, the multipath error grows very large and departs from its sinusoidal form. The target and its image combine to appear as a two-point, or dumb-bell-shaped, target with varying phase and amplitude ratio between the two sources (see Section 3.3). Under these circumstances, the elevation data from the radar is of little value, and the track may be lost completely even when the ground clutter remains well below signal level.

Antenna and Servo Torque Disturbances

The analytical evaluation of error terms arising within the tracking servo loop requires considerable knowledge of the open-loop transfer functions of the system. Curves such as that shown in Fig. 9.15 cannot be used by themselves to establish the response of the system to torques imposed on the antenna, or to friction and electrical noise within the servo loop. To supplement these curves, we need the type of transfer function shown in Fig. 10.8, which gives the magnitude of the torque applied to the antenna by the servo for a given position error, as a function of frequency. In most tracking servomechanisms, this torque function will result from the combined response of two loops: the direct error (or position) loop, and the velocity (or tachometer) loop. At low frequencies, the position loop response, dependent upon tracking error measured by the antenna, will provide torque to overcome disturbances introduced at the antenna or in the drive system. For high frequencies, the response of the position loop is superseded by the tachometer-loop response, which depends upon the velocity of the antenna and which bypasses the radar's position-error measurement circuits. However derived, the torque transfer function, designated $Y_p(f)$, may be used to calculate the motion of the antenna pedestal in response to loads imposed internally or at the antenna.

As one example of mechanical disturbance, consider the force of the wind on the antenna. The aerodynamic characteristic of the antenna may

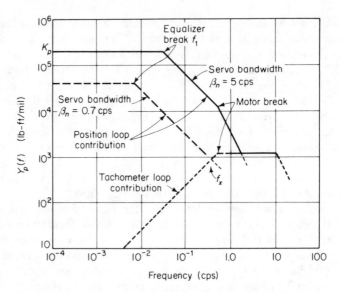

Figure 10.8 Antenna torque transfer function (AN/FPS-16 radar).

be summarized in terms of a constant K_w which relates the torque M around one of the tracking axes to the square of the wind speed v_w.

$$M = K_w v_w^2 \qquad\qquad \textbf{(10.15)}$$

This aerodynamic constant might typically be expressed in units of lb-ft per (ft per sec)^2, and a value of 0.2 has been experimentally determined for the 12 ft reflector of the AN/FPS-16 tracker (see Section 10.3), when the wind vector is $45°$ from the antenna axis. The torque for a wind speed of 70 ft per sec would, therefore, be about 1000 lb-ft. The bias error resulting from servo response to this steady wind torque would be

$$\epsilon_{ws} = \frac{M}{K_p} = \frac{K_w v_w^2}{K_p} \qquad\qquad \textbf{(10.16)}$$

Here, K_p gives the zero-frequency value of the torque transfer function: $K_p = Y_p(0)$, in appropriate units such as lb-ft/mil error. The response in Fig. 10.7 indicates a value $K_p = 2 \times 10^5$ lb-ft/mil in the wide-band mode, and would give a bias error ϵ_{ws} less than 5×10^{-6} rad for wind speeds up to 70 ft per sec.

Another component of tracking error due to torque on the antenna results from direct mechanical deflection of the antenna relative to its mounting surface. This is best described by an equivalent spring constant

K_s, which gives the deflection of the antenna axis per unit of applied torque.

$$\epsilon_{wd} = MK_s = K_w K_s v_w^2 \qquad \text{(10.17)}$$

Most tracking radar antennas are built to provide a very stiff mechanical structure, in order to obtain high resonant frequencies within the elements which form the servo loop. For the AN/FPS-16, with a 12 ft reflector, the spring constant is about 10^{-5} mils per lb-ft. The corresponding bias error ϵ_{wd} in a 70 ft per sec wind would be about 10 μrad. The same error would also result in elevation if the antenna structure were unbalanced by 1000 lb-ft due to added mechanical components, optical devices, ice loading, etc.

Variations in wind loading will cause variable deflection and servo errors. A typical gusty wind has been determined to have a standard deviation σ_v of a few ft per sec from the average velocity, and will produce a torque given by

$$M + \Delta m = K_w (v_w + \Delta v)^2$$

The variable component of torque will be approximately

$$(\Delta m)_{\text{rms}} = 2K_w v_w \sigma_v$$

This will cause a direct bending of the structure by an amount

$$\sigma_{\text{wd}} = 2K_s K_w v_w \sigma_v \qquad \text{(10.18)}$$

For the values given above, this error would be only about $\frac{1}{2}$ μrad. The servo error will be more serious. It is calculated from the spectrum of the wind gusts and the torque transfer function. The wind gust spectrum can be represented approximately by the form

$$W_g(f) = W_o \frac{f_a^2}{f_a^2 + f^2} \qquad (0 < f < f_{\text{max}})$$

(A Gaussian curve may also be fitted to most measured wind gust spectra). The power spectrum of the servo error is given by the torque spectrum divided by the square of the torque transfer function.

$$W_s(f) = 4K_w^2 v_w^2 \frac{W_g(f)}{Y_p^2(f)} \qquad \text{(10.19)}$$

The rms noise error owing to gusts may be found by integrating this spectrum.

$$\sigma_{\text{ws}}^2 = 4K_w^2 v_w^2 \int_0^{f_{\max}} \frac{W_g(f)}{Y_p^2(f)} \, df \qquad \text{(10.20)}$$

Asymptotic forms of the wind and error spectra are shown in Fig. 10.9. In general, adequate results may be obtained if these spectra are approximated by a series of straight-line segments, which may be integrated separately by using simple forms. Such a procedure, applied to the spectrum of Fig. 10.9, leads to an rms servo error of 0.012 mil (12 μrad) for $v_w = 70$ ft per sec. Most of the servo-error power appears near the frequency f_x in

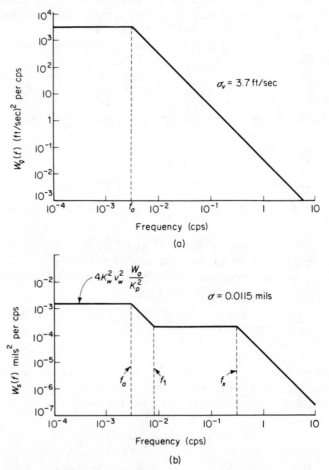

Figure 10.9 Spectrum of wind gusts and resulting servo error spectrum ($\beta_n = 0.7$ cps). (a) Spectrum of typical wind gusts. (b) Spectrum of servo error owing to wind gusts.

the spectrum, indicating the importance of wide bandwidth and high tachometer-loop gain in minimizing wind-induced error. Use of the wide-band-width mode would reduce the error considerably, since the torque transfer function is at least ten times higher over the region between f_1 and f_x, and remains higher up to $f = 2$ cps.

The problem of wind-induced error becomes rapidly worse as the size of the antenna is increased. The wind torque increases roughly according to the square of the antenna diameter, and the available gain in the servo system tends to be reduced sharply as a result of mechanical resonance and the requirements of loop stability.

Collimation and Drift Errors in the Electrical Axis

If the electrical axis of the radar is stable over long periods of time, the accuracy with which it can be collimated is dependent primarily upon the care and patience which are exercised in calibration. Comparisons of the electrical axis with boresight telescope observations on visible targets can be made over a period of time and over a range of operating conditions such that noise components of error are averaged to very low values. This process is especially accurate if photographic readings are taken from the telescope while a point-source target is being tracked at relatively high elevation angle, where multipath and propagation errors are minimized. The residual errors are caused by drift components which change too rapidly to be removed by calibration. These can be the result of variation in several operating parameters of the radar, and of environmental factors such as uneven heating of the radar components. In a complete error analysis, the variation in position of the axis must be determined as a function of the following:

1. Frequency of operation within the radar band.
2. Tuning of the system (center i-f frequency).
3. Phase or gain variations in the receiver(s).
4. Signal strength.
5. Temperature or intensity of solar (thermal) radiation.

When these effects are determined, it will be possible to devise calibration and collimation procedures, and to estimate the errors remaining in the radar output at various times after calibration. For example, consider the experimental curve of Fig. 10.10, which represents the shift in position of the null point in the r-f error pattern of the AN/FPS-16 as its operating frequency is varied over a 10 per cent band. No r-f tuning elements are included in the antenna system, so the error can be reduced during operation only by collimation at the frequency to be used (it can be reduced in data

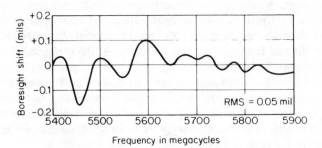

Figure 10.10 Typical boresight shift vs. r-f tuning.

processing by use of a calibration curve). If it is assumed that many different frequencies are to be used in the period between collimations, and that no calibrations are to be applied to correct for the shift, the error can be expressed as the rms value of the curve shown, or about 0.05 mil. This presumes that the frequency chosen for collimation gives an error near zero. When a different frequency is used for operation, a bias error will appear in elevation and traverse angle. The azimuth error Δ_a will be equal to the traverse bias Δ_{tr} at low elevation angles, but will increase as the target rises in elevation.

$$\Delta_a = \Delta_{tr} \text{ secant } E \qquad \qquad (10.21)$$

Although the azimuth error becomes infinite at zenith, the linear error in target position does not change.

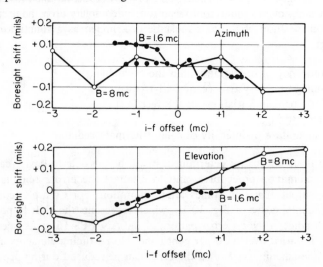

Figure 10.11 Typical boresight shift vs. i-f tuning.

A curve showing the sensitivity of a precision monopulse radar to i-f excursions (tuning error) is given in Fig. 10.11. Assuming that collimation is carried out with the system properly tuned, the shift in the tracking axis may be estimated for any operating condition which leads to tuning error by assigning to the tuning error a probability distribution and calculating the rms value of the corresponding error from this figure. The tuning error may arise from Doppler shift, drift in the transmitting frequency of a beacon, or uncorrected drifts in the radar oscillators. Let us assume that the tuning error in a particular case is normally distributed with a standard deviation of $\frac{1}{3}$ mc, and that the wide receiver bandwidth is used in the radar. For simplicity, we shall represent the tuning error distribution by a series of rectangular segments, and we shall calculate the rms error based on the boresight shifts occurring at the center of each rectangle. A typical calculation based on the elevation error curve [Fig. 10.11(b)] takes the following form

Tuning error (mc)	Boresight error x	Probability P	Variance Px^2
-1.0 to -0.67	-0.063	0.023	0.00009
-0.67 to -0.33	-0.37	0.136	0.00019
-0.33 to 0	-0.012	0.341	0.00005
0 to 0.33	0.012	0.341	0.00005
0.33 to 0.67	0.037	0.136	0.00019
0.67 to 1.0	0.063	0.023	0.00009
		1.000	$0.00066 = \sigma^2$

The rms borsight error for this case is 0.026 mil. The simplified procedure leads to a slight overestimation of the error, but is well within the accuracy required for error analysis. Results of the type shown above may be used to establish realistic limits for tuning accuracy of the receiver and transmitter system, to guide design of automatic-frequency-control circuits, or to set the allowable deviations in i-f amplifier phase shift and linearity over the center portion of the passband.

In an amplitude comparison monopulse system, the position of the tracking axis will also vary as a function of phase difference between the reference and error signals arriving at the error detector.* The extent of this variation depends upon the depth of null in the error pattern of the antenna, which in turn, depends upon the magnitude of the quadrature component remaining in the error channel after cancellation of the four horn outputs in the comparator (see Fig. 9.5). If the null depth (the ratio G_1/G_3 on the axis, in Fig. 10.5) is represented by G_n, then the quadrature component E_q will be equal to the reference channel voltage E_r divided

$$E_q = E_r / \sqrt{G_n}$$

* John H. Dunn and Dean D. Howard, "Precision Tracking with Monopulse Radar," *Electronics*, **35**, No. 17 (April 22, 1960), pp. 51–56.

by $\sqrt{G_n}$. When the reference and error phase shifts are exactly equal, the quadrature component is completely rejected by the phase-sensitive detector, and no boresight error will result. However, when the relative phase of reference and error signals arriving at the error detector is in error by an amount $\Delta\phi$, the projection of the quadrature component on the reference signal vector will not be zero, but $E_q \sin \Delta\phi$. The servo will drive the antenna until this component is balanced by an equal and opposite error signal, producing a boresight shift given by

$$\Delta_\phi = \frac{\theta E_q \sin \Delta\phi}{k_m E_r} = \frac{\theta \sin \Delta\phi}{k_m \sqrt{G_n}} \simeq \frac{\theta \, \Delta\phi}{k_m \sqrt{G_n}} \qquad (10.22)$$

where θ is beamwidth and k_m is tracking error slope. When the null depth is measured with respect to the maximum lobe in the error pattern rather than the reference pattern main lobe, the factor k_m in the denominator may be increased accordingly, and the factor 2.0 is sometimes used in place of k_m. The sensitivity of the axis to relative phase shifts is one of the factors which accounts for the interaction between tuning and boresight shift, since it is difficult to match the phase slopes of two or more receivers over a wide bandwidth. It also accounts for the sensitivity to signal strength, discussed below. If the null depth is 35 db, the error in the axis will be about 1 per cent of the beamwidth per radian of differential phase shift $\Delta\phi$. Phase tolerances of 5 to 15 deg are typical of monopulse receivers operating within their normal range of signal strengths and frequencies. The phase of local-oscillator signals at the mixers must also be controlled to avoid errors.

In the usual form of amplitude comparison monopulse, automatic gain control is used to normalize the error signal amplitude and to avoid overloading the reference channel. As the agc voltage changes the gain of the i-f amplifier stages, it also introduces phase shift through the "Miller effect," which changes the input capacitance of the amplifier tubes. Since it is impossible to match perfectly the characteristics of the three receiver channels, there will be some error introduced by this phase shift. A typical plot of resulting error is shown in Fig. 10.12. The error can be made very small over a considerable dynamic range, but it grows rapidly as the agc action brings one or more stages in the amplifier near cutoff. Since the receiver on which these data were gathered could not operate properly over a dynamic range much greater than 60 db, it was necessary to apply sensitivity time control (STC) to the preceding preamplifier, and to program the output of the transmitter, in order to hold the received signal levels within the operating range. Maintenance of good null depth is important in reducing these components of error.

Other sources of boresight shift may be important in tracking radars of

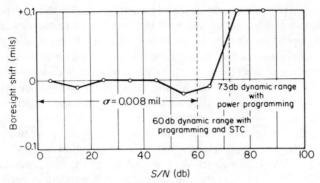

Figure 10.12 Typical boresight shift vs. S/N ratio.

various types. In each case, the rms value of the error must be evaluated over the range of operating conditions to be encountered, and an estimate must be made of the frequency components which make up the error. It is then possible to assign the error to the class of "bias" or to "noise," and to determine its effect on the radar's output data. The radar components which must be checked experimentally for possible tracking errors include the servo system, which may contribute electrical or mechanical noise; the antenna reflector and feed support structure, which may change dimensions as a result of solar heating or conduction of heat from operating mounted near critical elements; the antenna reflector and feed (in either monopulse or scanning-type radars), which may be sensitive to frequency or polarization of the received signal; and the receivers and error detectors, which may introduce unexpected noise and bias errors into the system. The preceding discussion, applicable specifically to one form of monopulse radar, is intended to serve as a guide to the procedures used in analysis of this class of error in any tracking system. Where the form of the tracker differs, there will be new sources of possible error, and some of those listed here may not appear.

Radar-Dependent Translation Errors

The ability of the radar to keep its tracking axis pointed at the target is only one part of the tracking problem. It is also necessary to translate this axis position into output data in a form usable by computers and similar devices, and properly related to a known coordinate system. A list of error sources which affect this translation was given in Table 10.1. The techniques for evaluating and expressing these error components are similar to those described above, but there is little general theory which can be applied in the absence of a specific example. Most of the "bias" errors listed will be functions of the angles at which the radar is pointed, whereas the noise components will vary in a more rapid or random fashion.

An example will serve to illustrate the analysis of errors of this type. Figure 10.13 shows a plot of error in the readings of a high-accuracy digital encoder, measured over both short intervals of time and angle and at a number of such intervals around a complete circle. Two types of error are evident: one is a rapid, noiselike variation in the indicated position when the encoder is fixed at any point; the second is a systematic error which rises and falls slowly as the encoder is rotated, and which is almost exactly repeatable with any single encoder unit. The distribution of total error is shown in Fig. 10.14, along with the distributions of the two separate components. Since the two errors are uncorrelated, it is possible to express the rms value of the total error as the square root of the sum of the two variances. The spectrum of the systematic component is proportional to the rate at which the target moves in angle, with spectral components at har-

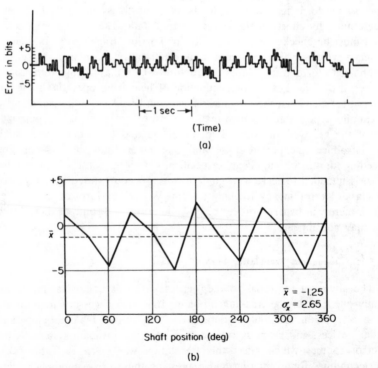

Figure 10.13 Digital encoder error (19-bit encoder, least bit = 0.012 mils). (a) Noise error with encoder shaft fixed. (b) Systematic error measured over one shaft rotation.

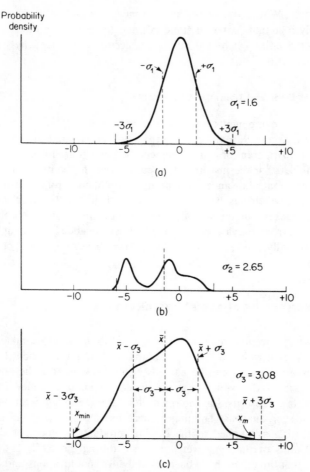

Figure 10.14 Distributions for encoder error. (a) Distribution of noise about short-term average. (b) Distribution of systematic error about mean value. (c) Distribution of total error about mean value.

monics of the rotation rate of the encoder. The noise component has been found to approximate white noise over the bandwidth determined by the sampling rate of the encoder. In other words, successive samples have uncorrelated noise components even at readout rates of 50 or 100 samples per sec. When tracking a nonmaneuvering target, it would be possible, in principle, to reduce the noise by taking an average (or smoothing) over a period containing many samples, and to reduce the systematic portion by careful calibration against some more accurate standard. This would leave a residual error considerably less than the rms value shown for the two components. With a device of this accuracy, however, it is difficult to find and apply a suitable external standard, and to reduce the other sources of error to the point where the additional accuracy provided by calibration would be useful.

Target Error and Nonradar Components

The error components listed as "target dependent" have been discussed in Chapters 3 and 9. Propagation errors will be covered in some detail in Chapter 15. Apparent or test-instrumentation errors are listed in Table 10.1 as a reminder that the instruments used to evaluate the radar error are not perfect, and that allowance must be made for their errors in using experimental data which has been gathered with their aid. In error analysis of an angle tracker, all the actual error components, along with the radar-dependent errors previously described, must be added in an appropriate fashion (usually rms), to find the total error. An example will be given to illustrate this process.

10.3 EXAMPLE OF ANGLE ERROR ANALYSIS

To show how the preceding theory is applied to arrive at estimates of total angle error of a radar, a specific example will be given. The AN/FPS-16 Instrumentation Radar can be used as an illustration. It is the only monopulse tracking radar which was developed on an unclassified basis, and which has been described in the open literature in enough detail to provide a basis for such analysis. The major characteristics of this radar are shown in Table 10.2, as applied to the original production model of the radar. Some 40 units of this type were produced during the 1958–1961 period, and since that time there have been a number of modifications and improvements placed in the field. Some of these improvements will be described later, insofar as they relate to measurement accuracy.

Errors of Fixed RMS Value

A number of the errors listed in Table 10.1 may be described as "fixed" in the sense that their rms values are nearly independent of the conditions under which the radar is operated. This is especially true of the radar-dependent translation errors (principally mechanical in nature), but it may apply also to some of the other errors over a broad range of conditions. As a first step in analysis of total error, we may list these fixed error values, as in Table 10.3, to set the minimum angle error attainable on any target. In some cases, the values shown represent the result of operating experience

Table 10.2 CHARACTERISTICS OF INSTRUMENTATION RADAR AN/FPS-16

Frequency	5400 to 5900 mc
Power output	1.0 Mw peak (5480 mc)
	250 kw peak (5450–5825 mc)
Pulse widths	0.25, 0.5, and 1.0 μsec
Repetition rates	160 to 1707 pps
Pulse codes	Up to 5 pulses
Duty cycle	0.001 at 1.0 Mw, 0.0016 at 250 kw
Antenna size	12-ft diameter reflector
Antenna gain	44.5 db
Beamwidth	1.1°
Monopulse feed type	4-horn, amplitude comparison
Receiver noise factor	11 db maximum
Receiver bandwidths	1.6 and 8.0 mc
Coverage	Azimuth 360°
	Elevation −10° to 85°
	Range 500 to 400,000 yd
Maximum rates	Azimuth 48°/sec
	Elevation 37°/sec
	Range 10,000 yd/sec
Servo bandwidths (β_n)	0.5 to 6 cps
Velocity error constant	300 (angle), 2500 (range)
Acquisition scans	Adjustable circle and sector
Data outputs	Digital, synchro, and potentiometer
Displays	Dual A-scope, dials, digital
Detection range (1.0 m²)	150 n miles
Accurate tracking range (1.0 m²)	75 n miles
Angle error (bias)	0.1 mil rms
(noise)	0.1 mil rms
Range error (total)	10 yd rms
Input power	75 kv, 208/115 v, 60 cps
Installation	Fixed building

(The AN/MPS-25 is a transportable version of the AN/FPS-16 with identical electrical characteristics.)

(e.g., collimation and alignment error), whereas others are based on typical measured values or those calculated in the preceding paragraphs. The boresight axis drift was calculated on the assumption that the system is calibrated within 50 mc of the operating frequency, tuned within $\frac{1}{3}$ mc, and that the differential phase shifts between receiver channels are within 5 deg rms. The tropospheric refraction terms are representative of typical conditions where the target is about 5° above the horizon.

Errors of Variable RMS Value

In order to proceed further with analysis of error, specific operating conditions must be assumed. For our example, we shall assume that the

Table 10.3   ERRORS OF FIXED RMS VALUE

Error source	Angle error in mils rms	
	Bias	Noise
Boresight axis collimation	0.025	–
Boresight axis drift	0.04	–
Wind forces (50 mph)	0.02	0.012
Servo noise and unbalance	0.01	0.02
Subtotal: Radar-dependent tracking errors	0.052	0.023
Leveling and north alignment	0.015	–
Orthogonality of axes	0.02	–
Mechanical deflections	0.01	–
Thermal distortion	0.01	–
Bearing wobble	–	0.005
Data gear error	–	0.03
Digital encoder error	–	0.025
Subtotal: Radar-dependent translation errors	0.023	0.04
Tropospheric refraction	0.05	0.03
Total "fixed" error	0.078	0.054

target is a satellite in orbit at 100 mi altitude, a corner reflector being used to provide extended radar tracking range. Radar operation at 1 Mw peak power, one μsec pulse width, and 340 pps will be assumed, with the target moving at 4 miles per sec along an orbit which crosses a point 40 miles in ground range from the radar. We shall evaluate the errors near crossover ($R = 110$ miles) and at a point near the maximum range of the ranging system ($R = 200$ miles).

As a starting point, signal-to-noise ratios are determined by using the radar range equation, and angular rates and accelerations found from Tables 9.4 and 9.5. The optimum servo bandwidths will then be calculated from Eq. (9.25). The results of these steps are shown in Table 10.4. Only the error components caused by multipath, glint, and scintillation remain to be estimated.

At this high elevation angle, the multipath error shown in Fig. 10.7 will drop to very low values, and may be ignored. When a corner reflector is used as a target, the glint term is reduced to zero, and any scintillation error is eliminated by the use of fast agc in the monopulse radar. As a result, the total angle error may be found by combining the "fixed" error level from Table 10.3 with the thermal-noise and lag terms from Table 10.4. It should be noted that some of the azimuth errors are larger near crossover

Table 10.4 THERMAL-NOISE AND LAG ERROR CALCULATION

		Azimuth		Elevation	
		$R = 110$ mi.	$R = 200$ mi.	$R = 110$ mi.	$R = 200$ mi.
S/N ratio	(db)	30	20	30	20
Angle sector	(deg)	60–90	12–18	60–70	25–35
Angle rate (max)	(rad/sec)	0.1	0.006	0.017	0.011
Angle acceleration (max)	(rad/sec²)	0.0065	0.0004	−0.003	0.0004
Effective target acceleration	(g)	54	14	−63	150
Optimum servo bandwidth β_o	(cps)	6.5	1.9	7.2	2.0
Thermal-noise error* σ_t	(mil)	0.09	0.078	0.043	0.078
Velocity-lag error (rms) ε_v	(mil)	0.30	0.02	0.057	0.037
Acceleration-lag error† ε_a	(mil)	0.05	0.04	−0.033	0.04
Total "variable" error	(mil)	0.32	0.1	0.1	0.11
Fixed error from Table 10-3	(mil)	0.095	0.095	0.095	0.095
Total error	(mil)	0.33	0.138	0.138	0.145
Position error of target	(ft)	80	140	92	175

* Calculated for bandwidth not exceeding 6 cps max, and includes secant term.
† Although the elevation lags tend to cancel before crossover, they add after crossover, and the totals shown include the sum of the two lags.

because of the high elevation angle, which causes the traverse error to be multiplied by the secant of elevation to give a larger azimuth error. The error in the position of the target is not increased, because the azimuth error is multiplied by the ground range, rather than by slant range, to find the linear error in target position. The rms values of error in both angular coordinates are summarized at the end of Table 10.4, which also shows the error converted to target position in feet. The most serious bias error components are those due to azimuth lag and to fixed error in elevation. The largest noise components are from thermal noise. Figure 10.15 shows how the error varies with time during the track from 200 miles to crossover, and also shows how thermal noise and total error would increase at long range if servo bandwidth were not reduced to the optimum value.

Use of a Beacon to Reduce Error

The large lag and thermal-noise components of error in the above example can be reduced by use of a beacon in the target. A relatively small beacon, providing a useful output of 400 w, increases the signal-to-noise ratio from 20 to about 36 db at 200 miles range. The angle servos may then be kept at their maximum bandwidth during the entire track without encountering excessive thermal noise, and the acceleration errors will be reduced when compared to those regions which required reduced bandwidth

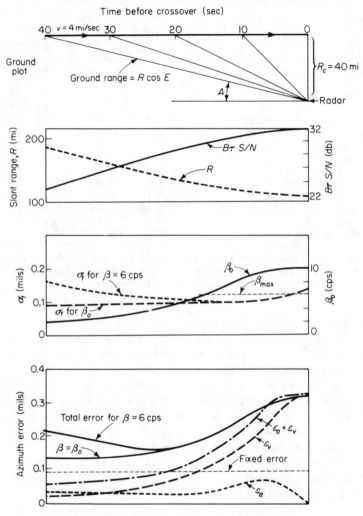

Figure 10.15   Calculation of error for crossing target.

in the echo-tracking case. Near crossover, however, the errors remain essentially the same, because the increased S/N ratio reduces the already small thermal-noise term, while the servos remain at their maximum bandwidth of 6 cps. If the satellite had not been originally assumed to carry a large reflector, the improvement due to use of the beacon would have been much greater. In fact, it might well have been impossible to establish a track at all without the beacon.

Effect of Open-Loop Data Correction

The preceding example in which a beacon was used shows the value of open-loop data correction. When the target approaches crossover, the optimum system bandwidth varies from 4 to about 11 cps, and the antenna servos are unable to provide this range. If open-loop correction in combination with a reasonably high servo bandwidth is used, the tracking error can be held well within the linear range of error slope, and large reduction in lag errors can be obtained. In addition, the velocity-lag error can be eliminated almost completely by the second-order response of the data corrector. This brings the over-all system error to about the level of the fixed error components, with a slight increase caused by thermal noise.

Since, in this case, the antenna lag errors are less than $\frac{1}{2}$ mil, the accuracy required of the open-loop correction circuits is not extreme, and no new error sources of significance need be introduced. Correction to about 10 per cent of the antenna lag is sufficient, and this implies that the error slope should be known to within 10 per cent of its true value. It has been shown* that calibrations of this accuracy can be maintained in monopulse radars of this type over a dynamic range greater than that required in our example.

---

* M. Korff, C. M. Brindley, and M. H. Lowe, "Multiple-Target Data Handling with a Monopulse Radar," *Trans. IRE*, **MIL-6**, No. 4 (Oct. 1962), pp. 364–65.

In this chapter we shall extend to the radar's range-measuring system the analysis of errors which was performed for the angle tracker. The basic processes of range tracking will be described, and techniques used in pulsed radar will be analyzed with respect to their ability to overcome thermal noise. The effect of target-dependent factors such as glint and lag will be considered, and the equation for optimum bandwidth in the range-tracking servo will be given. Other basic sources of error, such as uncertainty in the velocity of light, will be discussed and included in the complete error analysis. One example will be given to demonstrate how the analysis may be applied to estimating the over-all range-tracking error and to indicate what portions of the system could be improved to reduce this error in a specific tracking case.

11.1 DESCRIPTION OF RANGE-TRACKING DEVICES

Basic Processes of Range Tracking

Radar range measurement con-

RANGE-TRACKING SYSTEMS

11

sists of the determination of time delay between transmission and reception of a given signal. The process is basically one of correlation, as noted in Chapter 2. The received signal is multiplied by a stored and delayed replica of the transmission, and the delay is adjusted to obtain maximum output. Range tracking may be defined as the process by which the delay in the stored reference is kept matched to the delay of the selected signal, in order to provide continuous data on the range of the target. The basic steps in range tracking may be summarized as follows:

1. Generation of recognizable features in the transmitted wave (e.g., discrete pulses of low duty cycle).
2. Storage and delay of a reference signal (e.g., rectangular gate or strobe marker matched to signal width).
3. Measurement of the relative delay of reference and received signals (e.g., viewing of the two signals on an expanded A-scope).
4. Correction of the reference delay to obtain coincidence or maximum correlation between reference and received signals.
5. Measurement of reference delay.
6. Encoding and reading-out of reference delay to indicate signal delay and hence target range.

In a tracking radar, the target will normally remain in the antenna beam for prolonged periods, and the range-tracking system will process repeated samples of the echo signal to arrive at an accurate measurement of range. The velocity and acceleration of the target will place limitations on the bandwidth of the measuring circuits, and hence on the accuracy with which the target range may be determined. This accuracy will also be affected directly by the bandwidth of the transmission, the characteristics of the circuits used in storage and delay measurement, and by the propagation medium through which the measurement is made. The characteristics of the common types of range-tracking systems will be described briefly below, and this description will be followed by a detailed analysis of errors owing to radar and target parameters.

Manual Tracking

Most of the radars used in World War II contained manual-tracking systems for range measurement. In these systems, the echo (assumed to be a rectangular pulse) was displayed on one or more A-scopes, along with a marker indicating the stored reference delay. The marker was driven by manual controls such as handwheels or handcranks, through which the human operator could maintain alignment between marker and echo signal. The marker might consist of an electronic pulse or gate, indicated by intensity or deflection modulation on the display, or it might be a mechanical cursor scale placed in front of the display tube. In cases where angle tracking was to be carried out, the electronic gate served to exclude unwanted signals from the angle error detectors, as well as providing a reference for range measurement. By using linear sweeps or carefully calibrated circular or spiral sweeps, it was possible to translate motion of the gate or cursor into shaft rotations suitable for driving data output devices, such as synchros or potentiometers.

Any signals which could be detected by the operator, even intermittently, could be tracked in the manual system, with an accuracy of some fraction of the pulse width. By providing an expanded display which moved in delay along with the cursor, target visibility could be enhanced, and the full resolution capability of the pulse and receiver could be maintained. Depending upon the alertness of the operator and the effectiveness of his manual controls, it was possible to achieve tracking bandwidths approaching 1 cps, at least for short periods when the signal-to-noise ratio was high. The performance of continued tracks of high accuracy placed great burdens on the operator, and the response of the human-controlled system could not be depended upon in cases where high accuracy and bandwidth were required. Manual systems cannot be neglected even in today's advanced radars, however, and many of these systems could be improved by provision of adequate display and control elements for use by the operator. This is particularly true when intermittent signals must be acquired and tracked, or when interference or clutter contaminates the signal.

Aided Tracking

One advance over the pure manual tracking systems was provided by introducing "aided tracking," in which handwheel motion produced both displacement and rate control of the cursor. Proper coupling of the rate drive with the position drive made it possible for the operator to approach the correct rate setting with minimum overshoot or hunting, and greatly relieved the burden of following moving targets. Handwheel motion was required only during initial acquisition and in following target accelerations. During periods when the signal faded below the visible level, or when clutter or interference obscured the signal, the stored target rate permitted the operator to "coast" for several seconds before the target had an opportunity to leave the region around the cursor. The use of aided tracking remains an important technique whenever intervention by the operator is required in a tracking system.

Automatic Tracking

Modern tracking radars place great reliance on automatic (servo-controlled) tracking to achieve high accuracy and wide data bandwidth on well-defined targets. The automatic range tracker follows the same form as the manual system, except that the error in setting of the reference

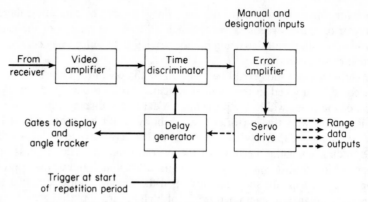

Figure 11.1   Basic elements of automatic range tracker.

delay is measured by a "time discriminator" instead of by eye, and the correction is applied by an instrument servo or electronic time-delay control loop (see Fig. 11.1). The most common type of time discriminator is the "split-gate" circuit, in which the signal is applied to a pair of rectangular gates which are staggered in time, and whose outputs are subtracted to indicate the direction and magnitude of the error. When the "center of gravity" of the echo signal coincides with the crossover point between the early and the late gate, the output error signal is zero. Each gate must be made slightly longer than one-half the pulse width to obtain maximum sensitivity to error.

The design of the follow-up servo for range is based on the theory outlined in Chapter 9 for angle tracking. The usual servo contains a drive motor which constitutes one stage of integration, and to this is added an equalizer or a second integration stage to increase the velocity error constant to the required value. Since the range servo need not drive a heavy antenna, its bandwidth is limited primarily by considerations of signal-to-noise ratio and desired smoothing of data. When the input signal is subject to fading, it is desirable to provide an effective rate memory, permitting the system to coast through temporary periods of low signal strength. The relationships between range error and bandwidth will be discussed in more detail later.

The Leading-Edge Tracker

In place of the "center-of-gravity" measurement performed by the split gate, it is possible to use a circuit which measures the range to the point of maximum slope on the leading or trailing edge of the echo signal. This portion of the pulse contains the spectral components of greatest frequency spread, and provides increased resolution in the system, although not necessarily increased accuracy. In one method, the received pulse is passed through a differentiating circuit, producing the waveform shown in Fig. 11.2. A very narrow tracking gate is then placed under the first output pulse of the differentiator, producing a null in error signal when the gate is centered exactly on the maximum point, which is the point where the slope of the echo pulse is greatest. When the target consists of an extended group of reflectors, each of about the same amplitude, it is possible to maintain track on the nearest (or farthest) reflector and to exclude the intermediate echoes from the range- and angle-tracking circuits. Better performance could also be obtained by reducing the pulse width and gate width in the split-gate type of tracker.

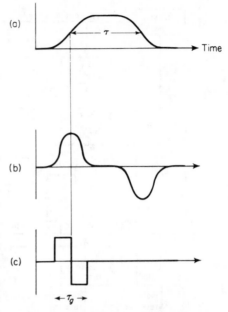

Figure 11.2 Leading-edge tracking. (a) Video pulse. (b) Differentiator output. (c) Tracking gate.

Range System Data Outputs

The range trackers described above generate internal time delays or mechanical shaft rotations which correspond to the range of the target. Data outputs are provided by coupling these indications to some form of transducer, such as a synchro generator, a resolver, a potentiometer, or a digital encoder. In the systems where a shaft is driven by the operator's control or by the range servo, it is a simple matter to couple the chosen output devices to this shaft, gear trains being used to provide two-speed or three-speed outputs of high accuracy. The block diagram of a high-accuracy, shaft-driven system is shown in Fig. 11.3. The basic time reference is provided by a crystal-controlled oscillator, whose period is equal to the range-delay time corresponding to 2000 yd. The variable delay for the internal reference pulse is generated in three steps. A sinusoidal output from the reference oscillator is passed through a precision phase-shifter and shaped to produce a series of pulses at 2000 yd intervals which are locked within 1 yd (0.006 μsec) of the range indicated by the resolver

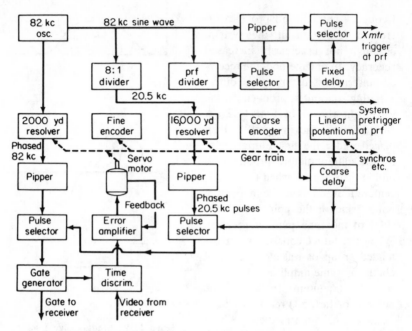

Figure 11.3 Precision electromechanical range tracker.

shaft. In order to select a single pulse from this train, at the desired range delay, a single-speed potentiometer is used to control a linear time-delay generator of the phantastron type, which covers the entire range of the radar (1 million yd in this case). This coarse delay in turn selects a pulse from the 16,000 yd series, which is generated and phase-shifted in synchronism with the original 2000 yd pulses. The tolerance on the phantastron must be such as to assure exclusion of the 16,000 yd pulses before and after the selected one, and these pulses must in turn assure that only the desired 2000 yd pulse is selected as the reference. Under these circumstances, the only error to appear at the output of the reference system is the error in generation and phase-shifting of the 2000 yd pulse train, and the output covers the entire one million yards under control of the servo-driven shaft.

The selected 2000 yd pulse initiates the split-gate time discriminator, whose output is used by the servo to control the gear train. On this gear train are mounted the phase-shifters, the coarse-delay potentiometer, and the data output devices. The most accurate indication of target range appears at the shaft of the 2000 yd resolver, and this shaft is coupled closely to the fine digital encoder and the fine synchro output. Other (coarse) outputs are driven by appropriate shafts on the gear train, but with relaxed mechanical tolerances. In addition to the circuits shown in the block diagram, there are a number of auxiliary systems which provide for initial designation of the tracking gate, for automatic detection and lock-on, and for generation of supplementary gates for displays, beacon tracking, etc.

Digital Ranging Systems

As digital computers became available for processing of radar data, the requirement for analog outputs became less important, and all-electronic ranging systems of high accuracy were developed. In these systems, there are no shafts or similar mechanical components, and inertia is eliminated completely. A modern ranging system of this type is shown in Fig. 11.4. The basic functions remain the same as before, but the instrumentation differs. The internal reference delay is now generated under control of the

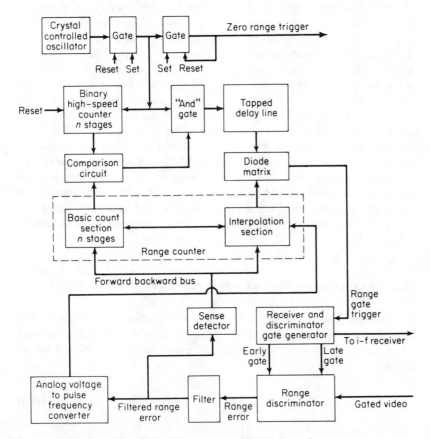

Figure 11.4 Basic block diagram of digital ranging unit.

number stored in the "range counter," which selects a given pulse from the output of the crystal-controlled oscillator and adjusts its delay to the exact figure required. Operating at a frequency of 1.64 mc, these clock pulses are separated by a delay equivalent to 100 yd. In order to interpolate between clock pulses, the selected pulse is sent through a delay line with precisely placed taps, and a diode matrix selects the output to the nearest 3.25 yd, for initiation of the split-gate time discriminator. From this point, the operation is as in the electromechanical system, except that the servo is entirely electronic, controlling the setting of the range counter to select the reference pulse and fine delay. At any instant, the number in the range counter represents this reference delay, and hence the target range (to the nearest 3.25 yd). By averaging the contents of the counter over a few repetition periods, the output granularity error is reduced to a few feet or less.

Digital systems which are designed for high accuracy retain the analog time-discriminator circuit because they must measure to the center of gravity of echo pulses which are immersed in noise, and must average over many pulses to obtain the desired output. The analog discriminator and servo-equalizer circuits are still well suited to this function, and do not limit the performance of the system. Early attempts to measure directly on the leading edge of the echo signal proved unsuccessful, because it was impossible to operate successfully on signals near noise level. When long pulses are used, it would be possible to apply digital processing to the measurement of the center of gravity, or to instrument a more exact solution to the correlation process, if this were desired. Such a process would require that many samples be taken over the pulse width, so that interpolation within the pulse width could be performed accurately. Another approach to this technique involves the use of a matched filter in the i-f system, so that the output pulse can be sampled at its maximum amplitude to indicate delay.

11.2   THERMAL AND TARGET NOISE IN RANGE MEASUREMENT

The fundamental limitations in range measurement owing to thermal noise were discussed in Chapter 2, where it was shown that the ideal accuracy for a single rectangular pulse was expressed by a time-delay error as follows

$$(\sigma_{r_t})_{\text{opt}} = \frac{1}{2B\sqrt{S/N}} \qquad (B\tau \gg 1) \qquad \textbf{(11.1)}$$

The above result was shown to be consistent with Woodward's theory and with approximate analysis based on the combining of leading-edge and trailing-edge measurements for a trapezoidal pulse. This theory will now be compared with the performance of practical range-tracking systems, and the effects of target noise (glint, scintillation and lag errors) will be discussed. The limitations in tracking at low S/N ratio will also be explored, and methods will be given for calculating optimum servo bandwidths and i-f filter characteristics.

Noise Output of the Split-Gate Tracker

Let us consider a range tracker in which a pair of adjacent, rectangular gates is applied to the i-f or video signal, the difference in average voltages during the gates being used as the error signal. If it is assumed that the gains of the two gated channels are matched, the long-term average output will tend toward zero, but the output measured during a single repetition period will have some finite value. This output will be proportional to the amplitude of noise in the receiver, and also dependent upon the width of the gate and the bandpass characteristics of the receiver. We shall express the noise output from a single repetition period as

$$E_{no} = \int_{\substack{\text{Early} \\ \text{gate}}} \frac{E_n}{\tau_g}\, dt - \int_{\substack{\text{Late} \\ \text{gate}}} \frac{E_n}{\tau_g}\, dt \qquad\qquad \textbf{(11.2)}$$

The ratio of noise output E_{no} to noise input E_n is shown in Fig. 11.5 as a function of $B\tau_g$, the product of i-f bandwidth and total gate width. Curves are shown for four different types of i-f filter. It is seen that the output approaches $(B\tau_g)^{-1/2}$ for large $B\tau_g$ in all cases. Large $B\tau_g$ implies that the noise voltages are essentially independent during most of the gate period. However, as the width of the gate approaches $2/B$, the two voltages become significantly correlated, and the output drops below the straight-line relationship. Below $B\tau_g = 1$, the output tends to fall with $B\tau_g$, and to follow the line $0.44B\tau_g$ for very low values of this parameter in the rectangular-filter case. With the matched filter, the output falls more slowly along a curve given by $\sqrt{B\tau_g/6}$. The single-pole filter follows generally the curve for the matched filter, approaching $\sqrt{B\tau_g/3}$ for low values of $B\tau_g$. Cascade-filter systems will follow curves close to that of the rectangular filter, falling off approximately as $0.5B\tau_g$.

The process described by Eq. (11.2) and Fig. 11.5 is a close approximation of that used in many actual split-gate trackers, which operate on the difference between the average video voltages during the early and late gates. The video output of the receiver is first multiplied by the bipolar gating function sketched in Fig. 11.5. The resulting bipolar video wave form is

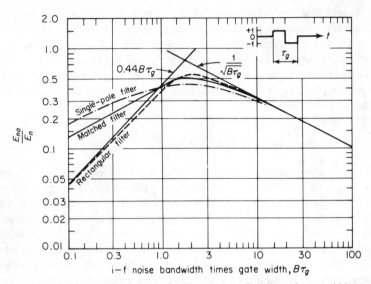

Figure 11.5   Noise output of split gate as a function of gate width and filter characteristics.

applied to an RC integrating circuit whose time constant is long relative to τ_g. After the conclusion of the gate, the integrated error voltage is held in a "boxcar" or "linear holding circuit,"* and amplified to provide the appropriate scale factor (depending upon the ratio RC/τ_g). The output voltage then represents the difference between the integrals of video output during the two halves of the gate, weighted by $1/\tau_g$, as expressed in Eq. (11.2). For wide gates, the difference in integrated video noise (as represented by the charge on the storage capacitor) increases only as the square root of the gate width τ_g, or the number of independent noise samples $B\tau_g$. This leads to a noise output voltage which varies inversely as $\sqrt{\tau_g}$, when bandwidth is held constant and weighting of $1/\tau_g$ is applied.

Calculation of split-gate response in the presence of noise which is correlated over the width of the gate starts with the impulse-response function $f(t)$ of the i-f filter. This is the transform of the filter frequency-response function $F(j\omega)$, and represents the output envelope waveform following an impulse of noise applied to the filter. If a linear envelope detector at the receiver output is assumed, the split-gate output voltage $g(\tau)$, resulting from an impulse which occurs at a time τ with respect to the

---

* Mischa Schwartz, *Information Transmission, Modulation and Noise* (New York: McGraw-Hill Book Company, 1959), pp. 178–80.

center of the gate, can be found from the convolution integral* of $f(t)$ and the gating function $h(\tau - t)$.

$$g(\tau) = \int_{-\infty}^{\infty} f(t)h(\tau - t)\,dt$$

The mean-square error voltage out of the system is then related to the i-f noise voltage E_n by

$$E_{no}^2 = \frac{E_n^2 B}{\tau_g^2} \int_{-\infty}^{\infty} g^2(\tau)\,d\tau$$

Figure 11.5 was calculated by performing these integrations, the impulse functions of typical and idealized i-f filters being used.

Identical results can be obtained by considering the frequency response of the split-gate circuit, which is shown in Fig. 13.6(g), for the same time function used as a differentiating filter for data. The response is assumed to apply symmetrically to frequencies on either side of the i-f center frequency, and a composite response function is found by multiplying the gate response by the filter response. When this is multiplied by the noise spectral density N_o and integrated over all positive frequencies, the result is the mean-square noise output of the split-gate system. The square root of this, when divided by the gate width τ_g, will yield values of E_{no} identical to those shown in Fig. 11.5.

It should be noted that use of a linear envelope detector has been assumed in these derivations, and this is consistent with the usual practice in design of range trackers. Use of a square-law detector can be expected to give similar results, but with some differences in performance when the gate extends well beyond the signal and when the S/N ratio is large.

Response of the Split Gate to Signals

The signal output in the presence of an arbitrary signal wave form may be calculated as shown in Fig. 11.6. It is assumed that the gate has been centered initially on the point of maximum amplitude E_m. A displacement by Δt will add to the area within the first gate the quantity $(E_m - E_1)\,\Delta t$, and will subtract from the second gate the area $(E_m - E_2)\,\Delta t$. A relative voltage slope k_t may be defined in terms of the total change in integrated voltage during the gate, divided by the signal voltage E_s which would have been present in the absence of filtering by the i-f amplifier.

* Schwartz, *Ibid.*, pp. 64–73.

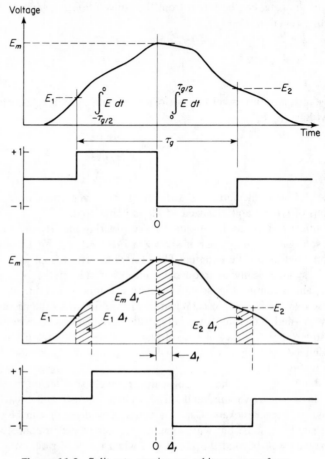

Figure 11.6 Split-gate ranging on arbitrary wave form.

$$k_t = \frac{1}{\Delta t E_s}\left(\int_{\substack{\text{Early} \\ \text{gate}}} E \, dt - \int_{\substack{\text{Late} \\ \text{gate}}} E \, dt\right) = \frac{2E_m - E_1 - E_2}{E_s} \qquad \textbf{(11.3)}$$

When the gate is wide enough to include the entire received pulse, including the rise and fall times which result from filtering by the i-f amplifier, this error slope will be equal to $2E_m/E_s$, or approximately two, for all systems with sufficient bandwidth to pass the full amplitude of the pulse.

The output error signal in the presence of an error Δt will be

$$\Delta E_o = k_t E_s \frac{\Delta t}{\tau_g} \qquad \textbf{(11.4)}$$

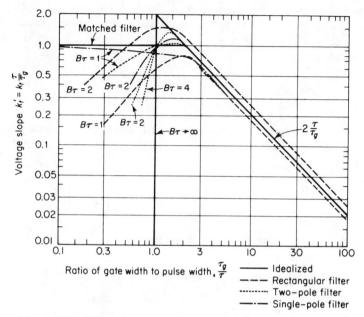

Figure 11.7 Discriminator voltage slope vs. gate width for different filter characteristics.

since it is the voltage averaged over the gate that has been defined as error in the previous paragraphs. Fig. 11.7 shows a plot of the factor k_t times the quantity τ/τ_g, as a function of the relative gate width τ_g/τ and the i-f filter characteristics. In general, this normalized error slope will approach $2\tau/\tau_g$ for wide gates. The curve for a rectangular filter with $B\tau = 2$ lies above the idealized curve by a factor 1.18, corresponding to the 18 per cent voltage overshoot at the center of the pulse for this filter.* Similarly, the curve for $B\tau = 1$ falls 13 per cent below the idealized curve, since the pulse remains this far below its unfiltered amplitude. When the gate width is reduced below the pulse width, the curves tend toward zero in most cases. Exceptions are the matched filter, and the single-pole filter with $B\tau = 1$, where an inflection point in the output waveform leads to a near-constant normalized error slope in the narrow-gate region. The significance of this point will be discussed below.

Range Noise Output

We may find the time delay error of the gate which corresponds to the

---

* See, for instance, Mischa Schwartz, *Information Transmission, Modulation and Noise* (New York: McGraw-Hill Book Company, 1959), p. 45.

thermal noise output by setting $(\Delta E_o)_{rms} = E_{no}$, and solving for $(\Delta t)_{rms}$, using Eqs. (11.2) and (11.4). Since the noise output represents an rms level, with zero mean value, we designate $(\Delta t)_{rms}$ as σ_{r_1}, indicating noise on a single-pulse basis.*

$$\sigma_{r_1} = \frac{\tau_g}{k_t(E_s/E_{no})} = \frac{\tau_g}{k_t(E_n/E_{no})(E_s/E_n)} = \frac{\tau_g}{k_t(E_n/E_{no})\sqrt{2S/N}} \tag{11.5}$$

To find the error in terms of the pulse width, we replace the relative error slope k_t with the factor $k_t' = k_t\tau/\tau_g$, previously plotted in Fig. 11.7, obtaining

$$\sigma_{r_1} = \frac{\tau}{k_t'(E_n/E_{no})\sqrt{2S/N}} = \frac{\tau}{k_r\sqrt{2S/N}} \tag{11.6}$$

The absolute error slope factor k_r is defined by

$$k_r \equiv k_t'\frac{E_n}{E_{no}} = \frac{\tau E_n}{\Delta t E_{no}} \times \frac{1}{E_s\tau_g}\left(\int_{\substack{\text{Early} \\ \text{gate}}} E\,dt - \int_{\substack{\text{Late} \\ \text{gate}}} E\,dt\right) \tag{11.7}$$

This slope is exactly analogous to the angular error slope k_m used in Eq. (9.9). In ranging, however, k_r is a function of gate width and filter characteristics, as shown in Figs. 11.5 and 11.7.

In comparing Eq. (11.6) with optimum performance, as expressed in Eq. (11.1) for $B\tau \gg 1$, we note that the error is larger by the factor $B\tau\sqrt{2}/k_r$. Let us assume that the pulsewidth τ has been fixed by system considerations, and explore the means of minimizing range error. A rectangular pulse will be assumed for the present, and we have three parameters to

---

* In the following discussion, use of a linear envelope detector will be assumed. Assume the signal-to-noise ratio is considerably greater than unity and the change in noise level resulting from presence of signal during only a portion of the gate is neglected. The resulting values of noise are slightly high for the case where $\tau_g > \tau + 1/B$, because the actual noise level drops when the signal is absent. The drop in level of the a-c component is expressed by the factor $\sqrt{2 - \pi/2}$, given by Mischa Schwartz, (op. cit.), p. 403. This effect is the opposite of that attributed to the detector by A. N. Shchukin, "Dynamic and Fluctuation Errors in Guided Vehicles," *Izdatel'stvo Sovetskoye Radio*, Moscow (1961), as translated by the Foreign Technology Div., Air Force System Command, (ASTIA Document AD 401, 738), pp. 153-54 and p. 163. Shchukin assumed an a-c noise voltage which was greater by a factor of $\sqrt{2}$ when the signal was absent, and arrived at an unnecessarily large value of error for $\tau_g > \tau$. The noise level remains constant, as noted by Shchukin, when a synchronous or coherent detector is used. However, the level for this case will be identical with that for the linear envelope detector, not greater by the factor $\sqrt{2}$ as assumed by Shchukin.

adjust: gate width τ_g, receiver bandwidth B, and i-f filter shape. First, since received energy and noise density are constant, Eq (11.6) may be rewritten in terms of the energy ratio $\mathcal{R}$.

$$\sigma_{r_1} = \frac{\tau\sqrt{B\tau}}{k_r\sqrt{\mathcal{R}}} = \frac{\tau}{M_r\sqrt{\mathcal{R}}} \tag{11.8}$$

The dimensionless "tracking merit factor" M_r for this tracker will then be defined as

$$M_r = \frac{k_r}{\sqrt{B\tau}} = k_t' \frac{E_n}{E_{no}} \times \frac{1}{\sqrt{B\tau}} \tag{11.9}$$

This factor is plotted in Fig. 11.8. The solid line represents the asymptotic performance of all split-gate trackers for $B\tau \gg 1$. This function has zero value for gates shorter than the pulsewidth τ, and rises abruptly to $M_r = 2$ at $\tau_g = \tau$. For wider gates, it drops according to the curve $M_r = 2\sqrt{\tau/\tau_g}$. The basic performance is independent of bandwidth and filter shape, and assumes complete independence of the noise appearing in the two gates $(E_n/E_{no} = \sqrt{B\tau_g})$.

An inspection of the M_r curves for specific i-f filters leads to some important conclusions:

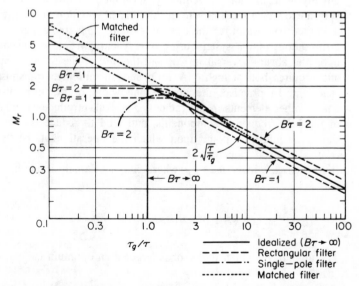

Figure 11.8 Tracking merit factor M_r as a function of gate width and filter characteristics.

1. In the split-gate tracker, there is no advantage to be gained from use of an i-f bandwidth greater than optimum for detection, as the output noise of the range tracker is almost independent of bandwidth for a given pulse energy and noise temperature. A minor exception appears in the case of a rectangular filter with $B\tau \cong 2$, which causes maximum overshoot at the center of the pulse.

2. Best over-all results in most cases are obtained when the filter is approximately matched to the pulse spectrum $(1 < B\tau < 2)$, and the gate width is matched to the pulse width $(\tau < \tau_g < 2\tau)$. This is close to the condition of "correlation processing," and leads to maximum values of $M_r = 2$, or $k_r = 2.5$.

3. Optimum performance for the split-gate tracker under the above conditions is then

$$\sigma_{r_1} = \frac{\tau}{2.5\sqrt{2S/N}} \qquad (B\tau \cong 1.4, \quad \tau_g/\tau \cong 1.4) \qquad \textbf{(11.10)}$$

4. The increased tracking performance shown for matched and single-pole filters with narrow gates $(B\tau_g < 1)$ is predicted in the theory of Woodward, and results from the fact that narrow gates act as differentiators on the input signal, increasing its rms spectral spread. When the signal spectrum is assumed infinite (as would be true for a rectangular pulse) and the i-f filter response falls off in voltage directly as a function of frequency, the differentiating action of the very narrow gate tends to cancel the filter attenuation, leading to an unlimited bandwidth for the over-all system. Thus, the apparent low error output of the system using a very narrow gate is dependent upon the abrupt reversal in slope at the end of the received pulse, and requires both a high S/N ratio and an unlimited transmission spectrum. In practice, the limited bandwidth of the transmission and of other elements in the radar system will introduce attenuation of the high-frequency components, and tend to reduce the discontinuity or inflection point upon which the operation of the "slope reversal tracker" depends.

5. In the cases where $B\tau \gg 1$, where Skolnik's analysis applies, we find that

$$k_r = M_r\sqrt{B\tau} = 2\tau\sqrt{\frac{B}{\tau_g}} \qquad \textbf{(11.11)}$$

The amount by which the error is larger than optimum is

$$\frac{\sigma_{r_1}}{(\sigma_{r_1})_{\text{opt}}} = \frac{\sqrt{2}\,B\tau}{k_r} = \sqrt{\frac{B\tau_g}{2}} \qquad \textbf{(11.12)}$$

Thus the split gate gives optimum performance for $B_{\tau_g} \leq 2$, and this confirms the earlier indication of optimum conditions, $B_\tau \cong 1.4$ and $\tau_g/\tau \cong 1.4$.

6. When we consider the reduced noise level which accompanies the absence of signal, values of M_r are increased in the region $\tau_g > \tau + 1/B$. Thus, when $B_\tau \gg 1$, values of M_r near 2 will persist for $1 < \tau_g/\tau < 2$, and the idealized curve represented by the solid line in Fig. 11.8 will approach an asymptotic line given by $3.2\sqrt{\tau/\tau_g}$ (instead of $2\sqrt{\tau/\tau_g}$ as shown). Curves for the matched filter, single-pole filters, and for rectangular filters with $B_\tau \gg 1$ will all merge with the modified line at $\tau_g/\tau = 3$, where $M_r \cong 1.5$.

7. The results shown in Fig. 11.8 are consistent with the only other published analysis of the split-gate tracker,* which also indicates performance equivalent to a value $M_r = 1.5$ for the case $\tau_g = 2\tau$, $B_\tau = 1$, with a rectangular filter and constant assumed noise level over the entire gate.

Leading- and Trailing-Edge Tracking

When the pulses received from the target are trapezoidal in shape, with rise and fall times which are short relative to the pulse width, the split-gate tracker performance is far from optimum in the high-S/N case. Noise received during the middle of the pulse contributes to the output, but there is no corresponding error sensitivity to the signal in this region.[†] Signal displacement can be sensed only in regions where the time derivative of the waveform is not zero (i.e., during the leading and trailing edges of the trapezoidal pulse). Various systems can be used to extract this information, while the region in between is ignored. One of the more simple approaches involves the use of two narrow gates, each of width equal to the rise time of the pulse, or $1/B$. The gates are separated by an amount equal to the pulse width [see Fig. 11.9(c)]. In this system, the error slope as previously defined in Eq. (11.7) would be approximately $k_r = 2$, and the noise output would be $E_{no} = \sqrt{2}\,E_n$ (assuming zero correlation over

---

* F. L. Warner and R. L. Ford, "Errors Due to Noise in the Measurement of Range, Range Rate and Range Acceleration When Using an Automatic Strobe," *Royal Radar Establishment*, Malvern, England, Tech. Note 657 (January 1960), ASTIA Document AD 237, 223.

† The analogous situation in angle measurement has been described by R. Bernstein, "An Analysis of Angular Accuracy in Search Radar," *IRE Conv. Record* (1955), part 5, p. 71. His maximum-likelihood weighting functions drop to zero when the radar beam passes directly across the target, because there is no information to be gained from that region. The weighting functions follow generally the derivative of the antenna pattern, as our gates follow the slope of the signal.

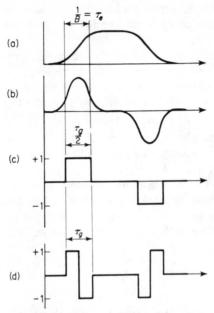

Figure 11.9 Leading- and trailing-edge tracking. (a) Video pulse. (b) Differentiated video pulse. (c) Separated gate used with video pulse. (d) Double gate used for derivative tracking.

the period between gates). The resulting thermal-noise error, for the "separated-gate" system, is

$$\sigma_{r_1} = \frac{2E_n \tau_g}{k_r E_s} = \frac{1}{2B\sqrt{S/N}} \qquad \text{(11.13)}$$

This represents an improvement by a factor $\sqrt{B\tau/2}$ when compared to the split-gate tracker for $B\tau \gg 1$, which is the full potential improvement predicted by theory. Tracking accuracy on extended targets with glint may be very poor, however.

Another way to realize the full improvement factor is to use the leading and trailing edges of the signal in a derivative tracker, as shown in Fig. 11.9(d). Here the video signal is passed through a high-pass filter whose time constant t_d is short compared to the rise time of the pulse. The output will be equal to t_d times the slope of the input signal, or $E_s B t_d$ in the trapezoidal case. The filter output is then applied to a split-gate tracker with total width τ_g equal to the rise time. The error slope of such a tracker will be approximately $2B t_d$, whereas the noise output will be $E_{no} = 2E_n B t_d$.* The output noise for a single pulse will be

$$\sigma_{r_1} = \frac{\tau_g E_{no}}{k_r E_s} = \frac{1}{B\sqrt{2S/N}} \qquad \text{(11.14)}$$

This is better than the split-gate tracker by the factor $\sqrt{B\tau}/2$. In order to appreciate the full predicted improvement, two split gates must be used to measure both the leading and the trailing edges of the pulse, providing the extra factor of $\sqrt{2}$.

The three types of tracking gate described above are seen to be equiva-

---

*Schwartz, *op. cit.*, pp. 46–47, shows that the actual voltage slope for a rectangular pulse passed through a rectangular filter has an average value $\pi E_s B/3.94 = 0.8E_s B$. However, the noise output of the rectangular filter is actually near $1.6E_n B$, so the results are the same.

lent when $B\tau \le 2$. Since the best tracking results are achieved when the system is approximately matched, the split-gate type of tracker is the one most generally used in practical radar. The leading-edge tracker finds use in special cases where thermal noise is of less importance than resolution of multiple targets or targets in clutter. Its operation is such as to ignore the flat center portion of the echo signal, and it makes no use of the energy contained therein. If this type of operation proves inadequate because of this waste of energy, and if it is not feasible to transmit the equivalent energy in a narrower pulse of higher peak power, then some form of pulse compression should prove useful.

Range Tracking on Compressed Pulses

When the radar operates with pulse compression, the video pulse leaving the second detector will already have been compressed to its narrow width. Although it may have associated with it the "time side lobes" generated in the expansion-compression networks (similar to the side lobes of an antenna pattern), the split-gate range tracker will operate on the main lobe of the compressed pulse in much the same way as it would on a narrow rectangular pulse. The value of τ used in the equations for thermal noise should be that of the compressed pulse, or approximately $1/\Delta f$, where Δf is the total bandwidth of the transmission. The width of the tracking gate should also be determined with respect to this bandwidth. The effective signal-to-noise ratio will be calculated on the basis of total pulse energy, yielding a result which would have been obtained if the peak power of the transmission had been increased by the compression ratio, while τ was held at the compressed value. The net effect of pulse compression is to increase the range precision by an amount equal to the compression ratio, as compared to a conventional matched system using the same pulse energy. When it is compared to a leading-edge technique with equivalent bandwidth, the effect is to improve the signal-to-noise ratio by the compression ratio, giving twice the amount of improvement that would be obtained by leading-edge tracking.

Thermal-Noise Output of a Servo System

Measurements made over n repetition periods will combine in a tracking servo or similar filter to provide an output whose noise is $1/\sqrt{n}$ times that of a single-pulse measurement. This is true because the noise terms are completely uncorrelated between widely separated pulses. In terms of the equivalent noise bandwidth β_n of the servo or filter, we have

$$n = \frac{f_r}{2\beta_n}$$

For the split-gate system, then, the noise output will be

$$\sigma_r = \frac{\tau}{k_r\sqrt{(S/N)(f_r/\beta_n)}} \tag{11.15}$$

For the optimum (leading-edge or separated-gate) system, with $B_T \gg 1$, we have

$$(\sigma_r)_{\text{opt}} = \frac{1}{B\sqrt{2(S/N)(f_r/\beta_n)}} \tag{11.16}$$

Tracking at Low S/N Ratio

In the preceding discussion, it has been assumed that the single-pulse S/N ratio was high, so that the predetection relationships between signal and noise amplitudes were preserved in passing through the second detector. As the S/N ratio approaches unity, the effect of detector loss becomes significant, and the preceding equations must be modified to account for this. As in the case of the angle tracker (Section 9.3), the gain of the tracking loop will be reduced by the progressive failure of agc action, due to reduced S/N, and the equivalent noise bandwidth of the system will vary directly with S/N in the region below $S/N = $ unity. As a result, all equations containing the factor $(S/N)(f_r/\beta_n)$ will hold approximately for any value of S/N, if β_n is defined as that value of bandwidth which would be observed in testing at high S/N. The actual bandwidth β_s for low S/N will be approximately $\beta_n S/(S + N)$.*

The deterioration in performance of the tracking loops at low S/N adds to the advantage of the matched, split-gate type of tracker. Obviously, the detector loss C_d will have a larger value if the i-f bandwidth is greater than optimum. Furthermore, the ability of the gate to maintain track will depend upon the presence of an extended region of linear error slope on each side of the tracking null. If the instantaneous error in the loop reaches a value such that there is little or no error signal to restore tracking (as would happen if the error slope dropped rapidly), tracking will be lost. This loss occurs if the error slope becomes negative within the region extending $2\sigma_r$ on each side of the tracking point. The extent of the positive-slope region is directly proportional to the width of the tracking gate, favoring the split-gate tracker over the types using narrow gates on the leading and trailing edges of the pulse. However, the gate cannot be made much wider than the pulse without increasing the output noise and reducing the ability of the agc system to maintain gain in the tracking loop.

* See Section 9.3, Eq. (9.6).

If an approximately matched system is used, with $B\tau \cong 1.4$ and $\tau_g \cong 1.4\tau$, we see from Eq. (11.15) that

$$\sigma_r = \frac{0.4\tau}{\sqrt{(S/C_d N)(f_r/\beta_s)}} = \frac{0.4\tau}{\sqrt{(S/N)(f_r/\beta_n)}} \qquad (11.17)$$

If it is assumed that the error slope is positive for displacements up to 0.4τ on each side of the tracking point, the rms error may reach a level $\sigma_r = 0.2\tau$ before loss of target occurs. This level is reached when the quantity within the radical in Eq. (11.17) equals about 7. Since the ratio f_r/β_n is normally quite high, the system should be capable of tracking at signal-to-noise ratios well below unity, and this has been observed in practice.

The practical limit in tracking at low S/N ratios is set by two factors: the ability of the tracking loop to follow target accelerations; and its ability to overcome small unbalanced output from the circuit which compares the two gate outputs. If the target acceleration is low and the balance is maintained, the bandwidth may be reduced to avoid noise errors which might carry the gate beyond the region of positive error slope. In the absence of friction or similar nonlinearities in the tracking loop, the track can be maintained when the signal is so low as to preclude positive identification of its presence. The usual problem in such cases is to know when the system is still tracking and when it has lost the target. Little credence can be placed on radar data which represent a linear extrapolation of an established track, unless there is some evidence of signal, since the velocity memory circuit would provide the same output in the absence of a signal. If the target is known to be following such a course, the radar track conveys no further information. In order to be useful, the radar data must at least add to our confidence that the target is in its assumed position. For this reason, the lower limit to use of radar data will depend not upon whether the range system is tracking, but on whether the operator *knows* that it is tracking. This limit occurs, with 90 per cent confidence, when the integrated signal-to-noise ratio is about 6 or 8 db. At this point, if the bandwidth of the tracker and the detection circuits are similar, the range error will be less than $\tau/4$, and the peak error will still be within the region of positive error slope. Only in cases where the tracker bandwidth is greater than the detection bandwidth, and especially when a fluctuating target is being tracked, will the operator be able to observe the signal leaving the tracking gate.

Range Glint and Scintillation

The physical origin of target glint error for both range and angle tracking was discussed in Chapter 3. An experimental study was cited which established an rms glint error in radial measurement equal to about one-fourth the radial extent of the target.* The spectral distribution of range glint is the same as for angle glint, and may be expressed in the form of Eq. (9.15). Typical values for the half-power frequency of this spectrum on aircraft in straight flight may be read from plots in the cited reference, and are in the neighborhood of one or two cps. As a result, most of the glint error will lie within the bandwidth of a typical range tracker.

Amplitude scintillation will have the same effect on the range servo as in the case of angle tracking, causing a modulation in the tracking-loop gain and hence in lag error. The curves given in Figs. 9.12–9.14 may be applied directly to the range tracker. Since sequential-gate sampling systems are seldom used in range, the larger errors of the conical-scan systems will not generally have a counterpart in range tracking. Also, since the range servo bandwidth may be made wider than most angle-tracker bandwidths, the lag errors will be less, and modulation of them will not be as significant.

Radial Velocity and Its Derivatives

The tracking rate, acceleration, and higher derivatives in range are dependent upon the true motion of the target and on geometrical components, as was the case in angle. True components are found simply by projecting the target velocity, etc., on the line of sight of the radar.

$$\dot{R} = v_t \cos \alpha \qquad\qquad \textbf{(11.18)}$$

$$\ddot{R} = a_t \cos \beta \qquad\qquad \textbf{(11.19)}$$

$$\text{etc.}$$

Here, α and β are the angles between the target velocity and acceleration vectors and the radar beam. The geometrical components of acceleration and higher derivatives may be expressed in terms of the target velocity v_t, the angle α, and the slant range at crossover R_a.†

$$\ddot{R} = -\frac{v_t^2}{R_a} \sin^3 \alpha \qquad \left(\text{max:} \ \frac{-v_t^2}{R_a}\right) \qquad \textbf{(11.20)}$$

---

* See also D. D. Howard and C. F. White, "External and Internal Noise Inputs to the Radar System," Sect. 8–2 in *Airborne Radar*, by Donald J. Povejsil, Robert S. Raven, and Peter Waterman. (Princeton, N.J.: D. Van Nostrand Company, Inc., 1961), pp. 395–401.

† Expressions and plots of these derivatives as functions of R/R_a will be found in Charles F. White's discussion of servos, "Servo System-Theory," Chap. 7 of *Guidance*, A. S. Locke, ed. (Princeton, N.J.: D. Van Nostrand Company, Inc., 1955), pp. 260–61.

$$\dddot{R} = -\frac{3v_t^3}{R_a^2}\sin^4\alpha\cos\alpha \qquad \left(\text{max: } \pm 0.85\frac{v_t^3}{R_a^2}\right) \qquad \textbf{(11.21)}$$

$$\ddddot{R} = -\frac{3v_t^4}{R_a^3}(5\cos^2\alpha - 1)\sin^5\alpha \qquad \textbf{(11.22)}$$

Plots of the derivatives of R are shown in Fig. 11.10. The crossing range R_a is the minimum slant range, as opposed to the ground range R_c used in calculation of angular rates.

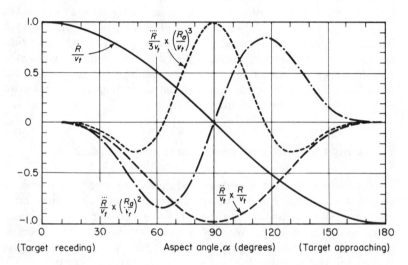

Figure 11.10 Range derivatives for pass course.

   The servo error coefficients are calculated as in the case of angle tracking, by using Fig. 9.15 and Tables 9.1 through 9.3 without change. Lag error components are then found by dividing the range derivatives by the corresponding error coefficients.

$$\epsilon_r = \frac{\dot{R}}{K_v} + \frac{\ddot{R}}{K_a} + \frac{\dddot{R}}{K_3} + \cdots \qquad \textbf{(11.23)}$$

Where both true and geometrical acceleration are present, the two terms must be added directly, before the preceding equation is used. Using the relationships between servo noise bandwidth β_n and acceleration error constant K_a, we may arrive at an expression for optimum bandwidth in the range servo.

$$(\beta_n)_{\mathrm{opt}} = \left[\frac{a_r^2 k_r^2 f_r (S/N)}{0.39 \tau^2 c^2} \right]^{1/5} \qquad (11.24)$$

Here, the quantity in the denominator is equivalent to 1.56 times the square of the range equivalent to the pulse length. The term k_r^2 may also be replaced by $M_r^2 B_\tau$ [see Eq. (11.9)].

For example, assume that a radar operates with a pulse width of 1 μsec (500 ft equivalent range), repetition rate 400 pps, and with $k_r = 2.5$. A target whose acceleration is 10 g (320 ft per sec²) is tracked with a signal-to-noise ratio of 10. The optimum range servo bandwidth would be 5.8 cps, giving a thermal noise error of 7.5 ft and a lag error of 4.0 ft. The bandwidth and errors would be independent of the range to the target, so long as this S/N ratio could be achieved. The presence of range glint would not change the optimum bandwidth appreciably, unless there were significant frequency components near the upper limit of the servo bandpass. Even then, a slight reduction of bandwidth might increase the lag error more rapidly than it would decrease glint.

11.3 ERROR ANALYSIS OF RANGE-TRACKING SYSTEMS

The procedures used in describing and combining errors from various sources were discussed in the preceding chapter, as related to angle error analysis. The same mathematical models and procedures may be applied to analysis of range errors. We shall list and describe the most important sources of error which affect the conventional types of radar ranging system, and give an example based on errors in a typical system.

Table 11.1 lists the errors most frequently encountered in radar measurement of range. The form of the listing is similar to that used in the preceding chapter for angular errors, and the definitions of the several classes of error will not be repeated here. Some of the more significant error components will be discussed in more detail below.

Velocity of Light

Since the basic range measurement process of the radar is dependent upon conversion of round-trip time delay in the signal to an equivalent target range, there will be an error component proportional to the error in the velocity of electromagnetic propagation. Two factors enter into this velocity error: the determination of the constant c, representing the vacuum velocity of light; and effects of the atmosphere on the particular path over which the radar measurement is made. We shall consider here the first factor, and leave the discussion of the second to another chapter.

Table 11.1 SOURCES OF RANGE ERROR

Class of error	Bias components	Noise components
Radar-dependent tracking error	Zero range setting Range discriminator shift Receiver delay	Thermal noise Multipath Servo noise Variation in receiver delay
Radar-dependent translation error	Range oscillator (velocity of light) Data takeoff zero setting	Range resolver Internal jitter Data gearing Data takeoff Range oscillator stability
Target-dependent tracking error	Dynamic lag Beacon delay	Dynamic lag Glint Scintillation Beacon jitter
Propagation error	Average tropospheric refraction Average ionospheric refraction	Variation in tropospheric refraction Variation in ionospheric refraction

As of 1962, the National Bureau of Standards estimated that the velocity of light was known to within about one part per million, the best accepted value being*

$$c = 299{,}792.5 \text{ km per sec} \pm 0.2 \text{ km per sec}$$

or

$$c = 983{,}569{,}200 \text{ ft per sec} \pm 650 \text{ ft per sec}$$

Determinations of this constant are necessarily made by using electromagnetic paths of limited extent, and it is difficult to simulate the free-space vacuum conditions under which the wave would actually propagate with this velocity. The error in c is of little importance in earth-bound radar measurements, since it will be less than that introduced by the atmosphere through which the measurements are made. When the radar measurement extends into the space, however, the bias error $R\sigma_c/c$ may exceed by a large factor the error introduced in passage through the limited portion of the path which lies within the atmosphere. Eventually, experiments will be performed in space with accuracies exceeding that of the present value of c. Meanwhile, there will be a scale-factor problem in relating electro-

* A. G. McNish, "Velocity of Light," Appendix A to *Report of Ad Hoc Panel on Basic Measurements*, National Academy of Sciences, Dec. 8, 1961.

magnetic measurements of lunar and planetary ranges to the standard meter bar.

The solution to this problem lies in the use of the atomic standard for length, combined with the atomic time or frequency standard. The frequency of the caesium resonance, in terms of ephemeris time (ET) has been measured as 9,192,631,770 ± 20 cps, and makes available a time or frequency standard with an accuracy of ± 22 parts in 10^{10}.* Assuming that the wavelength of this resonance line can be related to the standard of length with the same accuracy, it would be possible to define an atomic standard for the vacuum velocity of electromagnetic radiation $c_o = \lambda_o f_o$ which would be almost 1000 times more accurate than the present experimental value of c. Radar measurements made through millions of miles of empty space would then be consistent with the standards, and the scale-factor error would become evident only when these measurements were to be related to lengths on the earth. In such cases, the error would be properly attributable to the earth-bound measurement processes, rather than to the accurate and consistent measurements made through space.†

In summarizing the results of propagation error studies, covered by Chapter 15, we may state that the initial error (for microwave systems) is due to the troposphere, and amounts to a few tens of feet for paths above about 5°. The error may be reduced to about 2 per cent of its initial value, by using profiles of refractivity based on meteorological data, or in some cases by using surface data alone. Thus the error may be held to about 1 or 2 ft if care is taken in applying the tropospheric correction. The ionospheric error will be in the order of a few feet for radars operating in S-band (3000 mc), and will vary inversely as the square of frequency. In the region between 5000 and 10,000 mc the tropospheric and ionospheric errors will each contribute one foot or so to the final error.

Time-Interval Measurements

Within the radar, the limitations on accuracy of time-interval measurements are relatively few. Thermal-noise errors have been discussed earlier. Some of the errors introduced by practical circuit limitations will be discussed below. First, however, let us consider the basic processes by which time intervals are measured. Most systems depend upon local secondary standards for generation of time and frequency signals. In the tracking radar, the usual standard consists of a crystal-controlled oscillator operating in a temperature-controlled, vibration-free chamber. The day-by-day stab-

---

* L. Essen, "Frequency and Time Standards," *Proc. IRE*, Vol. 50, No. 5 (May 1962), pp. 1158-63.

† M. J. E. Golay, "Velocity of Light and Measurement of Interplanetary Distance," *Science*, **131**, No. 3392 (Jan. 1, 1960), pp. 31-32.

ility of such oscillators may be in the order of one part in 10^{11}, although this accuracy is seldom required.* The accuracy of the oscillator is dependent upon the time interval since the previous calibration. If needed, oscillators can be built with a drift rate as low as one part in 10^9 per month,* but the availability of atomic standards and standard-frequency signals from nearby laboratories makes this stability unnecessary in many cases. At most points where precision radar may be used, it is relatively easy to assure an accuracy in time-interval measurement of one part in 10^9. An accuracy of one part in 10^8 may be maintained over an indefinite period, even at remote points on the earth.

Once the local standard of frequency has been established, conventional techniques of frequency multiplication and synthesis may be used to obtain signals at any frequency up to the transmission frequency of the radar. These frequencies, in turn, are available to control the generation or measurement of internal time delays, with almost the same accuracy and precision as those which apply to the local standard oscillator. Examples of practical procedures of time-delay generation have already been given, in which precision resolvers and tapped delay lines were used to generate tracking gates whose delay could be expressed by the data output devices of the range tracker. The practical error levels of such devices, which are set more by convenience and economy than by fundamental considerations, are typically in the order of one to a few feet. Figure 11.11 shows an error

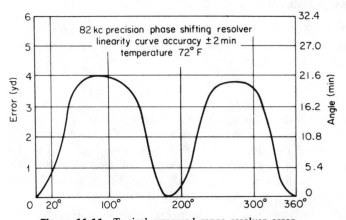

Figure 11.11 Typical measured range resolver error.

curve for a typical precision resolver, operating in this case at 82 kc and 4000 yd per revolution (two cycles of output are produced per revolution). The error is approximately sinusoidal, with a peak-to-peak value of 12 ft

* Essen, *op. cit.*

and an rms of 4.3 ft. This is not the ultimate in resolver accuracy, but is typical of what can be attained in the field with such units. An rms error of 4.3 ft in the 12,000 ft revolution corresponds to an error of 0.035 per cent, which is representative of many analog data devices. The error will appear in the output data at a frequency determined by the target velocity.

$$f_c = \frac{\dot{R}}{2000 \text{ yd}} \tag{11.25}$$

Gear errors and encoder or other takeoff errors are easily held below the level set by this resolver, by using standard components.

If the more advanced digital ranging system described earlier is used, the error in the delay line used for interpolation between clock pulses (100 yd intervals) will be smaller than the resolver error of the electromechanical system, and will appear at the higher frequency given by $\dot{R}/100$ yd. In many cases, this error will have the characteristics of random noise, owing to sampling of data at a rate below this frequency. This will subject the error to effective smoothing in the computer. Where it is warranted by the thermal noise and propagation limits, the internal error in time-delay measurement can be reduced to 1 ft or better, by using available circuits and components.

Range Multipath Error

The ray which is reflected from the surface of the earth, arriving at the receiver through side-lobe response of the antenna, is delayed by an amount equivalent to a range error

$$\Delta R = h \sin E \tag{11.26}$$

Since the phase of this signal varies relative to that of the direct ray, the reflected component may add to or subtract from the direct component in the region of overlap, as with the two-element target of Fig. 3.11. The center of energy for the combined signal will tend to oscillate about the range of the direct component, with a peak deviation equal to the delay ΔR multiplied by the amplitude ratio of the reflected to the direct component. This amplitude ratio, from Fig. 10.5, is $\rho/\sqrt{G_{sr}}$, where ρ is the surface reflectivity and $G_{sr} = G_1/G_2$ is the power ratio of main (reference) lobe response to side lobe response in the direction of the reflected ray. The resulting range error has an rms value given by

$$\sigma_{rm} = \frac{\rho h \sin E}{2\sqrt{G_{sr}}} \tag{11.27}$$

This oscillating error may be accompanied by a bias error approaching $\Delta R/\sqrt{G_{sr}}$ if the reflected signal lies entirely within the range gate ($\tau_g \geq \tau + 2\Delta R$). If the gate is only slightly longer than the echo pulse, the multipath error will be symmetrical about zero, as shown previously in Fig. 3.11.

Since the gate will exclude those portions of the reflected signal which are received more than $(\tau_g - \tau)/2$ after the end of the direct signal, the maximum bias error is limited to

$$(\Delta r)_{max} = (\tau_g - \tau)\frac{\rho\, c/2}{2\sqrt{G_{sr}}} \qquad (11.28)$$

Also, the maximum cyclic component will occur when the reflected signal is delayed by exactly one-half the pulse width.

$$(\sigma_{rm})_{max} = \frac{\rho\tau\, c/2}{\sqrt{8G_{sr}}} \qquad (11.29)$$

There will, of course, be no error at all when the delay ΔR exceeds the gate width (more exactly, when it exceeds the quantity $(\tau_g + \tau)/2$). In such cases, the first reflected signal energy will arrive after the end of the late gate, and will be excluded from the system. This limit, plus the relatively short pulse lengths normally found in the precision tracking radar and the 20 db or more of side-lobe rejection, will normally reduce the range multipath error to values of a few feet or less. For example, if the pulse length is 1 μsec, the surface reflectivity 0.3, and the side lobe ratio 20 db, the maximum cyclic error would be about 6 ft rms, and the maximum bias error about 2 ft. These errors would be encountered only when the radar was sited high above the reflecting surface, such that low-angle reflections could arrive with delays $h \sin E$ approaching 250 ft.

Range Servo Noise and Unbalance

The tracking noise and bias errors generated within the range servo system are dependent entirely upon the care exercised in circuit design, and may be held to almost any desired level. Since the range-tracking servo need not drive any heavy output mechanism, the best instrument servo techniques may be employed, or all-electronic systems may be used to avoid mechanical sources of error. Total servo error less than 1 ft is a common goal, and may be maintained with reasonable calibration procedures. The same applies to zero-setting of the range discriminator and correction of slow drifts which may occur as a result of temperature changes and aging of electronic components.

Receiver Delay

The signal delivered to the range tracker will be delayed in passing through both the passive r-f elements and the active i-f amplifier stages of the receiver. A large portion of this delay is constant, and may be removed in the initial calibration of the radar. Other portions are temperature-sensitive, and must be held within tolerances by minimizing the cable lengths within the radar or by maintaining the temperature of these elements within a prescribed range. The largest contributor to delay will usually be the i-f amplifier, consisting typically of 8 to 12 tuned stages with interstage coupling transformers or multipole filters. Since the delay in each stage is approximately equal to the reciprocal of the stage bandwidth, and the over-all bandwidth is about one-third the single-stage bandwidth, the total delay will be about three times the reciprocal of the i-f system bandwidth. This corresponds to about twice the pulse width for a typical system. This delay will change as a function of tuning of the receiver, and also as a function of i-f gain or bias level. Figure 11.12 shows a measured curve of delay variation with signal strength, covering a range of about 100 db. Over the center of this range, the delay changes by less than 1 ft for a 50 db change in input. Near noise level, a 10 ft change was measured, owing to actual receiver changes or to noise unbalance in the discriminator. Above 60 db, the apparent delay increased by 1 ft per decibel, owing to overloading of the receiver. In this case, bandwidth was 8 mc, and the total delay variation was about 10 per cent of the total delay. Errors of this sort can be minimized by careful design, by use of r-f attenuation in the transmitter or prior to the receiver, or by calibration of delay as a function of signal strength.

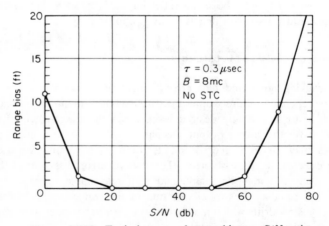

Figure 11.12   Typical measured range bias vs. S/N ratio.

Target-Dependent Errors

In addition to the lag and glint errors discussed in the preceding section, the target may contribute to range error through scintillation or through the characteristics of a transponder beacon, if one is used. The scintillation error is computed as some fraction of the lag error, as in the angle case, and is normally insignificant. Range errors produced by beacons, however, may be large compared to other errors in the system. A typical beacon will have an internal delay equal to twice the width of the interrogating pulse, or about 1 or 2 μsec. Although this delay can be calibrated in preflight testing, it is subject to variation as a function of interrogation signal level, tuning, and environmental conditions. If there is an initial delay equivalent to 1000 ft, it is difficult to hold the variation within 10 ft, even over limited intervals of signal strength. Delay variations of 0.05 μsec (25 ft) are usually encountered when the interrogation signal varies with change in range and antenna aspect. Special beacons have been designed using repeater techniques to assure phase stability of the response, and these beacons are also much more stable in regard to envelope delay. Variations as low as 0.01 μsec have been attained with repeater beacons.

Example of Range Error Analysis

Using again the AN/FPS-16 radar as an example of error analysis procedure (see Chapter 10), we may list the sources which contribute errors of fixed rms value, as shown in Table 11.2. These add to an rms error of

Table 11.2 ERRORS OF FIXED RMS VALUE

Error source	Range error in feet RMS	
	Bias	Noise
Discriminator zero setting	1.0	–
Discriminator drift	2.0	–
Servo noise and unbalance	0.5	0.5
Receiver delay	2.0	1.0
Subtotal: Radar-dependent tracking errors	3.0	1.3
Digital encoder error	1.0	1.2
Data gear error	–	1.0
Resolver cyclic error	–	4.3
Internal time jitter	–	2.0
Subtotal: Radar-dependent translation errors	1.0	5.0
Tropospheric refraction	2.0	0.5
Total "fixed" error	3.8	5.2

only 6.4 ft for echo tracking, divided equally between noise and bias. If we take the same satellite target which was used in the angle-tracking example ($h = 100$ miles, $v = 4$ miles per sec, $R = 40$ miles), the variable errors will be as shown in Table 11.3. The important components here are the velocity and acceleration lag errors, which total about 10 ft over this entire tracking interval. Thus the over-all error of the radar will be slightly above 10 ft when all errors are added.

Table 11.3 VARIABLE RANGE ERROR COMPUTATION

		$R = 110$ mi	$R = 200$ mi
Signal-to-noise ratio S/N	(db)	32	22
Aspect angle α	(deg)	60 to 90	30
Range rate $\dot{R}$	(ft per sec)	12,000 to 0	20,000
Effective target acceleration $\ddot{R}$	(ft per sec²)	550 to 880	110
Optimum servo bandwidth $(\beta_n)_{opt}$	(cps)	19 to 23	6.5
Thermal-noise error $\sigma_r{}^*$	(ft)	0.67	2.1
Velocity lag error $\epsilon_v{}^*$	(ft)	6 to 0	10
Acceleration lag error $\epsilon_a{}^*$	(ft)	6 to 10	1.2
Range oscillator error $R \times 10^{-7}$	(ft)	0.7	1.2
Ionospheric propagation error	(ft)	0.2	0.3
Total variable error	(ft)	8.5 to 10	10.3

* Noise and lag errors are calculated for maximum servo bandwidth $\beta_n = 6$ cps. Assumed parameters: pulsewidth $\tau = 1.0\ \mu$sec, bandwidth $B = 1.6$ mc, repetition rate $f_r = 340$ pps, $k_r = 2.5$, $K_v = 2000$.

In this example, it may be noted that the electromechanical range tracker, whose servo bandwidth is limited to 6 cps, is responsible for much of the error. The optimum bandwidths for the signal-to-noise ratio of 20 to 30 db vary between 6.5 and 23 cps, and use of the wider bandwidths would reduce the variable error considerably. It was this consideration, in part, which led to development of all-electronic ranging systems, whose bandwidth is not limited by mechanical elements. The use of an infinite velocity error constant is also an advantage in this case, provided that the tracking can begin early enough to permit settling of the initial transients after lockon. Calculations using the optimum bandwidths, infinite K_v, and the all-electronic tracker show a reduction in variable error to about 2 ft rms, with an additional fixed component of only 4 ft rms.

A relatively new type of radar tracker, evolved since the end of World War II, is the Doppler tracker, which provides output data on the radial velocity of the target by measuring the frequency shift between transmitted and received signals. This type of radar was used first in airborne applications, where frequency resolution was required to distinguish desired targets from clutter. In that application, however, the Doppler data were not generally furnished as an output, being used primarily to select the target to be tracked in the three spatial coordinates. We shall be concerned here primarily with tracking channels which are intended to provide a useful velocity measurement, using either coherent or noncoherent processing. The added resolution of the Doppler tracker, an all-important consideration in search radar, will play a secondary role in tracking, and may actually prove disadvantageous in some applications where acquisition is difficult. Several systems using Doppler measurement will first be described, following

DOPPLER-TRACKING SYSTEMS

12

which we shall analyze the effects of thermal noise and other sources of error. An outline of system error analysis procedure will be presented, as a guide to estimating the accuracy to be expected from this type of system.

12.1 DESCRIPTION OF DOPPLER-TRACKING TECHNIQUES

The techniques used in Doppler tracking have evolved from radar MTI systems. In tracking radar, the target is kept within the beam for a protracted period of time, and a range gate is generally used to assist in excluding undesired echo signals. This simplifies the processing task, as compared to search radar, and permits greater emphasis to be placed on accurate measurement. Five different types of Doppler tracker will be described below, representing the techniques used in this type of radar measurement.

C-W Systems

The earliest Doppler trackers used a pure c-w transmission, and depended entirely upon angular and frequency resolution to reject undesired targets. An example of this type of radar was the AN/TPS-5, used during the late 1940's for measurement of projectile and missile velocity at various ordnance proving grounds. With this radar, the antenna was pointed manually or by remote control, and automatic tracking was carried out only in the velocity coordinate. A narrow-band tracking filter was used to help reject noise and crosstalk from the transmitter. This filter was followed by a discriminator, which provided an error signal to an automatic-frequency-control (afc) loop (see Fig. 12.1). The transmitted and received spectra (see Fig. 12.2) indicate that the operation of this system is not limited by any type of ambiguity, but is dependent upon the elimination of transmitted sidebands, which may spill over into the region in which signals appear.

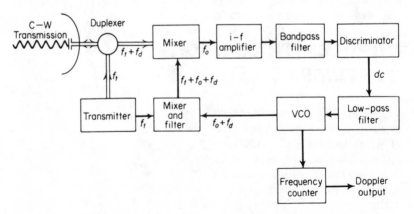

Figure 12.1 Simple c-w Doppler-tracking system.

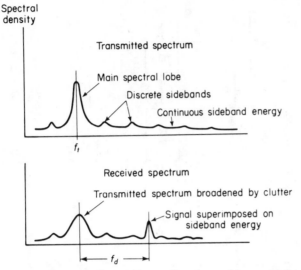

Figure 12.2   Spectra in c-w Doppler-tracking system.

There will also be a minimum target velocity, below which the transmitter spillover and clutter echo power will obscure the target.

Noncoherent Pulsed Systems

In order of increasing complexity, the noncoherent pulsed system should be considered next. This radar is operated as a conventional pulsed system, but the width of the pulses is sufficient to provide useful measurements based on the envelope of the transmitted spectrum (see Fig. 12.3). The width of this spectrum, and of the Doppler-shifted echo spectrum, is broader than in the c-w case, and the target velocity must be comparable to the spectral width $B_t = 1/\tau$ if it is to be measured with accuracy. Range gating may be used, along with angular and frequency resolution, to exclude undesired targets. This permits measurement of the Doppler shift even when the shift is not sufficient to provide resolution in the frequency domain. There is no ambiguity in the measurement, and no upper limit on the velocity which may be measured and tracked. Tracking is carried out by using a conventional frequency discriminator and afc loop, following the range-gated i-f amplifier. It is also possible to use a bank of filters on the output of the i-f amplifier, covering the region of expected Doppler shifts, and to track the target by comparing the outputs of adjacent filters. This method permits simultaneous tracking of two or more targets, separate range gates and interpolation circuits being used at the output of the

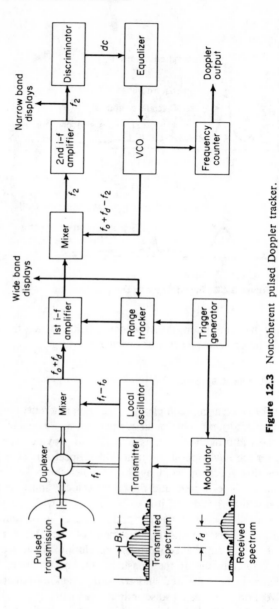

Figure 12.3 Noncoherent pulsed Doppler tracker.

Doppler filter bank. As will be shown below, the accuracy of tracking is directly dependent on the width of the spectral envelope of the transmission, and a compromise between range and Doppler accuracy is required.

When the long pulse is transmitted with phase or frequency modulation, such as that used for pulse compression, the spectrum is broadened and Doppler accuracy is decreased. However, upon reception, the signal can be applied to the inverse filter which restores the original phase relationships and the original Doppler resolution of the long pulse. In fact, when pulse compression is used, the sensitivity of the system to Doppler shifts may present serious problems in acquiring and tracking high-speed targets. By proper design of the pulse compression circuits, an error signal may be developed to permit accurate Doppler tracking in an afc loop. By using an additional frequency-conversion stage prior to the inverse filter in the i-f amplifier, the echo signal may be converted to the proper frequency for pulse compression, and the frequency of the mixing oscillator provides Doppler data.

Coherent Pulsed Systems

A coherent radar system, using either the pulsed-amplifier or pulsed-oscillator with COHO, furnishes an i-f signal which contains the discrete line spectrum structure used in MTI (see Fig. 7.1). A simple form of coherent Doppler tracker, developed to exploit this line spectrum, is shown in Fig. 12.4. In this system, normal tracking-radar operation is established

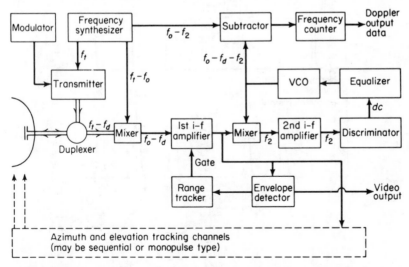

Figure 12.4 Coherent pulsed Doppler tracker used on conventional radar.

in the range and angle channels, and the target is resolved in one or more spatial coordinates without reliance on the fine structure of the spectrum. The Doppler tracker operates as a separate servo loop, controlling its own coherent mixing oscillator to keep the gated signal within the narrow-band filter of a separate i-f amplifier. Since range gating occurs in the wide-band i-f amplifier, the resolution in range remains that of the normal system. After gating, the signal may be converted to a lower i-f and passed through a narrow-band filter, which spreads it in time and eliminates adjacent lines in the fine spectrum (see Fig. 12.5). After amplification to make up for the loss of energy owing to rejection of all but one line, the signal is applied to a narrow-band discriminator, whose output controls the voltage-controlled oscillator (VCO) to close the frequency-tracking loop. In this system, the width of the narrow-band filter and discriminator is determined not by the width of the transmitted pulse but by the spectral spreading of the discrete line structure. This depends upon the stability of the radar oscillators and amplifiers, and the spread in velocity between reflecting surfaces on the target. With no limit imposed by motion of the beam across the target, it is possible to achieve line widths in the order of 1 cps over most of the microwave spectrum.

In the conventional pulsed radar, as noted earlier, there may be

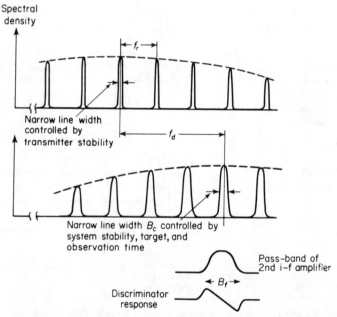

Figure 12.5 Spectra in coherent pulsed Doppler tracker.

hundreds or thousands of lines in the main lobe of the transmitted spectrum, and adjacent lines are virtually indistinguishable on the basis of relative amplitude. The tracker described here will, in fact, operate on any one of these lines, once it has acquired a signal. In order to assure that the output data represents the Doppler shift between the central line of the transmission and that of the received signal, the voltage-controlled oscillator must initially be designated to a frequency within one-half the filter bandwidth of the central line in the received spectrum (allowing of course, the proper offset to convert this line to the center frequency of the filter). In a radar which has already acquired the target and established a track in the three spatial coordinates, this designation is not difficult. Range data is differentiated over a period of one to several seconds, and the resulting range rate is converted to a form which sets the VCO at the proper frequency. When the VCO is held for a brief period at a frequency which converts the signal to pass through the filter, the discriminator output takes over and locks the VCO. The output data from the Doppler loop has all the characteristics of the c-w system output, with a frequency accuracy determined by the width of the narrow-band filter and the line width. Depending upon the location of the VCO and its mixer, the system can be designed to track targets of any velocity from zero to orbital velocities. By placing the mixer in the wide-band portion of the i-f system, the Doppler loop can be used to correct the tuning of the system for all tracking channels.

Four-Dimensional Resolution

The final step in pulsed-Doppler tracking involves the use of narrow-band filters in all four tracking channels of the radar (see Fig. 12.6). The Doppler-tracking loop operates in the manner described above, and the three spatial-coordinate tracking channels also operate on signals converted by the VCO and passed through filters. With range gating in the wide-band portion of the i-f system, the performance of the position channels is similar to that of the noncoherent radar. A major advantage is that bandwidth reduction now takes place before the final detection circuit, eliminating the detector loss factors C_d and C_a which previously degraded the performance in the low-S/N case. The agc system remains effective until the signal-to-noise ratio in the narrow-band filter approaches unity, and all servo bandwidths are maintained until this point is reached.

In order to find the signal-to-noise ratio $(S/N)_f$ applicable to the narrow-band filter channel, we must apply to the single-pulse S/N ratio two factors, one given by the ratio of the repetition rate to the filter bandwidth B_f, and a second equal to the factor $B\tau$ which applied to the wideband i-f.

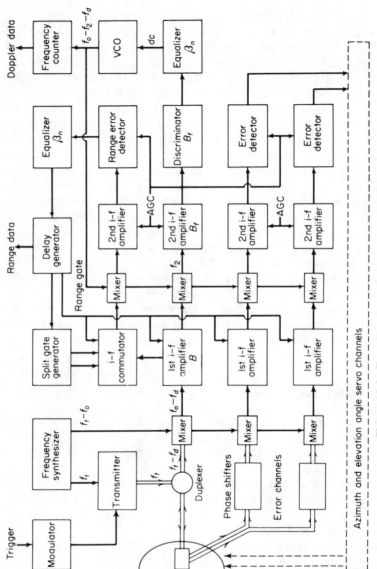

Figure 12.6 Four-dimensional tracker (monopulse type).

$$\left(\frac{S}{N}\right)_f = \left(\frac{S}{N}\right)\left(\frac{f_r}{B_f}\right)B_T = \mathcal{R}_1\frac{f_r}{2B_f} \qquad (12.1)$$

This follows from application of the same process used in establishing signal-to-noise and signal-to-clutter relationships in MTI, and tracking noise response for the angle tracker (Fig. 9.7). A further advantage of the system is that it permits resolution of targets in the Doppler coordinate, even though they may occupy the same spatial resolution element for a brief time. For example, in tracking one stage of a missile through separation from a booster, separation velocities of 2 or 3 ft per sec are sometimes found. It would require periods in the order of a minute for the two targets to be resolved in range, when 1 μsec pulses are used. At a frequency of 5600 mc, the two signals would separate immediately by 25 to 35 cps, and could be resolved in a system using $B_f < 20$ cps. A disadvantage of the four-dimensional system is that it requires designation in all four coordinates simultaneously, if it is to acquire a target. Special signal processing at i-f is required to display targets in the range-Doppler plane, in order to permit the operator to select a particular target. Such a radar was placed in operation during 1962 for special missile instrumentation applications at Kwajalein Island in the Pacific.*

Phase-Locked Loops

The four Doppler-tracking systems described above are controlled by frequency discriminators to maintain the received signal in the center of a given i-f filter. In the c-w or coherent pulsed systems, it is also possible to close the tracking loop by using a phase-sensitive discriminator, which will hold the internal oscillators in exact step with the phase shift produced by motion of the target. This actually results in a range-tracking system, since the phase of the echo signal corresponds to radial range and its derivative to radial velocity or Doppler frequency. The r-f phase information is highly ambiguous, and attempts are seldom made to convert the phase data to range. However, such loops are excellent devices for measuring Doppler, since there is no possibility of velocity bias error. This can be an advantage over the frequency discriminator, whose zero position depends upon the relative response of different circuit elements to closely spaced frequencies.

In order to use the phase-locked loop, the internal VCO must be held at all times within a fraction of a cycle of the signal phase. This requires that the initial designation be closer than in the frequency-locked case,

---

* J. T. Nessmith, "New Performance Records for Instrumentation Radar," *Space/Aeronautics* (Dec. 1962), pp. 86-94.

and that the radar and target be more stable. The systems shown in Figs. 12.4 and 12.6 could use phase-locked loops if these stability requirements were met. The frequency discriminator in each case would be replaced or supplemented by a phase detector, whose output would control the VCO in such a way as to hold its phase in step with the input signal. A disadvantage in this system, when pulsed transmissions are used, is that the phase-locked loop must have sufficient bandwidth to follow the input within about 45° in phase at all times. When the target has a high acceleration, the lag may become excessive unless a very wide loop bandwidth is available, and the pulsed system is limited to bandwidths below $f_r/2$. For example, in a system operating at 3000 mc, the tolerance of 45° in phase corresponds to a lag error of 1.25 cm or 0.041 ft. If the repetition rate is 400 pps, a loop bandwidth of about 100 cps might be attained, providing an acceleration error constant of 25,000. The maximum target acceleration would then be 1000 ft per $\sec^2$, or about 30 g. In the frequency-locked system, there is no limit on the acceleration of targets which may be tracked, since the loop responds to acceleration as "velocity" of the input quantity (frequency). Bandwidth restriction in the frequency-locked loop affects the response of the tracker to the third derivative of target range.

12.2 THERMAL AND TARGET NOISE IN VELOCITY MEASUREMENT

The fundamental limitation in frequency measurement owing to thermal noise was established in Chapter 2.

$$(\sigma_f)_{\text{opt}} = \frac{1}{\alpha\sqrt{\mathcal{R}}} \qquad\qquad \textbf{(12.2)}$$

Here, the parameter α represents 2π times the rms duration of the signal, measured with respect to its center of energy, and $\mathcal{R}$ is the energy ratio defined in Chapter 1. For a signal whose power level is constant over an observation time t_o, the value of α is $\pi t_o/\sqrt{3}$, giving

$$(\sigma_f)_{\text{opt}} = \frac{1}{1.8 t_o \sqrt{\mathcal{R}}} \qquad\qquad \textbf{(12.3)}$$

The time interval to be substituted for t_o in this equation depends upon the nature of the signal and the measurement process, since it is required that the relative phase of the signal and of the internal reference in the measuring system remain constant over this interval. Thus, in a noncoherent radar, the pulsewidth τ sets the limit of time over which coherence is possible. When τ, usually measured in microseconds, is placed in Eq.

(12.3), the result is an accuracy measured as some fraction of a megacycle. In a coherent system, there will be a time period t_c over which the target retains its phase, and this will often place a limit on the accuracy of the system. Only in the ideal system, with no target or radar instability, will t_o represent the total observation time.

Discriminator Noise Output and Error Slope

The performance of frequency discriminators may be analyzed by a process similar to that which was used for the split-gate range tracker. Instead of two displaced gates, we have two narrow-band filters whose outputs are subtracted to form the error signal. The signal spectrum replaces the wave form, and the filter frequency response is analagous to the wave form of the gate. Error-slope curves for three types of discriminator are shown in Figs. 12.7 and 12.8. Both figures are based on the assumption of a rectangular pulse, which applies to the discriminator a spectrum of the $(\sin x)/x$ type, where the bandwidth $B_t = 1/\tau$ gives the distance between the center frequency and the first null. Figure 12.7 applies when the

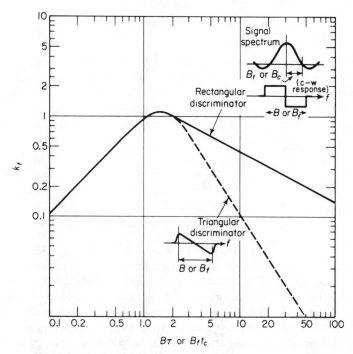

Figure 12.7 Error slope vs. normalized discriminator bandwidth.

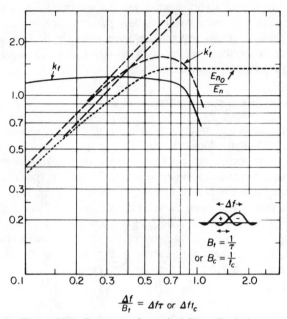

$$\frac{\Delta f}{B_t} = \Delta f\tau \text{ or } \Delta f t_c$$

Figure 12.8 Response of matched-filter discriminator.

over-all response is limited to the bandwidth B, with rectangular or triangular discriminator response inside this band. (The response shapes referred to here are those which would be measured with a c-w test signal. When a signal with broad spectrum is applied, these response curves would be broadened.) The error slope k_f has been defined in a way analogous to that used for the "tracking merit factor" in range.

$$k_f \equiv \frac{B_t}{\sigma_f \sqrt{\mathcal{R}}} \tag{12.4}$$

$$\sigma_{f_1} = \frac{B_t}{k_f \sqrt{\mathcal{R}}} = \frac{B_t}{k_f \sqrt{2B_\tau(S/N)}} \tag{12.5}$$

Here, σ_f is the rms error in frequency measurement, and σ_{f_1} refers to the value measured over a single sample of duration τ. The effect of both coherent and noncoherent addition of many such samples will be discussed later.

A third type of discriminator, whose performance is shown in Fig. 12.8, consists of two matched filters with frequencies of peak response separated by Δf. The difference in voltage output of these two filters, averaged over the duration of the received pulse, forms the output. One curve shows the variation of this output for an input consisting of white noise of uniform spectral density, where E_{no}/E_n represents the ratio of output voltage to that which would appear at the output of a single filter whose noise bandwidth is B_t. A second curve shows the relative output voltage slope for a signal, where k'_f represents the ratio of output voltage to signal voltage, per unit error in signal frequency (in units of B_t). The curve for k_f represents the product $k'_f(E_n/E_{no})$, and gives the efficiency of the filter in the presence of thermal noise. This curve shows that the performance remains constant for separations Δf smaller than about $0.75B_t$, dropping sharply for greater separations. It is interesting to note that the error slope k_f which applies to two closely spaced matched filters is almost exactly the same as the maximum value of slope for the other discriminators (Fig. 12.7). The response of this discriminator is equal to the derivative of the matched filter or signal spectrum curve, and this would be expected to give optimum performance. However, these two figures show that the same performance can be obtained from any one of several filters, and that the shape of the discriminator response or signal spectrum will not greatly affect the output until it departs radically from that assumed in the construction of these curves.

When measurements are made over extended periods of time in a coherent system, the discriminator response must be matched to the width B_c of the fine lines in the spectrum of the received signal. Equation (12.4) may be modified to apply to interpolation of frequency within the bandwidth B_c, and the curves of Fig. 12.7 remain applicable. The value of $\mathcal{R}$ must now be evaluated over the period of signal coherence $t_c = 1/B_c$. The error in such a measurement is

$$\sigma_{f_c} = \frac{B_c}{k_f\sqrt{\mathcal{R}}}$$

$$= \frac{B_c}{k_f\sqrt{2B_T(S/N)(f_r/B_c)}} \tag{12.6}$$

$$= \frac{B_c}{k_f\sqrt{2(S/N)_f(B_f/B_c)}}$$

Ideally, the line width B_c will be equal to the reciprocal of the observation time t_o during which the target remains within the radar beam. However, since this period may extend for many seconds or even for minutes, the line width will normally be set by instabilities in the radar system or in the target itself. Similarly, the bandwidth of the narrow-band filter B_f should be matched to B_c for best results, Fig. 12.7 indicating an optimum near $B_f = 1.4B_c$. Practical considerations may make it impossible to use such a narrow filter bandwidth.

Equation (12.6) shows that the original i-f bandwidth B is not important in determining the frequency error of the system. Only the signal-to-noise ratio within the narrow-band filter enters into the final result. When Fig. 12.7 is used for the coherent case, the parameter $B_f t_c = B_f/B_c$ is used in the abscissa. Instead of a "single-pulse" error in frequency measurement, used in the noncoherent radar, we obtain a "single-sample" error which applies for each interval $t_f = 1/B_f$. This error may be further reduced by passing the output through a servo or filter with bandwidth β_n narrower than $B_f/2$.

Thermal-Noise Output of the Doppler Loop

When the frequency measurements are made on a noncoherent basis, with results of several pulses averaged in the frequency-control servo, the single-pulse error σ_{f_1} will be reduced by the factor $\sqrt{n}$, where $n = f_r/2\beta_n$. The output error is

$$\sigma_f = \frac{B_t}{k_f \sqrt{B\tau(S/N)(f_r/\beta_n)}} \tag{12.7}$$

As in the three spatial coordinates, the effect of low S/N ratios may be expressed as a reduction in S/N ratio owing to the detector loss factor $C_a = (S + N)/S$. Also, unless special means are employed to hold the servo loop gain constant, the servo bandwidth will be reduced by this same factor. As before, this reduction permits Eq. (12.7) to be used without modification for all values of S/N, provided that we define β_n as the servo bandwidth which would be measured at high S/N. It is to be understood that this "adaptive" effect, where β_n is reduced automatically by the reduction in S/N ratio, may lead to large lag errors or to loss of the target when the frequency is subject to rapid change or when the servo loop is subject to unbalance.

In the coherent case, the frequency-tracking loop will be limited in bandwidth by the narrow-band filter in the i-f system. The servo bandwidth β_n, defined as a single-sided response relative to zero frequency, may approach but not exceed $B_f/2$. No loss in signal power or error slope will occur when the servo bandwidth is reduced to value much smaller than $B_f/2$, but lag errors may develop in the tracking loop if the signal frequency is changing rapidly. The noise output of the servo loop will be

$$\sigma_{f_c} = \frac{B_c}{k_f \sqrt{B\tau(S/N)(f_r/\beta_n)}} = \frac{B_c}{k_f \sqrt{(S/N)_f(B_f/\beta_n)}} \tag{12.8}$$

We can see that three different bandwidths are now involved in determining loop performance, not including the original i-f bandwidth, which drops from the calculation. The fine-line signal spectrum bandwidth $B_c = 1/t_c$ establishes the frequency interval over which the discriminator can best operate, or the time interval over which coherence could be maintained in a perfect radar. The fine-line or narrow-band filter bandwidth B_f, ideally matched closely to B_c to achieve maximum $(S/N)_f$ and k_f, drops out of the result when the product $(S/N)_f(B_f/\beta_n)$ is considered (this product is just $f_r/2\beta_n$ times the single-pulse energy ratio $\mathcal{R}_1$). Finally, the servo bandwidth β_n indicates the frequency interval over which noncoherent filtering of the output is carried out. The result may be compared to Eqs. (9.10b) and (9.14) for angle tracking, and it is seen that the form of the equation for thermal noise error is the same, with the spectral width B_c replacing beamwidth θ, and with a different error slope.

The form of the equations for frequency error has been chosen to show the dependence of error upon the signal bandwidth B_t or B_c, and the corresponding time duration τ and t_c. It should be emphasized, however, that this dependence is modified by the error slope factor when nonoptimum circuits are used. Consider, for example, the conventional (triangular) discriminator, in the case where its bandwidth B extends well beyond the spectral width B_t. Figure 12.7 shows that the error slope is approximately $3.46(B_t/B)^{3/2}$ for this case, leading to a single-pulse error given by

$$\sigma_{f_1} = \frac{B}{4.9\sqrt{S/N}} \qquad (B \gg B_t) \qquad \text{(12.9)}$$

Comparing this with Eq. (12.5), we see that the new form implies an error proportional to i-f or discriminator bandwidth B, for fixed S/N, or proportional to the square root of this bandwidth for a given signal (since S/N varies inversely with B). If the more optimum rectangular discriminator were used, the result would be

$$\sigma_{f_1} = \frac{B_t}{2\sqrt{S/N}} \qquad (B \gg B_t, \text{ rectangular discriminator}) \qquad \text{(12.10)}$$

This is still poorer than the optimum theoretical performance by the factor $\sqrt{2B\tau}$, but is better than the performance of the triangular discriminator.

Frequency Tracking on Compressed Pulses

It was shown during development of the Tradex radar* that it was possible to combine pulse compression with coherent Doppler tracking. The pulse compression circuits modify the envelope of the pulse spectrum,

---

* Nessmith, op. cit.

but need not destroy the structure of the fine lines within this spectrum. In Figs. 12.4 or 12.6, we may place pulse expansion and compression networks in the transmitter and broad-band i-f portions of the radar, and deliver to the Doppler measuring and filtering circuits the same signal that would result from use of short, high-amplitude pulses.

The requirements for processing Doppler signals do impose limitations on the type of pulse compression networks which may be incorporated in the radar. These networks must remain phase-stable over the period of coherent processing t_c, rather than over a single repetition period, in order to ensure that all pulses within the processing period have identical phase characteristics. If the delay or phase shift in these networks changes, the width of the fine spectral lines will be increased, and the useful period of coherent measurement t_c will be reduced. However, the frequency-tracking loop makes it possible to compensate for target Doppler shift before the received signal reaches the compression filter, eliminating such effects as range-Doppler crosstalk, when "chirp" is used, and simplifying filter design in other systems. The major design problem in simultaneous operation of pulse compression and Doppler tracking originates in the high resolution of the combined system, which leads to difficulties in acquisition (see Chapter 14). Like the other four-dimensional systems, designation is required in all four coordinates, and when pulse compression is used, the designation must be more accurate than in conventional systems. Multichannel processing of some sort is, therefore, required in order to display targets at different locations in the range-Doppler plane.

Signal Spectrum Spread

In the preceding discussion, it was assumed that the signal power remained essentially constant during the period of radar observation, so the spectral width was determined by the stability of the radar. Thus, if the observation period were doubled, the frequency error could be reduced by $2^{3/2}$, corresponding to reduction in bandwidth and increase in the ratio $(S/N)_f$ by a factor of two. In the actual case, there are a number of factors which set a lower limit on the width of the signal spectrum, and hence on the interval over which observations may be considered coherent. Some of these factors were considered in Chapter 7, in connection with MTI. Internal modulation, superimposed on the transmitted pulses or receiver oscillators, can broaden the width of the spectral lines or introduce spurious lines within the bandwidth of the discriminator. Motion of the antenna beam relative to the target will also produce modulation components in the received spectrum, and these must be minimized by accurate angle tracking. If conical scan or sequential lobing is used, there will be some residual energy at the scan frequency (see Fig. 9.7).

A more fundamental limitation originates in the motion of the target itself. Unless the target is a perfect sphere or is stabilized with respect to the radar line of sight, its echo signal will exhibit amplitude and phase modulation, corresponding to the scintillation and glint errors in spatial tracking. The extent of the Doppler shifts introduced by target motion and reflectivity changes was described in Chapter 3, and an example was given showing how the frequency spectrum could be found for an arbitrary target configuration. Generally, we may expect the width of the target spectrum to be as given in Eq. (3.8).

$$B_c = f_{max} = \frac{2\omega L}{\lambda}$$

This is the width measured between the extreme of the spectrum of signals reflected from a target of length L rotating at an angular rate ω around an axis normal to the radar line of sight. Inspection of the reflectivity patterns (e.g., Fig. 3.2 for a cylinder) shows that this is also the frequency at which reflectivity lobes pass the line of sight.

Internal processing of the signal in the radar may cause a further spreading of the lobe width. For example, if a wide band agc loop is used to normalize the receiver output (as measured with an envelope detector), the gain will be increased sharply as the target signal enters a null, leading to generation of harmonics of the lobing frequency. These harmonics will not correspond to any real Doppler shift generated by motion of the target, since such shifts are limited by the radial velocity of the end of the target.

As an example of line-width calculation, consider the case of a radar operating at S-band (3000 mc), tracking a missile whose length is 50 ft. If this target crosses at a slant range of 50 miles, with a velocity of 10,000 ft per sec, it will appear to be rotating at an angular rate $\omega = 0.033$ rad per sec. The corresponding frequency spread will be

$$f_{max} = \frac{2 \times 0.033 \times 50 \times 30.48}{10} = 10 \text{ cps}$$

A reduction in narrow-band filter bandwidth B_f below this value would exclude portions of the signal from the system and reduce signal-to-noise ratio. Use of a much narrower bandwidth would result in tracking of selected reflecting elements on the target, rather than the target as a whole. As noted in the discussion of error analysis below, the error introduced by exclusion of signals from some portions of the target can nullify any advantage gained through use of the narrower filter.

Lag Errors in the Doppler Loop

We can analyze the frequency-tracking loop as a servo, using the procedures we have applied earlier to the position-tracking loops. In this case, the basic input coordinate is radial velocity, and the lags in the output of the servo will be proportional to higher derivatives. The derivatives of range R were given by Eqs. (11.18) through (11.22). The frequency tracker operates on $\dot{R}$, and its output rate is given by $\ddot{R}$, the target acceleration term. The frequency lag error, in terms of servo error constants, will be

$$\epsilon_f = \frac{\dot{f}}{K_v} + \frac{\ddot{f}}{K_a} + \frac{\dddot{f}}{K_3} + \cdots = \frac{2}{\lambda}\left(\frac{\ddot{R}}{K_v} + \frac{\dddot{R}}{K_a} + \frac{\ddddot{R}}{K_3} + \cdots\right) \quad \text{(12.11)}$$

Since the value of K_v can be increased without limit, and is infinite for a second-order servo, the performance of the frequency tracker is determined by the third derivative of radial range and the bandwidth β_n of the servo loop, upon which K_a depends [Eq. (9.24)].

A second-order loop for frequency tracking, or a first-order loop with very high K_v can maintain a linear rate of change of VCO frequency with negligible error. When the incoming signal from an accelerating target is mixed with the VCO output, the spectral width of the signal can be reduced to zero (in the absence of instabilities), making it possible to track accelerating targets with extremely narrow filter bandwidths. Only when the acceleration of the target changes will there be a lag error in the loop, and the width of the spectrum reaching the filter will be broadened in proportion to the fourth derivative of range. The minimum output error in the case of a target with third derivative may be calculated on the assumption that the servo loop bandwidth is limited to a value somewhat less than one-half the narrow-band filter bandwidth. The expression for optimum noise bandwidth of the servo loop is then

$$(\beta_n)_{\text{opt}} = \left[\frac{2.56\dddot{R}^2 k_f^2 B_T f_r(S/N)}{B_c^2 \lambda^2}\right]^{1/5} \quad (2\beta_n < B_c) \quad \text{(12.12)}$$

In this case, the error slope may be optimized at $k_f = 1.18$, with $B_f \cong 1.5 B_c$. Where the resulting optimum servo bandwidth is greater than the width of the received spectral lines, the performance of the discriminator must be compromised in order to widen the servo bandwidth. Then, assuming the rectangular discriminator response, we shall have $k_f = \sqrt{B_c/\beta_n}$, and

$$(\beta_n)_{\text{opt}} = \left[\frac{1.23\dddot{R}^2 B_T f_r(S/N)}{B_c \lambda^2}\right]^{1/6} \quad (2\beta_n > B_c) \quad \text{(12.13)}$$

The equivalent expressions for the triangular discriminator are

$$k_f = 1.22(B_c/\beta_n)^{3/2}$$

and

$$(\beta_n)_{\text{opt}} = \left[\frac{0.96\dddot{R}^2 B_c B_T f_r(S/N)}{\lambda^2} \right]^{1/8} \quad (2\beta_n > B_c) \qquad \textbf{(12.14)}$$

As an example, consider the missile target crossing at a range of 50 miles, for which we have calculated a spectral width $B_c = f_{\text{max}} = 10$ cps. Assume that the radar operates at $f_r = 400$ pps, and has a single-pulse S/N ratio (in a matched receiver) equal to 10. Using Eq. (11.21), we may calculate the geometric component of $\dddot{R}$ to be 9.5 ft per sec³. A trial using Eq. (12.12) shows that the optimum bandwidth is near 10 cps, or twice as great as will permit optimizing of the discriminator. A new trial using Eq. (12.13) yields the optimum bandwidth $\beta_n = 6.0$ cps, indicating a required filter bandwidth $B_f = 12$ cps. The error slope k_f for a rectangular discriminator would then be about 1.0, found from Fig. 12.7 by setting $B_f t_c = B_f/B_c = 1.2$. The thermal noise error will be 0.21 cps and the lag component 0.64 cps, for an over-all error of 0.68 cps rms.

As the target departs from the region of maximum $\dddot{R}$, the spectral width is reduced according to the sine of the angle α, whereas $\dddot{R}$ varies approximately as $\sin^4 \alpha$. As a result, the optimum bandwidth drops rapidly to the point where the discriminator may be optimized to the signal spectrum, following which the servo bandwidth may be further reduced to obtain noncoherent averaging of the noise.

12.3 ERROR ANALYSIS IN DOPPLER-TRACKING SYSTEMS

The errors owing to thermal noise and dynamic lags, discussed above, may be combined with other components to estimate the accuracy of the complete Doppler measurement loop. The procedure is similar to that used previously in the case of range and angular coordinates, and some of the errors have already been covered in the discussion of range errors (e.g., error in knowledge of the velocity of light). In some cases, the Doppler error will be found simply by differentiating the corresponding component of range error, whereas in others the Doppler error will not be affected by the source of range error. The following discussion will cover the most significant components of Doppler measurement error, as shown in Table 12.1.

Radar-Dependent Tracking Errors

This class of errors is defined in the same way as before: it is the type

Table 12.1   SOURCES OF DOPPLER ERROR

Class of error	Bias components	Noise components
Radar-dependent tracking errors	Discriminator zero setting and drift Gradient of receiver delay	Thermal noise Multipath Servo noise Variation in receiver delay
Radar-dependent translation errors	Transmitting oscillator frequency	VCO frequency measurement Radar frequency stability
Target-dependent tracking errors	Dynamic lag	Dynamic lag Target rotation
Propagation error	Gradient of atmospheric refraction	Fluctuation in atmospheric refraction

of error which causes the servo loop output (in this case the VCO frequency) to depart from that which would correctly indicate the target's radial velocity, and which results primarily from characteristics of the radar itself. One example is the error in tuning or zero-setting of the discriminator. If the crossover point where the two filters produce equal outputs is not set accurately to the offset frequency defined by the fixed coherent oscillators, or if it shifts with the passage of time after calibration, the entire Doppler loop will operate with a bias error. For this reason, the filters used in forming the discriminator are often of the crystal-controlled type, having inherent stabilities of the order of 1 cps or less in the 100 kc region. If higher accuracy is required, the frequency of the final i-f must be reduced, or the phase-lock type of system must be used to eliminate the bias completely. Any noise introduced within the VCO or its control circuitry will appear at the output as a frequency deviation. This type of servo noise is difficult to control in an electronic VCO having a wide tuning range, and it may prove the limiting factor in the system.

Variation in delay through the receiver will appear as a Doppler error, if it is the result of phase shift of the signal carrier. Thus, a slow drift in phase owing to progressive detuning of the local oscillator or to use of agc on the i-f amplifiers would appear as a shift in the average frequency of the Doppler signal. In some frequency-tracking systems, the local oscillator is closely controlled to eliminate this component of error, but in other cases the tracking oscillator operates only on signals which have passed through the first few i-f stages, which are made wide enough to pass all Doppler shifts to be tracked. In those systems which have been designed to provide very small range delay errors, the derivative of the range error will also be low, and the receiver elements will not contribute significantly to the frequency error.

The multipath error in Doppler appears as a second signal component, whose amplitude, relative to that of the direct component, may be written

$$\frac{E_2}{E_r} = \rho\sqrt{\frac{G_2}{G_1}} = \frac{\rho}{\sqrt{G_{sr}}}$$

[see Fig. 10.6(a)]. The side lobe attenuation in the reference pattern is represented by G_{sr}, as in the case of range multipath error. The response of the discriminator to the combined signal is best described in terms of the vector diagram of Fig. 10.6(a), which shows the reflected component as a small vector rotating about the end of the direct component. The direct signal itself rotates about the origin at a rate (relative to the transmitted signal) corresponding to the Doppler shift. The presence of the reflected ray causes the combined signal alternately to lead and lag the direct signal by a phase shift

$$\Delta\phi_m = \frac{E_2}{E_r}\sin\phi = \frac{\rho}{\sqrt{G_{sr}}}\sin\left(\frac{4\pi h}{\lambda}\sin E + \psi\right) \qquad (12.15)$$

This phase will change as the elevation angle changes, and for low elevation angles we may write the frequency error of the combined signal, in cps, as $1/2\pi$ times the derivative of Eq. (12.15).

$$\Delta f_m = \frac{2h\dot{E}\rho}{\lambda\sqrt{G_{sr}}}\cos\phi \qquad (12.16)$$

The rms value of the multipath error in frequency is

$$\sigma_{fm} = \frac{\sqrt{2}\,h\dot{E}\rho}{\lambda\sqrt{G_{sr}}} = \frac{\sqrt{2}\,h\rho v_t\sin\gamma}{\lambda R\sqrt{G_{sr}}} \qquad (12.17)$$

The angle γ represents the angle in the vertical plane between the radar beam and the target velocity vector.

Assume, for example, a low-elevation target at a range of 50 miles, with a vertical velocity of 10,000 ft per sec. The elevation angle rate will be 0.033 rad per sec. If the antenna height is 50 ft, the wavelength 10 cm, the ground reflectivity $\rho = 0.3$, and the side-lobe attenuation 25 db, the rms error in Doppler will be

$$\sigma_{fm} = \frac{\sqrt{2}\times 50\times 0.033\times 0.3}{0.33\times 17.8} = 0.12 \text{ cps}$$

This error corresponds to a velocity of 0.02 ft per sec, and indicates that the multipath error should not be very troublesome in a narrow-beam system with good side-lobe rejection. As the antenna height is increased, the fm sidebands representing the reflected component may lie outside the bandwidth of the discriminator. These sidebands are separated from the direct signal by the multipath cyclic frequency $f_m = 2h\dot{E}/\lambda$ [see Eq. (10.14)]. In the above example, $f_m = 10$ cps, placing the sidebands within the width of the usual tracking filter. If the separation increases beyond this, or if very narrow bandwidths are used, the error may be reduced still further.

Translation Errors

The ability to hold the radar's transmitter frequency at the intended value is limited only by the accuracy and stability of the basic oscillators. Short-term stabilities of one part in 10^{11} are not difficult to obtain in crystal-controlled oscillators, and the same figure applies to the daily drift of the best oscillators of this type. Using an "ultrastable 1 mc/sec crystal oscillator" of commercial manufacture, workers at MIT's Lincoln Laboratory made measurements of lunar surface reflectivity with an over-all resolution of 0.1 cps at 400 mc, based on a 9 sec pulse train.* The fact that the resolution was as good as theoretically predicted for this signal duration indicates that the oscillator, transmitter, and receiver stability was at least as good as the 2.5 parts in 10^{10} resolution figure, and the time over which this stability was maintained was at least 10 sec. By combining the short-term stability of the crystal oscillator with the accuracy of an atomic standard, the transmitted frequency can be kept within 0.1 cps of its nominal value over the entire radar spectrum.

Characteristics of the target and its environment will have greater influence on the stability and accuracy of the Doppler indications than any elements of the radar. The translation of Doppler shift into velocity data is dependent upon the phase velocity of propagation in the medium surrounding the target. This can be known to no better accuracy than the vacuum velocity of light c. As noted in Chapter 11, this basic constant was determined only to about one part per million, as of late 1962. Targets associated with the earth, which are limited to velocities below about 40,000 ft per sec, may be subject to errors up to 0.04 ft per sec from this uncertainty alone, until a better standard is measured or defined. In addition, the refractive index of the medium must be known to one N-unit or better (see Chapter 15) for this accuracy to be achieved.

Probably the most serious component of translation error is that ori-

* Gordon H. Pettengill, "Measurements of Lunar Reflectivity Using the Millstone Radar," *Proc. IRE*, **48**, No. 5 (May 1960), pp. 933-34.

ginating in the circuits which measure the VCO output frequency. An ideal VCO will match the target signal shift on a cycle-by-cycle basis, relative to the selected offset frequency of the final i-f units. If its frequency is measured by a simple cycle-counting process, the accuracy will be given by $0.58/t_m$, where t_m is the period over which the count is accumulated (the peak-to-peak error will be $2/t_m$ cps, and the rms value of the sawtooth error distribution will be 0.29 times this amount, from Fig. 10.3). Since output data is usually needed more frequently than once per second, t_m will be small, and the resulting error will be larger than 1 cps. For this reason, a simple counting process will not generally be adequate. The VCO output must be multiplied to obtain a higher frequency, suitable for direct counting with the required accuracy, or a measurement system must be devised which can interpolate within one cycle of the fundamental output. In either case, phase stability of the VCO and its measuring circuits will be critical. Since the VCO must change frequency rapidly to follow accelerating targets, there will be a lag error associated with any system of measurement, limiting the accuracy to the order of 0.1 cps in most cases.

Target-Dependent Errors

The dynamic lag errors owing to limited servo gain and bandwidth have been discussed above. The only other target-dependent error of consequence is caused by rotation or internal motion of the target. If the reflecting surfaces are not distributed evenly over the target, the strongest reflections may originate from a region which is moving with respect to the center of gravity of the object, and the observed Doppler data will vary in a cyclic fashion with respect to the "true" velocity of the object. As an extreme example, we may use the dumbbell-shaped object, one end of which consists of a lightweight corner reflector and the other end of a dense sphere. If the object is in a ballistic trajectory or orbit, the center of gravity (located near the dense end) will move in a regular path, and the radar target will rotate around it. On the average, the Doppler signal will represent the velocity of the object, but the instantaneous measurement will be in error by

$$\Delta v \cong \omega L \sin (\omega_t + \phi) \sin \gamma$$

where ω is the rotational rate, L the length, ϕ the phase angle of the rotation, and γ the angle between the axis of rotation and the radar line of sight. The problem becomes one of defining the target, and the radar must depend on what portions of the object produce reflections (the same problems arise in position measurements).

This error need not be as great as the derivative of the glint error in range. To understand the distinction, we may again consider a dumbbell-shaped target, but with two reflecting ends of such size and shape that their signals are strongly dependent upon aspect angle. A small oscillation in the aspect angle of this target will cause the center of reflection to oscillate rapidly between the two reflectors, producing large range glint derivatives. The Doppler tracker would respond to the true radial velocity of the reflectors, not to the shifts in the center of reflection of the signal envelope. This type of operation has been observed when pulsed Doppler radars are used to measure sea clutter. The change in range of discrete target reflections follows the velocity of the wave crests, whereas the Doppler signal indicates the velocity of particles at the water surface. Thus the differentiated range data can exceed the velocity given by the Doppler channel by a large factor.

If the bandwidth of the Doppler channel is narrow relative to the components generated by internal motion or rotation of the target, the errors from this source may be reduced by the filtering effect of the system. In this connection, it is important to define properly the spectrum of the target signal, or to measure it over a period which is appropriate for the system. For example, in the earlier case of the target containing a reflector and a sphere, the rotational rate might be very low (e.g., one cycle per minute). When averaged over several cycles, the frequency spectrum of this target would contain all components between the two extreme Doppler frequencies, covering a spectral width up to several cps. When observed with a system whose bandwidth was in the order of 1 cps, however, the servo would follow accurately the "instantaneous" Doppler shift over its deviations of several cps in the course of the 1 min period, reproducing the velocity of the reflector within a small fraction of 1 cps. A target of smaller length, rotating more rapidly, might appear to have the same spectral distribution, but the servo would average its Doppler shift. Close inspection of the two signal spectra would disclose a difference in the form of a fine structure (at the rotational frequency), not apparent in some measurements.

In the preceding chapters we have described the characteristics of tracking and search radars, and the types of error which appear at their outputs. Errors attributable to the output data take-off devices have been included in the analysis, but we have not been concerned with the system elements which receive and process the radar data. In this chapter we shall discuss some of the basic processes by which radar data are delivered to other portions of the system in usable form. Although a detailed treatment of computer processing is beyond the scope of this book, we shall attempt to follow the the radar data through a typical processing system and to relate the errors of the radar to those which will appear at the final system outputs. The effects of sampling, analog-digital conversion, coordinate transformation, smoothing, and differentiation will be described briefly, and procedures will be given for estimating the errors after each of these processes.

PROCESSING OF RADAR DATA

13

13.1 SAMPLING, CONVERSION, AND TRANSMISSION OF DATA

Radar data are obtained in spherical coordinates, giving the position of the target in range and two angular coordinates. In some cases, Doppler frequency shift is also obtained to indicate the radial velocity of the target. The data appear initially in the form of antenna shaft angles, time intervals between pulses, and oscillator frequencies within the radar, sometimes supplemented by error voltages and similar analog signals. In order to transfer these data to remote points with reasonable accuracy, they must first be converted to an electrical form which is suitable for use over cable links or through standard communication circuits. For short-range cable transmission, analog devices are often used, and outputs in the form of synchro, resolver, or potentiometer voltages are found in most radars. During the 1950's, the trend was in the direction of digital data processing, which gave higher accuracy both in mathematical manipulation and in long-range transmission. Digital data were obtained initially by conversion of shaft rotations to digital form, the same output shafts being used which had originally operated the analog devices. More recently, direct digitization of electrical signals has been used where possible, as in digital range-tracking and Doppler-tracking systems. In some of these cases, it is now necessary to convert from the basic digital format back to analog data in order to operate dials and subsidiary systems of reduced accuracy.

There is still considerable freedom in choosing between analog and digital processing of certain signals, and in some cases the digital computer has been used to replace the conventional analog elements within the tracking servo channels. In an extreme example, it has proved possible to digitize the i-f signals prior to detection, and to perform within an associated digital computer all the steps of error detection, filtering, servo equalization, range and Doppler resolution, tracking, and adaptive control of the radar. As a result of this approach, it has become quite difficult to draw a line between the functions of the radar and those of the associated computing devices.

Analog Processing

The conventional devices used for processing of radar data in analog form are the synchro, the resolver, and the potentiometer. The synchro serves primarily as a means of transferring shaft rotation data over short distances. When it is used in two-speed or three-speed systems, with gear ratios in the order of 16:1 or 36:1 between fine and coarse shafts, it is possible to repeat the original data at a remote point with very high accuracy. If the input data are the angles of an antenna shaft, the synchro system, operating with an instrument servo, is capable of repeating the antenna position with negligible lag error, since the mass and inertia within the instrument servo are very small compared to those of the antenna. The use of synchros is not as convenient in range and Doppler channels, since the input data must first be converted to shaft rotation form, and the bandwidth available from instrument servos may be less than that of the basic data. The same considerations apply to the use of resolvers for transmission of data out of a radar. Either type of device is capable of operating into a servo with a bandwidth approaching 20 or 30 cps, and with an accuracy in the order of 1/10,000 of a revolution of the input shaft.

The resolver, in addition to serving as a data transmitter, may be used as a computer element for performing trigonometric operations on the radar data. When it is used with suitable isolation amplifiers, input data representing range or distance may be multiplied by the sine or cosine of an angle, again with an accuracy approaching one part in 10,000. When combined with the precision potentiometer (linear or with special functional relationship of resistance to shaft angle), the resolver system may perform almost any mathematical operation on the radar data. Since many elements are involved in the usual process, the output accuracy is usually limited to about one part in 1000, or 0.1 per cent of the full-scale range of the system.

During the course of World War II, the art of d-c analog computation was brought to an advanced state, largely as a result of radar data-processing requirements. The circuits and techniques of the M-9 fire-control computer served as a basis for many postwar computer and simulator systems. The major advantage of the d-c analog computer lies in its ability to perform integration, smoothing, and differentiation in simple electrical circuits, thus aiding the computation of predicted positions, velocity vectors, and similar quantities. As in the case of the a-c analog systems, the output accuracy is generally limited to about 0.1 per cent of the full-scale range of the system, although higher accuracies are available if the number of operations performed on the data is limited and if the best components are used. The bandwidth of the most accurate systems is limited by the characteristics of the instrument servos used, with operation up to 20 cps representing a practical limit for electromechanical systems. Systems with lower accuracy may operate without using mechanical elements, thus overcoming the bandwidth limitations at the expense of more error in computation.

The outputs of analog data-handling systems may be used in many ways, with or without human links. In many cases, the output data are presented on large plotting boards, which reproduce on maps or overlays the positions of targets with respect to some reference system or with respect to other targets. In automatic systems, such plotting devices are used to monitor the performance of the system, permitting human intervention when required. Radar data may be transferred from a search or tracking radar to designate targets for other trackers, on a fully automatic basis. The primary operations performed on the radar data in this case are coordinate conversions and parallax corrections, translating the original spherical coordinates to rectangular, shifting the reference to a second station, and reconverting to spherical coordinates with respect to that station. The controlled device may also be a gun or missile launcher, or a missile in flight. Examples of such systems are described in such books as Locke's *Guidance** and the Air Force's *Fundamentals of Aerospace Weapon Systems.*†

Analog-Digital Conversion

In order to avoid the errors which are inherent in analog processing of data, tracking-radar systems of very high accuracy will normally convert the target position and velocity data to digital form as soon as possible. Antenna position is converted directly from shaft rotation to digital signals representing the elevation and azimuth angles (or other suitable coordinates). Either two-speed or single-speed encoding devices may be used, depending upon whether the errors of gear-drive units needed for the former are tolerable. The state of the art in shaft-position encoders is such that negligible error is involved in the process of converting to digital form, provided that the cost, size, and weight of the most accurate encoders is accepted.

In digitizing range and Doppler data, the analog input is usually in the form of oscillator frequencies or time intervals between given portions of a wave form. In both cases, conversion to digital form may be carried out by using counting techniques. A digital range-tracking system was described in Chapter 11, in which the tracking gate delay was controlled by the output of a counter or storage register. Counting of individual cycles in a Doppler-tracking oscillator represents an even more direct approach to the digitization of data, and involves little opportunity for error in the conversion process.

Other radar data are generated initially as voltages or voltage ratios, as in the case of tracking error signals or indications of target signal strength. The accuracy of these voltages is limited to about 0.1 per cent of full scale by the nature of the electronic circuits and components used, but it is often convenient to convert the data to digital form to permit transmission, computation, or recording along with the more accurate coordinate data. Analog-to-digital converters for this purpose are available in a variety of forms, with accuracy and sampling rates appropriate to the original data. In some cases, this same type of voltage-to-digital converter may be used on the output of an analog computer which processes the radar coordinate data, in order to prepare data of medium accuracy for transmission. Such systems are often used in transferring tracking data to other radar sites

---

* A. S. Locke, *Guidance* (Princeton, N. J.: D. Van Nostrand Co., Inc., 1955).

† Department of the Air Force, *Fundamentals of Aerospace Weapon Systems* (Washington, D.C.: U.S. Government Printing Office, 1961).

for purposes of designation. The optimum accuracy of the tracking radar may not be required for this, and both sampling rates and digitizing accuracy may be adjusted to preserve only that accuracy needed for designation.

Errors in Analog-Digital Conversion

The important characteristics of any analog-digital conversion process are its sampling rate and its accuracy. Included in the latter are the granularity, or least-bit size, of the digital data; the random noise associated with conversion; and any systematic error which may be present. The sampling rate f_n establishes an upper limit on the bandwidth of the data, since no information can be transmitted at frequencies above $f_n/2$.* If the bandwidth of the servo channel prior to digitization is β_n, the sampling rate required to pass essentially all the data is at least $f_n = 2\beta_n$ and may often be two to four times this value. When the sampling rate exceeds this critical value, it will be possible to reproduce at any point in the computing process the frequencies of the original data, with small error. In the case of the radar servo whose bandwidth is in the order of a few cps, sampling rates of 10 to 20 per sec are used to preserve all the data. In some cases, the servo outputs and error signals are sampled at the radar repetition frequency, passing into the digital processing system all the information received by the radar (and all the noise), for later filtering or smoothing.

When the sampling rate must be set below the value needed to pass all the output frequencies of the radar servo or filter, there will be an irretrievable loss of information in the system. This may or may not be important in the use of the data, but the effect should be evaluated in establishing radar data bandwidths and sampling rates. Consider the typical radar angle-tracking error spectrum of Fig. 10.4, which is reproduced in Fig. 13.1. Five components of error were distinguished in this spectrum, extending from the zero-frequency "true bias" up to (and slightly beyond) the bandwidth β_n of the servo system. If the sampling rate f_n is set near or below this bandwidth, the rms value of the output error will not change, but frequency components above $f_n/2$ will disappear. Figure 13.1 shows where this energy goes when it leaves the high-frequency region of the spectrum. Each component above $f_n/2$ appears as a "mirror-image" or "foldover spectrum" in the region below $f_n/2$. The new frequency corresponds to the beat note between the original error frequency and the nearest harmonic of the sampling rate. The amplitude distribution and rms value of the error remain unchanged in this process, unless the granularity is such as to alter these statistics.

---

* See, for instance, Mischa Schwartz, *Information Transmission, Modulation and Noise* (New York: McGraw-Hill Book Company, 1959), p. 169.

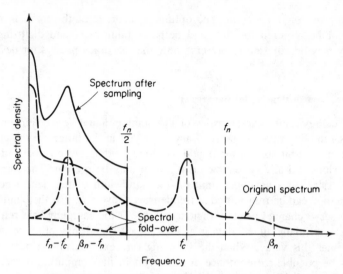

Figure 13.1 Modification of spectrum by sampling.

Thus we may distinguish four effects produced by the sampling or digitizing process. First is the adding of an error component due to granularity. When the input data are relatively smooth, this error takes the form of a sawtooth wave (Fig. 10.3) whose peak-to-peak amplitude is equal to the size of the least significant bit Δx, and whose rms value is, therefore, 0.29 Δx. Second, there are other errors associated with the particular device used to convert the data to digital form, which must be evaluated in each special case (e.g., cyclic or systematic error due to the mechanical construction of shaft encoders). Third, there may be a shift of noise energy into the low-frequency regions of the spectrum, caused by too low a sampling rate. This will make it more difficult to remove the noise errors by subsequent smoothing processes, since the high-frequency noise errors may now appear superimposed on the signal components in the low-frequency region of the spectrum. Lastly, in the case of an accelerating target, the limited bandwidth of the subsequent processing equipment may not be sufficient to pass all those components of the signal needed to reconstruct the target motion, and lag errors will result which are larger than those of the radar itself. These lag errors may be found by using the over-all system bandwidth (approximately $f_n/2$) instead of the servo bandwidth β_n, in cases where the digital portion of the system sets the upper frequency limit.

Two further points should be made in regard to error in sampling and digitizing. The spectral distribution of the quantizing error depends both on the sampling rate and on the relationship between the size of the least bit and the other noise in the system. When noise is absent, and the input quantity x changes very slowly, the sawtooth component representing quantizing error may appear at a low frequency given by

$$f_q = \frac{\dot{x}}{\Delta x} \tag{13.1}$$

In the presence of noise whose amplitude is greater than Δx, the quantizing error will assume a new value on each independent noise sample, and its spectral distribution will be broadened to that of the noise itself, extending all the way to the upper limit set by the sampling rate. In certain cases, it may be desirable to introduce noise or "dither" into the system at high frequency, in order to ensure that the quantizing error will not appear in the low-frequency regions of the spectrum where the signal is located. In order to take advantage of this error-reducing procedure, we must sample at a high rate and follow the sampling with a smoothing process which will attenuate both the noise and the distributed quantizing error components. This requires additional processing capacity, but it makes possible the virtual elimination of quantizing error from the system, even when the least bit is several times greater than the allowable output error. However, when the sampling rate must be low for reasons of simplicity or economy of equipment, the system error can be minimized by reducing the bandwidth of the radar servos or filters (below the optimum value β_o, if necessary), to ensure that no large error components will reach the digitizer in the spectral region above $f_n/2$.

Coordinate Conversion

Radar data in spherical coordinates, referred to the radar site, are of limited value in most systems. One of the first steps taken in processing these data is usually that of conversion to some other coordinate system, in which data from several radars or other instruments may be combined or compared. The desired coordinate system is often a Cartesian system (rectangular, or xyz, system) referred either to the local horizontal plane, to the axis of the earth, or to inertial coordinates. Earth-centered spherical coordinates are also used, as in navigation, with latitude, longitude, and a radius from the center of the earth as the three coordinates. Other special systems of coordinates are required in some applications. In any case, the radar data must be transformed from two angles and slant range (and range rate, when available) into the chosen system, without undue degradation of accuracy. In this process, it is also possible to apply various corrections and calibrations to the data, so that the output data represent the best possible estimates of the target position in the new coordinate system. Examples of this would be the correction for propagation errors, local gravity anomalies, and for other errors whose magnitude and direction may be predicted from the radar data and other available measurements.

There are several advantages to processing radar data in rectangular or earth-centered coordinate systems. We have seen in the preceding chapters that the accelerations seen by the radar servo channels are largely the result of "geometrical" components, originating in the radar's local spherical coordinate system, but not characteristic of the target's motion in inertial space. As will be shown in a later example, the radar may be operated with wide servo bandwidths and the data may be smoothed after conversion to rectangular form, reducing the lag errors of the smoothing filters to those contributed by real target accelerations. The allowable smoothing times will be greatly increased, and satisfactory system operation may be possible out to the limit of the tracking range of the radar, in spite of relatively high error levels in the radar outputs. An extreme example of this is encountered in the determination of satellite orbits, where the acceleration may be assumed to be known with great accuracy over periods of many days. Radar data may be fitted to a hypothetical elliptical orbit in such a way that noise terms from the radar are reduced to negligible importance in the final result.

In practice, it is necessary to balance the sampling rates and bandwidths in various portions of the system to prevent overloading of a critical element with unnecessary amounts of data and noise. In tracking a ballistic trajectory or orbit, the radar bandwidth might have to be set to such a high value (to eliminate lag errors) that loss of track due to thermal noise would take place. Obviously, the servo bandwidths would have to be reduced to assure continuance of the track. Open-loop error correction processes could then be used to extend the bandwidth of the radar data to the desired value, if this were beyond the operable radar servo bandwidth. Conversion of data to rectangular form would then take place in a relatively wide-band device, either by analog computation or in a digital system with

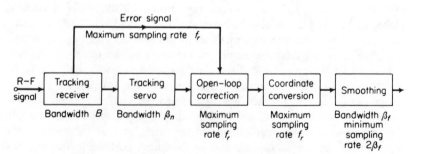

Figure 13.2 Block diagram of optimum processing system.

high sampling rate (see Fig. 13.2). Following this conversion, the bandwidth or sampling rate could be sharply reduced again, making it possible to transmit the target position over standard communication circuits or to perform further processing at a more economical rate.

Digital Transmission

One of the most valuable properties of digital data is the ease with which they can be transmitted over conventional communications circuits, on an error-free basis. Each sample of the target position output of a precise tracking radar can be represented by some 60 binary bits, and search-radar data will require perhaps half as many. In order to transmit the tracker data at a rate of 20 complete samples per sec, the channel must carry 1200 bits per sec. This speed is within the bandwidth capabilities of high-quality voice channels, on radio or wire circuits. With refined transmission techniques, or with a slight reduction in sampling rate, the data may be carried on a single voice channel of average quality. The design principles of equipment for such transmission will not be discussed here, but a few of the system characteristics are of interest to the radar systems engineer.

First, it is apparent that the data to be transmitted must be made available to the communication circuit at a uniform rate, with provisions for synchronizing the radar or processer timing cycle with that used in transmission and reception. The format used in transmission will usually be optimized with respect to the communication circuit, and may well be different from that used within the radar and its local processing equipment. Second, since the data will, in most cases, be transmitted serially, there will be a delay in assembling the sample at the receiving end of the channel. This delay will be at least equal to the sampling period $(1/f_n)$ plus the propagation time required by the signal in traversing the channel. Normally, this leads to a requirement that the data be accompanied by a "time tag" giving the exact time at which the data were taken from the radar. (We may note that the radar data are delayed by an amount at least equal to the one-way propagation time of the signal exchanged with the target.) The time tag may be attached at the time the radar encoders are sampled, or it may be supplied at a later step, on the assumption that delays between that point and the radar sampling are fixed and known to the required accuracy. Finally, the system must have some means of verifying the correctness of the data, since even the best transmission equipment has opportunity for gross error owing to the presence of noise or signal gaps in the channel. This last requirement is met by the editing process described below, or by a more refined error-correcting procedure applied to the channel.

When the transmission channel is severely limited in its bandwidth, as when teletype channels are used, the data must be adapted to minimize the degradation caused by the low sampling rate. It will be recalled that the effect of low sampling rate is to modify both noise and signal components which occur in the original data at frequencies above one-half the sampling rate. If the output of a tracking radar is to be carried on a single teletype channel whose capacity is 30 bits per sec, the maximum data rate will be about one sample per sec. In order to achieve optimum performance at this rate, it is almost essential to perform coordinate conversion and smoothing at the radar site before transmission of the data. The resulting system bandwidth will be limited to $\frac{1}{4}$ cps, and the corresponding value of acceleration error constant will be 0.16 or less. This leads to an acceleration lag error of about 6 ft for each ft per sec^2 of target acceleration, or almost 200 ft per g. The effect can be reduced if the bandwidth limitation applies only to rectangular coordinate data, or, better yet, to data which are referred to the coordinate system in which the target actually moves. For a ballistic missile or satellite, for instance, the motion can be described in terms of orbital elements which are essentially constant over periods of many minutes, hours, or days. Data in this form can be transmitted over a link at relatively low sampling rates with small lag error, unless the target is subject to accelerations which are outside the theory of orbital motion.

Editing of Digital Data

In the process of computing and transmitting thousands of samples of radar data, each consisting of 60 or more bits of binary data, it is almost inevitable that some errors will occur. Early systems using digital data were sometimes thrown into violent confusion by the occasional loss or addition of a data pulse in one of the more significant bit positions. When such an error is analyzed on the usual rms basis, it becomes virtually impossible to operate an accurate digital system, since a single misplaced pulse out of thousands or millions can introduce an error far greater than the worst tracking error of the radar itself. If digital data are to be used effectively, they must be subjected to a checking or editing process to minimize the effects of such errors. It is possible, of course, to use error-correcting codes* greatly to reduce the occurrence of these errors. In processing of radar data, however, it is more common to take advantage of the great redundancy inherent in the continuous tracking and sampling process, which permits the deletion of many random samples without substantial degradation of the data output. Each sample is checked to see if it lies within a reasonable distance from the previous point or from the expected

* See, for instance, W. Wesley Peterson, "Error-Correcting Codes," *Scientific American*, **206**, No. 2 (Feb. 1962), pp. 86-108.

new target position. If the position indicated by the new sample is beyond a certain limit, it is rejected completely. A dummy point may be inserted by the processing equipment, based on the expected value of the coordinate for that time, or the using equipment may be programmed to ignore that sample in the affected coordinate. In either case, it is necessary to obtain valid data on most of the samples, in order to be sure that the "expected" value actually follows the target position, and that real deviations of the target from its planned course are not overlooked. Depending upon the certainty with which the target is known to be limited in velocity or acceleration, it may be possible to operate successfully when as many as half of the received data points are rejected as erroneous. There will also be errors which are too small to be rejected on the basis of departure from th expected value of the data, and these will introduce noise errors into the edited output data. Unless the transmission system and digital processing equipment are hopelessly inadequate for their task, however, the error added to the final data by misplaced pulses in the transmission can be reduced to a negligible point by using one of the relatively simple editing procedures. It is those systems which have been designed without such procedures which have been subject to large error components owing to digital and transmission links.

13.2 DATA SMOOTHING

The output data from the radar result from successive conversion of r-f signals to intermediate frequency and then to a narrow band of signals near zero frequency. The bandwidth of the radar output is described by the servo or data bandwidth β_n, which represents the highest frequency at which oscillation in the target position can be followed by the radar with acceptable accuracy. Procedures have been given for choosing β_n to minimize the rms error of the radar output, but it has been noted that the optimum bandwidth cannot always be used, for various practical reasons. Rather than lose information through premature restriction of bandwidth in the radar, it is often preferable to operate at a greater bandwidth than needed, and to depend upon the data-processing system to perform the final reduction in data bandwidth, through use of smoothing techniques. If the operation of the radar and data circuits is linear, it makes no difference where the smoothing is applied. However, when coordinate conversions are to be applied, it is often more convenient to perform most of the smoothing at a point where the data bandwidth is at a minimum. We shall consider here the characteristics of some of the smoothing filters commonly used in both analog and digital equipment.

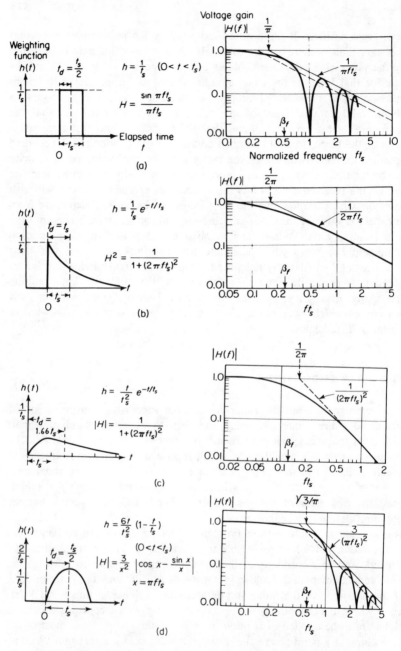

Figure 13.3 Response of smoothing filters. (a) Rectangular weighting. (b) Exponential weighting. (c) Cascaded exponential weighting. (d) Parabolic weighting.

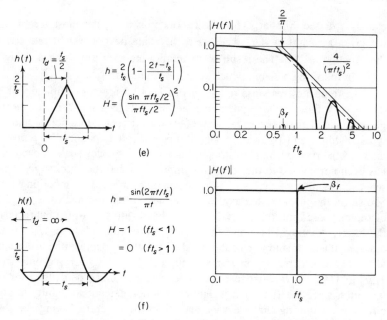

Figure 13.3 cont. (e) Triangular weighting. (f) $(\sin x)/x$ weighting.

Weighting Functions and Frequency Response

The response of a smoothing circuit may be described by using the procedures developed for any linear filter. One such description is the weighting function or impulse response of the filter, $h(t)$, which is a time function. An equivalent description is the frequency response $H(f)$ of the filter. The two functions are Fourier transforms of each other, and a complete specification of either will permit derivation of the output signal for any input.* Figure 13.3 shows plots of both response functions, for several common types of smoothing filter. The weighting function is plotted as a function of the time which has elapsed since the occurrence of a given input sample. Thus, the rectangular-weighting filter forms an average, with uniform weight, of all data received during the past t_s sec, giving an estimate of the input signal at the center of the weighting period. If the input is changing at a uniform rate, the delay t_d between input and output will be

* See, for instance, Mischa Schwartz, *Information Transmission, Modulation and Noise* (New York: McGraw-Hill Book Company, 1959), pp. 65-73.

Also see the classical papers by L. A. Zadeh and J. R. Ragazzini, "An Extension of Weiner's Theory of Prediction," *Journal of Applied Physics*, **21**, No. 7 (July 1950), pp. 645-55; "Optimum Filters for the Detection of Signals in Noise," *Proc. IRE*, **40**, No. 10 (Oct. 1952), pp. 1223-31.

exactly $t_s/2$ sec. Similarly, for the exponential filter, the most recent data are given a weight of $1/t_s$, whereas earlier data have a weight reduced by the function e^{-t/t_s}. The result is an infinite memory time, and a delay of t_s sec in following a signal which varies at a uniform rate.

The frequency response functions of Fig. 13.3 are not complete, but indicate only the absolute magnitude of the voltage gain, $|H(f)|$, ignoring phase angle. This is the usual way of plotting transfer functions for many types of filter, and by using a logarithmic scale it is possible to represent many of the relationships by straight lines. The frequency scale in each case has been normalized to the smoothing time t_s, so that the abscissa ft_s represents the number of cycles of input occurring during this smoothing time. Shown on the plots are the asymptotic values for each straight-line segment of the response, and the equivalent noise bandwidths β_f which characterize the filters. Although the noise bandwidths vary from $1/8t_s$ to $1/t_s$ for the different filters, a simple calculation shows that the bandwidths are almost identical when expressed in terms of the delay time t_d, varying only from $1/\pi t_d$ to $1/4t_d$ for the first five filters. No delay time can be defined for the last filter, since the $(\sin x)/x$ weighting function extends to infinity in both directions, requiring that we wait an infinite time before having the data required to form the estimate at any given point.

The two filters using exponential-weighting functions are those most often encountered in analog systems. In d-c analog data processing, the exponential weighting of Fig. 13-3(b) may be provided by a simple RC circuit, as shown in Fig. 13.4(a), or by the operational amplifier circuit of Fig. 13.4(b). Two such circuits in cascade give the weighting function of Fig. 13.3(c). By using more complex circuits, it is possible to approach the parabolic-weighting function of Fig. 13.3(d), although there will always be some tendency in the analog system to preserve the long "tail" of the exponential function, often with a superimposed sinusoidal oscillation. In digital systems, the weighting functions may be made closely to approximate any of the forms shown. The finite sampling rate and granularity of the digital system place a series of steps on any weighting function used, but

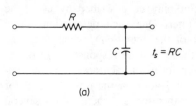

(a)

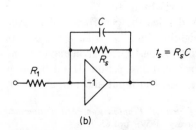

(b)

Figure 13.4 Basic d-c analog smoothing circuits. (a) Simple RC filter. (b) Smoothing circuit using operational amplifier.

these steps may be made small by frequent sampling and small bit size. The discontinuous frequency-response function, punctuated by lobes of a $(\sin x)/x$ form, is characteristic of the digital filter. When the detailed structure of this type of response is not of interest, we may replace the lobes with a straight line representing the rms response, as shown by dashed lines in Fig. 13.3. This response will normally be at a level 0.707 times that of the asymptotic-response line shown on the plot.

The noise error at the output of the smoothing filter will be the combined result of radar and filter characteristics. Consider the case of a white-noise component, described by a uniform spectral density W_n as in Fig. 10.4. The radar output noise will have a variance $\sigma_n^2 = W_n \beta_n$, as a result of the limited bandwidth of the receiver and servo circuits. If the noise bandwidth of the smoothing filter exceeds β_n, the output noise from the filter will be governed primarily by the radar bandwidth, and no useful smoothing will be obtained. In general, the filter bandwidth will be made much smaller than β_n, and the noise output will be approximately $\sigma_f^2 = W_n \beta_f$. When the radar and filter bandwidths are comparable in value, the composite response may be found as the product of the two frequency-response functions (or the convolution of the corresponding weighting functions). Thus, for example, if the radar response is represented by a cascaded exponential circuit whose time constant t_{sr} is $1/8\beta_n$, the rectangular-weighting filter would produce an over-all response as shown in Fig. 13.5. The over-all weighting function will be close to the original rectangle provided by the filter, but with rounded edges. The over-all time delay will be the rss sum of the individual delays.

$$t_d^2 = \left(\frac{1}{4\beta_n} \right)^2 + \left(\frac{t_s}{2} \right)^2$$

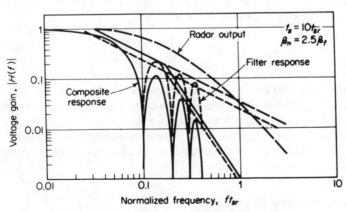

Figure 13.5 Composite response of radar and filter.

Radar error components which do not appear as uniform-density noise over the servo bandwidth must be considered separately. The cyclic components are easily evaluated by using the frequency response of the filter. A radar error whose magnitude is σ_c, appearing at a frequency f_c, will appear at the output of the filter with a magnitude $\sigma_c|H(f_c)|$. Errors of relatively high frequency are greatly attenuated by the filters shown in Figs. 13.3(c), (d), and (e), where the asymptotic response falls off with the square of frequency. In cases where the error frequency cannot be determined exactly, or where the error is spread over a narrow band of frequencies centered at f_c, it is convenient to evaluate the output by using the rms response for that region of the spectrum, if the actual response contains a lobing structure. For example, let us assume that a multipath error in elevation is encountered, with an rms magnitude of 1 mil and a frequency of 2 cps. Using the rectangular-weighting filter, with $t_s = 1.5$ sec, we would find exactly three cycles of error during the smoothing period, and the output error would be zero $[(ft_s = 3$ in Fig. 13.3(a)$]$. If, however, we were not sure of the exact error frequency, it would be more prudent to use the rms value $H(f) = 0.707/3\pi = 0.07$, which would apply over a range of frequencies in this general region.

Radar errors of the low-frequency type [Markoffian spectrum, Eq. (10.7)] may be considered as white noise which has been passed through an exponential filter, with $t_{sa} = 1/2\pi f_a$. If the smoothing time is much greater than the error time-constant t_{sa}, the output will be the same as for white noise of spectral density W_o. If t_s is much less than t_{sa}, the smoothing will have no effect, and the output error will be the same as that measured at the radar. For intermediate cases, the output error will be found by integrating the product $W_a(f)|H(f)|^2$ between the limits zero and infinity. The exact form for $|H(f)|$, or the rms value of the asymptote may be used. For example, if the smoothing filter uses a simple exponential-weighting function, the output error will be

$$\sigma_a^2 = \frac{W_o}{4(t_s + t_{sa})}$$

The error may also be found by integration in the time domain, by using the filter weighting function and the autocorrelation function of the input error, as described by Zadeh and Ragazinni.

Bias errors and "apparent bias," whose frequency components fall within the passband of the smoothing filter, will appear without change at the output of the system.

Lag Errors Owing to Smoothing

The use of smoothing filters on the radar data will reduce noise error, but at the expense of introducing additional lag error for moving targets. The effective time delay t_d of each smoothing filter is shown in Fig. 13.3. This indicates the amount by which the filter output will lag behind the input, in the case of an input signal which changes at a constant rate. Thus, if the target is moving along a path that gives a velocity component v in a given coordinate, the lag error at the output of the filter will be equal to vt_d. Recalling that the radar itself was described in terms of error coefficients K_v, K_a, etc., we may similarly characterize the filter as having a velocity error coefficient $K_v = 1/t_d$. The delay in the output, for constant target velocity, will equal the rss sum of the delay in the radar, $t_{dr} = 1/K_{vr}$, and the filter delay t_d. It should be noted that the K_v of the radar may be made very large (typically 100 to 1000 per sec) by use of a properly equalized servo amplifier, so that the output of the radar will lag a constant-velocity target by only a few milliseconds. Similarly, in the data-processing system, it is possible to "update" the data, the delay t_d being eliminated by adding to the output data a term equal to vt_d. The value of velocity v must be determined from the output of a differentiating filter, as described below, and will be subject to an additional noise error. However, if the target is following a path with constant velocity in the smoothed coordinate, there will be no lag error in the corrected or updated output.

There will remain in the output a lag error owing to target acceleration in the smoothed coordinate, unless this too can be estimated and corrections applied in the updating process. The smoothing filters shown in Fig. 13.3 may all be characterized by an acceleration lag coefficient K_a equal to about $-20\beta_f^2$, or to $-1.4/t_d^2$. The negative value of this error coefficient indicates that the velocity lag correction vt_d is too great to compensate acurately for the actual lag in the data, when the target has positive acceleration and the correction is based on actual, up-to-date velocity. In other words, the use of the symmetrical weighting functions shown in Fig. 13.3 leads to an accurate mid-point estimate of position, when the target has constant velocity. When this mid-point estimate is updated with current velocity, on an accelerating target, the output will lead the actual target position. However, if the velocity estimate used in updating is also based on smoothed values, over the data span t_s the correction will be smaller and will result in a lag in the output data, proportional to acceleration (K_a positive). A thorough analysis of this process has been made, for the general case in which the input data may be represented by a polynomial of any arbitrary degree.* Equations have been derived for the optimum weighting functions to be used in each case, depending on the noise spectrum, the degree of the polynomial which represents the input data,

---

* Irving Kanter, "The Prediction of Derivatives of Polynomial Signals in Additive Stationary Noise," *IRE Wescon Record* (1958), Part 4, pp. 131-46; "Some New Results for the Prediction of Derivatives of Polynomial Signals in Additive Stationary Noise," *IRE Wescon Record* (1959), Part 5, pp. 87-91.

and the type of output data desired (position, velocity, or acceleration, with mid-point or updated estimates). It is shown that the mid-point estimates are far less subject to noise errors than are updated values, as would be expected. The updating process requires that the input data be differentiated, in order to reduce the lag by correction for velocity, acceleration, and higher derivatives, and this increases the noise bandwidth of the processing system. Another interesting result of Kanter's study is his finding that the weighting functions for each derivative may be determined separately, and combined to give outputs representing both mid-point and updated estimates of the input data and their several derivatives. His optimum weighting functions consist of combinations of impulse and uniform weighting for position data, and derivatives of these for velocity, acceleration, and higher derivatives of the input.

13.3 DIFFERENTIATION OF RADAR DATA

Unless Doppler data are available in the desired coordinate, the target velocity must be determined by differentiation of the radar's position data outputs. Velocity in three coordinates can be obtained from one radar only by differentiation. The process of differentiation results in an increase in noise, as compared to position data, and differentiation must always be accompanied by smoothing. The filter transfer functions commonly used in differentiation of radar data are shown in Fig. 13.6. The impulse weighting function $\delta(t)$ represents the sampling of an instantaneous value of position data at the time t, and is indicated on the plots by a pair of closely spaced line with a symbol showing extension of these lines to infinity. The lines represent an infinitesimal sampling period, and the area under the impulse function is equal to unity. Two such impulses, one positive and one negative, separated by an interval t_s sec, give the difference in position over this interval, and this difference, divided by t_s, is the average velocity over this interval.

$$\bar{\dot{x}} = \frac{x(t) - x(t - t_s)}{t_s} \qquad \text{(13.2)}$$

The weighting functions h' (understood to be functions of elapsed time t) are plotted on a scale which is normalized to t_s, so that the product $h't_s$ can be represented by unit impulse functions. Similarly, the frequency response is plotted as the product $|H'(f)|t_s$.

The ideal differentiation would have a normalized weighting function $h't_s$ given by two opposing unit impulses with infinitesimal separation, and its frequency response $|H'(f)|$ would be equal to $2\pi f$ for all frequencies [its normalized value would be $|H'(f)|t_s = 2\pi f t_s$, which is shown by a solid line in Fig. 13.6]. A circuit corresponding to this ideal differentiator is shown in Fig. 13.7(a), consisting of an ideal operational amplifier, whose gain is -1 at all frequencies, with a capacitive input and resistive feedback. Such a differentiator cannot be built or used, in practice. The filter functions shown in Fig. 13.6 are used or approximated in actual digital and analog systems. The impulse filter, with finite time separation t_s, corresponds to the simple, two-point sampling process described by Eq. (13.2). It can be approximated in digital systems, where the sampling time is very short, and will provide a useful output in cases where the noise bandwidth of the input data is limited. If the rms error in each position sample is given by σ_x, the error in velocity output of the filter will be

$$\sigma_{\dot{x}} = \frac{\sqrt{2}\,\sigma_x}{t_s} \qquad\qquad \textbf{(13.3)}$$

For an input consisting of white noise, the output error is infinite, since the lobed response $|H'(f)|$ continues to infinity without reduction in lobe amplitude. Its average (rms) value is $\sqrt{2}\,/t_s$, when integrated over a number of lobes. If it is assumed that the input noise is limited to a band $\beta_n \gg 1/t_s$, we find that the variance of the output noise is simply the integral of $|H'(f)|^2$ between the limits zero and β_n, or

$$\sigma_{\dot{x}}^2 = \frac{2W_n\beta_n}{t_s^2}$$

This agrees with Eq. (13.3), since $\sigma_x^2 = W_n\beta_n$. In practice, the sampling impulses will consist of short exponentials, as in Fig. 13.6(b), or of narrow rectangles, as in Fig. 13.6(f). These filters have the basic response of the impulse differentiator, but are limited at high frequencies by the smoothing inherent in the width of the two sampling periods t_1. If the frequency response of the impulse filter is multiplied by the smoothing response of Fig. 13.3(b) (t_1 being used as the smoothing time), we have the response of the stretched-impulse filter, shown in Fig. 13.6(b). Similarly, the rectangular-weighting function of Fig. 13.6(f) has a frequency response equal to the product of the responses of the impulse differentiator and the rectangular-smoothing filter of Fig. 13.3(a).

If the ideal differentiator is placed in series with a simple RC smoothing filter, the response is as shown in Fig. 13.6(c). This will be recognized as identical to that of a simple RC differentiating circuit, Fig. 13.7(b). The weighting function is simply the derivative of the exponential smoothing function. In fact, the weighting function of the ideal differentiator, in series with any smoothing filter, is just the derivative of the weighting function of

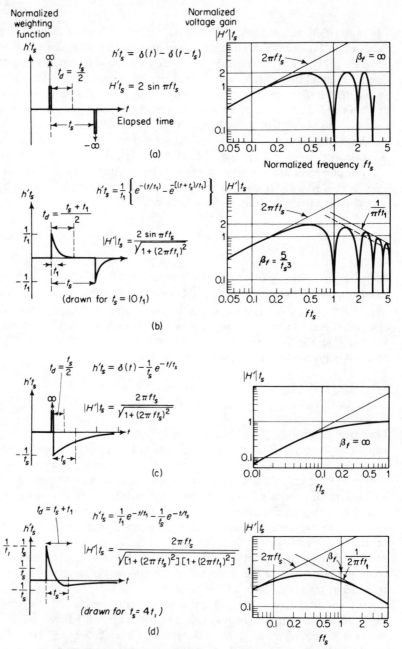

Figure 13.6 Response of differentiating filters. (a) Impulse weighting. (b) Stretched-impulse weighting, (c) Exponential weighting. (d) Cascaded exponential weighting.

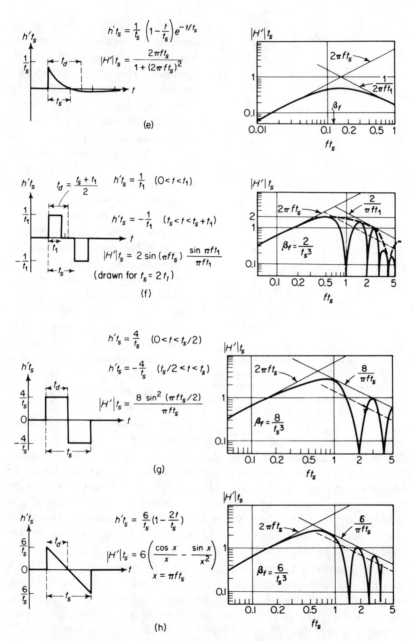

Figure 13.6 cont. Cascaded exponential weighting ($t_1 = t_s$). (f) Rectangular weighting ($t_s > t_1$). (g) Rectangular weighting ($t_1 = t_s/2$). (h) Triangular weighting.

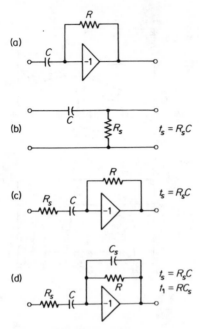

(a)

(b) $t_s = R_s C$

(c) $t_s = R_s C$

(d) $t_s = R_s C$
$t_1 = R C_s$

Figure 13.7 Simple differentiating circuits. (a) Ideal differentiator ($RC = 1$). (b) Simple RC differentiator (exponential weighting). (c) Smoothed differentiator (exponential weighting). (d) Differentiator with cascaded exponential weighting.

that smoothing filter. Thus, the impulse differentiator represents the derivative of the rectangular smoothing filter. The simple exponential differentiator also has an infinite noise bandwidth, although it serves as an effective differentiator only up to the frequency equal to $1/2\pi t_s$. In order to limit the amount of high-frequency noise, another low-pass filter must be used in cascade, providing the response shown in Fig. 13.6(d). When an operational amplifier is used, the second time constant may be provided by inserting a single capacitor across the feedback resistor, as shown in Fig. 13.7(d), or by use of a separate RC smoothing filter in cascade. The additional filter converts the impulse of Fig. 13.6(c) to a short exponential. As the second time constant is increased, the response approaches that shown in Fig. 13.6(e).

In digital systems, the filter weighting functions are more likely to be of the discrete type, such as the rectangular functions o Fig. 13.6(f), (g), or (h). In all these cases, the noise bandwidth is limited by the sampling interval t_1 or the smoothing t_s, and the envelope of the high-frequency response varies as $1/f$. Also, in each case, there is a lobing structure in the high-frequency region, and this may be replaced by an rms response, as indicated by the dashed lines. In the case of the rectangular functions, the rms response is at one-half the level of the asymptote, whereas for the triangular function it is at 0.707 of this level.

It is interesting to note that the rectangular weighting function, with $t_1 = t_s$, represents the derivative of the triangular smoothing function, whereas the triangular differentiator function is the derivative of the parabolic smoothing function. The parabolic function is often used as a model for the ideal smoothing filter, and is approximated by refined analog smoothing networks. The triangular function was shown by Kanter to be the optimum for estimation of velocity data in the presence of white noise. The noise output of each differentiating filter may be described by its noise bandwidth β_f, as indicated in Fig. 13.6. For the triangular function, for instance, the noise bandwidth of the normalized response is equal to $\beta_f t_s = 6/t_s$. In the presence of band-limited white noise, where $\beta_n \gg \beta_f$, the noise output may be written either as a function of the spectral density W_n or of the position error σ_x.

$$\sigma_{\dot{x}}^2 = \frac{3W_n}{4t_s^3} = \frac{2\sigma_x^2}{t_s^2} \times \frac{3}{\beta_n t_s} \qquad \textbf{(13.4)}$$

The expression is divided in the above manner to show a direct comparison with the output of the impulse differentiator, and it may be seen that the variance has been changed by the factor $3/\beta_n t_s$. This represents a substantial reduction in output error, since $\beta_n t_s \gg 1$.

The relationship of Eq. (13.3) is often used to relate the error in velocity data to that of the position data from which they are derived. The preceding discussion has shown that the error may be reduced to a much smaller level in some cases, provided that the input data have been measured with excessive bandwidth and that an optimum smoothing procedure is applied to the differentiator. A convenient way of finding the improvement possible over the performance indicated by Eq. (13.3) is to plot the ratio of the output velocity error to the quantity $\sqrt{2}\,\sigma_x/t_s$, for different types of input noise. This has been done in Fig. 13.8, where the

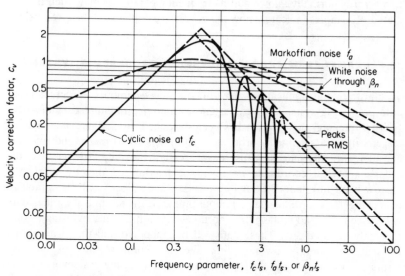

Figure 13.8 Velocity correction factor vs. noise frequency (for triangular-weighting function).

correction factor C_v represents this ratio, and the use of the triangular differentiating function has been assumed. The characteristics of the input noise are indicated by normalized frequency parameters: $f_c t_s$ for cyclic noise, $f_a t_s$ for Markoffian (low-frequency) noise, or $\beta_n t_s$ for band-limited white noise. The difference between the curves for Markoffian noise $f_a t_s$ and band-limited white noise $\beta_n t_s$ is due to the fact that f_a represents a half-power bandwidth, whereas β_n is the equivalent noise bandwidth, greater by a factor $\pi/2$ for Markoffian noise. Over a substantial range of frequency parameters, the velocity correction factor is close to unity, indicating that Eq. (13.3) may be applied without change to a wide variety of actual radar cases where the frequency parameter itself is between 0.1 and 3.0. The greatest departure comes when the noise is cyclic, in which case the correction factor follows the form of the differentiator frequency-response function, Fig. 13.6(h). In practice, unless the radar error spectrum is known to extend over a frequency range much larger or much smaller than $1/t_s$, it is difficult to arrive at a better estimate of velocity error than is given by Eq. (13.3).

When the differentiator has a weighting function other than triangular, curves similar to Fig. 13.8 may be prepared for different types of input error. Inspection of Fig. 13.6 shows that the curves for the rectangular and cascaded exponential integrators $(t_1 = t_s)$ will be very similar to Fig. 13.8, differing primarily in the location of the high-frequency asymptote.

Curves of this type are useful, because the magnitude of the radar position error is normally known and its frequency parameter may often be estimated. Examples indicating how the curves are applied will be given below.

13.4 EXAMPLES OF ERROR ANALYSIS OF PROCESSING SYSTEMS

To illustrate the application of error analysis techniques to actual radar data processing systems, let us assume as a source of input data the AN/FPS-16 tracking radar, which has been described in Chapter 10. A typical target will be a satellite in orbit at an altitude of 100 miles, and we shall consider two alternate systems for processing the data. In the first, the radar's spherical coordinate data will be transmitted directly through a narrow-band link, and the satellite position and velocity will be computed in rectangular coordinates at a remote point. In the second, the radar data will be converted to rectangular coordinates at a high rate, by using a computer at the radar site. These data will be smoothed and then transmitted through the same narrow-band link. We shall compare the noise and lag errors for the two cases, and show how the processing system will affect the ability of the radar to optimize its tracking parameters without degrading the final output data of the system.

Target Position Error

We shall assume that the target is equipped with a beacon whose effective power output (in the direction of the radar) is in the order of 10 w, and that the radar operates at a repetition rate of 160 pps to obtain an unambiguous range of about 500 miles. Figure 13.9(a) shows the variation in radar position coordinates as a function of time from crossover. The curves are identical before and after crossover, except that the azimuth angle is assumed to start at 0°, increasing from 90° at crossover to 180° at the end of the track.

In this example, we shall calculate the error in target position and velocity owing to azimuth error of the radar. Similar calculations may be made for elevation and range components. The azimuth error results in target position and velocity error components in a direction normal to the

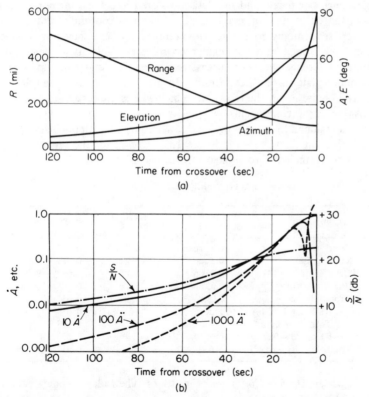

Figure 13.9 Target coordinates and S/N ratio vs. time. (a) Target position in spherical coordinates. (b) S/N ratio and azimuth derivatives.

radar beam and lying in a plane parallel to the horizontal plane through the radar site. The magnitude of these error components is equal to the azimuth error (in radians or radians per second) times the ground range of the target. It may also be calculated in terms of the traverse error times the slant range of the target. Figure 13.9(b) shows the variation in signal-to-noise ratio (assuming an approximately matched system with $\tau = 1$ μsec), and the first three derivatives of azimuth angle. The problem to be solved by the radar and its associated processing system in this example is to arrive at estimates of target position and velocity for each 6 sec interval during the track. We shall assume that the radar's encoders may be sampled at any rate up to 10 samples per sec, and that the data are to be transmitted to a central control point at a rate of one sample per 6 sec. The two cases to be discussed will use the following steps in processing:

Case I   Remote Computer (Fig. 13.10). The radar encoders are sampled once per 6 sec, and the data are stored in a local memory unit along with the exact time of sampling. The spherical-coordinate data are transmitted sequentially to the remote computer, where they are reassembled and transformed into rectangular coordinates. Velocity is determined by taking differences between successive position readings (and dividing by six to obtain data in feet per second). In order to limit the magnitude of the thermal noise error, the radar servo bandwidths are restricted to 1 cps. The position error components will be as shown in Fig. 13.11(a). Azimuth bias is assumed to be 0.08 mil, and low-frequency components, independent of target range, are listed in Table 13.1. Except for the region near crossover, the principal error is due to thermal noise.

Table 13.1 LOW-FREQUENCY ERROR COMPONENTS

Component	Width of spectrum	Correction factor C_v	RMS error in mils
Wind gusts	$f_a = 0.3$ cps	0.9	0.012
Servo noise	0.3 cps	0.9	0.02
Gear error	0.05 cps	1.1	0.03
Encoder error	0.05 cps	1.1	0.02
Tropospheric	0.05 cps	1.1	0.03
RMS total of low-frequency error			0.05
Total bias error			0.078
Bias + low-frequency error			0.093 mil
Total velocity error for $t_s = 6$ sec			0.0125 mil per sec

Case II   On-Site Computer (Fig. 13.2). The radar encoders supply data to the local computer at a rate of about 10 samples per sec, and transformation to rectangular coordinates takes place at this rate in the computer. Smoothing is applied, covering a 6 sec period, for both position and velocity

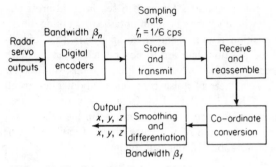

Figure 13.10 Data processing with remote computer.

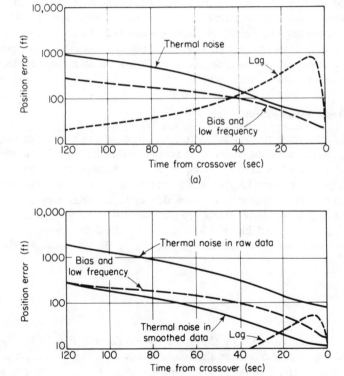

Figure **13.11** Radar position errors vs. time. (a) Errors for Case I: $\beta_n = 1$ cps. (b) Errors for Case II: $\beta_n = 4$ cps.

data. The position estimates for each 6 sec interval are transmitted to the central point, where velocity is again obtained by taking differences between successive position readings and dividing by six (a choice is available as to whether the position data are to be updated prior to transmission). The radar is operated with servo bandwidths of 4 cps, and the smoothed position data have a thermal noise component about one-seventh as great as the raw data [Fig. 13.11(b)]. Low-frequency error is reduced only slightly by smoothing, and bias remains the same as in the other case. Acceleration lag error is one-sixteenth of the value shown for Case I, whereas velocity lag is reduced to about 10 per cent of its original value by open-loop correction.

Comparison of Results for Position

The use of the on-site computer results in position errors about one-half to one-third as great as those of Case I. The data can be improved slightly in Case I, however, by varying the bandwidths of the radar servos to maintain minimum error as the target approaches and then recedes. It may also be improved by fitting the 6 sec points to an orbit whose general form is known in advance. Such a procedure tends to mask the effects of any maneuver during the track.

Target Velocity Error

In calculating the velocity error, the effect of the radar bias may be ignored. The remaining components are due to thermal noise, low-frequency noise, and variation in dynamic lag error. The thermal-noise error may be found by using Eq. (13.3), since the position error is essentially uncorrelated over 6 sec intervals. By coincidence, the low-frequency error components are distributed in frequency such that the velocity correction factor C_v is near unity, and Eq. (13.3) may also be used for these errors. The velocity error owing to dynamic lag is simply the derivative of the lag error itself.

$$\epsilon_v = \dot{\epsilon}_A = \frac{\ddot{A}}{K_v} + \frac{\dddot{A}}{K_a} + \cdots \tag{13.5}$$

These error components are shown for both systems in Fig. 13.12. Except for the region near crossover, the velocity error owing to thermal noise will predominate, and the system using on-site computation is about four times better than the one using remote computation.

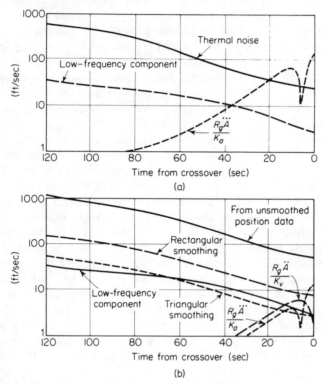

Figure 13.12 Comparison of velocity errors. (a) Case I: 1 cps band-width and remote computer. (b) Case II: on-site computer and smoothing.

In Case I, if the servo bandwidth is further reduced to minimize thermal-noise error, the lag term $R_g \dddot{A}/K_a$ will increase very rapidly, since K_a varies with the square of β_n. Hence, it will not be possible to obtain much improvement by varying bandwidth during the track. In Case II, reduced servo bandwidth will not improve the performance at all, since the computer smoothing already eliminates most of the noise appearing at the radar output. An increase in servo bandwidth will reduce the term $R_g \ddot{A}/K_a$, and the term $R_g \ddot{A}/K_v$ will be reduced by a factor of ten by the open-loop error correction process. Thus, it is possible to avoid the lag errors near crossover almost entirely, by using the open-loop corrector, and to hold the thermal-noise level down by smoothing in the rectangular coordinates. Differentiation may be performed in either spherical or rectangular coordinates, with similar effects.

Shown on Fig. 13.12(b) is a curve for triangular smoothing of velocity data at the radar site [see Fig. 13.6(h)]. With about 50 independent noise samples available in the computer during each 6 sec interval, the use of optimum weighting can reduce the thermal component of velocity error by a factor of two relative to the rectangular-weighting function assumed in the basic process [Fig. 13.6(g)]. However, use of this type of smoothing would make necessary a separate transmission of velocity data over the narrow-band link to the control center, and the over-all improvement might not warrant this use of the communications capacity. More important in the practical case would be the use of properly derived velocity data to update the position estimates being furnished to the control center. By operating the radar with wide servo bandwidths and forming position and velocity estimates in the computer, it is possible to eliminate any variation in the velocity or acceleration lags of the transmitted position estimates, and to update them with greater accuracy than would be possible at the remote station. Thus, although the limited bandwidth of the transmission circuit prevents the use of a high data rate, it would be possible, at least, to assure that the data which were transmitted represented estimates of the target position at precisely the times indicated on the time tags. Furthermore, the radar operators would be more free to adjust their controls to obtain the best tracking data, without projecting into the data system a variation in time delay and bandwidth of the data used for transmission.

Three-Coordinate Analysis

In the above example, only one coordinate has been examined, and the errors have been expressed in a direction determined by this coordinate. A complete analysis would cover all three radar coordinates, and would transform the errors into rectangular coordinates with a fixed orientation, relative to the earth. However, the results would be approximately the same, if expressed in terms of the major axis of the ellipsoid of error in any coordinate system. This error ellipsoid, for a precision microwave tracking radar, takes the form of a circular disc, whose short axis lies along the radar line of sight. The range error σ_r is almost always small relative to the other two error components, whose linear dimensions are given by $R\sigma_e$ and $R_g\sigma_a$. As the track progresses through crossover, the ellipsoid or disc changes its orientation with respect to a fixed coordinate system or to the target path, and hence the errors in rectangular coordinates will be more variable during the track. The minimum error will be equal to the range error σ_r, whereas the maximum will be equal either to the azimuth or elevation error times the appropriate range. In our example, the azimuth error is expected to be the larger of the two angular components, owing to the larger derivatives of angle in the azimuth coordinate. Hence, the linear error owing to azimuth will appear as the major axis of the ellipsoid of error in space. In other cases, it might be necessary to analyze all three components before selecting the one with the largest error. Having made this selection, however, it is generally unnecessary to go into detail in the other two coordinates, unless there is one particular direction in space along which an error is of particular interest.

A special application of search-radar theory is concerned with problems of target acquisition by the narrow-beam tracking radar. This chapter will review the use of search-radar theory for this special case, considering the type of designation data which has been made available to the tracking radar, the errors in this data, and the requirements for reliable target acquisition. Different acquisition procedures will be described, and a multistep procedure will be given for establishing the type of scan, if any, which should be used in any given case. A set of modified curves giving acquisition probability as a function of signal-to-noise ratio will be derived from the basic detection curves of Chapter 1. Some of the practical requirements of acquisition will also be described, and examples will be given to show the application of the theoretical and practical procedures derived. Limited experimental verification for the theory will also be presented.

TRACKING-RADAR ACQUISITION

14

14.1 APPLICATION OF SEARCH THEORY

During the process of acquiring a designated target, the tracking radar must operate in a search mode over a limited volume of space. The theory and procedures developed in Chapters 5, 6, and 8 may be applied to this mode of operation, in order to arrive at criteria for acquisition procedures and estimates of the probability of acquiring a given target. We shall consider first the different means by which the target is designated to the tracking radar, and the effects of errors in the designation data.

Sources of Designation Data

Ideally, the tracking radar target should be designated in all four radar coordinates, with sufficient accuracy to permit the tracker to center its resolution element exactly on the target before commencing to track. In such a case, the search problem is trivial, and consists of waiting for the echo signal to rise above the detection threshold. More generally, however, the designation data are neither complete nor accurate, and consist of two or three coordinates with some finite distribution of errors, describable on a statistical basis. The following are examples of the usual sources of designation data used with tracking radar:

1. *Optical designation.* When the target is visible to the naked eye, or when it can be located optically with a telescopic device, the radar antenna may be pointed by a nearby optical tracking instrument. Azimuth and elevation data (or other suitable coordinates) are transmitted to the radar, but no range or frequency information is given. The range interval to be searched is determined in advance by knowledge of the target characteristics, or the entire range of the radar may be searched. In the case of echo tracking, when a short-pulse radar is used, the Doppler shift will normally be small relative to the bandwidth of the receiver, and tuning of the system may be based on maximizing the return from nearby surface targets. In other cases, where the probable path of the target is known, an estimate of the Doppler shift may be sufficient data on which to tune the receiver. When a beacon is to be tracked, its transmitter may have been allowed to drift from its assigned frequency, and search may be required over many megacycles each side of the nominal value before the signal is located. To prevent this, a recent measurement of beacon frequency by some other radar or monitoring station is often made available to supplement the angular designation of the optical device.

2. *Search-radar designation.* When the target has been detected initially by a conventional search radar, azimuth and range data from that radar may be transmitted for designation. These may sometimes be used directly by the tracker, but when the two radars are separated by any substantial distance it is necessary to apply a parallax correction in order to point the tracker properly. In either case, the ground

projection of the target position is known, and the elevation angle must be found by scanning with the tracker. The scan may be made more efficient by using some a priori knowledge of target position and other characteristics to limit the elevation search sector. In other cases, a height-finder is used as part of the search system, and three-coordinate designation to the tracker is possible. With range data available, it is possible to derive the Doppler shift of the target as an aid in tuning. Beacon transmissions will be subject to the same errors as those applied to optical designation, unless the search radar operates in the same frequency band and can obtain a reading of the actual beacon frequency, which is unusual.

3. *Chain-radar designation.* In many cases, the target is already being tracked by another radar, which may be used as a source of accurate, four-coordinate designation data for the tracking radar which is acquiring its target. Three-coordinate position data will generally be converted to rectangular coordinates, transmitted over communications circuits of a standard type in digital form, corrected for parallax at the receiving station, and reconverted into spherical coordinate form for designation. Doppler shift can then be estimated from the range rate, if such data are important in acquisition. Beacon frequency can be transmitted along with the other data.

4. *Designation from orbital elements.* In satellite or ballistic-missile tracking, the path of the target is highly predictable, and may be stored in the form of orbital elements. At any given time, these elements are then converted to spherical coordinates with respect to the latitude and longitude of the radar site, yielding accurate, four-coordinate designation.* By performing this conversion continuously or at a high rate, continuous designation data may be made available to the tracking radar. The accuracy is dependent upon the correctness of the original orbital elements, the amount of drift which may have occurred since they were measured (in satellite orbits, drag and solar pressure may cause the elements to drift, and irregularities in the earth's gravitational field may also cause a shift), and on the degree of refinement used in the conversion process.

5. *Direction-finder designation.* A target which is emitting radio signals may be located in angle by a radio direction-finder, and two such instruments may be used to provide a position fix in two or three coordinates. The resulting designation data will sometimes be in the form of a ground projection, as in search-radar designation, and at other times will consist of two local angles, as in optical designation.

* Stanley J. Macko gives a simplified description of this process in *Satellite Tracking* (New York: John F. Rider, Publisher, Inc., 1962).

Although less accurate in most cases than radar or optical data, the radio d-f data may be obtained under all weather conditions and at long range.

6. *Manual designation.* The tracking radar operator is sometimes expected to acquire a target on the basis of an expected flight path and an estimated time of arrival at some point on this path. The rough information defining this path may be relayed over a communications link, or it may be part of a plan which is agreed to well in advance of the operation. At times, only an approximate sector is identified for search by the tracker. This procedure is best fitted to acquisition of large, slow targets, but it has been applied to a wide variety of cases where the resulting scan and detection requirements of the human operator are not excessive.

Designation Errors

Regardless of the method used for target designation, there will be some error in the information received by the radar, or in the center of the interval chosen for search. If the actual distribution of error is known, an appropriate scan pattern can be established to optimize the acquisition process. Otherwise, it is necessary to assume some distribution based on past experience or perhaps on a mere guess as to the adequacy of the designation. In the absence of specific information to the contrary, it may be assumed that each coordinate of the designation data is characterized by a normal distribution or error, with zero mean value when averaged over long periods of time (many acquisition attempts) and with standard deviation σ_x. The radar scan should then be centered at the designated value, and should extend to each side of this value by an amount which will provide the needed probability of scanning across the actual target position on this attempt. Rules for setting the scan limits are given below.

Estimation of the magnitude of designation error is relatively easy for those coordinates which are based on observation of the target by another radar. The radar error analysis procedures given in the previous chapters, combined with experimental data, may be applied to search radars and to systems which relay search or tracking information between radar sites. Optical and radio devices may be analyzed in a similar fashion, the same basic equations often being used. Most of these systems will be characterized by either an rms error or a "peak" error equal to about three times the rms value.

A more difficult problem arises when the designation is based on general knowledge of target characteristics. For example, if an aircraft has been detected by a conventional search radar, its elevation angle may be assumed to lie within the elevation coverage of that radar, and within a sector which extends from the horizon to a maximum value set by the measured range and the operational altitude of the types of aircraft believed to be in the area. A distribution of aircraft altitudes may be used as a guide, but this assumes that the new target is a member of the population from which the distribution was obtained. In most cases, search is extended to a "maximum" altitude, above which aircraft are assumed never to fly. There may be times when even this extension is insufficient, as when an unusual type of aircraft is designed to fly above the assumed ceiling. In designing the acquisition scan for this case, we should also decide whether the tracker is expected to acquire high-altitude balloons, missiles, and possible meteors, in case they are detected by the search radar. These considerations do not necessarily dictate a scan covering the entire 90° elevation sector, but they may lead to an extension of the scan beyond the limit which would normally be applied to aircraft alone.

In addition to knowledge of the amplitude distribution of designation error, it is helpful to know how the error varies with time. For example, if the target is known to pass through a given sector at a uniform rate, the error with respect to the center of that sector will be a sawtooth wave whose period is equal to the sector width divided by the target rate. In other cases, where continuous designation is received, it is important to know whether the error is constant over the period allotted to acquisition, or whether it varies significantly during this time. If variable, the rate of variation or the correlation time would provide a helpful guide to establishing the scan pattern, as will be shown below.

Probability of Acquisition

In the earlier discussions of detection and search processes, we used the concept of "probability of detection" to measure the performance of the radar. In the present case, the measure of performance will be the "probability of acquisition," designated P_a. This probability is defined as the summation, over all resolution elements covered by the radar, of the detection probability P_{di} for each element times the probability P_{vi} that the target actually lies in that element.

$$P_a = \sum_{i=1}^{i=n_v} P_{vi} P_{di} \qquad (14.1)$$

If the scan and detection processes are continuous, the summation in the above equation may be replaced by an integration over the scan volume. (We shall describe the scan coverage and each element in terms of a volume in space, although in special cases it may actually represent an element of one to four dimensions, including frequency.)

Equation (14.1) may also be used to define an average detection probability $\overline{P_d}$ over this volume, such that

$$P_a = \overline{P_d P_v} \qquad (14.2)$$

where P_v is the probability that the target lies somewhere within the volume searched, or the integral of P_{vi} over this volume. It can be seen that $\overline{P_d}$ represents an average probability of detection for each of the elements, weighted in accordance with the probability that the target is present in each element. The number of elements in the volume is given by

$$n_v = n_s \eta n_d \qquad (14.3)$$

where n_s is the number of beam positions to be searched [Eq. (5.1)], η is the number of range elements per beam position, and n_d is the number of Doppler elements. The latter two quantities are given by

$$\eta = \frac{R_{max} - R_{min}}{\tau_g} \times \frac{2}{c} \qquad (\eta \geq 1) \qquad (14.4)$$

$$n_d = \frac{f_{max} - f_{min}}{B} \qquad (n_d \geq 1) \qquad (14.5)$$

where R_{max} and R_{min} define the limits of range search, τ_g is the length of the range gate or other detection resolution element, f_{max} and f_{min} are the limits of frequency search, and B is the bandwidth of the receiver channel used.

If we represent the search interval in any given coordinate by the quantity $2X_m$, where X_m is the peak excursion from the designated position or scan center, the probability P_v that the target lies within this interval may be found from a table of areas under the normal distribution curve, as in Table 14.1. The entries for P_x^2, etc., will represent the probability P_v that the target lies within the search volume for the rectangular scan in two, three, and four dimensions, when each coordinate is scanned out to the same value of X_m/σ_x. In the case of angular scanning, we may find it more convenient to generate a circular or elliptical scan, omitting the corners which apply to the rectangular scans assumed in Table 14.1. The resulting probability P_r that the target lies within the radius of the outermost scan circle or ellipse is tabulated in Table 14.2, with a comparison to the two-dimensional rectangular case P_x^2. It can be seen that there is little penalty paid for omitting the corner regions of the scan, should this prove convenient. The entries in Table 14.2 may be derived from Fig. 1.8 (the single-pulse threshold-crossing probability curves), where the curve for noise

Table 14.1 PROBABILITY OF TARGET'S LYING
WITHIN SEARCH VOLUME

X_m/σ_x	P_x	P_x^2	P_x^3	P_x^4
0.25	0.20	0.04	0.008	0.0016
0.50	0.38	0.14	0.055	0.02
1.0	0.68	0.46	0.31	0.21
1.5	0.87	0.76	0.65	0.58
2.0	0.955	0.91	0.87	0.83
2.5	0.988	0.976	0.964	0.95
3.0	0.997	0.994	0.991	0.988
3.5	0.9995	0.999	0.9985	0.998
4.0	0.9999	0.9998	0.9997	0.9996

(Assumes search of rectangular volume extending by an amount X_m from the designated point in each coordinate, for one to four dimensions.)

alone gives the probability that the absolute value of a two-dimensional random vector will exceed a given radius.

The probability P_{vi} that the target lies within a given resolution element may be calculated from the location of the element relative to the center of the scan in each of the one to four coordinates which apply, by using the normal distribution curve. For example, if a one-dimensional scan (as used in elevation when range and azimuth are known to the required accuracy) is assumed, the elements or beam positions may be numbered consecutively from one end of the scan to the other. If the scan extends over $n_s = 20$ beam positions ($X_m = 10\theta$, where θ is beamwidth) and covers a

Table 14.2 PROBABILITY OF TARGET'S LYING
WITHIN SEARCH ELLIPSE

X_m/σ_x	P_r (see Fig. 1.8)	P_x^2 (from Table 14.1)
0.25	0.03	0.04
0.5	0.10	0.14
1.0	0.35	0.46
1.5	0.66	0.76
2.0	0.86	0.91
2.5	0.95	0.976
3.0	0.99	0.994
3.5	0.998	0.999
4.0	0.9997	0.9998

(Assumes search of elliptical volume extending by an amount X_m in each angular coordinate from the designated point, as compared to a rectangle with the same peak dimensions.)

region extending to the 3σ limit of designation error $(X_m = 3\sigma_x)$, then the probability of the target's lying within each of the 20 beam positions is as shown in Table 14.3. The second part of this table gives the corresponding probability for a two-dimensional case, where the elements are defined in terms of rings centered on the designated point. The probabilities P_{vi} in the first case correspond to the Gaussian distribution, whereas in the second they follow the Rayleigh curve.

Table 14.3 PROBABILITY OF TARGET'S LYING WITHIN TYPICAL
GIVEN LIMITS AND RESOLUTION ELEMENTS

X/σ_x	One-dimensional case			Two-dimensional case		
	Beam position	P_x	P_{vi}	Ring number	P_r	P_{vi}
0.3	10, 11	0.236	0.118	1	0.044	0.044
0.6	9, 12	0.451	0.108	2	0.164	0.120
0.9	8, 13	0.632	0.090	3	0.333	0.169
1.2	7, 14	0.770	0.069	4	0.513	0.180
1.5	6, 15	0.866	0.048	5	0.675	0.162
1.8	5, 16	0.928	0.031	6	0.802	0.127
2.1	4, 17	0.964	0.018	7	0.890	0.088
2.4	3, 18	0.984	0.010	8	0.943	0.053
2.7	2, 19	0.993	0.0045	9	0.974	0.031
3.0	1, 20	0.997	0.0021	10	0.989	0.015

(Applies to discrete beam positions or rings, extending ± 3.0 σ_x from the designated point. P_{vi} represents the probability that the target lies within that particular element whose outer limit is given by X.)

14.2 ACQUISITION PROCEDURE

The general theory given above will now be applied to design of scan patterns for acquisition with a pencil-beam tracking radar. We shall first establish the volume to be scanned, based on the estimated errors in designation. The detection procedure will then depend upon the extent of the scan and on the availability of suitable display or automatic detection devices. Finally, methods of covering the required volume with antenna scans and range or frequency search processes will be investigated.

Defining the Search Volume

The search volume should be set in such a way that the cumulative probability of acquisition is maximized within the time available for search. The cumulative probability of acquisition is defined along the same line as the cumulative probability of detection [Eq. (5.12)], but the acquisition probability P_a in place of the detection probability P_d is used. For the case where both P_d and P_v are independent from scan to scan, the cumulative probability of acquisition is

$$P_{ca} = 1 - (1 - P_a)^j \qquad (14.6)$$

This represents the probability that the target will have been acquired prior to the completion of j scans, each giving an acquisition probability P_a, during the search time t_s. Each scan will occupy a period $t_1 = t_s/j$ sec, and detection on any scan is sufficient for acquisition.

It is obvious that the single-scan probability of acquisition P_a will be increased if the scan is conducted as slowly as possible over the volume, so that just one scan is completed in the available time. On the other hand, if the target is fluctuating or if the integration time of the detection system is less than the available time on target for one slow scan $(t_n = t_s/n_s)$, then the system performance will be improved by making two or more scans over the entire volume and depending upon the increase in cumulative acquisition probability to meet the required performance. In either case, the first step is to define the volume to optimize the final acquisition probability. This is not a straightforward process in most cases, as it involves a delicate balance between speed of scan, overlap of successive scans, signal-to-noise ratio, and probability that the target will lie within the scan volume. A multistep procedure for arriving at the proper scan volume and pattern will be given, consisting of the following major steps:

1. The problem is first checked for over-all feasibility; approximate values for maximum range and scan angle (such as 3σ limits) are used, and the search radar equation used in Chapter 5 is applied. Uniform distribution of energy over the volume is initially assumed.
2. If step 1 indicates that acquisition is possible over such a volume of scan, a specific pattern is chosen to cover the volume uniformly, and the probability of detection is found by using the procedures of Chapters 1 and 8, or by using Blake's method.
3. If a marginal level of acquisition probability results from uniform distribution of search energy, the scan volume may be subdivided into several regions, following contours of equal probability P_v. The available energy or scan time is allocated in proportion to the probability that the target lies within each such subdivision, and the detection calculation is applied separately to each, in order to arrive at a second estimate of acquisition probability.
4. If the acquisition probability has been improved by the previous step, the required scan pattern of the antenna is checked against the

mechanical capabilities of the antenna pedestal, to determine whether the system has approached or exceeded the limits set by velocity and acceleration. If further capability remains, the previous steps may be repeated until maximum improvement is obtained. If the antenna is unable to scan the required volume in the available time, the pattern must be simplified to ease requirements on the pedestal, even if this results in reducing the acquisition probability.

5. Having arrived at a satisfactory pattern for the antenna the fluctuating characteristics of the target must be checked to assure optimum performance. If the time allotted to acquisition is not sufficient to assure independence of target echo amplitude between scans for a fluctuating target, the fluctuation loss must be increased accordingly.

6. If the required scan is of a very limited extent, special detection curves may be required to arrive at an estimate of acquisition probability for the searchlighting and small-scan cases. The effects of scan-to-scan correlation of designation error and of target signal strength must be evaluated carefully in this case, by using the curves given later in this chapter.

Initial Estimate

We are given the errors in designation σ_x for each coordinate, and a required probability of acquisition P_a. Using Tables 14.1 and 14.2, we may estimate the extent of the scan in each coordinate, so that the probability of not acquiring $(1 - P_a)$ is equally attributable to failure to detect $(1 - P_d)$ and to failure to scan past the target $(1 - P_v)$. Thus, we have

$$(1 - P_v) \cong \sqrt{1 - P_a} \qquad \text{(14.7)}$$

For example, if the acquisition probability is to be 90 per cent, we should choose the initial scan to obtain $P_v = 95$ per cent. This requires the scan to extend between 2σ and 2.5σ on each side of the designated point, depending upon how many coordinates are to be searched. Assuming that all range and Doppler elements are observed simultaneously, we may then apply the search-radar equation, with due allowance for losses, to determine whether the required value of P_d (equal to P_v) is available with the radar, at the maximum range of the target. If the integrated $(S/N)_i$ ratio is not adequate by a wide margin, the problem must be redefined by reducing the range, increasing the acquisition time, or some other change. If the available signal is more than sufficient, it will permit wide latitude in choice of scans, or in sequential search of range or Doppler elements.

Refining the Energy Distribution

Let us assume that the results of the first step have shown marginal performance, and that the extent of the scan covers a number of beam positions in reaching the 2.5σ limit (e.g., $\sigma_x = 2\theta$, $n_s = 100$). We may divide the volume into rings whose radii differ by, say, $\theta/2$ or $\sigma_x/4$. From Tables 14.2 or 14.3, the relative energy distribution for the ten rings (starting with a circle whose diameter is one beamwidth) will be as shown in Table 14.4. Also shown are the energy distribution for the uniform search case, that for the nonuniform case based on P_{vi}, the detection probabilities for each ring, and the resulting acquisition probability. If the integrated $(S/N)_i$ ratio were 10 db for uniform search, the value of P_d would be 12 per cent for the entire volume. With nonuniform distribution, the average $\overline{P_d}$ will be increased to 63 per cent for the same radar power. When the original value of P_d for uniform search is high, further concentration of energy is not usually advantageous, and it may be possible to improve P_a by extending the scan over a larger area.

Table 14.4 COMPARISON OF UNIFORM AND NONUNIFORM ENERGY DISTRIBUTIONS

Ring number	Uniform distribution Per cent area in ring P_{vi}	Nonuniform distribution P_{vi} or per cent power	Db above uniform dist.	P_d	$P_d P_{vi}$
1	1	4.4	+6.4	0.9999	0.044
2	3	12	+6	0.9999	0.120
3	5	16.9	+5.3	0.996	0.168
4	7	18	+4.1	0.94	0.169
5	9	16.2	+2.5	0.62	0.100
6	11	12.7	+0.6	0.18	0.023
7	13	8.8	−2.7	0.01	0.009
8	15	5.3	−4.6	0.001	0.000
9	17	3.1	−7.4	0.000	0.000
10	19	1.5	−11	0.000	0.000
$P_a = P_d P_{vi}$	0.12				0.633

(Slightly higher values of P_a are available if power is redistributed to match P_d to P_{vi}, rather than matching power to P_{vi}.)

Establishing the Scan Pattern

The path followed by the tracker beam in covering the solid angle of scan must now be determined. The choice will generally be limited to one of the following patterns:

1. *Circle or spiral scan.* When the scan angle is small relative to the

beamwidth, the tracker antenna may be left fixed or perturbed slightly to follow a circular path around the designated point. This circle scan is similar to the conical scan used in some tracking radars, and it may be extended to a radius of about one beamwidth before the hole in the center of the pattern becomes intolerable. Using the criteria of Tables 14.1 and 14.2, one should use the circle scan whenever the rms error in angle designation exceeds about one-fourth beamwidth. When the error exceeds about one-half beamwidth, the circle scan should be expanded into a spiral scan or concentric-circle scan, so that the outer edge extends to a radius near the 3σ value of error and the center remains filled. The number of concentric circles or revolutions in the spiral is not too critical, and may be established by using the beamshape and scanning-loss curves of Figs. 5.5 and 5.7. Generally, a spacing of about one-half beamwidth in radius of successive scans is used, and this may be increased to one beamwidth when a large area is being scanned to locate a strong signal. As shown in Chapter 5, the loss increases rapidly when the number of scans or the number of pulses exchanged per beamwidth drops toward unity.*

In our previous example, where the scan volume extended to a radius of five beamwidths, a spiral scan or concentric-circle scan would be indicated. This type of scan permits the energy distribution to be varied as a function of the scan radius either by modifying the scan frequency or by changing the spacing between successive scans. Merely by maintaining a constant rotational frequency, as the scan radius increases, we are able to taper the energy distribution to approximate that shown in Table 14.4, although the inverse variation with radius may not be precise enough to take full advantage of the potential improvement in detection probability. If the time permits completion of 10 rotations at increasing radius and constant frequency, without exceeding the servo limitations, this type of scan would be chosen for our case. With additional time available, two or more complete scans would be carried out in this period. Difficulties arise if the antenna cannot be scanned at the required rate. If this condition applies, it is necessary to increase the spacing between scan rotations and to decrease the frequency, adding to the beamshape loss (see Fig. 5.5). Mechanical limitations generally protect us from scanning at speeds high enough to reach one pulse per beamwidth, but the spacing of successive scans by one beamwidth in radius is quite possible. Use of an inter-

---

* Daniel Levine, "Volume Scanning with Conical Beams," *Proc. IRE*, **38**, No. 3 (March 1950), pp. 287-90.

laced scan, or an equivalent effect produced by noise perturbation of the designation data permits us to use spacings of one beamwidth or more between scan revolutions, so long as the scan pattern is repeated two or more times during the acquisition period.

2. *Sector scan.* The sector scan is used when data are obtained in only one angular coordinate, or when the error in one coordinate is much greater than in the other. Examples are the use of search-radar designation in azimuth and range, or of flight-plan designation, when the time of arrival is in doubt and azimuth scan is used at a fixed elevation. The sector width is determined from the one-dimension column of Table 14.1. The fixed angular coordinate is set at the designated value, or at a value such that the target will pass through the pattern within the detection range of the radar. This type of scan was covered by the discussion on search radars in earlier chapters.

3. *Raster scan.* When large errors are present in both angular coordinates, a raster scan may be used to distribute energy in the desired way over the scan volume. Considerations governing choice of scan limits and spacing are the same as those described above for spiral scan. Motion of the target between scans may eliminate the need for a closely spaced pattern of scanning lines, and the acceleration required to reverse the antenna at the end of each line will often limit the area which may be covered at a given time. For this reason, the spiral scan is preferred in most cases, except when the volume to be scanned is bounded by the horizon. The raster scan is particularly recommended when a fluctuating target is expected to rise from the horizon within a defined azimuth sector at an unknown time. In this case, a single search fan, as produced by sector scan in azimuth alone, may permit the target to pass unnoticed, unless the fan is placed at high elevation to achieve a short range. The raster scan provides some probability of detecting the target on signal peaks as it rises from the horizon, and still maintains coverage at the higher elevation angles and shorter ranges, in case earlier detection fails.

Target-Fluctuation Effects

If a high probability of acquisition is required, it is necessary to scan the volume at least twice to acquire targets whose amplitudes fluctuate. The curves given in Chapter 5 may be applied here to determine the optimum number of scans and the fluctuation loss to be expected. In cases where a very large number of independent looks is available during the acquisition period, the range of detection may extend well beyond that calculated for the steady target, as a result of the "fluctuation gain" applicable to low values of single-scan P_d. An example of this will be given later.

The Small-Scan Case

When the rms designation errors are less than one beamwidth, it becomes difficult to apply the general search theory used above. A modified curve for acquisition probability in this case is shown in Fig. 14.1, based on the detection-probability curves of Fig. 1.9. Solid curves show the value of P_a for a single look with a fixed beam, as a function of integrated signal-to-noise ratio $(S/N)_i$ applicable to the center of the beam. Designation errors from zero to one-half beamwidth are covered, with the acquisition probability remaining below 60 per cent for errors of one-half beamwidth at all values of S/N ratio. If higher values of P_a are needed, a scan must be used, or two or more looks must be taken with independent designation error. The number of independent looks available in the search time t_s is given by $j = t_s/t_c$, where t_c is the correlation time of the designation error. For this case, the value of $(S/N)_i$ applicable to t_c should be used to enter Fig. 14.1, and the cumulative probability of acquisition should be found by using Eq. (14.6). For example, assume that a 10 sec period is available, and that t_c is 1 sec. At a repetition rate of 300 pps, the available integration gain in 1 sec is about 18 db. If the single-pulse S/N ratio (after allowance for losses other than beamshape and "integration loss") is unity, we may enter Fig. 14.1 at $(S/N)_i = 18$ db, and find a single-look acquisition probability of about 40 per cent, with no scan and $\sigma_x = 0.5\theta$. After 10 sec, the cumulative probability of acquisition would be 99.4 per cent. The antenna would be moved at random over a region near the target, in response to the varying designation errors, and use of one of the regular scans would be either redundant or harmful.

In cases where the error remains fixed during the acquisition period, the curves for the circle-scanning case may apply. Although these fall below the fixed-beam curves in the low-signal region, they continue to higher values of acquisition probability when the signal-to-noise ratio increases, and they make it possible to acquire with reasonable probability when the rms error is in the order of one beamwidth. For an error of one-half beamwidth, they cross at $(S/N)_i = 18$ db, above which value the scan should be used.

Figure 14.2 describes the performance of a pencil-beam system which acquires on the basis of the cumulative probability evaluated on 10 successive looks, with the designation error constant over this period. For each look, the probability of a false alarm is made one-tenth as great as in the previous figure, to provide a basis for direct comparison. The use of circle scan is now indicated at lower values of $(S/N)_i$, since the amount of integration has been reduced some 8 db by distributing the energy over 10 separate looks. The results of this and the previous figure confirm the rules which were established on an intuitive basis earlier in this discussion, as to when a scan is required for acquisition. The circle scan considered here is a rotation of the center of the beam in a circle whose radius is one-half beamwidth, giving a constant loss of 6 db for targets on the axis. The variation in loss during the scan cycle for other target locations is shown in Fig. 14.3(a), whereas the average loss for a signal which is integrated over more than one scan cycle is shown in Fig. 14.3(b).

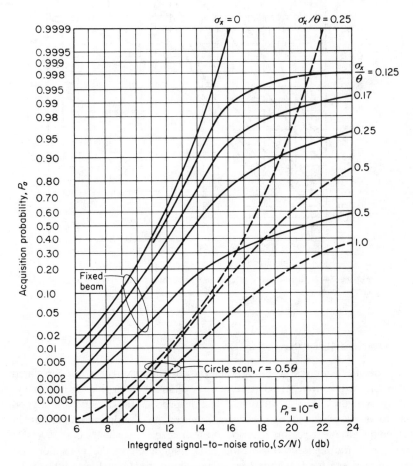

Figure 14.1 Acquisition probability for single look with continuous integration.

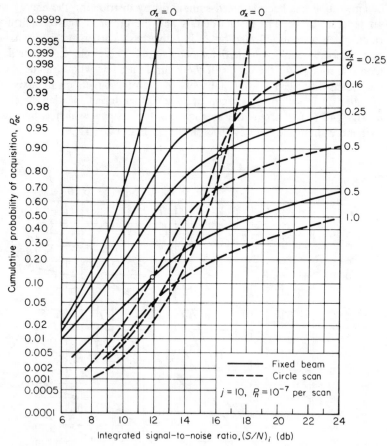

Figure 14.2 Cumulative acquisition probability with fixed beam and circle scan.

Error and Signal Correlation

The effect of circle scanning when the designation error is not correlated over the acquisition period is somewhat different from that shown in the previous figures. The cumulative detection process, for 10 scans, is a good approximation of the process actually used by the human operator in scanning an A-scope for a period of 1 or 2 sec. In such a case, $(S/N)_i$ represents the signal-to-noise ratio after integration in the phosphor and the human eye for 0.1 or 0.2 sec. Thus, the ordinate in Fig. 14.2 can be used as the acquisition probability after 1 or 2 sec of observation, and the

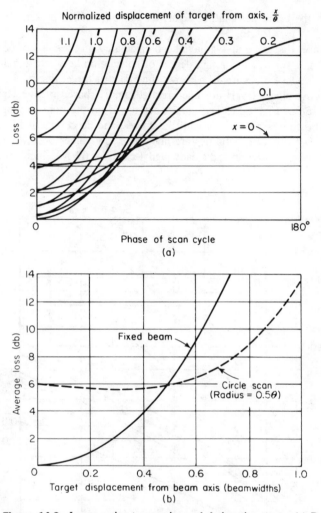

Figure 14.3 Losses owing to scanning and designation error. (a) Relative signal loss during circle scan. (b) Average loss over scan cycle.

results of several such periods may be combined on a cumulative basis when the error is uncorrelated over periods longer than a few seconds. For example, in 10 to 20 sec, with $\sigma_x = \theta$ and $(S/N)_i = 14$ db at beam center, the cumulative probability would rise in 10 looks to 75 per cent, based on a single-look probability of 13 per cent. This is well above the value which would be obtained during a single look of 10 times the duration, with the same designation error, as can be verified by entering Fig. 14.1 at $(S/N)_i = 22$ db.

The cases discussed above have been calculated on the basis of constant signal strength, or at least a high degree of correlation between signal levels on successive looks. Figure 14.4 shows the effect of target fluctuation when the successive signal levels are uncorrelated, and is based on the Rayleigh characteristic of Fig. 1.12. The curves are based on the nonscanning case, and may be compared directly with the corresponding curves of Figs. 14.1 and 14.2. Detection probabilities are higher for low values of $(S/N)_i$, and fall off rapidly, relative to the steady-signal case, as $(S/N)_i$ increases. The

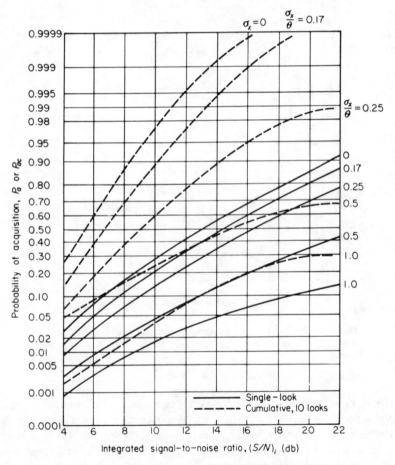

Figure 14.4 Single-look and cumulative acquisition probability for fluctuating targets.

curves for cumulative probability of acquisition increase quite rapidly in the region below $(S/N)_i = 14$ db, as a result of detection on signal peaks.

Detection Procedure

In the foregoing discussion, the nature of the detection process has been considered only indirectly. The results of Fig. 14.1 were based on any process using integration to achieve a single-look value of $(S/N)_i$ as given in the abscissa. In Fig. 14.2, the cumulative detection process was assumed with no storage of information from one look to the next. Somewhat better performance could have been obtained in the latter case by using continuous integration over the scan cycle, but it is unusual for the radar to use an integrator which covers such a long period (in the order of 1 sec). The most common practical procedure for integration and detection is based on the human operator and oscilloscope display, often a simple A-scope. Blake's curves for A-scope visibility factor V_o were derived from the experiments of Lawson and others, who used a presentation time of about 3 sec. It can be shown that these results agree with the cumulative probability of detection obtained in an automatic detector which responds to samples integrated over periods of about 0.1 sec and repeats the process 30 times in 3 sec. When the human operator can concentrate his attention on a small number of resolution elements, he may improve his performance by mentally correlating the observations over several 0.1 sec intervals, but as the number of observed elements is increased [Fig. 6.8(d)], he is reduced to the simple cumulative detection process with 0.1 sec of integration provided by the eye-phosphor system.

Range and Frequency Scan

Unlike the angular coordinates, the dimensions of range and frequency may be covered simultaneously in search by a single radar, by using such devices as a two-dimensional display on the output of a wide band receiver system.* As a result, there need be no loss of energy or time in scanning whatever volume is required by the designation errors in these coordinates. Most tracking radars, however, lack the equipment needed to realize fully the two-dimensional display, and aquisition may fail as a result of errors in range and frequency designation. This is especially true in the case of beacon acquisition, where the frequency of the beacon cannot be predicted accurately. Collapsing losses due to use of long-range search intervals and wide receiver bandwidths can reduce the range of acquisition, when these

---

* Such a system is described in J. T. Nessmith's article, "New Performance Records for Instrumentation Radar," *Space/Aeronautics*, (Dec. 1962), pp. 86-94.

intervals are extended in an attempt to assure coverage of the target signal.

The proper procedure to be used in frequency acquisition of beacons is not always understood, as a result of uncertainties as to the distribution of frequency designation errors and beacon signal amplitudes. The normal procedure is to use a receiver bandwidth wide enough to accomodate the expected tolerance in beacon frequency, and to consider a greater shift as a failure of the beacon. However, when the beacon signal strength is subject to variation as a result of antenna patterns on the vehicle, and when angle search in one or two coordinates is combined with the range and frequency acquisition process, there are many possible search routines and many opportunities for error by the radar operator. The best procedure in a specific case can be determined only with respect to the probabilities of various designation errors and system failures, as illustrated in the following example.

Assume that an incoming target is to be acquired by beacon signal, at a range beyond that of useful echo tracking. A nominal beacon frequency is known to the radar operator, along with a three-coordinate designation in space and an estimated signal strength. The acquisition scan is set to give 99 per cent probability that the target will lie within the volume scanned, and the detection probability over the first 30 sec of the scan is calculated from the signal strength to be 99 per cent. What is the proper procedure to follow if no signal has been seen within this period? There are several possible reasons for this failure, with different consequences in operating procedure:

1. The target is present in the search volume, but this is one of the one-in-a-hundred cases when the signal fails to cross the detection threshold. Continued scan in the same pattern is required.
2. The target is the one in 100 which lies outside the search volume. An extension of the scan over a wider interval is required.
3. The beacon has failed completely. Acquisition attempts should be abandoned, or shifted to echo signal acquisition if this is even remotely possible.
4. Some element of the radar has failed without the knowledge of the operator. Quick repair or readjustment should be made if possible.
5. The beacon frequency has shifted out of the bandwidth of the radar receiver. Scan should be continued with added frequency search.
6. The beacon receiver frequency has shifted so that radar interrogations are not received. The radar transmitter should be retuned, and scan continued until the proper frequency is found.
7. One of the initial assumptions as to probable error or signal strength was wrong. The acquisition procedure should be corrected to match a new estimate.

Unless the operator comes into possession of some new information from an external source during the 30 sec, he must quickly verify the operating status of his own radar, and then proceed to the next most probable difficulty and the corresponding remedy. He may have time for only one choice, since the extension of search in any dimension (including time) will consume all the resources of the average system. Thus, the a priori probability of acquisition, which was calculated to be about 98 per cent in 30 sec, may change abruptly to a very low value when the target fails to appear. Since there is such a relatively short time for decision, the proper procedure is one which will provide a reasonable "second chance," based on a re-evaluation of the original assumptions. In order to make this possible, the probable performance of all elements in the system must be evaluated as accurately as possible before the operation begins. An over-all acquisition probability of 99 per cent requires confidence levels well above 99 per cent on many different elements in the system. These performance levels should be re-evaluated at intervals during the operation, so that the acquisition procedure may be optimized without delay when one of the factors changes. Maximum use should be made of procedures worked out in advance for various contingencies, to relieve the operating personnel of distractions which may reduce the detection performance of the system.

14.3 TRANSITION FROM SEARCH TO TRACKING

Following successful acquisition of a target in the search mode, there remain several steps which are necessary to convert the operation to the tracking mode. A failure in any of these steps will lead to loss of the target, and the search process must begin again.

Error Sensing

In the course of scanning for the target, the antenna may have been placed in rapid motion relative to the designation coordinates. When the target signal crosses the detection threshold, the antenna will then have moved to a new position, as much as several beamwidths away from the point where the target signal was first received. The acquisition process must provide some means of storing the location at which the target actually appeared, enabling the radar to return to this point for tracking. A two-step process is often used for this purpose. After initial detection, the scan is modified to concentrate energy in the general area where detection took place. Antenna scan rates are reduced to the point where a second detection may stop the antenna within one beamwidth of the target position. Alternatively, the position of the antenna relative to the designated point may be measured when first detection takes place, and the designation may be corrected by that exact amount to bring the antenna to the proper point without further scanning. In either case, if the target is moving at an appreciable rate, the radar antenna must remain under the control of the designation data, and corrections must be applied to those data. Otherwise, the target may move far enough during the transition time to defeat attempts to enter the automatic tracking mode.

Rate Matching

Where continuous designation data are generated or received at the radar site, the antenna and range gate will be in motion at an average rate which is close to that of the target. When the scan is stopped, these rates will still approximate those of the target, and tracking may begin without a large transient. This is important if the tracking data are to reach their rated accuracy in a short time. In other cases, as when a sector scan is used for acquisition, there is no basis for estimation of the target rate, and the antenna must be brought to a stop before tracking begins. If the target rate is high, it may prove impossible to accelerate the antenna or range gate to the proper rate on the basis of error signal measured in the tracking loops, before the target passes from the beam or gate. Again, it may be necessary to provide a two-step procedure to make a successful transition from the search to the tracking mode. A form of rough, manual tracking, using the acquisition displays as error sensors, may be used to permit the operator to establish the initial tracking rates.

An initial error in velocity of the servo will produce an error in position, which decays slowly over the early part of the tracking period [see discussion following Eq. (9.24)]. Manual tracking during the transition process is unlikely to reduce this initial rate error to the point where accurate tracking is possible immediately after lockon of the automatic loop, and to improve on the accuracy of manual tracking would require too great a period of time in transition. Hence, it is necessary to use the widest possible bandwidth, with minimum integration in the automatic tracking loop, until the target rate has been matched by the servo. As mentioned in Chapter 9, the use of open-loop correction offers one way out of this initial transient problem, making it possible to start accurate tracking even before the servo loop has been closed.

Reacquisition

Even if the best transition procedures are used, or if lockon has been successful, the target may be lost as a result of fading or similar causes. The tracking radar and its designation system should have a means of rapidly reacquiring such a lost target, based not only on the original designation data from outside sources but on any information obtained during the previous acquisition process and any subsequent tracking. Depending upon the capacity and flexibility of the equipment associated with the radar, the available data may be combined to correct the original errors in designation and to provide a scan, if needed, appropriate for the expected error in redesignation. In the case of a satellite in orbit, for instance, a single measured point is enough information to generate corrected orbital elements which can be much more accurate than those used in the original search process. It has also been shown* that a relatively short track with a precision radar can be used to generate orbital elements for reacquisition of a satellite on its next revolution, $1\frac{1}{2}$ hr later.

14.4 EXAMPLES OF ACQUISITION CALCULATIONS

Two types of acquisition calculation will be given, one based on search radar data in acquisition of an aircraft, and a second covering a typical satellite problem. Limited experimental data will be presented to confirm the fact that reasonable values can be found by using the procedures described earlier in this chapter.

Search Radar Designation

Assume that a radar such as the AN/UPS-1, with cosecant-squared antenna, is to be used as a source of designation for the AN/FPS-16 tracker (see Chapters 8 and 10 for data on these radars). Range and azimuth data are furnished on each target, with a rated accuracy of $\pm 1°$ and 1 mile plus 3 per cent of range (see Table 8.4). A review of the accuracy limitations for search radar shows that the thermal and fluctuating noise error for this radar should be less than $\frac{1}{2}°$ rms, and that the rated accuracy of $\pm 1°$ is probably a 2σ value, including some bias due to misalignment. The azimuth scan limits for the tracker should, therefore, be approximately $\pm 1.5°$, or $3\sigma_a$. In elevation, we have no information from the search radar, except that the target lies below the $40°$ upper limit of coverage. However, the fact that we are interested in conventional aircraft allows us to set an upper limit to the elevation scan of the tracker which is based on a ceiling of about 50,000 ft, and is dependent on the designated range. Inspection of the vertical-coverage chart (Fig. 8.5) shows that a target at 150-mile range would be below $3°$ elevation, whereas at 100 mi it would be below $5°$. This relationship is fortunate, as it permits us to limit the extent of

---

* Sidney Shucker, "Results of Space Tracking and Prediction with Precision Radar," *Proc. 1962 Natl. Symposium on Space Electronics and Telemetry*, IRE-PGSET.

the elevation scan on the longest-range targets, where the signal is the weakest. In fact, it may prove increasingly difficult to acquire a given target as it comes closer to the radar, owing to the greater elevation sector which must be searched and the mechanical limitations on antenna scan.

In range, the extent of the scan is not too important if the human operator is to view an A-scope. For the maximum range of 200 to 250 miles needed in tracking of conventional aircraft, the entire range can be displayed with moderate collapsing loss, even when the pulse width is 1 μsec. Even this loss can be avoided if the search radar data are used to center an expanded-sweep A-scope display, extending somewhat beyond the ±3 per cent of target range which was given as the rated accuracy (this may again be assumed to represent approximately a 2σ error). Similarly, in the frequency coordinate, the short-pulse tracking radar will require no correction in tuning for Doppler shift in this case.

Based on the considerations above, we may set the outside limits of the search volume for acquisition to include an azimuth sector of $3°$, an elevation sector extending from the horizon to a limit (below $40°$) which is given roughly by $E_{max} = \arcsin (8/R)$, where R is in miles, and a range interval of about 10 per cent of the range or 3 miles (whichever is greater). The basic scan will be an elevation sector scan, on which will be super-imposed a slow variation in azimuth to cover the $3°$ sector (alternatively, a rapid azimuth scan over $3°$ could be superimposed on a slow elevation sector scan, but the resulting accelerations would be higher). If it is found experimentally that the azimuth error of the search radar is variable from scan to scan, or if a track-while-scan unit is used to couple the data from a remote search radar to the tracker, then the azimuth scan may be dispensed with, and the variable error may be used to assure coverage of the error volume.

Probability of acquisition will be very high for the radar combination used in this example, on any target detected by the search radar. At maximum range of 150 miles (1.0 m² target in the reflection lobe of the vertical pattern), the search volume covers only a $3 \times 3°$ solid angle, and the tracking radar S/N will be near unity. In a 10-sec period, at a repetition rate of 320 pps, there will be a total of 3500 pulses, each with unity S/N ratio, available for integration, providing 350 pulses per look at one look per sec. We may enter Fig. 14.2 with $(S/N)_i = 15$ db (allowing about 3 db for collapsing loss), and read P_{ac} from the curve for $\sigma_x = \theta$. The 10 sec value for this case will be $P_{ac} = 18$ per cent, a circle scan being used, and the same value will apply to fluctuating targets, read from Fig. 14.4, with no scan. If we consider a maximum approach velocity of 1000 ft/sec, the target will penetrate about 1.5 miles per 10 sec acquisition period, and an over-all acquisition probability of 90 per cent will be achieved within about 18 miles, after 2 min, on those targets seen by the search radar, at maximum range. In the absence of ground reflections, the acquisition probability of the tracker would be a good deal higher, and the acquisition time shorter, at the reduced search-radar range.

Satellite Acquisition

As a second example, let us consider the case of satellite acquisition based on three-coordinate designation from orbital elements. Tracking data from another distant site or from a previous passage of the satellite will be assumed to have provided elements which locate the satellite to the following accuracies:

$$\text{Position along orbit} \quad \sigma_x = 20 \text{ mi (4 sec in time)}$$
$$\text{Position normal to orbit} \quad \sigma_y = \sigma_z = 5 \text{ mi}$$

The altitude of the satellite will be assumed to be 100 mi, and its radar cross section for echo tracking will be assumed to be 400 m², based on the use of a corner reflector.* The corresponding value of R_o (for unity single-pulse S/N ratio) will be 600 miles, with an allowance for practical losses and propagation. A steady signal will be expected from the reflector, owing to its smooth pattern over much of the solid angle around the vehicle.

At a range of 500 miles, the rms designation error in angle will be equal to one-half beamwidth, and we shall start the multistep acquisition calculation with an assumed scan angle covering four beam positions. By using Eq. (5.11), with $t_s = 10$ sec and an assumed search loss $L_s = 20$ db, we find an initial estimate of $(S/N)_i$ for the 500 mile range.

$$(S/N)_i = + 25 \text{ dbw} + 20.5 \text{ db(ft)}^2 + 10 \text{ dbs} + 26 \text{ db(m}^2) - 108 \text{ db(mi}^4)$$
$$+ 14.5 \text{ db(rad)} + 14.5 \text{ db} - 11 \text{ db} - 20 \text{ db} + 52 \text{ db}$$
$$= + 23.5 \text{ db}$$

Although this would appear more than adequate to assure a high probability of acquisition at this range, we must now check the loss figures to see whether the 20 db figure is enough for this case. We will recall that a minimum loss of 13 db, relative to the "ideal" search radar, was found in Chapter 5, and that the loss in a practical search radar was found in Chapter 8 to be near 20 db. In the present case, since the radar was designed primarily for accurate tracking, it turns out that its search performance is far below the optimum. A total search loss in excess of 30 db, relative to "ideal" performance, is shown in Table 14.5. The major factors which lead to this excess loss are

---

* This value is used to permit comparison with experimental data which were taken with a more powerful radar.

1. The large number of pulses exchanged in the 10 sec search period, which leads to a large value of "integration loss" L_i.
2. The lack of a suitable integration device to perform scan-to-scan integration or correlation.
3. The lack of an adequate display to avoid collapsing loss when searching a range interval of 100 miles or more.
4. The operator loss involved in scanning a large number of range elements over a period of many seconds.

Table 14.5 LOSSES IN ACQUISITION BY AN/FPS-16

Loss component		Loss in db
Antenna efficiency (L_n)		5.0
Matching loss (L_m)	$(B\tau = 1.6)$	2.5
Integration loss (L_i)	$(n = 200)$	7.0
Scan dist. loss (L_d)	$(j = 4)$	4.5
Beamshape loss (L_p)		3.2
Scanning loss (L_j)	$(n = 200)$	0.0
Fluctuation loss (L_f)	$(j = 4, \ P_c = 90\%)$	0.0
r-f line loss $(L_t + L_r)$		2.0
Propagation loss (L_a)	$(E = 10°)$	0.5
Collapsing loss (L_c)	$(\rho = 10)$	3.5
Operator loss (L_o)	$(t = 10 \ \text{sec})$	3.0
Total loss relative to ideal search radar		31.2 db

Minimum allowance: $L_n = 3.0, \ L_m = 1.0, \ L_i + L_d + L_p + L_j + L_f = 9.0$
In the case above: $L_n = 5.0, \ L_m = 2.5, \ L_i + L_d + L_p + L_j + L_f = 14.7$

Even with the large excess loss, however, the value of $(S/N)_i$ should be high enough to achieve a reasonable probability of acquisition by the time the target has approached to about 400 miles.

Calculation of Acquisition Probability

Since the amount of scan required in our example is relatively small, the second step in the process involves the computation of acquisition probability from the curves presented earlier in this chapter. We shall first compute the values of S/N and $(S/N)_i$ and the size of the designation error in units of beamwidth, as a function of target range. It will be assumed initially that the target is approaching the radar directly, so the relatively large error in position along the orbit does not introduce a significant component into the angle error. Figure 14.5(a) shows the error and signal-to-noise ratio as a function of range. For A-scope detection, the effective integration is limited to about 0.1 sec, and any longer period must be used

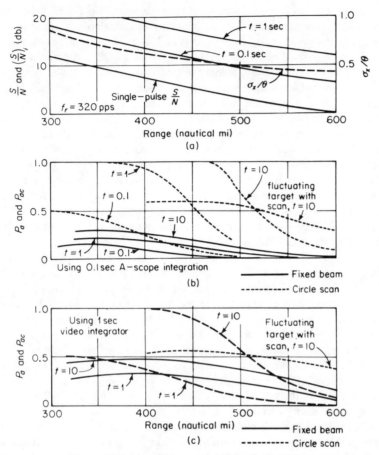

Figure 14.5 Acquisition probabilities for satellites.

by the observer to increase his cumulative probability of detection. An integrated value for $t = 1$ sec is included on the plot, and will be used to determine whether any advantage could be gained by use of such an integration period, if suitable equipment were added to the radar.

Figure 14.5(b) shows the results of the acquisition probability calculations, using Figs. 14.1 and 14.2 for the steady-target case, with video integration over 0.1 sec. The beam-center value of $(S/N)_i$ given in Fig. 14.5(a) includes an allowance of 5.5 db for this case, representing the sum of collapsing and operator losses for detection on the A-scope. No beamshape loss need be included, since the figures are based on beam-center values of signal-to-noise ratio. It will be seen that the three curves for acquisition with a fixed beam lie below about 25 per cent acquisition probability, whereas those for the scanning beam approach 100 per cent for cumulative probability over periods of 1 and 10 sec. The reason for the failure of the fixed-beam curves to rise more rapidly, compared to the single-look value ($t = 0.1$), is that the target has a high probability of lying just outside the beam in this case. It has been assumed here that the designation error is constant over each 10 sec period, and hence that the cumulative probability of acquisition cannot be found from Eq. (14.6). Instead, a cumulative probability of detection P_c must be found for each possible value of designation error, and the acquisition probability must be found by averaging P_c, weighted by the distribution of error. This leads to much lower values of P_{ac} in the nonscanning case. In the scanning case, where the primary reason for failure to acquire lies in low $(S/N)_i$, the curve for $t = 10$ sec represents more nearly the cumulative probability expressed by Eq. (14.6), since independent opportunities for detection exist on each of the successive scans.

Figure 14.5(c) shows the results which would be obtained if a video integrator were made available for continuous integration over $t = 1$ sec. The results would be better for the nonscanning case, approaching a 50 per cent acquisition probability for the 10 sec period at 450 miles. However, when scanning is used, the curves fall below those obtained from the cumulative probability with the 0.1 sec integration, because the short-period integration takes full advantage of the stronger signals which occur when the scan carries the beam directly over the target. The primary advantage of video integration devices in this type of problem lies in their making possible the use of more efficient displays, without the 5.5 db loss owing to collapsing and operator factors.

Dotted lines on both plots show the effects of a fluctuating target, giving higher probabilities of acquisition at long range and lower values at shorter range. The correlation time of the signal fluctuation is assumed to be less than 0.1 sec in Fig. 14.5(b), and less than 1 sec in Fig. 14.5(c). In all cases, it has been assumed that the designation error is completely correlated. If the error were to vary over the 10 sec period, a true cumulative probability of acquisition could be calculated according to Eq. (14.6), and much better results would be obtained. If this error is correlated over each 10 sec period, but uncorrelated over successive periods, then an over-all cumulative probability may be calculated from the plotted values of P_{ac} at 10 sec intervals (40 mile intervals in range). This over-all cumulative probability will rise steeply above the 10-sec values of P_{ac} shown.

In summary, for our particular example, we may conclude the following:

1. Acquisition will probably take place between 500 and 600 miles, on a steady target, and near 600 miles on a fluctuating target.

2. Circle scan should be used, with a radius of about one-half beam-width and a scan rate of about 1 cps. This is well within the capabilities of the radar servos.
3. The system is far from optimum for this type of search task, owing to collapsing loss and operator loss as well as to the large number of pulses available for noncoherent integration.

Experimental Results

It is difficult to obtain quantitative data to confirm the theoretical analysis given above, but there have been a number of operations on which reliable information is available. In particular, a series of satellite tracks were undertaken during 1962, using a modified version of the AN/FPS-16 which achieved unity S/N ratio at a range of about 550 mi on targets whose mean cross section was about 8 sq m. Designation data were made available from a nearby long-range tracking radar, and the accuracy of the data in angle was such that no scan was required ($\sigma_x < 0.2\theta$). The range designation errors were 20 to 40 miles, requiring an operator to scan a considerable portion of the scope. All targets were subject to wide variation in signal strength. The results of these operations, in terms of range and $\overline{S}/N$ ratio observed after lockon, are shown in Table 14.6. In these cases, acquisition by the human operator actually occurred tens of seconds prior to lockon by the tracking servos, and it is estimated that the average value of $\overline{S}/N$ required for operator detection was very close tő unity, as would be predicted by theory.

One interesting aspect of these experiments was the development by the radar operator of a two-step acquisition process. At long range, the

Table 14.6 SATELLITE ACQUISITION RESULTS (MODIFIED AN/FPS-16 RADAR)

Date	Satellite	Tumble period (sec)	Radar lockon	
			Range (miles)	S/N (db)
2/14/62	1961 Lambda 1	22	300	7
2/14/62	1960 Epsilon 1	44	450	4
2/14/62	1961 Alpha Kappa 1	7.6	400	6
2/14/62	1961 Lambda 1	22	300	7
2/14/62	1961 Epsilon 1		350	8
2/28/61	1962 Epsilon 1	*	400	6
2/29/62	1962 Epsilon 1	*	300	7
	Average lockon		360	6.5

* Stabilized in orbit

All targets had an average cross section of 8 m², for which radar R_o was about 550 mi, and a peak cross section approaching 3000 m².

signal appeared only periodically, during fluctuation peaks, which were seen twice per revolution of the target (see tumble rate data in Table 14.6). Upon detecting the first signal peak, the operator attempted to place the expanded A-scope sweep on the target, and to match the known rate of the incoming satellite. After the first two or three signal peaks, he was usually successful in this, and for the remainder of the acquisition process he was able to use the expanded display, avoiding the collapsing and operator losses assumed earlier. Thus, within a few seconds of the initial detection on rapidly tumbling objects, the system performance was upgraded by about 6 db, and the target was subject to almost continuous observation for the remainder of the operation, even when its instantaneous S/N ratio was below unity. The subsequent delay in achieving lockon of the automatic tracking circuits was caused primarily by difficulties with the controls during the transition process, and by the need to reacquire when the tracking gate was unable to coast through nulls of several seconds duration. To avoid the lost time of reacquiring, the operator retained his manual control until he was sure of the ability of the automatic circuits to lock on and track, and this caused a delay of several seconds in some of the operations.

The general theory of wave propagation, with applications to both radar
and communications systems, has
been the subject of several excellent
books,* and will not be covered
here. Instead, we shall survey the
material most directly applicable to
radar system design and analysis,
and shall present the results of some
recent studies which will assist the
radar engineer in solving the most
common types of problem.

The effects of propagation conditions on the performance of search
and tracking radars may be discussed under four general topics.
Atmospheric attenuation will cover
the departures from the free-space
radar equation results owing to absorption of the wave in the propagation medium: air, clouds and precipitation, and the ionosphere. The
background noise generated by these
interactions is also discussed as a
part of the attenuation phenomenon.
The section on surface reflection will
consider the effects of propagation
of radar waves over land and sea
surfaces, providing a basis for estimation of the reflectivity factor

SURVEY OF
PROPAGATION EFFECTS

15

which was used in calculating the
lobing patterns of search radar and
the multipath error of tracking sys-

* See, for instance, Donald E. Kerr
(ed.), *Propagation of Short Radio Waves*
(New York: McGraw-Hill Book Company, 1951).

H. R. Reed and C. M. Russel, *Ultra
High Frequency Propagation* (New York:
John Wiley & Sons, 1953).

tems. Tropospheric and ionospheric refraction will be covered in two separate sections, which will summarize the results of recent studies on the delay and angular deviation of transmissions through these two layers of the atmosphere. Although refraction will affect both search and tracking radars, the emphasis here will be on the magnitude of the errors introduced in tracking, and on the means available for correction of these errors.

15.1 ATMOSPHERIC ATTENUATION

In Chapter 4 we derived the basic radar range equation on the basis of free-space propagation of the transmitted and reflected radar waves. At the end of that chapter, various loss factors were listed which could be used to correct the calculations for the nonideal conditions actually encountered in most radar applications. The discussion below will summarize the results of calculations of this atmospheric attenuation loss L_a, and of the related contribution to the antenna noise temperature T_a. The material has been derived primarily from the published reports of L. V. Blake of the Naval Research Laboratory.

Attenuation in the Normal Troposphere

Calculations of attenuation in the troposphere are based on the original analysis of Van Vleck,* who gave equations for the absorption per unit distance, as a function of wavelength, pressure, temperature, and type of gas in the atmosphere (see Fig. 15.1). Having chosen a particular model for the troposphere, we may find the total attenuation over a given path by integrating the loss equations over the path. Figures 15.2 and 15.3 give the results of such integrations in a form directly applicable to most radar problems. Note that the values of L_a are given in decibels for the complete, two-way path, and that the one-way value in decibels will be just one-half of L_a. The method and constants used in constructing these curves are described in Blake's report.† About half the loss is seen to occur within the first 10,000 ft of the troposphere, as a result of the higher concentration of water vapor, as well as higher pressure, in this region. The relative

* J. H. Van Vleck, "The Absorption of Microwaves by Oxygen," and "The Absorption of Microwaves by Uncondensed Water Vapor," *Physical Review*, **71**, No. 7 (April 1, 1947), pp. 413–33.

   Donald E. Kerr, (ed.), *Propagation of Short Radio Waves* (New York: McGraw-Hill Book Company, 1951), pp. 646–64.

   † Lamont V. Blake, "A Guide to Basic Pulse-Radar Maximum-Range Calculation (Part 1)," Naval Res. Lab. Report 5868, Dec. 28, 1962, pp. 38–47; "Curves of Atmospheric-Absorption Loss for Use in Radar-Range Calculation," Naval Res. Lab. Report 5601 (March 23, 1961).

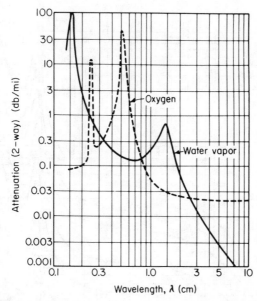

Figure 15.1 Theoretical values of atmospheric attenuation (after Van Vleck).

attenuation owing to water vapor and to the oxygen molecules was shown in Fig. 15.1.

Attenuation in Clouds and Rain

Theoretical values for attenuation of radar waves in clouds and rain were given by Goldstein,* whose curves are replotted in Fig. 15.4 in terms of two-way losses per mile of radar range. It can be seen that X-band (10,000 mc) radar can suffer an attenuation in the order of 1 db per mile of rain-filled path. The attenuation drops very rapidly with increasing wavelength, so that at C-band (5000 mc) the attenuation is only one-eighth as great. In clouds, the attenuation varies with the square, rather than with the cube, of frequency over this portion of the spectrum. By comparison with Fig. 15.1, we can see that the clear-troposphere value of loss for low-

---

* Herbert Goldstein, "Attenuation by Condensed Water," Sect. 8.6, *Propagation of Short Radio Waves*, D. E. Kerr, ed. (New York: McGraw-Hill Book Company, 1951), pp. 671–92.

H. E. Hawkins and O. La Plant, "Radar Performance Degradation in Fog and Rain," *Trans. IRE*, ANE-6, No. 1 (March 1959), pp. 26–30.

Raymond Wexler and Joseph Weinstein, "Rainfall Intensities and Attenuation of Centimeter Electromagnetic Waves," *Proc. IRE*, 36, No. 3 (March 1948), p. 353.

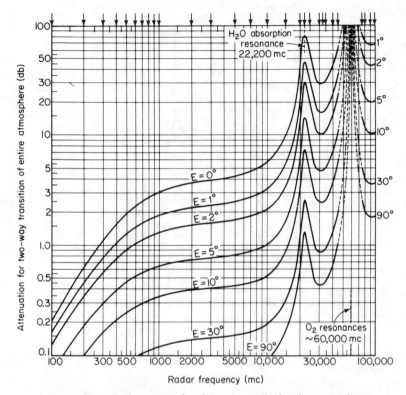

Figure 15.2 Radar attenuation for transversal of entire troposphere at various elevation angles, applicable for targets outside the troposphere. Does not include ionospheric loss, which may be significant below 500 mc in daytime. Arrows at top margin indicate frequencies at which calculations were made. (Courtesy L. V. Blake, Naval Res. Lab.)

altitude paths may be doubled by clouds or rain for S-band radars (3000 mc), increased by a factor of four for C-band, and by 20 or 30 for X-band (all in terms of decibel loss).

Ionospheric Attenuation

The attenuation produced by propagation through the ionosphere was studied by Millman,* whose results show that radars operating above 100 mc will not normally encounter losses greater than about 1 db, even

---

* G. H. Millman, "Atmospheric Effects on VHF and UHF Propagation," *Proc. IRE*, **46**, No. 8 (Aug. 1958), p. 1501 (Fig. 16).

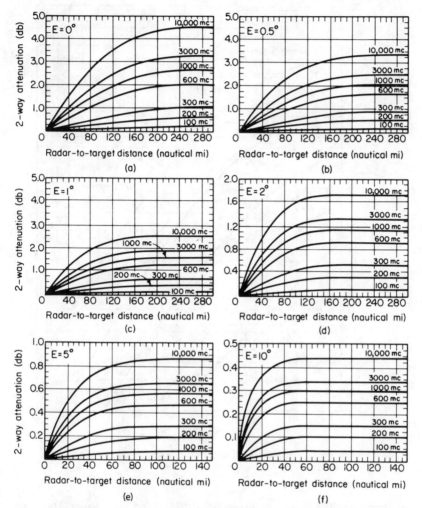

Figure 15.3 Radar atmospheric attenuation for different elevation angles. (Courtesy L. V. Blake, Naval Res. Lab.)

under daytime conditions when the ionospheric density is greatest. Specifically, at 100 mc, the two-way loss for a ray at zero elevation angle can reach about 1.2 db in the daytime, and 0.07 db at night. Losses over vertical paths are reduced by a factor of three, and all losses vary with the square of wavelength, becoming negligible above about 300 mc.

An exception to this has been noted in the case of "blackout" from high-altitude nuclear blasts, which may create, for limited periods of time,

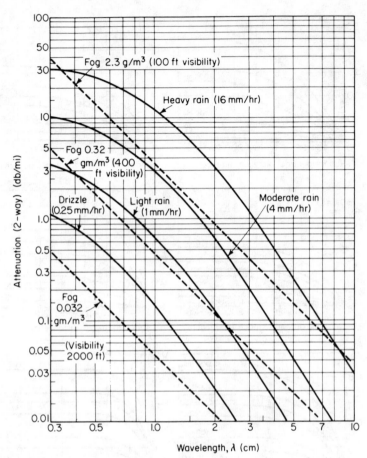

Figure 15.4 Theoretical values of attenuation in rain and fog (after Goldstein).

electron densities up to or exceeding 10^{13} per cc.* Densities 100 times greater than those of the normal F-layer (10^8 as compared to the normal 10^6 per cc) may prevail over paths of 200 or 300 miles. This ionization decays over a period of about an hour. Meanwhile, however, a radar operating at 100 mc could encounter attenuations of tens or even hundreds of decibels in looking through the region near the blast. As with other ionospheric effects, the attenuation in decibels varies with the square of the wavelength, so microwave radars will be affected for much shorter times,

* *The Effects of Nuclear Weapons*, Dept. of Defense and Atomic Energy Comm. (Washington, D. C.: U.S. Govt. Printing Office, April 1962), pp. 532–43.

at points closer to the blast site. Similar considerations apply to attenuation in the plasma generated by missiles and satellites as they re-enter the atmosphere, leading to use of the higher microwave frequencies in tracking and communication for such vehicles.

Atmospheric Noise Temperature

Just as the atmospheric molecule or particle is capable of extracting energy from the radar wave, it is also capable of emitting thermal radiation on the radar frequency. In Eq. (4.37) we related the temperature contribution of a lossy r-f transmission line, used in carrying the received signal, to the physical temperature of the line (assumed to be near reference temperature T_o) and the line loss L_r. The same relationship is used to find the temperature contribution T_p of the propagation loss in the troposphere.

$$T_p = T_t\left(1 - \frac{1}{\sqrt{L_{at}}}\right)$$

(15.1)

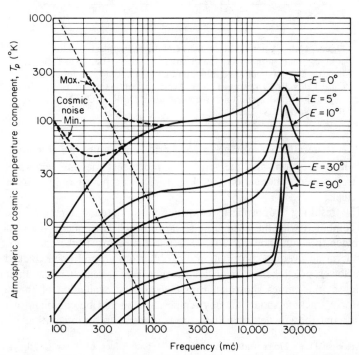

Figure 15.5 Background temperature from atmosphere and cosmic noise.

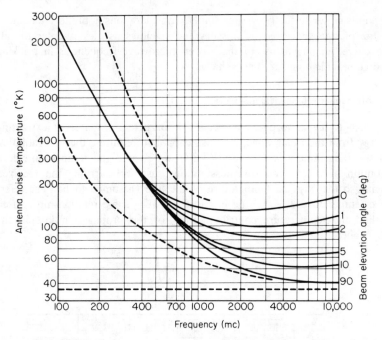

Figure 15.6 Antenna noise temperature for typical conditions of cosmic, solar, atmospheric, and ground noise. The dashed curves indicate the maximum and minimum levels of cosmic and atmospheric noise likely to be observed. The horizontal dashed line is the assumed level of ground-noise contribution (36°K). (Courtesy L. V. Blake, Naval Res. Lab.)

Here, T_t represents the average tropospheric temperature, weighted according to its absorption, and the term $\sqrt{L_{at}}$ represents the one-way value of total tropospheric attenuation (see Fig. 15.2), as a power ratio rather than in decibels.

The resulting tropospheric contribution to antenna temperature has been combined with the cosmic noise term, which is frequency-dependent, to provide the data shown in Fig. 15.5.* In his report, Blake added to this "sky" term an average contribution of 36°K from ground noise, picked up in the lower side lobes or main lobe response of a search radar antenna, arriving at the curves shown in Fig. 15.6. In the evaluation of the noise term for a tracking radar, the sky temperature should be combined with a ground contribution determined by measurement or by integration of the

* See also D. C. Hogg and R. A. Semplak, "The Effect of Rain on the Noise Level of a Microwave Receiving System," *Proc. IRE*, **48**, No. 12 (Dec. 1960), pp. 2024-25.

actual side-lobe pattern of the antenna, taking into account partial reflection of the side-lobe energy toward the sky. Careful designs of tracking antennas have led to values as low as 10°K for this term.

15.2 SURFACE REFLECTION

The processes of reflection from the earth's surface are discussed by Kerr.* His curves for reflection coefficient ρ, for smooth sea water, are reproduced in Figs. 15.7 and 15.8. Curves of this sort have been extended to values of depression angle up to 90°, showing that the coefficient for horizontal polarization remains near unity for all angles, whereas that for vertical polarization increases toward the same value as the angle increases above 10°. Values of ρ for rough sea are reduced considerably from those shown. The reflection coefficients for various types of ground have also been measured,† and have been found to vary irregularly from 0.1 to near unity, depending upon the type of terrain and the angle of incidence. The presence of vegetation tends to increase the absorption of the ground at all angles, whereas smooth and sandy terrain approaches a reflectivity of unity at low angles. It is difficult to state any general rules for estimating the coefficient in the case of land surfaces, but a value near $\rho = 0.3$ may be used in the absence of information which would justify some other specific value.

The effect of earth's curvature on surface reflection was discussed by Fishback,‡ who used the "divergence factor" D to describe the reduction in effective value of the reflection coefficient at long range and low angles of incidence. Generally, this factor reduces the effective value of ρ below unity in those cases where the region of surface reflection is far enough from the radar to introduce significant curvature in the surface, as when the radar is sited high above the sea. We must evaluate its effect when the angle of incidence is below a minimum value given by

$$E_{\min} = \frac{\sqrt{h_{ft}}}{4000} \text{ rad} \qquad \textbf{(15.2)}$$

Thus, if the antenna is at 100 ft altitude above the sea, the effect of the divergence factor will be negligible at angles of incidence above about

* Donald E. Kerr, *Propagation of Short Radio Waves* (New York: McGraw-Hill Book Company, 1951), Chap. 5.

† William T. Fishback, "Reflection Coefficient of Land," Sects. 5.9–5.10, *Propagation of Short Radio Waves*, pp. 430–34.

‡ William T. Fishback, "Methods for Calculating Field Strength with Standard Refraction," Sects. 2.13–2.16, *Propagation of Short Radio Waves*, pp. 112–140.

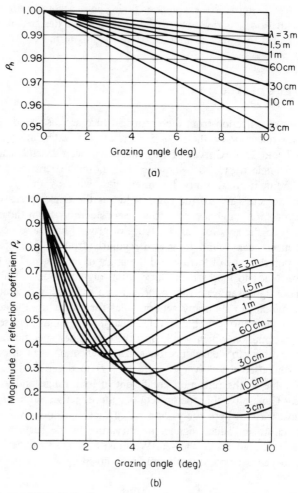

Figure 15.7 Reflection coefficients for smooth sea water. (a) Magnitude of reflection coefficient for horizontal polarization. (b) Magnitude of reflection coefficient for vertical polarization. (From Kerr.)

$0.15°$. In those cases where an exact lobing pattern is to be evaluated at low angles, the discussions in Kerr's book should be consulted.

15.3 TROPOSPHERIC REFRACTION

The effect of tropospheric refraction on radio communication paths is often described in terms of the "$\frac{4}{3}$ earth's radius" correction, which in-

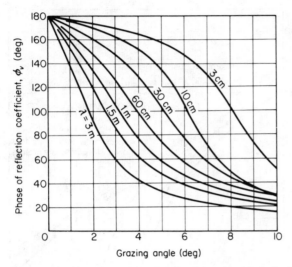

Figure 15.8 Phase angle of reflection for vertical polarization (from Kerr).

creases the radio line-of-sight pathlength beyond the geometrical limits. Early calculations on search radar coverage were made on the same basis, and the resulting coverage patterns were quite accurate for use with targets below 50,000 ft in altitude. In recent years, the subject has received further study, and most calculations are now made on a more accurate basis, using the "exponential reference atmosphere" defined by the National Bureau of Standards.* We shall present the data applicable to this atmosphere in several convenient forms, and then investigate the variations in atmospheric conditions which lead to deviations from this reference. Both short-term and long-term variations will be discussed, with most effects appearing only in the more accurate, tracking-type radar systems. The state of the art in correcting for the long-term variations will be discussed, the discussion being based on the work of the propagation personnel at the major missile test ranges.

Exponential Reference Atmosphere

The refractive index of the air, for frequencies below about 20,000 mc,

---

* B. R. Bean and G. D. Thayer, "CRPL Exponential Reference Atmosphere," Natl. Bureau of Standards Monograph No. 4 (Washington, D.C.: U.S. Govt. Printing Office, Oct. 29, 1959).

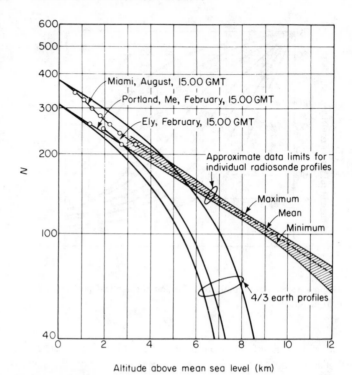

Figure 15.9 Refractivity vs. altitude for mean and extreme conditions, compared to 4/3 earth profiles (from Bean and Thayer).

can be expressed by using the "Smith-Weintraub constants"* in the equation

$$N \equiv (n - 1) \times 10^6 = \frac{77.6}{T}\left(P + \frac{4810p}{T}\right) \qquad \text{(15.3)}$$

where T represents the atmospheric temperature in degrees Kelvin, P is the total pressure of the atmosphere in millibars, p is the partial pressure of the water-vapor component, n is the refractive index, and N is a scaled-up value commonly known as the "refractivity." Typical values of refractivity at sea level are in the order of 300 to 350 "N-units", and the variation with altitude for mean and extreme profiles has been given in the form of Fig. 15.9. The approximate straight-line relationship of this data, when plotted on a logarithmic sale, suggests the use of an exponential form for

---

* E. K. Smith and S. Weintraub, "The Constants in the Equation for Atmospheric Refractive Index at Radio Frequencies," *Proc. IRE*, **41**, No. 8 (Aug. 1953), pp. 1035 37.

a mathematical model of the atmosphere, and the following form for refractivity as a function of altitude has been adopted in many discussions:

$$N(h) = 313.0 \exp(0.04385h) \tag{15.4}$$

with h expressed in kilofeet (for h in km, the constant will change from
0.04385 to 0.14386). This was the form of the refractivity profile used in
calculation of the coverage chart (Fig. 6.2) for search radar use.

Range and Elevation Errors

In addition to affecting the coverage of a radar for targets at a given
altitude, the refraction of the troposphere causes an extra time delay in
transmission of the signal, and an increase in the elevation angle measured
by the antenna system. The ray actually follows the path of minimum delay
to reach the target and return, and the errors are proportional to the refractivity of the air along the measurement path. Figures 15.10 and 15.11

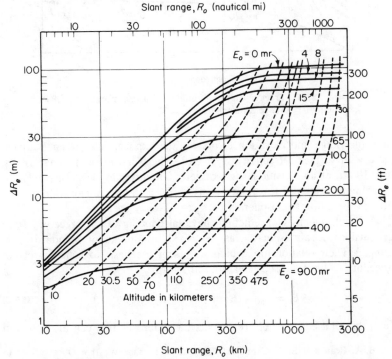

Figure 15.10 Range bias vs. range for CRPL exponential reference
atmosphere ($N_o = 313$).

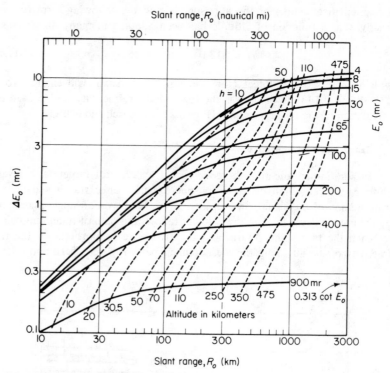

Figure 15.11 Tracker elevation angle error vs. range for CRPL exponential reference atmosphere ($N_o = 313$).

show the range and elevation errors for the exponential reference atmosphere with surface refractivity $N_s = 313$. The worst error in range is just over 300 ft, at an elevation angle of zero. The error in elevation angle has a maximum value just less than 1°, also for a path at zero elevation passing through the entire troposphere.

A simple means of predicting the total bending of rays received from far outside the atmosphere was described by Bureau of Standards personnel.* Their relationship, which will form the basis for one correction procedure to be discussed later, took the form

$$\Delta E_t = E_o - E_c = (bN_s + a) \times 10^{-6} \, \text{rad} \qquad \textbf{(15.5)}$$

Here, the total bending ΔE_t represents the difference between the observed

* B. R. Bean and B. A. Cahoon, "The Use of Surface Weather Observations to Predict the Total Atmospheric Bending of Radio Rays at Small Elevation Angles," *Proc. IRE*, **45**, No. 11 (Nov. 1957), pp. 1545–46.

elevation angle at which the ray arrives at the surface of the earth (E_o) and the true or corrected angle E_c at which it would have arrived had there been no atmosphere. The surface refractivity is given by N_s, and a and b are functions of the elevation angle E_o. At angles above about 5°, the parameter b is given approximately by the cotangent of E_o, and a is approximately zero, leading to

$$\Delta E_t \cong N_s \cot E_o \times 10^{-6} \text{ rad} \qquad (E_o > 5°)$$

Plots of a and b, normalized in each case by the function $\tan E_o$, are given in Fig. 15.12. These may be used to estimate the total bending which would apply to radiation from an astronomical source or a distant missile or satellite.

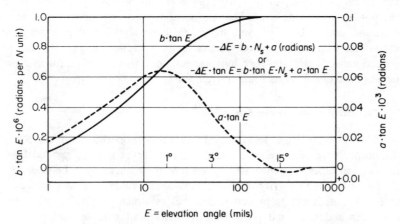

Figure 15.12 Refraction correction constants for radio-astronomy case (after Bean and Cahoon).

Variations in Refractivity

Results of a large number of observations of both surface refractivity N_s and slope of the refractivity profile have been published.* These show variations of surface refractivity at a number of locations, as a function of time of day and of season. Diurnal cycles with peak-to-peak changes in the order of 20 to 40 N-units are quite common, and the standard deviation of the seasonal change is in the order of 10 N-units. Thus, the refrac-

* B. R. Bean, J. D. Horn, and A. M. Ozanich, Jr., "Climatic Charts and Data of the Radio Refractive Index of the United States and the World," Natl. Bureau of Standards Monograph No. 22 (Washington, D.C.: U.S. Govt. Printing Office, Nov. 25, 1960).

tivity at a given site may change by 100 N-units or more during the course of a year. These changes will be reflected into values of tropospheric errors ΔR_c and ΔE_o, given by Figs. 15.10 and 15.11, the error varying approximately in proportion to the surface refractivity. In addition, geographical differences will affect the mean value of N_s and the percentage of change in this parameter during diurnal and seasonal cycles. The maximum values of N_s are associated with moist climates and sea-level pressures, as would be expected from Eq. (15.3). Dry, high-altitude sites will have low average values of N_s with smaller variations except during the relatively infrequent periods of heavy rain.

Extreme refraction effects are sometimes observed, in which the rays launched at low elevation angles are trapped in a "duct" and carried beyond the horizon to provide unusual coverage patterns. For the duration of the ducting situation, there will be a "hole" in the coverage, from which the rays have been diverted. Ducting phenomena do not normally occur at elevation angles above $\frac{1}{2}°$, and are chiefly important in that they vary the long-range coverage of search radars. During ducting, there may appear ground reflections from very long ranges, and also aircraft returns from targets beyond the normal line-of-sight limits.

Correction of Refraction Errors

The curves of Figs. 15.10 and 15.11 provide one basis for correcting the output data of a tracking radar. Similar curves may be plotted from the data on other reference atmospheres, or these curves may be scaled by the factor $N_s/313$ to arrive at an estimate for any value of surface refractive index. A more accurate procedure involves ray-tracing, by using a suitable computer and measured profiles of refractive index vs. altitude. Digital computers are normally used for this purpose, and a detailed discussion of the procedures followed has been published.* The estimated accuracy of such procedures is between 1 and 2 per cent of the initial tropospheric error in range, and perhaps 2 to 3 per cent in elevation angle, for paths above 5° elevation. Such accuracy requires that the absolute value of N_s be known to within about 1 per cent, which presents a major problem unless a radio refractometer is used to measure the local atmosphere. Profiles of refractivity as a function of height can be obtained from radiosonde data or from airborne refractometers which move along the path to be traveled by the ray.

When the facilities for complete measurement of profiles and ray-tracing computations are not available, or when the accuracy provided by these

---

* Preston Landry, "Atmospheric Sounding and Correction of Tracking Data," *Report of Ad Hoc Panel on Electromagnetic Propagation*, National Academy of Sciences, D. K. Barton, ed. (Feb. 1963), Sect. 5.

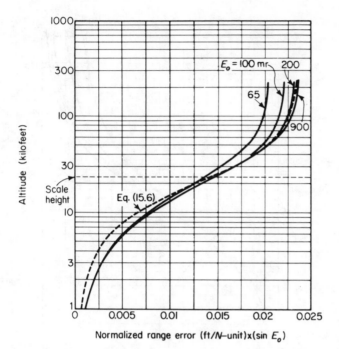

Figure 15.13 Comparison of approximate error corrections with range error for exponential reference atmosphere.

methods is not required, approximate methods of calculating the tropospheric correction terms may be employed. One such method, developed simultaneously at White Sands Proving Ground and at RCA,* will be described here. We recall that Bean and Cahoon arrived at a simple relationship which gave angular refraction as a function of surface refractive index and parameters related to elevation angle [Eq. (15.5)]. If the target is located within the troposphere or just beyond it, the angular refraction will be less than the total bending given by Bean and Cahoon, and we must allow for this by inclusion of a term which scales the refraction to target height. Figures 15.13 and 15.14 show the variation of range and angle errors as a function of target height, for different elevation angles.

* David K. Barton, "Final Report, Instrumentation Radar AN/FPS-16 (XN-I), Evaluation and Analysis of Radar Performance," RCA, Moorestown, N.J., Contract DA 36-034-ORD-151. Astia Document AD 212, 125.

Kermit E. Pearson, Dennis D. Kasparek, and Lucille N. Tarrant, "The Refraction Correction Developed for the AN/FPS-16 Radar at White Sands Missile Range," U.S. Army Missile Support Agency, White Sands Missile Range, N.M., Tech. Memo. No. 577 (Nov. 1958).

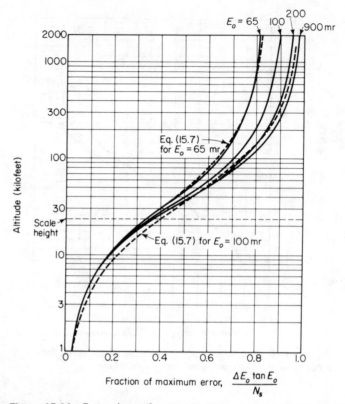

Figure 15.14 Comparison of approximate error corrections with elevation error for exponential reference atmosphere.

Both error plots have been normalized to the surface refractivity, and in addition the factors sin E_o and tan E_o have been applied to range and angle errors, respectively, further to compress the variation between the curves for different elevation angles. When this is done, a single curve can represent all angles between 200 and 1600 milliradians (12 and 90°), with an error of about 1 per cent. The same curve is within 8 per cent of the proper values for $E_o = 100$ mr (6° elevation).

For ease in calculation, the curves for range and angle error are fitted by simple polynomials in target altitude.

$$\Delta R_r = 0.0235 \frac{N_s}{\sin E_o} \times \frac{h + 0.1h^2}{50 + h + 0.1h^2} \tag{15.6}$$

$$\Delta E_o = (bN_s + a)\frac{h}{35 + h} \cong \frac{N_s}{\tan E_o} \times \frac{h}{35 + h} \tag{15.7}$$

Here, h is the altitude in kilofeet, and the constants are chosen to match the curves of error, as shown by the dashed lines on Figs. 15.13 and 15.14. Without use of the parameters a and b from Bean and Cahoon's study, the results are good to within 10 per cent of the maximum error at all elevation angles above 6 deg. Use of the two elevation-dependent parameters (Fig. 15.12) will reduce the error in elevation angle to within 5 per cent, all the way down to 2 or 3 deg elevation. Similar factors could be applied to the case of range error, but this is not generally as serious a problem as elevation error. In addition, the quality of the fit could be improved by including additional terms in the polynomial, but the residual errors owing to irregularity of the actual profiles would not appear to warrant the added complexity. In fact, Pearson limited himself to a simple, first-order fit in both range and angle equations, as we have done in angle.

Tropospheric Fluctuations

Temporal fluctuations in observed range, elevation, and azimuth of a target are caused by the drifting of tropospheric irregularities past the radar beam. These errors are in addition to the bias errors caused by the regular, stratified structure of the atmosphere, and the residual errors caused by departure of this structure from the exponential models adopted for mathematical simplicity. The nature of the errors owing to fluctuating components was discussed in several early papers,* and has since been the subject of much analysis and experimental study.† Following Muchmore and Wheelon's work, the standard deviation in range and angular measurement caused by these fluctuations may be expressed approximately in the following form:

$$\sigma_r = \sqrt{2l_o L \, \overline{\Delta N^2}} \times 10^{-6} \tag{15.8}$$

$$\sigma_a = \sqrt{\overline{\Delta N^2} L/l_o} \times 2 \times 10^{-6} \text{ rad} \tag{15.9}$$

---

* R. B. Muchmore and A. D. Wheelon, "Line-of-Sight Propagation Phenomena," *Proc. IRE*, **43**, No. 10 (Oct. 1955), pp. 1437–66.

J. W. Herbstreit and M. C. Thompson, "Measurements of the Phase of Radio Waves Received over Transmission Paths with Electrical Lengths Varying as a Result of Atmospheric Turbulence," *Proc. IRE*, **43**, No. 10 (Oct. 1955), pp. 1391–1401.

A. P. Deam and B. M. Fannin, "Phase-Difference Variations in 9350-Megacycle Radio Signals Arriving at Spaced Antennas," *Proc. IRE*, **43**, No. 10 (Oct. 1955), pp. 1402–11.

† K. A. Norton, et al., "An Experimental Study of Phase Variations in Line-of-Sight Microwave Transmissions," Natl. Bureau of Standards Monograph No. 33 (Washington, D.C.: U.S. Govt. Printing Office, Nov. 1, 1961).

David K. Barton, "Final Report, Instrumentation Radar AN/FPS-16 (XN-1), Evaluation and Analysis of Radar Performance," RCA, Moorestown, N.J., Contract DA 36-034-ORD-151. Astia Document AD 212, 125.

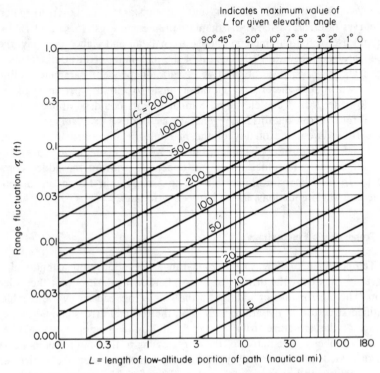

Figure 15.15   Range fluctuation vs. path length for different atmospheres ($C_r^2 = l_o \overline{\triangle N^2}$).

where l_o represents the scale length of the tropospheric anomalies causing the fluctuation, L the path length which contains the anomalies, and $\overline{\triangle N^2}$ the mean square refractivity variation of the anomalies. [In place of the constant 2 in Eq. (15.9), Muchmore and Wheelon used $\sqrt{2\sqrt{\pi}}$ or $\sqrt{3\pi/2}$, depending upon the exact distribution of refractivity which was assumed to represent the anomalies. The difference is not significant in view of the variability of the actual errors]. The magnitude of the errors is indicated in Figs. 15.15 and 15.16, in terms of affected pathlength, scale length, and refractivity variation.

More recent data have provided a somewhat more general description of atmospheric variation, in terms of a spectral distribution of refractivity and radio-pathlength change, shown in Fig. 15.17.* This distribution having

---

* M. C. Thompson, H. B. Janes, and R. W. Kirkpatrick, "An Analysis of Time Variations in Tropospheric Refractive Index and Apparent Radio Path Length," *Journal of Geophysics Research*, **65**, No. 1 (Jan. 1960), pp. 193-201.

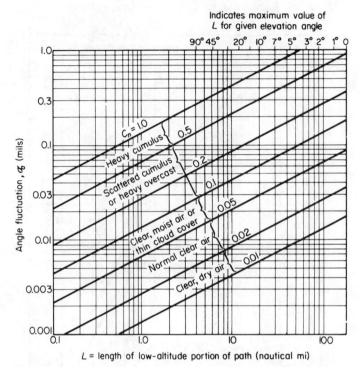

Figure 15.16 Angle fluctuation vs. path length for different atmospheres ($C_n^2 = \overline{\triangle N^2}/l_o$).

been found to be closely related to the distribution of refractivity variations in space, it became possible to arrive at relationships for both range and angle fluctuations, as functions of aperture size, relative velocity of the measurement rays and the troposphere, and amount of smoothing applied to the output data.* Thus, the range fluctuation in an atmosphere described by Fig. 15.17 is given by the integral of the spectral density from some chosen low-frequency limit to the upper limit set by temporal or aperture smoothing. The longer the observation is continued (reducing the low-frequency limit), the greater will be the observed range fluctuation. For short periods (e.g., 10 min) the value of the integral may be about 0.05 ft, for the 15 mile path at $E = 5°$ on which this spectrum was measured. For a period of a year or more, the fluctuation will increase to an rms value of about 1 ft, corresponding to an rms variation in refractivity of about 14

* David K. Barton, "Reasons for the Failure of Radio Interferometers to Achieve Their Expected Accuracy," *Proc. IEEE*, **51**, No. 4 (April 1963), pp. 626–27.

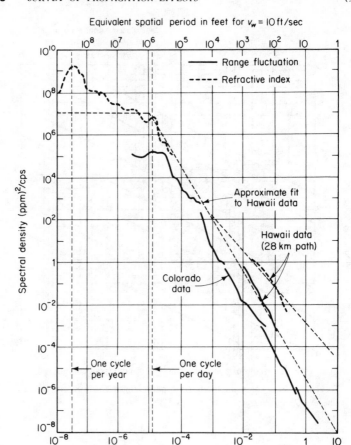

Figure 15.17 Spectra of refractivity and range fluctuations (after Thompson, Janes, and Kirkpatrick).

N-units. The value of range fluctuation for the shorter period is consistent with the use of the parameter $C_r^2 = l_o \overline{\Delta N^2} = 100$ in Fig. 15.15, whereas the larger value is consistent with the sum of diurnal and seasonal effects described earlier. It would appear that the magnitude of the error encountered is affected largely by the region of the spectrum to which the instrument is sensitive, and that the variation in actual atmospheric characteristics is not as variable as might be supposed from the scale of parameters given in the figures and in Table 15.1.

The angular error for the same spectral distribution is shown in Fig. 15.18, as a function of aperture or effective length of the measurement

Table 15.1 FLUCTUATION OF RANGE AND ANGLE IN TROPOSPHERE

Type of weather	$(\wedge N)_{rms}$	l_o (ft)	C_r	C_n	$(\sigma_a)_{max}$ (mil)	$(\sigma_r)_{max}$ (ft)
Heavy cumulus	30	3600	2000	0.5	0.7	2.0
Scattered cumulus	10	2500	500	0.2	0.3	0.5
Small scattered cumulus	3	1000	100	0.1	0.15	0.1
Clear, moist air	1	400	20	0.05	0.07	0.02
Clear, normal air	0.3	200	5	0.02	0.03	0.005
Clear, dry air	0.1	100	1	0.01	0.015	0.001

baseline. The mean values for the 10 foot aperture are in good agreement with Fig. 15.16, assuming that $C_n^2 = \overline{\wedge N^2}/l_o = 0.2$. Again, this is near the middle of the range of values given for this parameter, and the effect is probably more dependent upon the spectral sensitivity of the instrument than upon the particular atmosphere through which observations are made. These frequency components of fluctuation error are dependent upon the velocity with which the measurement rays or beams move through the troposphere, as well as upon the wind speed. An average crosswind component of 10 ft per sec was assumed in placing the upper scale on Fig. 15.17, which would make this figure apply directly to the case of a fixed beam in a moderate wind. When the beam moves to track a target, the frequencies will be shifted upwards. The average or effective beam velocity

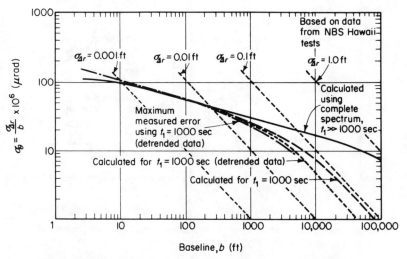

Figure 15.18 Tropospheric angle fluctuation vs. baseline length (t_1 is observation time).

will be determined by weighting the relative velocity at different altitude levels according to the intensity of refractivity variation at those altitudes. Since the refractivity changes are due, in large part, to the water vapor content of the air, the greatest weight should be placed on portions of the beam which lie below 10,000 or 12,000 ft in altitude.

15.4 IONOSPHERIC REFRACTION

When radar measurements are made on targets above 60 or 70 miles altitude, the effects of ionospheric layers must be considered. These effects are all dependent upon operating frequency, varying directly with the square of the wavelength. A brief summary of the errors produced by the ionosphere will be given, based on simplified models of the ionospheric profile for day and night conditions.

Ionospheric Profiles

Figure 15.19 shows typical day and night profiles of the ionosphere,*

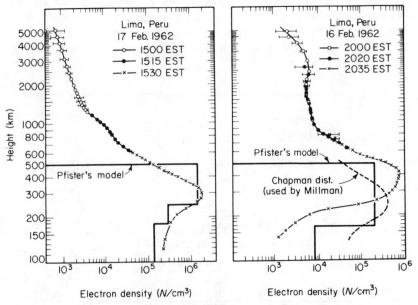

Figure 15.19 Comparison of ionospheric models with measured profiles for day and night conditions.

* Kenneth L. Bowles, "Lima Radio Observatory," Natl. Bureau of Standards Report 7201 (supplement) (April 30, 1961).

based on backscattering data, as compared to the "Chapman distribution" and with a simple, rectangular distribution used in an Air Force study of ionospheric effects* These profiles may be used to calculate the refractivity of the ionosphere as a function of frequency.

$$N_i = (n - 1) \times 10^6 = \left[\frac{N_e e^2}{\epsilon_o m \omega^2} \right]^{1/2} \times 10^6 \cong \frac{1}{2} \frac{N_e e^2}{\epsilon_o m \omega^2} \times 10^6 \quad \textbf{(15.10)}$$

Here, N_e is the electron density per m³, e is the charge of the electron, m is the electron mass, ω is the radian frequency, and ϵ_o is the permittivity of free space. Inserting the values of these constants, and applying the results to frequencies well above the "critical frequency" f_c, we have

$$N_i \cong 40 \frac{N_e}{f^2} \times 10^6 = \frac{1}{2} \left(\frac{f_c}{f} \right)^2 \times 10^6 \quad \textbf{(15.11)}$$

where N_e remains in electrons per m³, f is in cps, and $f_c = 9\sqrt{N_e}$ is the critical frequency of the medium. When N_e is expressed in electrons per cc, f and f_c will be in kc for the same result. These relationships are plotted in Fig. 15.20, which defines the limits of the operating region for accurate measurement systems (better than one part in 10^4 accuracy).

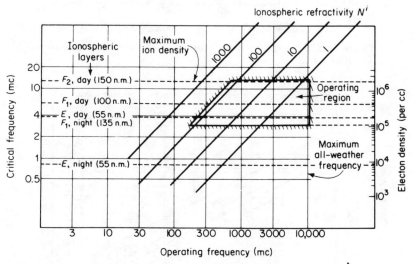

Figure 15.20 Ionospheric refractivity vs. operating and critical frequencies and electron density.

* W. Pfister and T. J. Keneshea, "Ionospheric Effects on Positioning of Vehicles at High Altitudes," Air Force Surveys in Geophysics No. 83 (March 1956), Cambridge Research Center, Astia Document AD 98, 777.

Ionospheric Errors

Curves for range and angular refraction errors, based on the approximate rectangular models of the daytime ionosphere, are shown in Figs. 15.21 and 15.22. These may be reduced by a factor of about three for nighttime conditions, and increased by a similar factor for periods of extreme ionospheric disturbance. The expected value of range fluctuation is shown in Fig. 15.23, for both normal and disturbed ionospheric conditions.

Corrections for ionospheric refraction errors have proved difficult to calculate. One very effective method is to make redundant measurements at

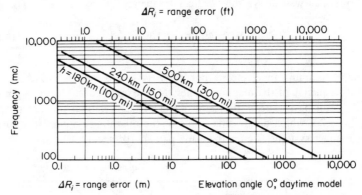

Figure 15.21 Ionospheric range error vs. frequency (after Pfister and Keneshea).

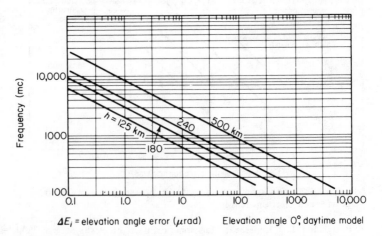

Figure 15.22 Ionospheric angle error vs. frequency (after Pfister and Keneshea).

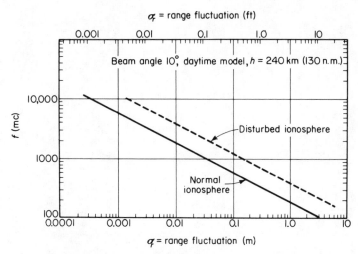

Figure 15.23 Ionospheric range fluctuation vs. frequency.

two frequencies, widely separated to assure that the ionospheric errors will be different. By combining the pairs of readings, it is possible to cancel out the effects of ionospheric refraction in both transmissions. This approach is not easy to apply to radar or to instruments measuring angles, but it has been used in a navigation system based on measurement of Doppler shift.* Correction procedures based on use of ionospheric profiles have not proved very accurate, and may not lead to any significant reduction of error. The estimates of error reduction run from slight to a factor of four (leaving 25 per cent of the initial error), but the improvement is quite dependent upon the stability of the ionospheric layers through which the measurement ray passes, and these layers are always in motion near dawn or dusk.

* W. H. Guier and G. C. Weiffenbach "A Satellite Doppler Navigation System," *Proc. IRE*, Vol. **48**, No. 4 (April 1960), pp. 507–16.

The previous chapters have considered the radar as an individual instrument for search and tracking of targets. The coverage of the individual radar was described in terms of a maximum range and azimuth and elevation sectors. In this chapter we shall review briefly the geographical considerations involved in use of radar, and shall consider the performance of radar networks for both search and tracking applications. The most obvious way of using additional radar sites is to extend the coverage beyond the maximum range limit of the single radar, as established either by line of sight or by radar power. However, there are several other benefits to be gained in operation of radar networks. The space-diversity feature of overlapping radar coverage provides insurance against loss of targets because of deep nulls in the reflectivity pattern, and adds to the over-all reliability of the radar system. Radars at separated sites are sometimes used as "bistatic" detection systems, making possible the transmission of c-w signals without

MULTISTATION RADAR NETWORKS

16

direct spillover of transmitter power into the receiver. Finally, in the case of tracking systems, the data from widely separated radars may be combined in a "range trilateration" solution to find target position more accurately than is possible with angle tracking. The interferometer tracking system is a variation of this approach.

16.1 SEARCH RADAR NETWORKS

The search-radar coverage problem depends greatly upon the way in which radar sites may be distributed to cover the total search area. In the earlier analysis it was assumed that the coverage required for a single radar could be defined in terms of maximum range and solid angle to be searched from one site. At times, however, the system engineer is given only the limits of the total geographical area to be searched or defended against penetration by incoming targets, and he is able to choose the number of radars to be used as well as the characteristics of the individual radar. Limitations are imposed by the geography of the area subject to search; suitable land sites may not be available at the points most desirable for radar operation, and it may be necessary to provide the coverage from nearby sites or to use ships or aircraft to fill in the coverage over some regions. Assuming, however, that the choice of sites represents primarily an economic compromise in system design, we shall discuss the considerations which establish the optimum number of sites for this type of problem, and outline some of the special cases in which it is desirable to use geographically dispersed radars in preference to a smaller number of long-range units.

Approach-Line vs. Area Coverage

Initial detection of targets (or "early warning") is often provided by a line of coverage, across which incoming targets must pass to enter the defended area. The range and solid angle required of an individual radar which forms part of this line of coverage may be reduced to the vanishing point by using a great many radar sites. The radars may then be made relatively small and economical, since the power-aperture product is proportional to the product of solid angle times the fourth power of range. However, when the costs of establishing the large number of radar sites are considered, the economic balance may dictate use of longer-range radars at less frequent intervals along the line. If the coverage must be extended to provide data on targets over a large area, the use of long-range radars is even more economical.

Examples of the warning-line approach to detection are the Distant Early Warning (DEW) Line, the Mid-Canada Line, and the Pine Tree Line, shown on Fig. 16.1. These radar lines were established during the late 1940's and the 1950's to alert the defense forces in North America to possible air attack from the north, and were supplemented* by over-water

* Philip J. Klass, "DEW Line Adds 100 Minutes Warning," *Aviation Week* (Sept. 16, 1957), pp. 100-13.

Herbert J. Coleman, "Navy Maintains Barrier in All Weather," *Aviation Week* (April 3, 1961), pp. 63-73.

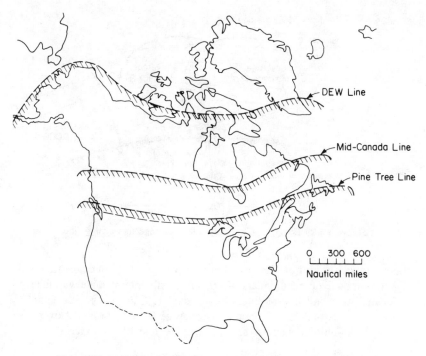

DEW Line

Mid-Canada Line

Pine Tree Line

0   300 600
Nautical miles

Figure 16.1 Radar warning lines in North America (late 1950's).

lines which extended from each end of the DEW line, using airborne search radars. Since the maximum altitude of the target aircraft was limited to about 50,000 ft, radars for use in the line could have been designed for maximum ranges as short as a few tens of miles, if it had been practical to use sites of this spacing over the entire 3000 mile line. In the case of the DEW Line, the initial cost and logistical support problems dictated the use of fewer sites, with radars of longer range. The coverage diagram of Fig. 16.2 is typical of the actual system used, in which overlapping long-range coverage of the main radars was supplemented by low-altitude "gap-filler" coverage from intermediate stations. The size of the remaining low-altitude gaps in the coverage was made as small as possible by using high towers for the gap-filler antennas (the drawing is not to scale). The lack of coverage in the zenith cones above each radar is not important, since the main radars scan continuously in azimuth, providing considerable depth of coverage in the direction normal to the search line. It is this depth that projects the warning line forward over the Arctic Ocean, as shown in Fig. 16.1.

In order to extend the coverage from a narrow line to a broad area, we

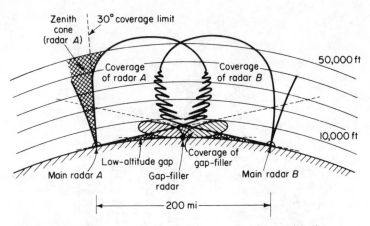

Figure 16.2 Long-range and gap-filler coverage in warning line.

must establish a grid of stations as shown in Fig. 16.3. Behind the initial line (composed of main radars *A*, *B*, etc., with intermediate gap-fillers) is located a second line of long-range radars (*A'*, *B'*, etc.), also supplemented with gap-fillers. If continuous coverage is to be provided over the area for low-altitude targets, an intermediate line of gap-filler sites, in-

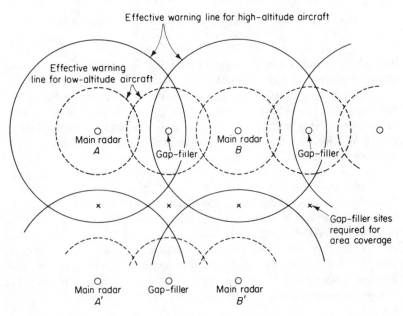

Figure 16.3 Extension from line to area coverage in aircraft search.

dicated by X's, must also be established, and similar lines must be repeated between each long-range radar line in the area. This double grid of long-range and gap-filler radar sites is representative of the Semiautomatic Ground Environment (SAGE) System, which provides continuous coverage over the continental United States, for both high- and low-altitude targets. Contrasted with this approach is that of the Federal Aviation Agency radar network, in which low-altitude coverage is limited to the regions near major airports. Continuous coverage is provided at the higher altitudes, long-range radar being used with the assistance of airborne transponders, to permit ground monitoring and control of large aircraft flying the airways. Limitation of continuous coverage to the higher altitudes, combined with the use of beacons for range-extension and altitude-reporting, makes it possible to cover the essential airways with a more economical system than would otherwise be required.

Economic Considerations

As shown above, the potential coverage of a search radar is dependent primarily upon the target altitude and the line-of-sight limit set by altitude. The minimum slant range which is useful for air-search applications is about 20 miles which will provide coverage up to 50,000 ft over a region whose radius is about 20 miles from the radar site. If the radar range were reduced to 10 miles, the coverage circle would extend only about 5 miles in ground range from the site. The maximum range which is useful for most targets is about 200 miles, beyond which the targets must exceed an altitude of 30,000 ft to remain visible. Between the limits of 20 and 200 miles, the choice of radar range for line or area coverage is dependent upon a number of economic considerations.

For line coverage, the radars will be much smaller and cheaper if they are limited to the minimum range and located at spacings of 30 to 40 miles. Table 16.1 shows a comparison of the requirements for search radars de-

Table 16.1 COVERAGE AND POWER REQUIREMENTS FOR SEARCH LINE

Maximum range R_m (miles)	Station spacing (miles)	Main beam elevation width E_m	Sin E_m	Relative search time t_s	Relative R_m^4	Relative $P_{av}A_r$	Relative number of radars	Relative System $P_{av}A_r$
20	35	35°	0.57	1	1	1	1	1
40	70	16°	0.28	2	16	4	0.5	2
80	140	8°	0.14	4	256	16	0.25	4
160	280	2.5°	0.044	8	4000	42	0.125	5.2

(Note that altitude of gap between radars increases from 50 ft for 35 mi spacing to 450 ft for 70 mile, 2500 ft for 140 mile, and 11,500 ft for 280 mile spacing.)

signed around ranges of 20, 40, 80, and 160 miles range, when used in a search line. As the range is increased, the fourth-power relationship of power-aperture product is offset, in part, by the reduction in required solid angle of search (to maintain a constant search line height). The relative coverage patterns are shown in Fig. 16.4. Search time is also increased for targets of given velocity. Use of the cosecant-squared type of pattern

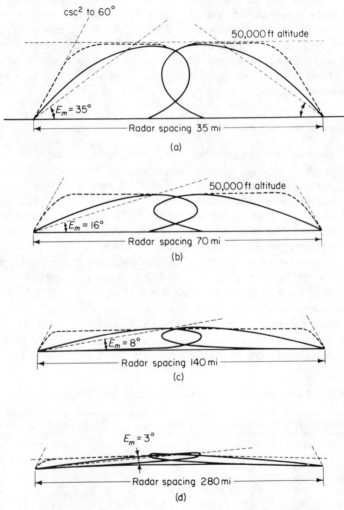

Figure 16.4 Relative vertical coverage for $R_m = 20$ to 160 miles. (a) $R_m = 20$ miles. (b) $R_m = 40$ miles. (c) $R_m = 80$ miles. (d) $R_m = 160$ miles.

in the longer-range radars is relatively inexpensive in power, as it restricts the long-range coverage to a relatively narrow solid angle.

In assessing the relative cost of using the different radars, the fixed costs of establishing each site must be added to the cost of the transmitter and antenna, which vary with radar range. The receiving equipment of the radar, along with communications terminals and many other types of facilities needed to operate the radar, will be almost independent of the radar range. In addition, the cost of developing and constructing the higher-power transmitters or larger antennas may not be related in any simple way to the power-aperture product. Curves of cost vs. power and cost vs. aperture areas, at different radar frequencies, are usually discontinuous, as a result of past development of components in specific bands with characteristics to match particular systems. Attempts to reduce the choice of the most economical radar system to an exact science have not proven successful in the past, and we shall present no such approach here. Beyond noting that the number of radars required will vary inversely with their range in the search line, and with the square of their range in area coverage, we shall omit economic formulas and proceed to two more technical factors.

Space Diversity

Overlapping coverage with considerable depth offers a number of advantages, which make it undesirable to limit the number of radars and the width of the search line to their minimum values. In a network of search radars with overlapping coverage, it is not necessary that any single station achieve a very high probability of detection. Furthermore, by allowing a reasonable depth to the warning line, each radar has many independent opportunities to detect the target. These two factors may lead to substantial reduction in the power-aperture requirements for each radar, especially when the target signal is subject to fluctuation. Two examples will illustrate this point.

In an area surveillance system, the track-while-scan device will be subject to lag errors which increase as the square of the time between data points (see Chapter 6). The system may be designed to tolerate loss of a single sample in a sequence, but may lose track when two successive samples are lost owing to failure of the search radars to detect the target. Let us assume that a track must be carried for 300 sec, with a data rate of 1 per 5 sec and with a probability of 90 per cent that the target will not be lost during the track. If the single-scan detection probability is given by P_d, the probability of not detecting on each scan is $(1 - P_d)$, and the probability of not detecting on two successive scans is $(1 - P_d)^2$, if there is no correlation between signal fluctuations over the scan period. With

60 possible opportunities for losing the target, we can write (approximately)

$$P_f = 60(1 - P_d)^2 = 0.10$$

where P_f is the specified 10 per cent probability of losing track during the 5 min period. The required value of P_d is thus 96 per cent, and the fluctuation loss for a Rayleigh target is 11 db (Fig. 1.13). By adding a second radar with an independent path to the target, the single-scan P_d required for each radar will be reduced to 80 per cent (it is assumed that data from either radar will be accepted by the track-while-scan device). The new fluctuation loss will be only 5 db, for a reduction of 6 db in individual radar power. If it were to be assumed that transmitter power were the determining factor in radar cost, and that cost were proportional to transmitter power, the overall cost of the system would be reduced by a factor of two through the use of two radars with space diversity. The same results could not be achieved merely by doubling the power of the single radar and scanning at twice the rate, since the target signal might well be correlated over periods of 5 or 10 sec, and the radar's failures in detection would tend to come in pairs. Similar considerations apply to cases where there is a fading of the signal owing to reflection or interference phenomena along the path.

As a second example of space diversity in radar detection, consider a missile defense system which is expected to warn of the approach of medium-range missile warheads when they are several hundred miles from the defended area. If the warhead is stabilized, it may present a very low, constant radar cross section in the solid angle around nose aspect (e.g., 0.001 sq m or less, depending on the frequency of radar operation).* Radars which can see the target only from nose aspect (see Fig. 16.5) will have to operate with very large power-aperture products. A network of search radars scanning overlapping regions from widely dispersed sites would provide a high degree of aspect diversity, and the applicable radar cross section for range computation might well rise to equal the optical cross section of the warhead (e.g., 0.5 to 1.0 sq m). The "diversity gain" of the dispersed system might be as high as 30 db, which would greatly outweigh other economic factors in radar design. The difficulty and cost of establishing the necessary sites and communication links might also be less than the radar cost for single-site operation of the system.

In general, whenever system specifications call for performance under some extremely disadvantageous condition of target cross section, propagation loss, reliability, or other statistical factor, some diversity scheme should

---

* E. M. Kennaugh and D. L. Moffatt, "On the Axial Echo Area of the Cone-Sphere Shape," *Proc. IRE*, **50**, No. 2 (Feb. 1962), p. 199.

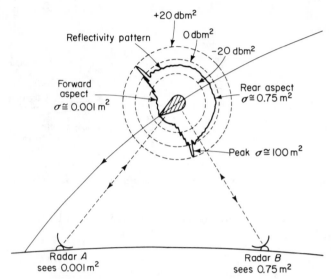

Figure 16.5 Increased target cross section using space diversity radar system.

be considered as an alternate to the "brute force" approach. In a system which has a "null point" in performance, it is almost always more economical to shift out of the region of the null, rather than to fill in this null with more power. Space diversity is one such technique, along with frequency diversity and equipment redundancy, both of which have received much discussion in connection with reliability and communications engineering.

Bistatic Search Radar

A special case of geographic diversity is the bistatic search radar, in which the transmitting and receiving sites are widely spaced and the target is forced to travel between the two sites. One advantage of this arrangement is that it eliminates the direct transfer of power from the transmitter to the receiver, thus permitting various c-w modulation schemes to be used which would be impractical in a monostatic radar. A second advantage arises when the two beams arrive at the target with an angle of intersection near 180° (see Fig. 16.6). For this case, the bistatic cross section may be increased by a large factor, as compared with the normal, monostatic radar cross section of the target.* The increase is due to the relatively

* Merrill I. Skolnik, "An Analysis of Bistatic Radar," *Trans. IRE,* ANE-8, No. 1 (March 1961), pp. 19–27.

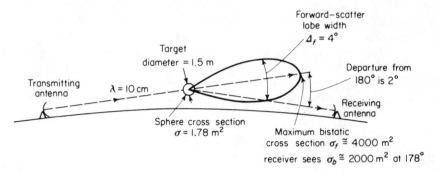

Figure 16.6   Bistatic cross section enhancement by forward scatter.

large "forward scatter" cross section of the target, shown by Siegel* to be equal to

$$\sigma_f = \frac{4\pi A^2}{\lambda^2} \tag{16.1}$$

where A is the projected area of the target and λ is the radar wavelength. Thus, we may regard the target as a flat antenna, oriented normal to the transmitting radar beam, reradiating energy in the forward direction (along the radar beam). As the wave front passes the target, the "hole" caused by the presence of the target will have the effect of negative illumination over this projected target area, and will lead to a lobe structure in the forward-scattered component identical to that of the equivalent flat antenna. The maximum, along the radar beam, will correspond to the cross section given by Eq. (16.1), and this main lobe will have a width (to the first null, or between half-power points) given by $\Delta_f = \lambda/d$ rad, where d is the diameter or length of the target measured in the plane in which Δ_f is defined.

In the example diagrammed in Fig. 16.6, the beams intersect at an angle of 178°, or 2° from the optimum. The lobe width Δ_f for enhanced bistatic operation must be in the order of 4°, if it is to cover the receiving site, and the enhancement of bistatic cross section σ_b when compared to the average backscattering cross section σ will be about 1000, or 30 db. This condition is met when the wavelength is about 1/15 of the target dimension. As the target rises in altitude, the beams will intersect at a smaller angle (farther removed from 180°), and the ratio of wavelength to target size must be increased in order to broaden the forward lobe. This will, of course, reduce the amount of gain available. As a result, this special ad-

* K. M. Siegel, "Bistatic Radars and Forward Scattering," *Proc.* 1958 *Natl. Conf. on Aero. Electronics,* Dayton, Ohio, pp. 286-90.

vantage of bistatic operation is available only in a limited class of search problems.

When looked at in the horizontal plane, the aircraft target may appear greatly elongated, leading to a narrow forward-scatter lobe. Bistatic operation will then provide great enhancement when the target crosses the line connecting the two radar sites.

16.2 TRACKING RADAR NETWORKS

Geographical limitations in station location play an important part in the design of systems which use tracking radar. Some of the considerations are the same as those discussed earlier for search radar, involving a compromise between the maximum radar range and the number of radars and sites required to cover a given line or area. The considerations of spatial diversity, for overcoming certain losses, may also apply to tracking systems. Networks of tracking radars have been used primarily for tracking of missiles, aircraft, or satellites during test and evaluation of these vehicles. We shall discuss the considerations which led to development of tracking radar networks, and some of the advantages which are gained when special network geometry is available.

Chain Radar Systems

The first chain radar system for missile tracking was established at White Sands Proving Ground in 1948, using two modified SCR-584 radars. Data were transmitted between the radars in the form of synchro voltages, multiconductor cable being used to cover a distance of about 50 miles. An analog computer was used for parallax correction of data, permitting one radar to designate targets to the second. The system was used to assure continuation of missile tracks which originated at a launcher near the first radar and terminated within the coverage of the second. Within a few years, the system was expanded to include five major radar sites, each with two or more radars. Data for operation of the chain could originate at any one of the sites, and could be used to plot the course of the target as well as to designate the target to the other radars. During an operation, the origin of the data could be switched to that site which was obtaining the best tracking data, and the remaining sites could continue to track independently or by following the chain designation data. Similar chains have since been placed in operation at the other major missile test ranges. One of the longest operates from Cape Canaveral, Florida, to the tip of South Africa, a distance of some 8000 miles, and is extended over the Indian Ocean by instrumentation ships. The National Aeronautics and Space Ad-

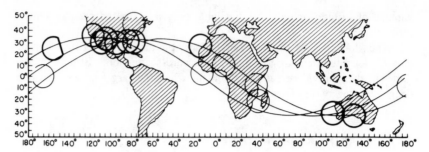

Figure 16.7 Mercury tracking radar and telemetry network—heavy circles show radar coverage, May 1962 (after Donegan and Jackson).

ministration, with cooperation from the military test ranges, established the Mercury Tracking Network (Fig. 16.7) for use with the first U. S. manned orbital flight tests. The operating principles and results of this network have been described in a widely circulated NASA report.*

The radar chain consists of a series of radar sites stretched in a line along the path of the test vehicle. There may be considerable overlap between stations, as at White Sands or the uprange portion of the Atlantic Missile Range, and there may also be gaps in the coverage, as in the region between the West Indies and Ascension Island in the Altantic, or between most of the Mercury stations. In general, data from one radar at a time are selected to operate the chain, and these data are used for display at the central control point. When adequate data-transmission and computing facilities are available, it may be possible to combine the data from two or more radars by simple averaging, improving the accuracy by a factor equal to the square root of the number of radars used. In other cases, weighting factors may be applied to the data from different stations or radars, tending to select the data which appears the smoothest or the most accurate at a particular time. These weighting factors will usually make a rapid transition from zero to unity and vice versa, so the best source of data dominates the output as in the case where simple selector switching is used to choose the source. The best method of combining data is one in which the individual target coordinate data are weighted according to estimated accuracy, permitting data from two or three stations to be combined with considerable improvement in accuracy. This approach leads to a "trilateration" solution, as described below.

* James J. Donegan and James C. Jackson, "Mercury Network Performance," Sect. 2 of NASA Report SP-6, *Results of the Second United States Manned Orbital Space Flight, May 24, 1962* (Washington, D.C.: U.S. Govt. Printing Office, 1962), pp. 15–26.

Radar Trilateration Systems

It has been recognized that radar range data are inherently more accurate than the angular data, at ranges beyond a few miles. This has led to numerous attempts to use the range data from two or three stations in deriving target position by a process of trilateration, in which the lengths of the sides of the measurement triangles serve as the input to the computer. A two-dimensional representation of the process is given in Fig. 16.8, showing the improvement which can be expected as compared to the range-angle solution. The range error remains essentially constant with increasing range, whereas the product $R\sigma_\theta$, representing the error component due to angular error σ_θ, increases in proportion to range. Whenever the target range exceeds the ratio σ_r/σ_θ, there will be an improvement in accuracy of measurement when an optimum pair of radar sites is used to determine position by trilateration.

When the two radar beams intersect at a right angle [Fig. 16.8(a)], the error in target position may be described as a circle of probable error

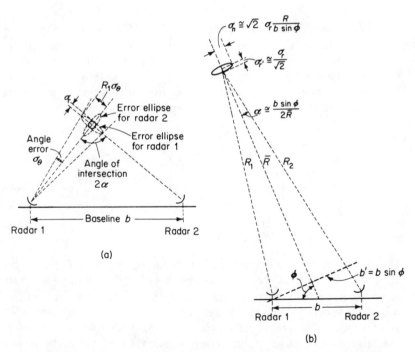

Figure 16.8 Radar range trilateration. (a) Comparison of trilateration with range-angle error at short range. (b) Trilateration error at long range.

(or of rms error), whose radius is equal to the error of the individual radar range measurement. In other cases, the trilateration error will appear as an ellipse whose major axis has an rms value given by the range error times two factors.

$$\sigma_n = \sigma_r C_c \times (\text{GDOP}) \qquad (16.2)$$

The first factor, C_c, depends on the correlation between the two range errors, and is equal to $\sqrt{2}$ for radars which are separated enough to have no error correlation. The second factor is the "Geometrical Dilution of Precision," which depends upon the angle at which the beams intersect. With reference to Fig. 16.8b, where the angle of intersection is represented by 2α, the exact expressions for the two components of position error are

$$\sigma_{r'} = \frac{C'_e \sigma_r}{2 \cos \alpha} \qquad (16.3)$$

$$\sigma_n = \frac{C_e \sigma_r}{2 \sin \alpha} \qquad (16.4)$$

When the beams intersect at a right angle ($\alpha = 45°$), both $\sin \alpha$ and $\cos \alpha$ become equal to $1/\sqrt{2}$, and the errors are equal to σ_r, as stated above.

As the radar range increases, the angular separation of the two beams decreases, amplifying the error component σ_n [see Fig. 16.8(b)]. We may now describe the system in terms of an equivalent range-angle measuring instrument, whose effective radial and normal errors are given by

$$\sigma_{r'} = \frac{C'_e \sigma_r}{2} = \frac{\sigma_{\Sigma r}}{2} \qquad (16.5)$$

$$\sigma_n = \frac{C_e \sigma_r}{2 \sin \alpha} \cong \frac{\sigma_{\Delta r}}{2 \sin \alpha} \cong \sigma_{\Delta r} \frac{\bar{R}}{b \sin \phi} \qquad (16.6)$$

$$\sigma_{\theta'} = \frac{\sigma_n}{R} \cong \frac{\sigma_{\Delta r}}{b \sin \phi} \qquad (16.7)$$

The term $\sigma_{\Sigma r}$ represents the rms error of the range sum ($R_1 + R_2$), whereas $\sigma_{\Delta r}$ is the rms error of the range difference ($R_1 - R_2$). When the errors are not correlated, both these terms are equal to $\sqrt{2}\,\sigma_r$.* As an example, consider a system where uncorrelated range errors in R_1 and R_2 are equal to 10 ft rms, and where the radars are located 50 miles apart (300,000 ft). For targets whose range is appreciably longer than 50 miles, the system

---

* The factor $C'_e = \sigma_{\Sigma r}/\sigma_r$ is related to u, the correlation coefficient of the two range errors by $C'_e = \sqrt{2(1 + u)}$, whereas $C_e = \sigma_{\Delta r}/\sigma_r = \sqrt{2(1 - u)}$.

radial error will be 7.07 ft, and the normal error will be equivalent to an angle error $\sigma_{\theta'} = 4.7 \times 10^{-5}/\sin \phi$ rad. When the target is in a direction normal to the baseline, the angle error will be 0.047 mil, or about half that of the best angle-tracking antennas. However, in order to obtain substantial improvement over angle tracking, the baseline must be increased further, or the range error reduced.

The principles discussed here for the two-dimensional case apply also to tracking in three dimensions, although it becomes more difficult to find three radar sites such that the beams intersect orthogonally. In general, it will be found that the target position error appears as an elongated ellipsoid, whose major axis is oriented in a direction parallel to the shortest radar baseline (see Fig. 16.9). The three-dimension case may be solved by breaking it down into a pair of two-dimensional triangles, in planes normal

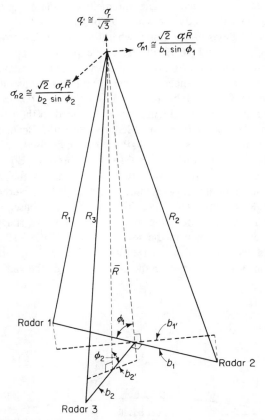

Figure 16.9 Approximate solution for three-dimensional trilateration errors.

to each other. The first plane is passed through any two stations and the target, and contains a baseline b_1 oriented at an angle ϕ_1 relative to the direction of the target. The second plane is passed through the target and the third station, at right angles to the first plane, and contains a baseline b_2 at an angle ϕ_2 from the target direction. The exact relationships for error are quite complex, but approximate values may be found by solving the two triangles separately, using Eqs. (16.3) through (16.6), and assuming that the first pair of radars is equivalent to a single radar at the point where the second baseline meets the first.

When the baselines are very long and the target is at an altitude less than the baseline, the angle 2α may approach $180°$ (see Fig. 16.10). In this case, error estimates may be made from Eqs. (16.3) and (16.4), or by assuming that one of the radars is moved to an effective position above the first, creating a short, vertical baseline. The errors are the same, regardless of the direction in which the measurement is made along the line BB'. Studies have shown that radar networks on baselines of 500 to 1000 miles can cover broad regions of the globe, when the targets lie above about 100 miles. Figures 16.11 and 16.12 illustrate one possible application of the radar trilateration technique for the region near Cape Canaveral. In this case, the region near the launcher is covered by two stations on the Florida coast and one on Grand Bahama Island. As the target rises and moves downrange, a station on Grand Turk Island is added to the network, along with the Bermuda station from the Mercury network. Beyond 600 miles from Cape Canaveral, the station on Puerto Rico is used, in conjunction with Grand Turk and Bermuda. The GDOP factors [see Eqs. (16.3) and (16.4)], calculated exactly on a digital machine, are in the order of two or three over most of the region, indicating that the measurement beams intersect at angles between $150°$ and $160°$ for the worst baseline. Resulting errors in target position, including both effects of radar range error and of station survey, are shown in the solid curves of Fig. 16.12.

The use of trilateration in a case such as this will provide moderate improvement in position data over the combination of the same radar sites

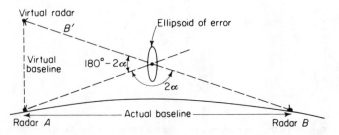

Figure 16.10 Trilateration solution for low-altitude targets.

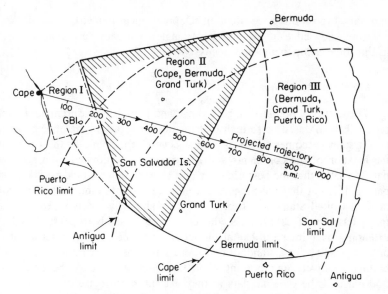

Figure 16.11 Proposed AMR trilateration network.

in an averaged solution based on range and angle data. A more significant improvement is obtained when velocity components are measured by using Doppler tracking. The dashed curves on Fig. 16.12 show the results obtained when the Doppler velocity data have errors of 0.1 ft per sec. This accuracy can be obtained with measurement times as short as a few tenths of one second. The velocity errors shown are more the result of station

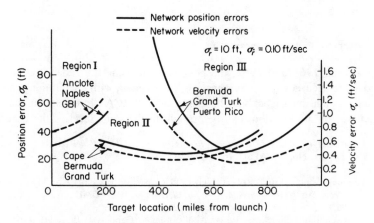

Figure 16.12 Accuracy of AMR trilateration network.

and target position error than of Doppler measurement, since lack of accurate knowledge of the position triangles will cause a systematic error in the three components of velocity.

Effect of Site Surveys

The accuracy obtained in both range and Doppler measurements, tracking radar being used, is so high that the accuracy of long-baseline trilateration systems becomes limited by knowledge of the shape of the earth and the location of the continents and islands. It was estimated by a scientific panel in 1961 that the best accuracy of a short baseline would be about one part in 10^7, but that the range between more distant points within the continental United States would be in error by about five parts in 10^6.* Thus, for a baseline of 2000 miles, the error would be about 60 ft. Baselines extending over intercontinental distances would be known to an accuracy of 100 to 300 ft, as a standard deviation. Thus, the radar range error would become relatively unimportant when the baseline extended beyond about 500 miles, if the geodetic data as they existed in 1961 are used.

The same scientific panel estimated that the accuracy of baselines within the continental United States could be improved by a factor of five, by using then-existing techniques and a great deal of time and effort. With some improvement in technique, it would be reasonable to assume that survey errors could be brought to the same level as the existing radar range error for baselines of 1000–3000 miles on a single continent. This error would take the form of a fixed bias, since the estimated drift of the land within a continent is less than a few feet over a period of many years. Major improvement in the acuracy of geodetic measurements appears likely during the 1960's, as a result of satellite measurements such as those outlined below.

Geodetic Radar Networks

During 1961 and 1962, the possibility of using the radar itself to improve the accuracy of site surveys was investigated, with encouraging results. Studies made by Lieber and his associates at RCA† have shown that the accuracy of site locations can be brought to the same level as the accuracy

---

* National Academy of Sciences, *Report of the Ad Hoc Panel on Basic Measurements*, Report No. ACAFSC 102, Contract AF 18 (600)-1895. Washington, D.C. (Dec. 8, 1961).

† Sidney Shucker and Robert Lieber, "Comparison of Short-Arc Tracking Systems," *Proc. 5th Natl. Conv. on Military Electronics*, IRE-PGMIL, Washington, D.C. (June 1961).

A. Reich and S. Shucker, "Radar Applications to Geodesy," *Proc. 9th East Coast Conf. on Aero. and Nav. Electronics*, Baltimore, Md. (Oct. 1962).

of radar range measurement, by using data obtained from coordinated tracking of a suitable calibration satellite by a world-wide radar network. Starting with a few well-surveyed sites in the continental United States, the network of coordinated sites would be extended through simultaneous tracking of the satellite (the method of "intervisibility"). The position of the satellite at each moment of the track would be established by range-only trilateration, and the location of the unknown station would be found by a process of "inverse trilateration" using the known target positions and the radar range data from the unknown station. Each new station would thus be located in three coordinates relative to the existing stations. Over a period of time, the orbital parameters would then be refined with reference to the center of the earth and the gravitational field, and the location of all stations would then be referred to the earth-centered coordinate system. In the process, the higher-order terms of the gravitational field would be determined. An important factor in the radar geodetic network would be that the relative location of stations would be measured directly, rather than deduced from the shape of the earth as determined by the gravity field.

The procedure described takes advantage of the sensitivity of the trilateration network to site location errors. Since these errors are bias errors, they may be evaluated by combining the results of several independent measurements of satellite position, some of which are based on stations whose relative locations are known. The satellite provides a desirable target for this network, both because it is high enough to assure intervisibility, and because its motion follows a regular pattern. Even when the higher-order perturbations in the orbital parameters are not known exactly, the motion over a short period is regular enough to permit the smoothing of radar noise errors. As the orbital perturbations become better known through continuous tracking, the radar range data may be fitted to the known orbit to adjust the site coordinates to even higher accuracy, any true bias in radar range measurement being eliminated and data being provided which are limited in accuracy only by the low-frequency components of radar range error. World-wide surveys to within one part in 10^6 are expected to result from use of this technique.

16.3 INTERFEROMETER RADAR

During the late 1940's and the 1950's, development was carried forward on several types of tracking system using the interferometer principle. This type of system is related to the radar trilateration system, but is usually operated with c-w signals provided by a cooperative transponder device. Measurement of range is based on phase-comparison of modulation frequencies which are superimposed on the c-w carrier, or equivalently on phase-comparison of two carriers whose frequency separation is varied in a sawtooth fashion. Angular measurements are based on comparison of the r-f carrier phase, rather than on modulation components. The resulting sensitivity to small changes in target location represents a considerable improvement over those tracking devices which measure angle of arrival by averaging phase over a small antenna aperture.* Figure 16.13 shows a simple interferometer.

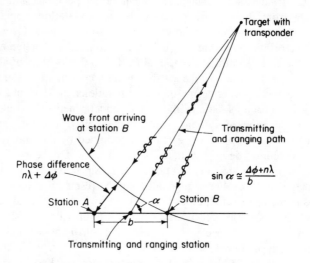

Figure 16.13   Interferometer system geometry.

Precision vs. Baseline Length

The theoretical precision of the interferometer was given by Eq. (2.17), in terms of aperture size b or interference lobe width $\Delta\alpha$.

$$\sigma_\alpha = \frac{\lambda}{\pi b \sin \alpha \sqrt{\mathcal{R}}} = \frac{\Delta\alpha}{\pi \sqrt{\mathcal{R}}}$$

* R. S. Grisetti and E. B. Mullen, "Baseline Guidance Systems," *Trans. IRE*, **MIL-2**, No. 1 (Dec. 1958), pp. 36–44.

D. C. Prim and L. N. Lawhead, "Quantitative Performance Evaluation of a Position and Velocity Range Measurement System," *IRE NEREM Record* (1960), pp. 72–73.

E. Gehrels and A. Parsons, "Interferometer Techniques Applied to Radar," *Trans. IRE*, **MIL-5**, No. 2 (April 1961), pp. 139–46.

John T. Mengel, "Tracking the Earth Satellite, and Data Transmission, by Radio," *Proc. IRE*, **44**, No. 6 (June 1956), pp. 755–60.

The parameter $\mathcal{R}$ represents the energy ratio of the signal measured over the total observation time for which σ_α is to be evaluated

$$\mathcal{R} = \frac{2E}{N_o} = \frac{2S_{av}t_o}{N_o}$$

The operation is equivalent to that of the range trilateration system at long range [Eq. 16.7)], where the range-difference error $\sigma_{\Delta r}$ is replaced by the term $(\lambda/2\pi)\sigma_{\Delta\phi}$. The phase-difference error $\sigma_{\Delta\phi}$ in radians is given by

$$\sigma_{\Delta\phi} = \sqrt{2}\,\sigma_\phi = \frac{\sqrt{2}}{\sqrt{\mathcal{R}_1}} = \frac{2}{\sqrt{\mathcal{R}}} \text{ rad} \qquad \textbf{(16.8)}$$

Thus, the single-station phase-measurement error σ_ϕ is simply $1/\sqrt{\mathcal{R}_1}$, where $\mathcal{R}_1$ is the energy ratio for one of the two stations, or one-half the system energy ratio $\mathcal{R}$. In a matched receiver system, the energy ratio may be equated to the signal-to-noise power ratio. When this ratio is reasonably high, the equivalent range-difference error will be a small fraction of the system wavelength, and great precision will be obtained.

Interferometers operating in the microwave region, with baselines of thousands of feet, provide a theoretical precision of a few microradians. Electrical circuits for measuring r-f phase to the required accuracy have been available for some time, and the interferometer technique has been believed to offer an attractive way to avoid the limitations of the smaller tracking antennas used in radar. In practice, for the reasons given below, the system accuracy of the interferometer seldom approaches the precision limit owing to thermal noise, as calculated above.

GDOP Factors for Interferometers

Most interferometer systems use a pair of orthogonal baselines, arranged in an X or L configuration. The angular sensitivity of such a system is constant as a function of azimuth angle, and geometrical dilution of precision (GDOP) is held to a minimum. However, the full accuracy of the system is available only for targets at zenith, and the effective length of the baseline for elevation measurement will decrease as the target nears the horizon. For a system using a pair of horizontal baselines of length b, the effective baseline for measurement of elevation will be

$$b' = b \sin E \qquad \textbf{(16.9)}$$

Thus, at an angle $E = 6°$, the effective length will be reduced to one-tenth the physical baseline b. The corresponding GDOP factor is equal to the cosecant of the elevation angle, and goes to infinity at the horizon. Addition of a vertical baseline will avoid this increase in error, but it is difficult to construct a vertical baseline of great length and to control its length with the accuracy required.

Limitations in Baseline Stability and Propagation

The first major problem to be overcome in using the precision microwave interferometer is caused by variation in length and orientation of the baseline. Survey of the mechanical elements of the baseline to within one part in 10^6 appears to be feasible, and construction of foundations for the instruments which will preserve this accuracy is dependent upon the stability of the ground at the instrument site. More difficult, but not impossible, is calibration of the electrical length of the baseline to this accuracy, and correction for variations which may occur during operation. Water-cooled wave guide, filled with dry air at a controlled pressure, has provided the answer in some cases. Microwave links have also been used successfully, when equipped with compensation systems to cancel the effect of variable propagation velocity over the path between the ends of the baseline. When the antennas at the ends of the baselines are moved to track the target, as is usually done in microwave systems with high-gain antennas, the phase-center of the antennas must be carefully controlled to avoid change in the baseline length. The error in target position owing to an uncorrected change σ_b in baseline length is given by

$$\sigma_n = \frac{R\sigma_b}{b \sin \phi} \tag{16.10}$$

where R is the target range and ϕ is the angle between the baseline and the direction of the target.

Other significant sources of error in interferometer systems, apart from equipment design limitations, are those caused by multipath and atmospheric refraction. The multipath term can be kept low by using directional antennas with low side lobes, and by extending the baselines. The refraction error was described in Chapter 15, where it was shown that the low-frequency components of refractivity change would place a severe limit on the accuracy of all systems for measurement of angle of arrival. The improvement in accuracy of interferometers over existing radars appears to be limited to a factor of about ten, and most such systems have provided an improvement of only two or three over tracking radar. As the length of the baseline is extended to overcome the tropospheric fluctuations, the distinction between the interferometer and the trilateration system using range and Doppler data becomes lost, and the two systems provide equivalent performance, limited only by the atmosphere and by the accuracy of site survey and equipment calibration. The actual interferometer equipment configuration is more complex than shown in Fig. 16.13, because of the need for resolving the ambiguities that exist in phase-difference data. In some cases, the errors in the system will be be large enough to prevent identification of the proper lobe in the pattern.

Our ability to predict the performance of a radar in search or tracking applications is no better than our knowledge of the important parameters of the radar: transmitter power, receiver sensitivity, antenna gain, etc. In this chapter, we shall review briefly the procedures which are available for measurement of these parameters, and the limitations in measurement accuracy. Then we shall consider the means by which the total performance of the radar may be evaluated, to check on the predicted results. Both search radar testing and tracking radar testing will be discussed, and examples will be given of successful test procedures which have been developed for these two types of systems.

17.1 RADAR PARAMETER MEASUREMENT

The first step in experimental evaluation of a radar system is to measure the important radar parameters on an individual basis, to check against the values given in the specification.

RADAR TESTING AND EVALUATION

17

This is important in many cases, because the characteristics of the various elements of the radar may be specified in ways which are difficult to relate to system performance. During the design and development process, these specifications may have been subject to interpretations, modifications, and compromises, many of which will not appear until the radar is tested as a unit. The following procedures illustrate the type of testing which is normally carried out on individual elements of the radar after they have been assembled into the system. The same tests should be repeated at intervals during the life of the equipment, to keep track of changes which occur as a result of component aging, wear and tear, and environmental changes.

Transmitter Parameters

Assuming that a successful design has produced a transmitter which is safe, reliable, and able to withstand the environmental conditions which are encountered at the radar site, the parameters which are of importance in system evaluation are as follows:

1. Peak output power P_t.
2. Pulse width τ.
3. Repetition rate f_r.
4. Average power P_{av}.
5. Spectral distribution of power.
6. Tuning range and stability.
7. Phase and amplitude stability.
8. Spurious radiation.
9. Internal losses of power.

Many radar transmitters include built-in test facilities for measurement of at least some of these quantities. A directional coupler, waveform monitor, and frequency meter are generally available, and additional external test equipment can be connected to perform the remaining tests. A typical transmitter test program would include the steps shown in Table 17.1. The calorimeter is preferred as a means of measuring output power, because it requires less fixed attenuation between the transmitter output and the measuring device. If a thermister or bolometer is used, it will usually be necessary to insert attenuation of 40 to 60 db at the input to the meter, introducing a possible error of 1 or 2 db. Direct measurement using a calorimeter capable of dissipating the full output of the transmitter will yield results to a few per cent, or about 0.2 db, in average power output. Some means of measuring spurious or harmonic output should also be provided.

In measurement of pulse width, some care is necessary to assure that the reference amplitude of the pulse is properly chosen. Figure 17.1 illustrates the process by which pulse width and reference amplitude are related to the peak power level calculated. The 50 per cent amplitude level is chosen as the point at which the pulse width will be measured, and in

Table 17.1 TEST PROCEDURES FOR TRANSMITTER PARAMETERS

Measured parameter	Instrumentation	Derived parameters
Average power P_{av}	Calorimetric power meter	Pulse energy $P_t\tau = P_{av}/f_r$
Repetition rate f_r	Oscilloscope	
Pulse width τ	Oscilloscope	Peak power $P_t = P_{av}/\tau f_r$
Spectral width (envelope) B_t	Spectrum analyzer or echo box	Pulse width $\tau = 1/B_t$
Spectral shape	Spectrum analyzer or echo box	
Width of fine lines B_c	Coherent analyzer	
Amplitude of spurious lines	Coherent analyzer	
Internal loss	Attenuation test set	System power output

most cases the r-f pulse will approximate the trapezoidal shape closely enough to eliminate serious error. Where the pulse has been passed through a narrow-band filter, or where special shaping techniques have been employed to eliminate sideband energy, the peak power calculation may be in error by a somewhat greater amount. For this reason, the check on

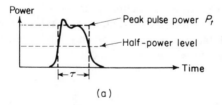

(a)

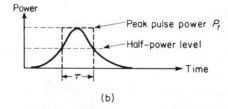

(b)

Figure 17.1 Relationships between pulse width and energy. (a) Wide band pulse waveform, $P_t\tau \cong$ pulse energy. (b) Narrow band pulse waveform, $P_t\tau <$ pulse energy.

pulse width using spectral width is valuable, as it tends to indicate the presence of error in the wave form measurement.

Equipment for measurement of the fine line structure in coherent transmissions has been available since about 1960. The test equipment scans the spectrum slowly, using a very narrow filter bandwidth (in the order of 1 cps). An associated recorder plots the spectral density, with appropriate logarithmic scales to present the width and relative amplitudes of the main sequence of lines and the spurious emissions at harmonics of the power-line frequency.

Measurement of internal losses may be necessary in cases where the basic power measurement is made at the output of the transmitting tube, rather than at the point where the power becomes available to the duplexer and antenna system. Several standard techniques are available which will hold the error in attenuation to the order of $\frac{1}{10}$ db or less, so long as the total attenuation is only a few decibels. One excellent method consists of direct power output measurements at the tube and at the duplexer or antenna, using the calorimetric power meter. This assures that the system will be well matched, and that only ohmic losses will be measured. As a supplement to this test, it may be desirable to measure separately the loss in power owing to reflection from unmatched antenna or other r-f line components. The test setup shown in Fig. 17.2, using a bidirectional coupler at the transmitter output terminal, provides for such measurements as well as for output measurements (with relatively low attenuation between the transmitter and the power meter). If the loss owing to the directional coupler is not excessive, the coupler may be left in place for frequent or even continuous monitoring of output and reflected power. If a 20 db coupler is used, the power loss will be only 1 per cent, and the attenuation can be measured to within about 0.2 db.

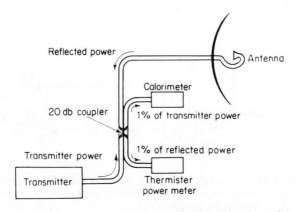

Figure 17.2 Power monitoring with bidirectional coupler.

Receiver Parameters

The radar receiver is usually characterized by its noise factor (or noise temperature) and its bandwidth. The distinction between the conventional noise factor and the "operating noise factor" used in the radar equation was discussed in Chapter 4, where the concept of receiver noise temperature was also introduced. As in the case of the transmitter, it is necessary to establish a particular point at which these parameters are measured, and to measure the attenuation between this point and the common point to which transmitter, antenna, and receiver are connected. From the standpoint of system evaluation, it would be best to make all measurements directly at that point. However, since the specification may require a certain value of "receiver noise factor," the measuring equipment is often connected in such a way as to by-pass much of the r-f line system and sometimes the duplexer as well. The tests required for receiver evaluation are shown in Table 17.2.

The wide-band noise source has come into widespread use for receiver evaluation, because it is insensitive to filter characteristics of the i-f amplifier. By adjusting the attenuation between the noise source and the receiver until the total noise output is double that of the receiver without the noise source, the noise factor of the receiver can be established with an accuracy of a few tenths of a decibel. When a low-noise receiver is used, it may be necessary to connect the receiver to a special "cold load" to establish its noise factor with this accuracy. The instructions which accompany the standard noise-factor instruments give procedures for elimination of errors owing to image response in the receiver, and various other procedures and techniques have been covered in the literature.*

Table 17.2 TEST PROCEDURES FOR RECEIVER PARAMETERS

Measured parameter	Instrumentation	Derived parameter
Noise factor $\overline{NF}$	Wide-band noise source	Input temp. $T_i = T_o(\overline{NF} - 1) + T_r + T_a$
Antenna temperature T_a	Radiometer	$\overline{NF_o} = (\overline{NF} - 1) + (T_r + T_a)/T_o$
i-f bandwidth B	c-w signal generator	Input noise $N = kT_o B\overline{NF_o}$
Input noise level N	c-w signal generator, i-f power meter	$\overline{NF_o} = N/(kT_o B)$ (check)
Minimum discernible signal $P_{\min}$	Pulsed signal generator	Input noise $N = P_{\min}/V_o L_c$

---

* See, for instance, A. J. Hendler, "Noise Figure Measurements, Definitions, Techniques and Pitfalls," *Topics in Noise*, Airborne Instruments Laboratory, Deer Park, N. Y (Aug. 1960).

IRE Standards on Receivers (1952).

Although the noise factor is of primary interest to the designer of the front end of the receiver, and is also the best indicator of deterioration of the crystal mixed or r-f amplifier, the systems engineer is equally concerned with the equivalent input noise power. This power level depends upon noise factor, antenna temperature, and i-f filter bandwidth. The check of i-f band-pass, using the c-w signal generator, provides the information from which the equivalent noise bandwidth B may be calculated. Typical curves for a single-pole and a stagger-tuned i-f amplifier are shown in Fig. 17.3, to illustrate the graphical procedure for determination of noise bandwidth.

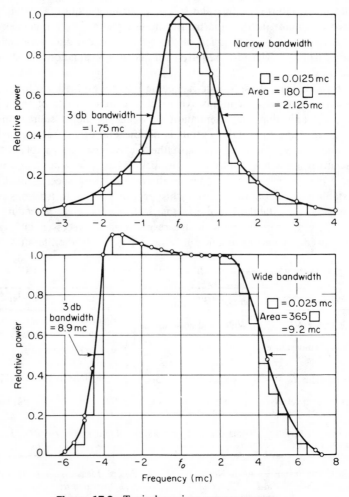

Figure 17.3 Typical receiver response curves.

The response is plotted on a scale where power and frequency are linear, and a reference level is chosen as representing unity power gain. The area under the curve is then found by counting the squares under the curve (including the portions which lie between the rectangular approximation and the actual curve) and multiplying by the value of the individual squares in units of relative power gain times frequency. In the examples shown, the noise bandwidths are wider than the half-power bandwidths, by a significant amount in the case of the narrow-band circuit and by a lesser amount in the stagger-turned case.

With the equivalent noise bandwidth of the receiver and the noise factor having been found, it is possible to calculate the equivalent input noise power. A direct check on this quantity can be made by using a calibrated r-f signal generator with a c-w output and an i-f power meter. The test setup and procedure are shown in Fig. 17.4. The power meter should be sensitive enough to measure the normal receiver noise output, without saturation of the amplifiers. By finding the level of the signal which doubles the output of the receiver, compared to noise alone, the equivalent input noise level may be measured directly.

A further test on the receiving and display system is provided by the "minimum discernible signal" test, in which a pulse signal is applied and adjusted until it is barely visible to the operator. Knowledge of the display visibility factor V_o (see Fig. 6.9) and of all collapsing losses may be used

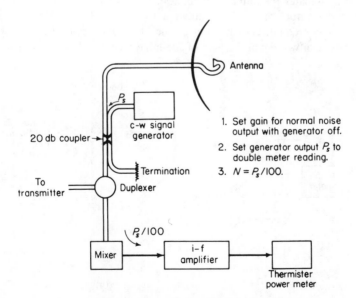

Figure 17.4 Input noise level measurement.

to compute the expected value of P_{min} as shown in Table 17.2. This will check for possible excess losses in the receiver and display circuits, if the noise level N is already known.

Both the noise level test and the MDS test may be applied to low-noise systems if a suitable directional coupler is used in the input transmission line. A 20 db coupler will introduce only about 3° K of additional noise temperature into the receiving system, and will cause a loss of 1 per cent in received signal. When carefully calibrated, the coupler will serve as an input channel for signal generator c-w or pulse signals, as used in these two tests.

Antenna Testing

The gain, pattern, and noise temperature of the antenna are of considerable importance in analysis of system performance. All these may be determined approximately from standard pattern tests, but in some cases it is desirable to perform more sensitive measurements of critical parameters. In fact, unless considerable care is taken in the test process, the results will be no better than design estimates of the various parameters. This is especially true of antenna gain measurements. Two methods will be described here.

Measurement of antenna gain can be carried out by comparison with the gain of a standard horn antenna, which is placed in the same field which illuminates the antenna under test. By inserting a calibrated attenuator in the output of the large antenna (see Fig. 17.5), the two outputs may be made equal, and the sum of the attenuation plus the gain of the standard horn will equal the gain of the large antenna. This method sounds

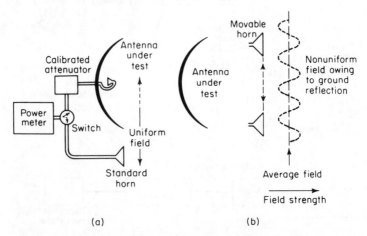

Figure 17.5 Antenna gain test techniques using standard horn. (a) Direct comparison gain test. (b) Method of obtaining average field.

simple and accurate, but it is subject to error because the field is not uniform over the large aperture. As a result of this, it is necessary to probe the field over the entire aperture and to average the results in order to obtain the output level from the standard horn. Astronomical sources are sometimes used to permit the antenna to be pointed upwards at a uniform field.*

The second measurement method is based on integration of the power gain over the entire sphere surrounding the antenna. Measured patterns used for this in one case are shown in Figs. 17.6, 17.7, and 17.8. Three

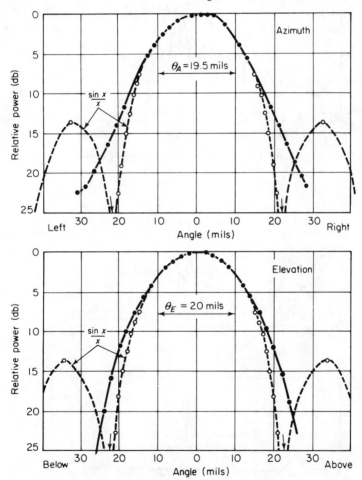

Figure 17.6 Typical main lobe antenna patterns.

* M. E. Armstrong, G. W. Swenson, R. L. Sydnor, and H. D. Webb, "The Use of Radio Noise from the Sun of Calibrating Radio Receiving Systems," *Proc. NEC*, **15** (1959).

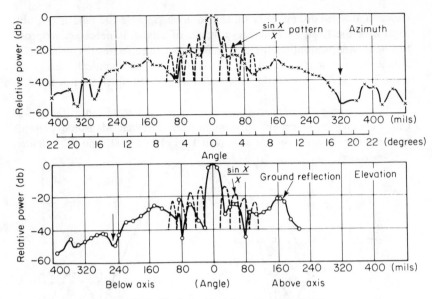

Figure 17.7 Typical side lobe antenna patterns.

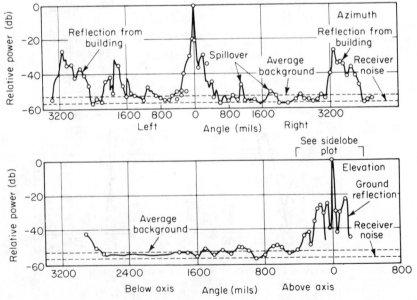

Figure 17.8 Typical far lobe structure, including reflections.

patterns were needed in this case to assure accuracy in measuring the solid angle of the various lobes of the pattern. Using the main-lobe power gain as a reference, we find the integral of gain over the sphere.

$$\psi_b = \iint_{4\pi} G_{\phi,\theta} \, d\phi \, d\theta$$

When the antenna pattern is circular in cross section, the integration may be carried out graphically in rings centered on the main lobe.

$$\psi_b = \sum_i 2\pi r_i G_i \, \Delta r_i$$

where r_i is the mean radius of the ith ring, Δr_i is the width of the ring, and G_i is the relative power gain in the ring. The absolute power gain at the center of the main lobe is given by

$$G = \frac{4\pi}{\psi_b}$$

When the patterns are plotted in mils, the conversion of mils^2 to steradians appears as a constant 0.96×10^{-6} in the equations for ψ_b. For the patterns shown, the radiated energy was distributed as follows.

Portion of pattern	Contribution to ψ_b (mils²)	Percentage of energy
Main lobe	440	63.3%
First side lobe (25 to 100 mils)	52	7.5%
Secondary lobes (25 to 250 mils)	144	20.6%
Spillover (near 1600 mils)	10	1.4%
Background level (-53 db)	50	7.2%
Total radiation pattern	696	100.0%

Main lobe gain $= \dfrac{4\pi}{696 \times 0.96 \times 10^{-6}} = 18{,}900$, or 42.6 db

The results of gain calculations using different formulas and of measurements using the standard horn and the integrated pattern may be summarized for our example as follows:

Antenna diameter $D = 12$ ft
Physical aperture $A = 113$ ft$^2 = 105{,}000$ cm^2
Wavelength $\lambda = 5.35$ cm
Ideal gain $= 4\pi A/\lambda^2 = 1{,}320{,}000/28.7 = 46{,}000$ or 46.6 db
Gain for 60% aperture efficiency $= 27{,}600$ or 44.4 db

Gain from measurements of beamwidth, using $G = 25{,}000/\theta_a\theta_e$
[or $k = 0.6$ in Eq. (4.41)]: 20,600 or 43.1 db
Measured gain (standard horn) $= 42.9$ db
Measured gain (pattern integration) $= 42.6$ db
Actual efficiency $= 40\%$

A major portion of the loss in gain for this particular antenna is accounted for by the relatively large secondary lobe structure beyond the first side lobe, which consumes over 20 per cent of the radiated power, even though the power density is below 30 db, relative to the main lobe. Further loss occurs because the beamwidth is somewhat broader than would be calculated from the relationship of $1.2 \, \lambda/D = 0.0175$ rad $= 1.0°$ [see discussion following Eq. (2.10)].

This example demonstrates some of the possible errors in measurement procedure. For instance, if the background level of the pattern had been increased from -53 db to -47 db, the contribution to ψ_b would be increased from 50 to 200 mils^2. This would reduce the gain by an additional 0.9 db. The only way in which this contribution can be evaluated is to have pattern records in which the main lobe lies at least 53 db above the noise level. The records shown here barely meet this requirement, so the figure of -53 db for average background level may be somewhat in error. However, the effect of small errors in this term on the over-all gain will not be large. Another source of possible error is in measuring the width of the main lobe, and its effective area. The data of Fig. 17.6 were taken at 1 mil increments, measured accurately by use of the precision pedestal on which the antenna was mounted. Had the measurements been subject to error of even 0.5 mil in beamwidth, the area and the gain would have been in error by about 5 per cent. Other errors can result from failure to account properly for reflected energy, which may appear as image lobes (see reflections from building, Fig. 17.8). These must be dropped from the calculation and replaced by the pattern which would have been observed in the absence of reflections. Presence of ground reflections also leads to error.

17.2 SYSTEM TESTING OF SEARCH RADAR

Knowledge of the individual parameters of the radar will permit fairly accurate prediction of over-all performance, but testing of the entire system against known targets will still be necessary to assure that all loss factors have been considered and to establish the actual coverage pattern. In search radar, the final performance will be measured by the ability to detect targets in specified regions of coverage, to make measurements of target position, and to eliminate undesired clutter and other competing signals.

Testing programs to verify the theoretical performance of the radar in these areas will be simplified if preliminary analysis has been performed to indicate the critical factors which must be evaluated experimentally. The tests can then be designed to isolate and measure the individual factors which are subject to question.

Search-Radar Loop Gain

The single-pulse S/N ratio has been used frequently in the preceding chapters as an intermediate quantity in prediction of system performance. One of the first steps to be performed in system testing is to see whether the actual S/N ratio obtained from a given target agrees with that value found from the radar equation. If agreement can be established, the remaining tests will be simplified. The problem of measuring S/N ratio in search radar is made difficult by the fact that it may be necessary to keep the antenna in continuous motion, and by the complex lobing structure resulting from surface reflections. In general, the indicators on the radar are not adequate for reading S/N ratio, and external test equipment must be set up for the test. The following illustrates one procedure which is applicable to scanning search radars, and which is even more accurate when applied to radars which may be controlled to point in the direction of a specific target.

The target used for evaluation should be of known cross section if the radar loop gain is to be measured. Lightweight, aluminum spheres have come into wide use in testing of tracking radars, and their application to search radar testing will lead to improved acuracy of results. The size and weight of the sphere must be held within strict limits, both to permit them to be carried aloft by small weather balloons and to avoid hazards to air safety when they are used. The standard 6 in sphere weighs a few ounces and has a radar cross section of 0.0183 sq m [modified over most of the radar bands by the resonance characteristic plotted in Fig. 3.4(a)]. The cross section is constant with aspect angle, and will not be influenced by the presence of the standard, 3 ft diameter weather balloon unless the surface of the balloon is wet. Other than its small cross section, the major limitation of the sphere-balloon target is that its path cannot be controlled except by adjusting the rate of rise and the release point. Figure 17.9 shows three typical paths, superimposed on the coverage pattern of a cosecant-squared search antenna. The rate of rise may be varied from a few hundred to 1000 ft per minute by inflating the balloon to different diameters, and the rate of lateral motion is dependent entirely upon wind speed. When suitable balloon paths cannot be obtained, or when the sphere cross section is inadequate, an aircraft will have to be used as a target. The highly variable cross section of the aircraft requires that many test runs be made,

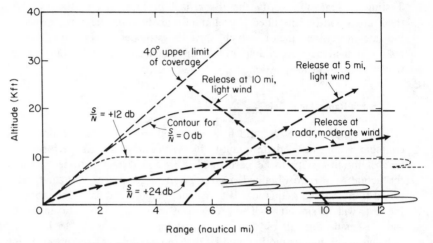

Figure 17.9 Typical balloon paths for search-radar testing (super-imposed on AN/UPS-1 coverage; see Fig. 8-5).

so that averages can be taken to approach the gain pattern of the radar itself. The accuracy of the tests is then dependent upon how well we know the average cross section of the aircraft over the aspect angles seen during the test runs.

The procedure for measuring S/N ratio is based on comparison with the output of a calibrated signal generator, connected in the same way as for noise level tests (see Fig. 17.4). The delay and amplitude of the signal generator pulse are varied to place the two signals side by side on an A-scope with matched amplitudes. In cases where an aircraft is used as a target, an attempt should be made to match the average value of the signal. The signal generator output level can be related directly to received power, or to S/N ratio when the noise level is known. In the region of the main lobe where reflection lobing is present, the variation in signal level should be plotted carefully as a function of elevation angle (determined from optical observations or known rate of rise of the target). This will permit an estimate of the free-space value of radar gain, as well as establishing the reflective properties of the surface under the radar path.

Coverage-Pattern Evaluation

Several sets of test data, covering a wide spread in elevation angle and surface conditions, may be combined into a plot of effective vertical coverage of the radar. Figure 17.10 shows such a plot, prepared from aircraft test runs in which the "blip-scan" ratio was observed. It is difficult to convert

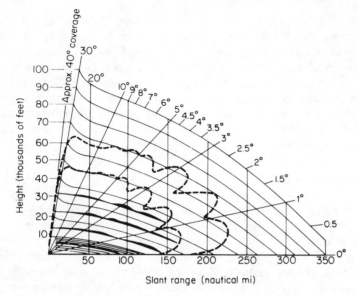

Figure 17.10 Typical results of aircraft flight tests plotted as relative blip-scan ratio.

these observations to S/N ratio with high accuracy, but they will serve to trace out the relative coverage pattern of the radar and will also indicate the actual detection performance of the entire system on the specific type of aircraft used. In the figure, the width of the shaded lines appearing at 15,000, 25,000, and 35,000 ft altitude is proportional to the blip-scan ratio, and the dashed plots of lobe structure have been estimated by fitting the theoretical reflection patterns to the data. The problem in deriving accurate coverage patterns from this type of data is that the signal strength on a given scan varies as a function of several parameters, whose effects must be isolated in data analysis. The aircraft cross section has a rapid, random variation in most cases, owing to small variations in aspect angle, propeller modulation, and similar effects. There will also be a systematic change in the average cross section as the elevation angle changes, owing to the fact that the aircraft flies at almost the same angle of attack, presenting a nose aspect at long range and changing to underside aspect as it flies overhead. Combined with these effects are the reflection-lobe structure, the statistical nature of the thermal noise and the operator's detection process, and the changing free-space pattern of the antenna. The reflection coefficient of the surface will vary with the type of surface beneath the path, with the surface contour, with time of day, and with moisture conditions. In the case of sea water, the height of the surface will follow a tidal variation, and the surface characteristics will depend on wind speed and direction,

534   RADAR TESTING AND EVALUATION   CH. 17

as well as distant weather which generates waves or swells. As a result of these many variable factors, it is important that the test program be set up in such a way that the results can be related to specific conditions, or that the conditions be varied over the same range of values to be encountered in field operation of the radar. The variable and often conflicting results of many search radar test programs are not surprising in the light of these factors, and the future use of spheres as targets is to be recommended as one way to narrow the uncertainties which are present in gain and coverage tests.

Probability of Detection

Regardless of the ability of the radar to achieve its expected S/N ratio on a given target, the most important measure of its performance is the ability to detect targets whose presence is not known in advance. Tests of single-scan detection probability or blip-scan ratio are often run on aircraft whose paths are under control of the test personnel. It should be recognized that such tests are not necessarily indicative of the performance of the radar and its displays in search and warning functions, since the attention span and search area are greatly reduced by concentration of the operator's attention on a small region known to contain the target. At the same time, the controlled tests may indicate the ability of the operator and equipment to carry tracks on aircraft whose presence has been detected. The difference in system loss can be estimated from the curves given in Chapter 6.

When automatic detection equipment is used, the probability of detection will give a consistent and accurate measure of radar performance. The approximate position of the target is known continuously, and the number of missed opportunities for detection may be established with accuracy. The false-alarm rate may also be measured with greater accuracy, over a prolonged period, and the parameters of the detection process are subject to better control during testing. For best results, the test data should be correlated with the vertical-coverage pattern to isolate the effects of target fluctuation and range from those of the reflection lobes.

Search-Radar Accuracy

The accuracy of range measurement in a search radar is usually measured on a subsystem basis by using a "range calibrator." Internal delays in the radar are relatively small compared to the errors in search radar displays or automatic tracking channels. The ability of the operator or automatic circuits to perform range measurements under various conditions of S/N ratio, scanning speed, and target fluctuation may be evaluated by using signal generators whose outputs are modulated in accordance with antenna patterns and other system parameters, or actual flight tests may be used. The fact that aircraft may be depended upon to fly along relatively stable flight paths, whereas most of the errors in the radar are independent from scan to scan, permits the deviations of the radar data to be compared with the average of the same data as a useful measure of error. When absolute error determinations are needed, the output of the search radar may be compared with an adjacent tracking radar, or with position measurements made optically.*

Azimuth measurements are best checked by comparison with optical data or tracking-radar outputs. Elevation data on aircraft are often obtained most accurately by relying on the aircraft's altimeter to determine height, and by calculating elevation from height and range data. In both azimuth and elevation measurements, it is necessary to synchronize the observations, and to control the conditions which affect the angular accuracy of the radar under test, in such a way that the test yields data on the particular processes which are under investigation. Otherwise, it may be impossible to extrapolate the test results to conditions encountered in field operations.

Clutter Rejection

Control of the clutter environment is so difficult that most MTI tests are run on a subsystem basis, using simulated signals for both clutter and moving targets. Field tests with the antenna fixed are capable of measuring the over-all stability of the transmitter, receiver, and cancellation circuits, if a genuinely stable target can be located at a range which provides a noise-free signal. Even in this case, if the radar is very stable, phase variations may be introduced by fluctuations in the propagation path, since this path will lie within the atmospheric layer immediately above the ground.

In testing against actual surface clutter, the relative performance of different modes of radar operation and of different radars may be established under nearly identical conditions. Absolute measurements of clutter rejection, however, are dependent upon the presence of "standard" clutter conditions, which are all but impossible to control. The same considerations apply to weather clutter, where there is no standard means of measuring the amount of turbulence or the distribution of velocities in the clutter spectrum. The over-all system test can be expected to yield qualitative data to confirm predictions made from theory, but considerable variation in absolute levels of performance can be expected, owing to local variation in clutter characteristics.

* E. V. Kullman, "A Precise Optical and Radar Tracking Range," *IRE Conv. Record* (1958), Part 5, pp. 142–49.

17.3 SYSTEM TESTING OF TRACKING RADAR

The task of system-testing a tracking radar is considerably more complex and difficult than testing a search radar, because it is more difficult to maintain a standard or reference whose accuracy is better than that of the radar under test. In the case of search radar, the primary data are in the form of detection of the target, which is easily verified. The accuracy expected of search radars is usually well within the capability of simple optical devices, and at worst will require checking against a tracking radar when long range or weather conditions rule out optical methods. In the case of tracking radar, the accuracies obtained are comparable to those of the best optical tracking instruments, and evaluation of radar accuracy may have to be carried out in several stages. We shall review the basic procedures which have been developed for testing tracking radars and give examples which will guide the formulation of test plans for future radars of this type.

S/N Ratio vs. Range

The loop gain of the tracking radar is verified more easily than that of search radar, because the antenna pattern is usually narrow and is kept on the target by the tracking process. Ground reflection effects will rarely influence the amplitude of the echo signal to any great extent, and continuous data on the target are available for averaging the effects of fluctuation, if present. Direct comparison with a signal generator pulse may be used, or the agc voltage of the receiver may be used to record continuous measurements of signal strength on a chart which has been calibrated with the signal generator. The standard sphere is generally used as a target. Results of a single sphere track in a typical case are shown in Fig. 17.11, plotted as received signal power and *S/N* ratio vs. slant range. A series of such tests will verify radar loop gain with almost any desired accuracy, since random errors in measurement are averaged out and only the systematic errors in calibration remain. The only appreciable external sources of error will be those in defining radar cross section of the sphere and balloon, and propagation loss in the troposphere. If the surface of the balloon is kept dry, there will be little contribution from this source, but tests made during rain storms may be in error from variable cross section and tropospheric attenuation errors.

Angular Accuracy

The discussion in Chapter 10 on angle-error analysis emphasized the great number and variety of error sources which may influence the output

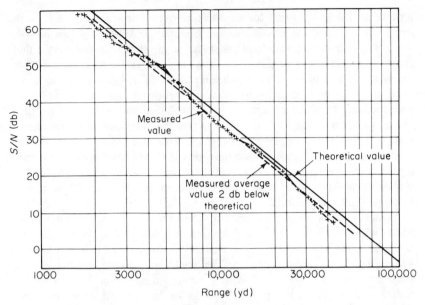

Figure 17.11 Typical plot of signal-to-noise ratio vs. range.

of a tracking radar. The classification of errors was arrived at in an attempt to aid the design of tests and the analysis of tracking test data. The two types of tracking test which are commonly run measure different types of error. In one test, a boresight telescope is used to establish departures of the antenna beam from the target, measuring the "tracking errors" as listed in Table 10.1. In the second test, radar angle outputs are compared with angular data derived from an independent source, measuring all components of radar error. In both cases, the errors of the reference instruments must be checked to ensure that the differences in data are due to "radar error" and not to "apparent error" arising in the reference system.

The standard tracking test used during the 1940's and 1950's used an aircraft as a target and the boresight telescope as a reference. The target was flown over a series of paths which would exercise the radar servos and would carry the target into the region where thermal noise was observed. Carried out in clear weather, the test depended on film images of the target, which could be observed at ranges of 40 or 50 miles. A reference point was established on the film by tracking a "boresight tower" which emitted a test signal from a small horn antenna, or the reference was based on the average as observed through the boresight telescope in several tracking runs. Any deviations of the target from this reference point were considered tracking error. Plots of this error as a function of time were prepared manually or by automatic equipment, and the same data were reduced to the form of standard deviations for each test, power spectral density plots, and amplitude distributions, as described in Chapter 10. When this type of test was run on aircraft, it was found that the target glint term dominated the error out to very long ranges. When the radar precision was improved to the order of 0.1 mil, a typical small aircraft (with rms glint of 15 to 20 ft along the wing or fuselage) was found to contribute this amount of error over most of the range of testing, or out to 25 or 30 miles. At twice this range, which was the limit of visibility on the telescope, the glint term remained the largest single component of error. Larger aircraft could be used to extend the range of testing, but their glint was also larger and continued to dominate the error. Thus, the use of aircraft proved to be a test of the ability of the radar to track aircraft and to minimize response to target glint and scintillation.

Tracking tests using the standard sphere as a target were first run in 1956, with results which showed for the first time the inherent tracking acuracy of the radar itself.* The first test runs showed that the sphere was an excellent point-source target, as would be expected, but that it was subject to oscillations in position as it swayed beneath the balloon in the turbulent atmosphere. When placed inside the balloon and suspended near the center with nylon cord, the sphere provided a relatively stable, point-source target for tracking tests. Even in this case, the position of the sphere was unknown within 2 or 3 in., and data gathered at very short range (within one mile) had to be discounted. Since the balloon could be photographed at ranges out to 10 or 20 miles (in the New Jersey atmosphere), this limita-proved to be minor. The sample of data shown in Fig. 10.1 was taken from boresight telescope film on a sphere test with the AN/FPS-16 radar, at a range of about 15 miles. By varying the inflation of the balloon and waiting for the right wind conditions, tests were arranged to cover a wide variety of conditions, including low-angle tracking and operation at different signal-to-noise ratios. Tests over a long period of time revealed the slowly changing bias components of tracking error, including such terms as antenna deflection due to solar heating. The major limitation of the sphere-tracking tests was the inability to exercise the servos for evaluation of lag error components, and the lack of data on the radar translation errors, which arise in conversion of antenna position to angular output data. Servo testing, however, was carried out with great success by using the boresight tower test signal to close the tracking loop, and inserting sinusoidal and sawtooth waves at the point where the error signal enters the servo amplifier (see Fig. 17.12).

* David K. Barton, "Final Report, Instrumentation Radar AN/FPS-16(XN-1), Evaluation and Analysis of Radar Performance," RCA, Moorestown, N. J., Contract DA-36-034-ORD-151. Astia Document AD 212, 125; "Accuracy of a Monopulse Radar," *Proc. 3rd Natl. Conv. on Military Electronics*, IRE-PGMIL, Washington, D.C. (June 28–July 1, 1959).

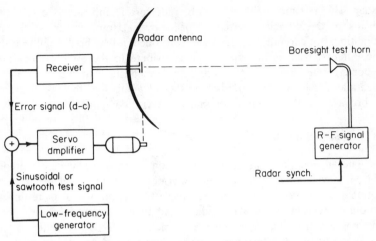

Figure 17.12 Servo test procedure.

The second type of tracking test for angular accuracy requires the presence of a reference instrumentation system to indicate the true position of the target. This type of test has been carried out at most of the missile test ranges, where well-established optical networks are available as reference system.* In most cases, the accuracy of the optical devices is little better than that of a radar at the same range, and the test is set up so that the radar operates at a much longer range (see Fig. 17.13). In this way, the errors of the reference system are held to a small portion of the radar error, when referred to linear error at the target or angular error at the

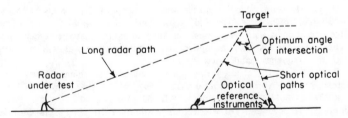

Figure 17.13 Optical reference for tracking test.

* Kermit E. Pearson, "Evaluation of the AN/FPS-16 (System No. 16) at White Sands Missile Range," U.S. Army Signal Missile Support Agency, White Sands Missile Range, N.M., Tech. Memo 606, Feb. 1959.

A.-E. Hoffman-Heyden and Vector B. Kovac, "Catalog of Static and Dynamic Errors in a Tracking Radar," *Proc. 6th Symposium on Ballistic Missile and Aerospace Technology*, Los Angeles, Calif. (1961).

radar. This test measures all types of radar error, and serves to verify the results of the complete system error analysis. However, unless the test conditions are carefully arranged and controlled, it may be difficult to isolate the different sources of error which contribute to the total. By operating a boresight telescope on the radar simultaneously with the other instrumentation, it is possible to combine the advantages of the two tests, and to break down the system error into its several components more easily.

Use of Astronomical Targets

The angular positions of the sun, moon, stars, and planets are predictable, if navigational tables and ephemerides which are available from the national observatories are used. All these targets are useful in testing the absolute accuracy of tracking radars, and they are used in several ways. In some cases, the radar may actually track the echo signal returned from the moon or planet.* In others, it may track passively on noise emitted by the target.† (Most modern radars can easily detect radiation from the sun, for instance, and many are also capable of receiving radio emissions from stars.) The stars are also useful for running optical tests with the boresight telescope to test the mechanical accuracy of the radar pedestal and the errors of the data system.‡ By combining the results of optical tests on stars with the tracking test data on spheres or aircraft, a complete evaluation of angular accuracy of the radar is possible.

Availability of artificial satellites has provided another means of calibrating the longer-range types of tracking radar. It has been shown§ that the orbits of high-altitude satellites can be determined by using range data only

* W. O. Mehuron and A. Rauchwerk, "Lunar Echo Boresighting of Large Antenna Systems," *Proc. IEEE International Conf. on Aerospace Support*, Washington, D.C. (Aug. 4-9, 1963).

† W. O. Mehuron, "Passive Radar Measurements at C-Band Using the Sun as a Noise Source," *Microwave Journal*, 5, No. 4 (April, 1962), pp. 87-94. If either active or passive operation is used, care must be taken to assure that the center of radiation or reflection coincides with the optical center of the target, as will normally be the case in the frequency bands used for radar.

‡ R. W. Davis, Jr., et al., "Radar Parameter Measurement Program," Interim Progress Report (April 1, 1960-Nov. 15, 1960), RCA, Moorestown, N.J., Contract DA 36-039-SC-783111, Astia Document AD 265, 336.

John H. Sandoz, "AN/FPS-16 Radar Evaluation Utilizing Star Observations," U.S. Army Signal Missile Support Agency, White Sands Missile Range, N.M., Tech, Report RA-101 (Sept. 1960).

§ Sidney Shucker, "Results of Space Tracking and Prediction with Precision Radar," *Proc. (1962) Natl. Symposium on Space Electronics and Telemetry*, IRE-PGSET.

Sidney Shucker and Robert Lieber, "Comparison of Short-Arc Tracking Systems," *Proc. 5th Natl. Conv. on Military Electronics*, IRE-PGMIL, Washington, D.C. (June 1961).

(after an initial estimate is made with range and angle data). The true angular position of the target may then be found from the orbital parameters and the known location of the radar site. By comparing these angles with the output of the radar angle system, it is possible to arrive at a check of radar accuracy which is at least as good as that obtained by the methods described earlier. The advantage of this method, shared with other active or passive radar tracking tests on astronomical targets, is that it provides an all-weather, day-or-night check of radar performance which can be run in the field by operating personnel with data reduced digitally at a remote point. There is no need for special optical observations or photography. In principle, at least, it would be possible to test a radar completely by remote control from a central site, by designating the target, receiving and recording the digital output data after lockon, and comparing them with ephemeris data or with orbital parameters obtained from the range data itself. A major limitation is that tracking range of the radar must be sufficient to permit it to acquire the satellite, and network provisions for designation and computation must be made. The use of a corner-reflector satellite has been proposed as a means of extending the range of coverage to most of the present tracking radars. The same satellite would serve as the calibration source for the radar geodetic network described in the previous chapter.

Range-Accuracy Evaluation

The measurement of radar range accuracy is one of the most difficult tasks in system testing, because it is difficult to obtain a source of data which is more accurate than the radar itself. The tracking tests against optical reference networks, mentioned above, can provide good data in cases where the target is close to the optical instruments. Satellite tracking tests are also capable of providing data on range errors, but it is difficult to evaluate the bias term unless the absolute orbital parameters of the target are known from continuous tracking of very high accuracy. For these reasons, the tracking radar is often evaluated on the basis of subsystem tests, in which each element of the range system is checked against internal test equipment. A final test on a few fixed calibration targets may be used to determine whether the bias error caused by internal delay has been calibrated out.

In calibration on a fixed target, the major limitation is the multipath error component and the possible presence of ground clutter. A typical target consists of a corner reflector mounted on a high, wooden pole at a range of several thousand feet from the radar. By raising the main beam of the antenna above the target, it is possible to place the first null of the antenna on the ground beneath the target, and thus to exclude clutter

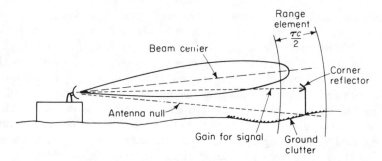

Figure 17.14 Use of reflector for range calibration.

signals. The height of the target pole must be sufficient to place the target on the shoulder of the main beam, if not in the center of it (see Fig. 17.14). The range to the target should not change as the antenna is moved slightly in azimuth or elevation. If a change is observed, it indicates that ground clutter or multipath reflections are affecting the range reading, and the test setup must be altered to eliminate this error. Radar fences may also be used to eliminate ground echoes and reflections which occur in the same range element as the fixed calibration target.

Acquisition Probability

Tests to determine the acquisition probability of a tracking radar as a function of designation error, signal strength, and other parameters may be designed with reference to Chapter 14. It was pointed out there that a large number of different factors will influence the acquisition process, and that it will be difficult to control all of them in full-scale system testing. A steady signal may be provided from a device such as a boresight test tower, and the antenna may be scanned with respect to this fixed point to simulate the actual acquisition of moving targets. By using suitable signal generators and modulators, it is also possible to simulate the fluctuation of typical targets.

In testing on actual targets such as aircraft or satellites, controlled conditions can be ensured by providing a second tracker which follows the target continuously and provides error-free designation. The designation data may then be perturbed by errors of known amplitude and frequency, as well as by the acquisition scan signals. The reference tracker will also provide a record of actual target signal strength, which may vary in an unpredictable way. As in other test problems, artificial satellites seem to offer a good source of signals, if the radar is powerful enough to track them.

Symbol	Meaning	Section Number
A	Area, physical aperture, amplitude, or azimuth angle	
A_b	Area of clutter within beam [Eq. (3.19)]*	
A_c	Area of clutter in resolution element	3.5
A_m	Azimuth search sector [Eq. (5.1)]	
A_r	Effective receiving aperture [Eq. (4.5)]	
A_τ	Area of clutter within pulse width [Eq. (3.21)]	
a	Barlow's clutter-stability parameter [Eq. (3.24)]	
a_t	Target acceleration	9.4
B	Equivalent noise bandwidth of receiver	1.2
B_c	Width of coherent spectral line [Eq. (12.6)]	
B_e	Effective bandwidth of detection system	1.3
B_f	Fine-line filter bandwidth	12.1
B_t	Spectral bandwidth of pulsed signal	2.1
B_v	Video amplifier bandwidth [Eq. (4.43)]	
B_1	Bandwidth of one i-f stage [Eq. (1.7)]	

LIST OF SYMBOLS USED

b	Length of interferometer baseline or Bean's correction constant [Eq. (15.5)]	15.3
C	Cancellation ratio	7.2
C_a	Detector loss factor $= (S + N)/S$	1.3
C_d	Detector loss factor $= (2S + N)/2S$ [Eq. (9.6)]	

* Equation numbers indicate where symbol is first used or defined.

Symbol	Meaning	Section Number
C_e	Range-difference correlation factor	16.2
C'_e	Range-sum correlation factor	16.2
C_n	Tropospheric parameter	15.3
C_r	Tropospheric parameter	15.3
C_v	Velocity error reduction factor	13.3
c	Velocity of light	1.1
c_o	Atomic reference velocity of light	11.3
D	Diameter of antenna	2.2
d	Diameter of cathode-ray spot [Eq. (4.42)]	
d_i	Diameter of rain droplet [Eq. (3.29)]	
dP	Probability density	1.2
E	Energy	1.2
	Elevation angle [Eq. (3.19)]	
E_c	Corrected elevation angle [Eq. (15.5)]	
E_e	Monopulse error-pattern voltage	9.1
E_m	Maximum search elevation angle [Eq. (5.1)]	
E_n	Amplitude of envelope of i-f noise [Eq. (1.2)]	
	Noise voltage (rms) in reference channel	9.2
E_{ne}	Noise voltage (rms) in error channel	9.2
E_{no}	Noise output of split gate [Eq. (11.2)]	
E_o	Observed elevation angle at surface [Eq. (15.5)]	
	Minimum search elevation angle [Eq. (5.1)]	
	Amplitude of pulse train [Eq. (9.1)]	
E_q	Quadrature voltage in monopulse tracker [Eq. (10.22)]	
E_r	Monopulse reference-pattern voltage	9.1
E_{ro}	Maximum possible reference-pattern voltage	9.2
E_s	Peak i-f signal voltage [Eq. (1.4)]	
E_t	Threshold voltage [Eq. (1.3)]	
ΔE_i	Elevation error from ionosphere	15.4
ΔE_o	Range error signal [Eq. (11.4)]	
	Elevation error from troposphere [Eq. (15.7)]	
ΔE_t	Total tropospheric bending [Eq. (15.5)]	
F	Propagation factor	4.4
f	Frequency	
f_a	Half-power frequency of spectrum [Eq. (3.14)]	
f_b	Reciprocal of time-on-target	3.2
f_c	Critical frequency of ionosphere [Eq. (15.11)]	
f_d	Doppler frequency	1.1
f_g	Half-power frequency of glint [Eq. (9.15)]	
f_m	Modulation frequency [Eq. (7.31)]	
	Multipath error frequency [Eq. (10.14)]	
f_n	Data sampling rate	13.1
f_o	Center-frequency of i-f amplifier	9.2

Symbol	Meaning	Section Number
f_p	Peak frequency deviation of reference oscillator	7.2
f_q	Quantizing noise frequency [(Eq. (13.1)]	
f_r	Repetition rate	1.1
f_s	Conical scan rate [Eq. (9.1)]	
f_t	Radar transmission frequency	1.1
f_{to}	Center frequency of transmission	2.3
$f_{0.5}$	Half-power frequency of clutter	3.5
Δf	Frequency separation [Eq. (2.5)]	
G	Antenna gain [Eq. (2.8)]	2.8
G_b	Beacon antenna gain [Eq. (4.28)]	
G_i	Gain of video integrator [Eq. (1.10)]	
G_n	Null depth in monopulse tracker	10.2
G_o	Antenna gain on the axis	
G_r	Receiving antenna gain [Eq. (4.34)]	
G_{se}	Relative antenna gain in side lobes of error pattern [Eq. (10.12)]	
G_{sr}	Relative antenna gain in side lobes of reference pattern [Eq. (11.27)]	
G_t	Transmitting antenna gain [Eq. (4.34)]	
G_1, G_2, G_3, G_4	Relative gains at different points in monopulse pattern	10.2
H	Wave height [Eq. (3.28)]	
$H_{1/3}$	Significant wave height	3.5
$H(f)$	Filter frequency-response function	13.2
$H'(f)$	Differentiating-filter frequency response	13.3
h	Height of antenna above surface [Eq. (6.1)]	
	Vertical dimension of antenna [Eq. (2.8)]	
$h(t)$	Filter weighting function	13.2
h_t	Target height above surface [Eq. (6.1)]	
$h'(t)$	Derivative of filter weighting function	13.3
I_o	Bessel function of first kind with imaginary argument [Eq. (1.4)]	
J	Jamming power in receiver [Eq. (4.32)]	
j	Number of search scans [Eq. (5.12)]	
K	Droplet refractivity parameter [Eq. (3.29)]	
	Relative difference slope	9.2
K_a	Acceleration error constant [Eq. (9.21)]	
K_g	Reference channel gain ratio	9.2
K_o	Position error constant [Eq. (9.21)]	
	Maximum possible difference slope	9.2

Symbol	Meaning	Section Number
K_r	Difference slope ratio K/K_0	9.2
K_s	Antenna spring constant [Eq. (10.17)]	
K_v	Velocity error constant [Eq. (9.21)]	
K_w	Aerodynamic constant of antenna [Eq. (10.15)]	
K_3, K_4	Error constants for third, fourth, and fifth derivatives	
K_5	[Eq. (9.21)]	
K_θ	Antenna beamwidth ratio	9.2
k	Boltzmann constant $= 1.37 \times 10^{-23}$ joules per °Kelvin [Eq. (4.8)]	
	Number of threshold crossings required for detection [Eq. (1.16)]	
k_c	Pulse compression ratio	2.1
k_e	Pulse voltage slope	2.1
k_f	Optimum detection criterion for fluctuating target	5.3
	Discriminator slope [Eq. (12.4)]	
	Frequency modulation rate	App. D
k_m	Monopulse error slope	9.1
k_o	Optimum detection criterion for steady target [Eq. (1.14)]	
k_r	Absolute range error slope [Eq. (11.7)]	
k_s	Conical scan error slope [Eq. (9.1)]	
k_t	Range discriminator slope [Eq. (11.3)]	
$k_{t'}$	Normalized range discriminator slope [Eq. (11.6)]	
L	Length of target [Eq. (3.8)]	
	Length of tropospheric path [Eq. (15.8)]	
	System loss factor [Eq. (4.14)]	
L_a	Atmospheric loss factor	15.1
L_{at}	Total atmospheric loss	15.1
L_b	Beacon interrogation or response loss [Eq. (4.28)]	
L_c	Collapsing loss	1.3
L_d	Scan distribution loss	5.3
L_e	Spacing loss for gates or filters	5.3
L_f	Fluctuation loss	1.2
L_g	Gain-adjustment or threshold loss	4.4
L_i	Loss from post-detector integration [Eq. (1.11)]	
L_j	Scanning loss	5.3
L_k	Conical-scan crossover loss	9.2
L_m	Filter-matching loss	1.2
L_n	Antenna efficiency loss	5.3
L_o	Integrator or operator loss	4.4
L_p	Pattern or beamshape loss	5.3
L_r	Receiving line loss	4.4
L_s	Total search loss [Eq. (5.6)]	
L_t	Transmitting line loss	4.4
L_x	Loss corresponding to propagation factor	4.4
$\mathcal{L}$	RMS aperture length	2.2

Symbol	Meaning	Section Number
M	Torque applied to antenna [Eq. (10.15)]	
M_r	Range-tracking merit factor [Eq. (11.9)]	
m	Number of extra noise samples [Eq. (1.12)]	
	Ratio of steady to random clutter power	3.5
	Complex refractive index [Eq. (3.29)]	
	Fractional modulation	9.2
m'	Fractional modulation after agc action	9.3
m_e	Fractional modulation in elevation channel [Eq. (9.3)]	
m_s	Fractional modulation due to scintillation [Eq. (9.16)]	
m_{tr}	Fractional modulation in traverse channel [Eq. (9.3)]	
N	Noise power referred to receiver input [Eq. (4.11)]	
	Atmospheric refractivity [Eq. (15.3)]	
N_e	Electron density [Eq. (15.10)]	
N_i	Ionospheric refractivity [Eq. (15.10)]	
N_o	Noise power density	1.2
N_r	Input noise power at reference temperature [Eq. (4.10)]	
N_s	Surface refractivity [Eq. (15.5)]	
$\overline{NF}$	Noise factor [Eq. (4.8)]	
$\overline{NF_o}$	Operating noise factor [Eq. (4.12)]	
ΔN^2	Mean-square refractivity variation [Eq. (15.8)]	
n	Number of signal pulses integrated	1.3
	Number of data samples [Eq. (10.1)]	
n_a	Total number of false alarms	1.2
n_d	Number of Doppler resolution elements [Eq. (14.5)]	
n_f	False-alarm number	1.2
n_f'	Modified false-alarm number [Eq. (1.13)]	
n_{fa}	Marcum's false-alarm number	1.2
n_i	Number of i-f amplifier stages [Eq. (1.7)]	
n_s	Number of angular resolution elements searched [Eq. (5.1)]	
n_t	Total number of pulses received	5.3
n_v	Number of elements in search volume [Eq. (14.3)]	
P	Power, probability	
P_a	Acquisition probability [Eq. (14.1)]	
P_{av}	Transmitter average power [Eq. (4.21)]	
P_b	Beacon peak power [Eq. (4.29)]	
P_c	Cumulative probability of detection [Eq. (5.12)]	
P_{ca}	Cumulative probability of acquisition [Eq. (14.6)]	
P_d	Detection probability after integration	1.3
P_e	Effective radiated beacon power $= P_b G_b$	3.4
P_n	False-alarm probability after integration	1.3
P_t	Transmitter peak power [Eq. (4.1)]	

Symbol	Meaning	Section Number
P_v	Probability of target being within searched volume [Eq. (14.2)]	
dP	Probability density	
p	Probability of threshold crossing on single sample of signal or noise [Eq. (1.15)]	
p_d	Single-pulse probability of detection [Eq. (1.5)]	
p_n	Single-pulse false-alarm probability [Eq. (1.3)]	
R	Range from radar to target [Eq. (4.1)]	
	Precipitation rate [Eq. (3.31)]	
R_a	Minimum slant range at crossover [Eq. (11.20)]	
R_b	Range at which geometrical and real acceleration components are equal [Eq. (9.23)]	
R_c	End of near zone for clutter [Eq. (3.28)]	
	Ground range at crossover	9.4
R_h	Line-of-sight range limit [Eq. (6.1)]	
R_m	Maximum range of search [Eq. (5.3)]	
R_o	Range for $S/N=$ unity [Eq. (4.15)]	
R_r	Range from receiver to target [Eq. (4.34)]	
R_t	Range from transmitter to target [Eq. (4.34)]	
ΔR	Range resolution element	3.5
ΔR_e	Range error due to troposphere	15.3
ΔR_g	Ground-range resolution element	3.5
ΔR_i	Range error due to ionosphere	15.4
$\dot{R}, \ddot{R}$ etc	Range rate, acceleration, etc. [Eq. (11.18)]	
$\mathcal{R}$	Energy ratio $2E/N_o$	
S	Signal power referred to signal input	1.2
S_b	Signal power at beacon receiver [Eq. (4.28)]	
S/N	Received single-pulse signal-to-noise power ratio [Eq. (4.14)]	
$(S/N)_{av}$	Average S/N during t_o [Eq. (2.5)]	
$(S/N)_f$	Signal-to-noise ratio within fine-line filter [Eq. (12.1)]	
$(S/N)_i$	Integrated signal-to-noise ratio [Eq. (1.10)]	
$(S/N)_o$	Signal-to-noise ratio at beam center [Eq. (2.12)]	
	Signal-to-noise ratio at receiver output [Eq. (4.8)]	
$\bar{S}/N$	Average S/N for fluctuating target	1.2
s	Cathode-ray sweep speed [Eq. (4.42)]	
T	Absolute temperature	
T_a	Antenna temperature [Eq. (4.11)]	
T_a'	Antenna temperature referred to receiver [Eq. (4.38)]	
T_e	Effective receiver temperature [Eq. (4.38)]	
T_i	Total input temperature [Eq. (4.11)]	

Symbol	Meaning	Section Number
T_i'	Input temperature referred to receiver [Eq. (4.38)]	
T_p	Propagation loss temperature [Eq. (15.1)]	
T_r	Transmission line loss temperature [Eq. (4.37)]	
T_t	Average temperature of troposphere [Eq. (15.1)]	
t	Time in seconds	
t_a	Target rotation period [Eq. (3.8)]	
t_b	Time-on-target with scanning beam	4.4
t_c	Period over which signal retains coherence	12.2
	Correlation time of designation error	14.2
t_d	Filter delay time	13.2
t_{fa}	False-alarm time	
$\overline{t_{fa}}$	Average false-alarm time	1.2
t_m	Mean wave period	3.5
t_n	Period of single scan	14.2
t_o	Observation time [Eq. (2.3)]	
t_p	Repetition period of pulse train	1.1
t_r	Range delay time	1.1
t_v	Video integration time [Eq. (4.44)]	
t_s	Search time [Eq. (5.5)]	
	Smoothing time	13.2
U	North's filter response factor	1.2
V	Instantaneous i-f noise voltage [Eq. (1.1)]	
V_c	Volume of clutter [Eq. (3.23)]	
V_o	Visibility factor	6.3
v_r	Target radial velocity	1.1
v_t	Target velocity	9.4
v_w	Wind velocity [Eq. (10.15)]	
$W(f)$	Power spectral density [Eq. (3.24)]	
W	Total weight of chaff [Eq. (3.34)]	
W_c	Spectral density of cyclic error	10.1
W_g	Spectral density of wind gusts [Eq. (10.19)]	
W_m	Spectral density of scintillation	9.3
W_n	Noise spectral density [Eq. (9.20)]	
W_s	Servo error spectral density [Eq. (10.19)]	
W_o	Spectral density at zero frequency [Eq. (3.24)]	
w	Width of antenna [Eq. (2.8)]	
X	Tangent $E_m = h_t/R_c$	9.4
x	Generalized variable	
$\overline{x}$	Mean value of x [Eq. (10.1)]	
x_i	Value of ith sample of x [Eq. (10.1)]	
x_{rms}	RMS value of x	

Symbol	Meaning	Section Number
x_1, x_2	Signal-to-noise ratios in coherent detector (after East)	
x_0	[Eq. (9.11)]	
Y	Complex transfer function	9.3
Y_a	Open-loop agc transfer function	9.3
Y_c	Closed-loop servo transfer function	9.3
Y_p	Servo torque transfer function	10.2
Y_s	Scintillation error factor	
Y_{11}	Open-loop servo transfer function	9.4
Z	Volume reflectivity factor [Eq. (3.30)]	
α	Generalized angular coordinate	2.2
	RMS time span of signal [Eq. (2.21)]	
	Normalized drop radius $= 2\pi a/\lambda$ [Eq. (3.29)]	
	Angle between target velocity and radar beam [Eq. (11.18)]	
$\Delta\alpha$	Interferometer lobe width [Eq. (2.17)]	
β	RMS frequency span of signal	
	Conical-scan squint angle	9.1
	Angle between acceleration vector and radar beam [Eq. (11.19)]	
β_f	Noise bandwidth of filter	13.2
β_n	Noise bandwidth of servo system	9.2
β_0	Optimum noise bandwidth of servo [Eq. (9.25)]	
β_s	Servo bandwidth after reduction of gain by small-signal suppression [Eq. (9.6)]	
γ	Integration efficiency factor [Eq. (1.10)]	
Δ	Width of reflection lobe [Eq. (3.2)]	
	Normalized off-axis angle	
	Fixed or bias error	
Δ_a	Azimuth bias error [Eq. (10.21)]	
Δ_0	Width of reflection side lobe [Eq. (3.4)]	
Δ_r	Range delay error [Eq. (2.1)]	
Δ_{tr}	Traverse bias error [Eq. (10.21)]	
Δ_ϕ	Boresight shift owing to phase error [Eq. (10.22)]	
ϵ	Tracking error [Eq. (9.1)]	
ϵ_a	Acceleration lag error	10.3
ϵ_e	Elevation tracking error [Eq. (9.2)]	
ϵ_f	Frequency tracking error	12.2
ϵ_n	Scintillation component of lag error [Eq. (9.19)]	

Symbol	Meaning	Section Number
ϵ_o	Bias component of lag error [Eq. (9.18)]	
ϵ_r	Range lag error [Eq. (11.23)]	
ϵ_t	Traverse tracking error [Eq. (9.2)]	
ϵ_v	Velocity lag error	10.3
ϵ_{wd}	Wind deflection error owing to gusts [Eq. (10.17)]	
ϵ_{ws}	Servo error owing to wind gusts [Eq. (10.16)]	
ϵ_1	Initial error owing to designation rate	9.4
η	Number of range elements during each repetition period	1.2
	Volume reflectivity of clutter	3.5
η_a	Antenna efficiency factor [Eq. (2.8)]	
θ	Angle, beamwidth (measured to one-way half-power points of pattern), phase angle	
θ_a	Azimuth beamwidth	
θ_c	COHO phase angle [Eq. (7.6)]	
θ_d	Bipolar video phase angle [Eq. (7.19)]	
θ_e	Elevation beamwidth	1.1
θ_i	Input angle to servo [Eq. (9.21)]	
	i-f signal phase angle [Eq. (7.17)]	
θ_o	Output angle of servo [Eq. (9.21)]	
θ_r	Received signal phase angle [Eq. (7.12)]	
θ_s	STALO phase angle [Eq. (7.4)]	
θ_t	Transmitter phase angle [Eq. (7.8)]	
λ	Radar transmission wavelength	
	North's energy ratio parameter	1.2
π	3.14159	
ρ	Collapsing ratio [Eq. (1.12)]	
	Surface reflectivity coefficient [Eq. (10.12)]	
σ	Radar cross section, standard deviation	
σ_a	Standard deviation of low-frequency error [Eq. (10.8)]	
σ_b	Standard deviation of apparent bias error	10.1
σ_c	Clutter spread in cps [Eq. (3.25)]	
	Standard deviation of cyclic error	10.1
σ_f	Standard deviation in frequency measurement [Eq. (2.20)]	
σ_{f_1}	Single-pulse error in frequency [Eq. (2.19)]	
σ_{f_c}	Error in coherent frequency measurement [Eq. (2.23)]	
σ_g	Standard deviation due to glint [Eq. (3.18)]	
σ_m	Standard deviation due to multipath [Eq. (10.13)]	
σ_n	Standard deviation of noise error [Eq. (10.9)]	
σ_r	Range delay error	2.1

Symbol	*Meaning*	*Section Number*
σ_{r_m}	Multipath error in range [Eq. (11.27)]	
σ_{r_1}	Single-pulse range delay error [Eq. (2.2)]	
σ_s	Scintillation error [Eq. (9.16)]	
σ_t	Thermal-noise error [Eq. (9.4)]	
σ_v	Clutter velocity spread [Eq. (3.26)]	
σ_{wd}	Standard deviation owing to wind gust deflection [Eq. (10.18)]	
σ_{ws}	Standard deviation of servo owing to wind gusts [Eq. (10.20)]	
	Wind gust velocity	10.2
σ_x	rms error in x coordinate [Eq. (10.5)]	
σ_α	Interferometer error [Eq. (2.17)]	
σ_θ	Angle measurement error [Eq. (2.14)]	
$\sigma_{\Delta r}$	Range-difference error across baseline	15.3
σ_{50}	Median radar cross section	1.2
σ^0	Surface reflectivity of clutter	3.5
σ_1	Single-pulse thermal noise error [Eq. (9.9)]	
σ_0	Bias error	
$\bar{\sigma}$	Average radar cross section [Eq. (1.8)]	
τ	Pulse length	1.2
τ_e	Pulse rise time	2.1
τ_g	Width of range gate [Eq. (4.46)]	
τ_r	Reference delay for correlation	2.1
τ'	Effective compressed pulse width	2.1
ϕ	Phase angle	
	Phase of error in conical scan [Eq. (9.1)]	
ϕ_c	COHO phase angle [Eq. (7.7)]	
ϕ_o	Transmitter oscillator phase angle [Eq. (7.11)]	
ϕ_s	STALO phase angle [Eq. (7.5)]	
ϕ_t	Transmission phase angle [Eq. (7.9)]	
$\Delta\phi$	Monopulse phase error [Eq. (10.22)]	
ψ	Phase angle at surface reflection [Eq. (10.11)]	
ψ_b	Solid angle of radar beam	
ψ_o	Mean-square i-f noise voltage [Eq. (1.1)]	
ψ_s	Solid angle covered by search radar [Eq. (5.1)]	
ω	Angular frequency or rate	
ω_a	Azimuth scan rate [Eq. (4.44)]	
	Intersection of servo transfer function with unity-gain axis	9.4
	Azimuth rate of target	9.4
ω_c	Frequency for unity servo gain	9.4

Symbol	Meaning	Section Number
ω_e	Elevation scan rate [Eq. (4.44)]	
ω_m	Modulation frequency [Eq. (9.18)]	
	Maximum azimuth rate	9.4
ω_n	Radian noise bandwidth [Eq. (9.24)]	
ω_s	Conical scan rate [Eq. (9.1)]	
ω_t	Target angular rate [Eq. (5.7)]	
ω_r	Intersection of extended 6 db/octave slope with unity gain axis	9.4
$\omega_1, \omega_2, \omega_3$	Corner frequencies in servo response	9.4

ANONYMOUS REFERENCES AND DICTIONARIES

Dept. of the Air Force, *Communications-Electronics Terminology, Definitions and Abbreviations* (Washington, D.C.: U.S. Govt. Printing Office, July 1, 1960).

————, Fundamentals of Aerospace Weapon Systems. Washington, D.C.: U.S. Govt. Printing Office, May, 1961.

Dept. of Defense and Atomic Energy Comm., *The Effects of Nuclear Weapons* (Washington, D.C.: U.S. Govt. Printing Office, April 1962). [15.1]

The International Dictionary of Physics and Electronics, Walter C. Michels (sr. ed.) (Princeton, N.J.: D. Van Nostrand Company, Inc., 1956).

Military Specification, Radar Signal Simulator SM-65( )/UP, MIL-15293 B(SHIPS) (Washington, D.C.: Bureau of Ships, Department of the Navy, March 28, 1955).

National Academy of Sciences, *Report of the Ad Hoc Panel on Basic Measurements*, Report No. ACAFSC 102, Contract AF 18 (600)-1895. Washington, D.C. (Dec. 8, 1961), Astia Document AD 286,086.

"The SCR-584 Radar," *Electronics* (Feb. 1946), p. 110. [9.1]

Institute of Radio Engineers, *IRE Dictionary of Electronic Terms and Symbols* (1961).

BIBLIOGRAPHY AND REFERENCES*

ALPHABETICAL LISTING OF AUTHOR'S REFERENCES

Adler, S. B., "Pulsed Radar Measurement of Backscattering from Spheres," *RCA Review*, **23**, No. 1 (March 1962). [3.1]

* Numbers in brackets indicate section in which the work is mentioned.

Armstrong, M. E., G. W. Swenson, R. L. Sydnor, and H. D. Webb, "The Use of Radio Noise from the Sun for Calibrating Radio Receiving Systems," *Proc. NEC*, **15** (1959). [17.1]

Ashby, R. M., V. Josephson, and S. Sydoriak, "Signal Threshold Studies," Naval Res. Lab. Report R-3007 (Dec. 1, 1946). [6.3]

Barlow, Edward J., "Doppler Radar," *Proc. IRE*, **37**, No. 4 (April 1949), pp. 340-55. [3.5]

Barton, David K., "Final Report, Instrumentation Radar AN/FPS-16(XN-1), Evaluation and Analysis of Radar Performance," RCA, Moorestown, N.J., Contract DA 36-034-ORD-151 (1957). Astia Document AD 212,125. [15.3, 17.3]

———, "Accuracy of a Monopulse Radar," *Proc. 3rd Natl. Conv. on Military Electronics*, IRE-PGMIL, Washington, D.C. (June 28-July 1, 1959). [17.3]

———, "Tracking Radars," Lecture 25 of special summer course, University of Pennsylvania (June 1961; to be published). [9.2]

———, *Report of the Ad Hoc Panel on Electromagnetic Propagation*, D. K. Barton, ed. National Academy of Sciences, Washington, D. C., Report No. ACAFSC 103, Contract AF 18 (600) 1895, (Feb. 1963), Astia Document AD 296,845 [15.3, 15.4].

———, "Reasons for the Failure of Radio Interferometers to Achieve Their Expected Accuracy," *Proc. IEEE*, **51**, No. 4 (April 1963), pp. 626-27. [15.3]

Battan, Louis J., *Radar Meteorology*. (Chicago: University of Chicago Press, 1959). [3.5]

Bean, B. R. and B. A. Cahoon, "The Use of Surface Weather Observations to Predict the Total Atmospheric Bending of Radio Rays at Small Elevation Angles," *Proc. IRE*, **45**, No. 11 (Nov. 1957), pp. 1545-46. [15.3]

Bean, B. R., J. D. Horn, and A. M. Ozanich, Jr., "Climatic Charts and Data of the Radio Refractive Index of the United States and the World," Natl. Bureau of Standards Monograph No. 22 (Washington, D.C.: U.S. Govt. Printing Office, Nov. 25, 1960). [15.3]

Bean, B. R. and G. D. Thayer, "Models of the Radio Refractive Index," *Proc. IRE*, **47**, No. 5 (May 1959), pp. 740-55. [15.3]

———, "CRPL Exponential Reference Atmosphere," Natl. Bureau of Standards Monograph No. 4 (Washington, D.C.: U.S. Govt. Printing Office, Oct. 29, 1959). [15.3]

Benner, A. H. and R. Drenick, "An Adaptive Servo System," *IRE Conv. Record* (1955), Part 4, pp. 8-14. [6.4]

Bernstein, R., "An Analysis of Angular Accuracy in Search Radar," *IRE Conv. Record* (1955), Part 5, pp. 61-78. [2.2, 11.2]

Blackman, R. B. and J. W. Tukey, *The Measurement of Power Spectra from the Point of View of Communications Engineering*. (New York: Dover Publications, Inc., 1958. [3.2, 10.1]

Blake, Lamont V., "The Effective Number of Pulses per Beamwidth for a Scanning Radar," *Proc. IRE*, **41**, No. 6 (June 1953), 770-74. [5.3]

———, "Interim Report on Basic Pulse-Radar Maximum Range Calculation," Naval Res. Lab. Memo. Report 1106, (Nov. 1960). Most of this report also appeared in the paper, "Recent Advancements in Basic Radar Range Calculation Technique," *Trans. IRE*, MIL-5, No. 2 (April 1961), pp. 154-64. [1.1, 1.2, 1.3, 3.2, 4.4, 6.2, 6.3, 8.2]

————, "Curves of Atmospheric-Absorption Loss for Use in Radar-Range Calculation," Naval Res. Lab. Report 5601 (Mar. 23, 1961), Astia Document AD 255,135. See also "Tropospheric Absorption Loss and Noise Temperature in the Frequency Range 100–10,000 Mc," *Trans. IRE*, AP-10, No. 1 (Jan. 1962). [15.1]

————, "Antenna and Receiving-System Noise-Temperature Calculation," Naval Res. Lab. Report 5668, Sept. 19, 1961, Astia Document AD 265,414. See also paper of same title, *Proc. IRE*, 49, No. 10 (Oct. 1961), pp. 1568–69. [4.4]

————, "A Guide to Basic Pulse-Radar Maximum-Range Calculation (Part 1)," Naval Res. Lab. Report 5868, Dec. 28, 1962, Astia Document AD 298,126. [8.2, 8.3, 15.1]

Bowles, Kenneth L., "Lima Radio Observatory," Natl. Bureau of Standards Report 7201 (supplement) (April 30, 1961). [15.4]

Brockner, Charles E., "Angular Jitter in Conventional Conical-Scanning, Automatic-Tracking Radar Systems," *Proc. IRE*, Vol. 39, No. 1 (Jan. 1951), pp. 51–55. [3.3]

Burrington, Richard S., *Handbook of Mathematical Tables and Formulas*, 3rd ed. (Sandusky, Ohio: Handbook Publishers, 1949). [1.2]

Bussgang, J. J. and D. Middleton, "Optimum Sequential Detection of Signals in Noise," *Trans. IRE*, IT-1, No. 1 (Dec. 1955), pp. 5–18. [5.1]

Bussgang, J. J., P. Nesbeda, and H. Safran, "A Unified Analysis of Range Performance of CW, Pulse, and Pulse Doppler Radar," *Proc. IRE*, 47, No. 10 (Oct. 1959), pp. 1753–62. [7.4]

Cheetham, Roger P. and Warren A. Mulle, "Enhanced Real-Time Data Accuracy for Instrumentation Radars by Use of Digital-Hydraulic Servos," *IRE Wescon Record* (1958), Part 4. [9.4]

Cohen, William and C. Martin Steinmetz, "Amplitude- and Phase-Sensing Monopulse System Parameters," *Microwave Journal*, 2, No. 10 (Oct. 1959), pp. 27–33; No. 11 (Nov. 1959), pp. 33–38. [9.1]

Coleman, Herbert J., "Navy Maintains Barrier in All Weather," *Aviation Week* (April 3, 1961), pp. 63–73. [16.1]

Cook, Charles E., "Pulse Compression—Key to More Efficient Radar Transmission," *Proc. IRE*, 48, No. 3 (March 1960), pp. 310–16. [2.1]

Cooke, Nelson M. and John Markus, *Electronics and Nucleonics Dictionary* (New York: McGraw-Hill Book Company, 1960).

Cutrona, L. J., W. E. Vivian, E. N. Leith, and G. O. Hall, "A High-Resolution Radar Combat-Surveillance System," *Trans. IRE*, MIL-5, No. 2 (April 1961), pp. 127–31. [6.1]

Davenport, Wilbur B., Jr. and William L. Root, *An Introduction to the Theory of Random Signals and Noise* (New York: McGraw-Hill Book Company, 1958). [4.1, 9.2]

Davis, R. W., Jr., D. K. Barton, D. R. Billetter, A. C. DuPont, M. H. Paiss, R. E. Stacy, and J. L. Foster, "Radar Parameter Measurement Program," Interim Progress Report (April 1, 1960–November 15, 1960), RCA, Moorestown, N. J., Contract DA 36-039-SC-78311, Astia Document AD 265,336. [17.3]

Deam, A. P. and B. M. Fannin, "Phase-Difference Variations in 9350-Megacycle Radio Signals Arriving at Spaced Antennas," *Proc. IRE*, 43, No. 10 (Oct. 1955), pp. 1402–11. [15.3]

Delano, Richard H., "A Theory of Target Glint or Angular Scintillation in Radar Tracking," *Proc. IRE*, **41**, No. 12 (Dec. 1953), pp. 1778–84. [3.3]

Delano, Richard H. and Irwin Pfeffer, "The Effect of AGC on Radar Tracking Noise," *Proc. IRE*, **44**, No. 6 (June 1956), pp. 801–10. [3.3, 9.3]

Develet, Jean A., Jr., "Thermal-Noise Errors in Simultaneous-Lobing and Conical-Scan Angle-Tracking Systems," *Trans. IRE*, **SET-7**, No. 2 (June 1961), pp. 42–51. [9.2]

Dickey, F. R., Jr., "Theoretical Performance of Airborne Moving Target Indicator," *Trans. IRE*, **PGAE-8**, No. 2 (June 1953), pp. 12–23. [7.3]

Donegan, James J. and James C. Jackson, "Mercury Network Performance," Sec. 2 of NASA Report SP-6, *Results of the Second United States Manned Orbital Space Flight*, May 24, 1962. (Washington, D.C.: U.S. Govt. Printing Office, 1962), pp. 15–26. [16.2]

Dunn, John H. and Dean D. Howard, "The Effects of Automatic Gain Control Performance on the Tracking Accuracy of Monopulse Radar Systems," *Proc. IRE*, **47**, No. 3 (March 1959), pp. 430–35. [3.3, 9.3]

——, "Precision Tracking with Monopulse Radar," *Electronics*, **35**, No. 17 (April 22, 1960), pp. 51–56. [9.1]

Dunn, John H., Dean D. Howard, and A. M. King, "The Phenomena of Scintillation Noise in Radar Tracking Systems," *Proc. IRE*, **47**, No. 5 (May 1959), pp. 855–63. [3.3, 9.3]

East, T. W. R., "The Coherent Detector with Noisy Reference Input," McGill University Physics Dept. Report No. 563, Contract P 69-8-442 DRB, 1955, Astia Document AD 90,783. [9.2]

Emslie, A. G. and R. A. McConnell, "Moving Target Indication," Chap. 16 in *Radar System Engineering*, L. N. Ridenour, ed. (New York: McGraw-Hill Book Company, 1947). [7.1, 7.2, 7.3, 8.1]

Essen, L., "Frequency and Time Standards," *Proc. IRE*, **50**, No. 5 (May 1962), pp. 1158–63. [11.3]

Fishback, William T., "Methods for Calculating Field Strength with Standard Refraction," Sec. 2.13–2.16 of *Propagation of Short Radio Waves*, D. E. Kerr, ed. (New York: McGraw-Hill Book Company, 1951). [15.2]

——, "Reflection Coefficient of Land," Sects. 5.9 and 5.10 of op. cit. above.

Fowler, C. A., A. P. Uzzo, Jr., and A. E. Ruvin,"Signal Processing Techniques for Surveillance Radar Sets," *Trans. IRE*, **MIL-5**, No. 2 (April 1961), pp. 103–8. [7.4, 8.1]

Gehrels, E. and A. Parsons, "Interferometer Techniques Applied to Radar," *Trans. IRE*, **MIL-5**, No. 2 (April 1961), pp. 139–46. [16.3]

George, Samuel F. and Arthur S. Zamanakos, "Multiple Target Resolution of Monopulse vs. Scanning Radars," *Proc. NEC*, **15** (1959), pp. 814–23. [9.1, 9.5]

Golay, M. J. E., "Velocity of Light and Measurement of Interplanetary Distances," *Science*, **131**, No. 3392 (Jan. 1, 1960), pp. 31–32. [11.3]

Goldstein, Herbert, "Sea Echo," "The Origins of Echo Fluctuations," and "The Fluctuations of Clutter Echoes," Sects. 6.6–6.21 in *Propagation of Short Radio Waves*, D. E. Kerr, ed. (New York: McGraw-Hill Book Company, 1951). [3.5]

, "Attenuation by Condensed Water," Sect. 8.6 of above. [15.1]

Goldstein, Herbert, Donald E. Kerr, and Arthur E. Bent, "Meteorological Echoes," Chap. 7 in *Propagation of Short Radio Waves*, D. E. Kerr, ed. (New York: McGraw-Hill Book Company, 1951). [3.5]

Grant, C. R. and B. S. Yaplee, "Back Scattering from Water and Land at Centimeter and Millimeter Wavelengths," *Proc. IRE*, **45**, No. 7, (July, 1957), pp. 976–82. [3.5]

Grisetti, R. S. and E. B. Mullen, "Baseline Guidance Systems," *Trans. IRE*, **MIL-2**, No. 1 (Dec. 1958), pp. 36–44. [16.3]

Grisetti, R. S., M. M. Santa, and G. M. Kirkpatrick, "Effect of Internal Fluctuations and Scanning on Clutter Attenuation," *Trans. IRE*, **ANE-2**, No. 1 (March 1955), pp. 37–41. [7.3]

Guier, W. H. and G. C. Weiffenbach, "A Satellite Doppler Navigation System," *Proc. IRE*, **48**, No. 4 (April 1960), pp. 507–16. [15.4]

Haeff, A. V., "Minimum Detectable Radar Signal and Its Dependence Upon Parameters of Radar Systems," *Proc. IRE*, **34**, No. 11 (Nov. 1946), pp. 857–61. [1.3]

Hall, W. M., "Prediction of Pulse Radar Performance," *Proc. IRE*, **44**, No. 2 (Feb. 1956), pp. 224–31. [1.2, 1.3, 4.4]

Hannan, Peter W., "Optimum Feeds for All Three Modes of a Monopulse Antenna," *Trans. IRE*, **AP-9**, No. 5 (Sept. 1961), pp. 444–61. [9.1]

Hansen, W. W., "C-W Radar Systems," Chap. 5 in *Radar System Engineering*, L. N. Ridenour, ed. (New York: McGraw-Hill Book Company, 1947). [2.1]

Hawkins, H. E. and O. La Plant, "Radar Performance Degration in Fog and Rain," *Trans. IRE*, **ANE-6**, No. 1 (March 1959), pp. 26–30. [15.1]

Hendler, A. J., "Noise Figure Measurements, Definitions, Techniques and Pitfalls," *Topics in Noise*, Airborne Instruments Laboratory, Deer Park, N.Y. (Aug. 1960). [17.1]

Herbstreit, J. W. and M. C. Thompson, "Measurements of the Phase of Radio Waves Received over Transmission Paths with Electrical Lengths Varying as a Result of Atmospheric Turbulence," *Proc. IRE*, **43**, No. 10 (Oct. 1955), pp. 1391–1401. [15.3]

Hicks, B. L., N. Knable, J. J. Kovaly, G. S. Newell, J. P. Ruina and C. W. Sherwin, "The Spectrum of X-Band Radiation Back-Scattered from the Sea Surface," *Journal of Geophysics Research*, **65**, No. 3 (March 1960), pp. 825–37. [3.5]

Hoffmann-Heyden, A-E. and Victor B. Kovac, "Catalog of Static and Dynamic Errors in a Tracking Radar," *Proc. 6th Symposium on Ballistic Missile and Aerospace Technology*, Los Angeles, Calif., (1961) (Also issued as AD 274,752). [17.3]

Hogg, D. C. and R. A. Semplak, "The Effect of Rain on the Noise Level of a Microwave Receiving System," *Proc. IRE*, **48**, No. 12 (Dec. 1960), pp. 2024–25. [15.1]

Hollis, Richard, "False Alarm Time in Pulse Radar," *Proc. IRE*, **42**, No. 7 (July 1954), p. 1189. [1.2]

Howard, Dean D., "Radar Target Angular Scintillation in Tracking and Guidance Systems Based on Echo Signal Phase-Front Distortion," *Proc. NEC*, **15** (1959), pp. 840–49. [3.3]

———, "Instrumentation for Recording and Analysis of Audio and Sub-Audio Noise," *IRE Conv. Record*, 1958, Part 5, pp. 176–82. [3.2]

Howard, Dean D. and C. F. White, "External and Internal Noise Inputs to the Radar System," Sect. 8-2 in *Airborne Radar*, by Povejsil, Raven, and Waterman (Princeton, N.J.: D. Van Nostrand Company, Inc., 1961).

Howard, Dean D. and B. L. Lewis, "Tracking Radar External Range Noise Measurements and Analysis," Naval Res. Lab. Report 4602 (Aug. 31, 1955). [3.3]

James, Hubert M., Nathaniel B. Nichols, and Ralph S. Phillips, *Theory of Servomechanisms*. (New York: McGraw-Hill Book Company, 1947). [9.4]

Kanter, Irving, "The Prediction of Derivatives of Polynomial Signals in Additive Stationary Noise," *IRE Wescon Record* (1958), Part 4, pp. 131–46. [13.2]

———, "Some New Results for the Prediction of Derivatives of Polynomial Signals in Additive Stationary Noise," *IRE Wescon Record* (1959), Part 5, pp. 87–91. [13.2]

Katzin, Martin, "On the Mechanisms of Radar Sea Clutter," *Proc. IRE*, 45, No. 1 (Jan. 1957), pp. 44–54. [3.5]

Keller, J. B., "Backscattering from a Finite Cone," *Trans. IRE*, Vol. AP-8, No. 2 (March 1960), pp. 175–82. [3.1]

Kennaugh, E. M. and D. L. Moffatt, "On the Axial Echo Area of the Cone-Sphere Shape," *Proc. IRE*, 50, No. 2 (Feb. 1962), p. 199. [3.1, 16.1]

Kerr, Donald E., ed., *Propagation of Short Radio Waves* (New York: McGraw-Hill Book Company, 1951). [3.1, 3.2, 15.1, 15.2]

King, Leonard H., "Reduction of Forced Error in Closed-Loop Systems," *Proc. IRE*, 41, No. 8 (Aug. 1953), pp. 1037–42. [9.4]

Kirkpatrick, G. M., "Aperture Illuminations for Radar Angle-of-Arrival Measurements," *Trans. IRE*, PGAE-9 (Sept. 1953), pp. 20–27. [9.1]

Klass, Philip J., "DEW Line Adds 100 Minutes Warning," *Aviation Week* (Sept. 16, 1957), pp. 100–13. [16.1]

Klauder, J. R., A. C. Price, S. Darlington, and W. J. Albertsheim, "The Theory and Design of Chirp Radars," *Bell System Tech. Journal*, 39, No. 4 (July 1960), pp. 745–820. [2.1]

Korff, M., C. M. Brindley, and M. H. Lowe, "Multiple-Target Data Handling with a Monopulse Radar," *Trans. IRE*, MIL-6, No. 4 (Oct. 1962), pp. 359–66. [9.4, 10.3]

Kullman, E. V., "A Precise Optical and Radar Tracking Range," *IRE Conv. Record*, (1958), Part 5, pp. 142–49. [17.2]

Landry, Preston, "Atmospheric Sounding and Correction of Tracking Data," Sect. 5 of *Report of Ad Hoc Panel on Electromagnetic Propagation*, D. K. Barton, ed., National Academy of Sciences, (Feb. 1963). [15.3]

Lawson, James L. and George E. Uhlenbeck, *Threshold Signals*. (New York: McGraw-Hill Book Company, 1950). [9.2, 9.3]

Levine, Daniel, "Volume Scanning with Conical Beams," *Proc. IRE*, 38, No. 3 (March 1950), pp. 287–90. [14.2]

Lisicky, A. J., "Digital Ranging System," *Proc. 2nd Natl. Conv. on Military Electronics*, IRE-PGMIL, Washington, D.C., (June 16–19, 1958), pp. 219–24. [11.1]

Macko, Stanley J., *Satellite Tracking* (New York: John F. Rider, Publisher, Inc., 1962). [14.1]

Manasse, Roger, "Range and Velocity Accuracy from Radar Measurements," Lincoln Lab. Report 312-26 (Feb. 3, 1955), Astia Document AD 236,236. [2.3]

——, "An Analysis of Angular Accuracies from Radar Measurements," Lincoln Lab. Group Report 32-24 (Dec. 6, 1955). [2.2, 9.2]

——, "Summary of Theoretical Accuracy of Radar Measurements," Mitre Corp. Technical Series Report No. 2 (April 1, 1960), Astia Document AD 287,563. [2.3]

Manger, W. P., "General Design Principles for Servomechanisms," Sects. 4.2-4.9 of *Theory of Servomechanisms*, James, Nichols, and Phillips, eds. (New York: McGraw-Hill Book Company, 1947). [9.4]

Marcum, J. I., "A Statistical Theory of Target Detection by Pulsed Radar," RAND Corp. Research Memo, RM-754 (Dec. 1, 1947).

——, "A Statistical Theory of Target Detection by Pulsed Radar: Mathematical Appendix," RAND Corp. Research Memo. RM-753 (July 1, 1948). [1.2, 1.3, 4.4] [The two reports cited above were reprinted in *Trans. IRE*, Vol. **IT-6**, No. 2 (April 1960).]

McNish, A. G., "Velocity of Light," Appendix A, *Report of Ad Hoc Panel on Basic Measurements*, National Academy of Sciences (Dec. 8, 1961). [11.3]

Meade, John E., "Target Considerations," Chap. 11 in *Guidance*, A. S. Locke, ed. (Princeton, N.J.: D. Van Nostrand Company, Inc., 1955). [3.3]

Mehuron, W. O., "Passive Radar Measurements at C-Band Using the Sun as a Noise Source," *Microwave Journal*, 5, No. 4 (April 1962), pp. 87-94. [17.3]

Mehuron, W. O. and A. Rauchwerk, "Lunar Echo Boresighting of Large Aperture Systems," *Proc. IEEE International Conf. on Aerospace Support*, Washington, D.C. (Aug. 4-9, 1963). [17.3]

Mengel, John T., "Tracking the Earth Satellite, and Data Transmission, by Radio," *Proc. IRE*, 44, No. 6 (June 1956), pp. 755-60. [16.3]

Merrill, Grayson, ed., *Dictionary of Guided Missiles and Space Flight*. (Princeton, N.J.: D. Van Nostrand Company, Inc., 1959). [3.3]

Millman, G. H. "Atmospheric Effects on VHF and UHF Propagation," *Proc. IRE*, 46, No. 8 (Aug. 1958), pp. 1492-1501. [15.1]

Muchmore, R. B. and A. D. Wheelon, "Line-of-Sight Propagation Phenomena," *Proc. IRE*, 43, No. 10 (Oct. 1955), pp. 1437-66. [15.3]

Nergaard, L. S., "Amplification—Modern Trends, Techniques and Problems," *RCA Review*, 21, No. 4 (Dec. 1960), pp. 485-507. [1.1]

Nessmith, J. T., "New Performance Records for Instrumentation Radar," *Space/Aeronautics* (Dec. 1962), pp. 86-94. [12.1, 12.2, 14.2]

Nichols, Nathaniel B., "General Design Principles for Servomechanisms: Applications," Sects. 4.18 and 4.19 of *Theory of Servomechanisms*, James, Nichols, and Phillips, eds. (New York: McGraw-Hill Book Company, 1947). [9.4]

North, D. O., "An Analysis of the Factors Which Determine Signal/Noise Discrimination in Pulsed Carrier Systems," RCA Labs. Tech. Report PTR-6C (June 25, 1943). [1.2, 1.3] [The preceding report has been reprinted in the *Proc. IEEE*, **51** (July 1963), pp. 1015-27).]

Norton, K. A. and associated authors, "An Experimental Study of Phase Variations in Line-of-Sight Microwave Transmissions," Natl. Bureau of Standards Monograph No. 33. (Washington: U.S. Govt. Printing Office, Nov. 1, 1961). [15.3]

Pachares, J., "A Table of Bias Levels Useful in Radar Detection Problems," *Trans. IRE*, **IT-4** No. 1 (March 1958), pp. 38–45. [1.3]

Pearson, Kermit E., "Evaluation of the AN/FPS-16 (System No. 1) at White Sands Missile Range," U.S. Army Signal Missile Support Agency, White Sands Missile Range, N.M., Tech. Memo. 606 (Feb. 1959). [17.3]

Pearson, Kermit E., Dennis D. Kasparek, and Lucile N. Tarrant, "The Refraction Correction Developed for the AN/FPS-16 Radar at White Sands Missile Range," U.S. Army Missile Support Agency, White Sands Missile Range, N.M., Tech. Memo. No. 577 (Nov. 1958). [15.3]

Peters, Leon, "Accuracy of Tracking Radar Systems," Ohio State U. Research Foundation, Report 601-29 (Dec. 31, 1957), Astia Document AD 200,027. [9.5]

Peterson, W. W., *Error Correcting Codes*. (New York: John Wiley and Sons, 1961). [13.1]

Peterson, W. Wesley, "Error-Correcting Codes," *Scientific American*, **206**, No. 2 (Feb. 1962), pp. 96–108. [13.1]

Pettengill, Gordon H., "Measurements of Lunar Reflectivity Using the Millstone Radar," *Proc. IRE*, **48**, No. 5 (May 1960), pp. 933–34 [12.3]

Pfister, W. and T. J. Keneshea, "Ionospheric Effects on Positioning of Vehicles at High Altitudes," Air Force Surveys in Geophysics No. 83 (March 1956), Cambridge Research Center, Astia Document AD 98,777. [15.4]

Povejsil, Donald J., Robert S. Raven and Peter Waterman, *Airborne Radar*. (Princeton, N.J.: D. Van Nostrand Company, Inc., 1961). [4.1]

Prim, D. C. and L. N. Lawhead, "Quantitative Performance Evaluation of a Position and Velocity Range Measurement System," *IRE NEREM Record* (1960), pp. 72–73. [16.3]

Probert-Jones, J. R., "The Radar Equation in Meteorology," *Proc. 9th Weather Radar Conference* (Oct. 23–26, 1961). [3.5]

Purcell, E. M., "The Radar Equation," Chap. 2 in *Radar System Engineering*, L. N. Ridenour, ed. (New York: McGraw-Hill Book Company, 1947). [1.2, 1.3]

Raven, R. S., "The Calculation of Radar Detection Probability and Angular Resolution," Chap. 3 in *Airborne Radar*, by Povejsil, Raven, and Waterman. (Princeton, N. J.: D. Van Nostrand Company, Inc., 1961). [1.3]

Reed, H. R. and C. M. Russell, *Ultra High Frequency Propagation*. (New York: John Wiley and Sons, 1953). [15.1]

Reich, A. and S. Shucker, "Radar Applications to Geodesy," *Proc. 9th East Coast Conf. on Aero. and Nav. Electronics*, Baltimore, Md., (Oct. 1962). [16.2]

Reintjes, J. F. and G. T. Coate, *Principles of Radar*, 3rd ed. (New York: McGraw-Hill Book Company, 1952). [1.1, 4.1]

Rhodes, Donald R., *Introduction to Monopulse*. (New York: McGraw-Hill Book Company, 1959). [3.3, 9.1, 9.5]

Rice, S. O., "Mathematical Analysis of Random Noise, Parts I and II," *Bell System Tech. Journal*, **23**, No. 3 (July 1944), pp. 282–332.

———, "Mathematical Analysis of Random Noise, Parts III and IV," *Bell System Tech. Journal*, **24**, No. 1 (Jan. 1945), pp. 46–156. [These papers are reprinted in *Selected Papers on Noise and Stochastic Processes*, N. Wax, ed. (New York: Dover Publications, Inc., 1954).] [1.2]

Ridenour, L. N., ed., *Radar System Engineering*. (New York: McGraw-Hill Book Company, 1947). [1.2, 2.1, 2.2, 8.1, 9.1]

Roberts, Arthur, ed., *Radar Beacons*. (New York: McGraw-Hill Book Company, 1947). [3.4]

Rubin, W. L. and S. K. Kamen, "SCAMP—A Single-Channel Monopulse Radar Signal Processing Technique," *Trans. IRE*, MIL-6, No. 2 (April 1962), pp. 146–52. [9.5]

Sandoz, John H., "AN/FPS-16 Radar Evaluation Utilizing Star Observations," U.S. Army Signal Missile Support Agency, White Sands Missile Range, N.M., Tech. Report RA-101 (Sept. 1960), Astia Document AD 243,921. [17.3]

Schlesinger, Robert J., *Principles of Electronic Warfare* (Englewood Cliffs, N. J.: Prentice-Hall, Inc., 1961). [2.1, 3.5]

Schooley, Allen H., "Some Limiting Cases of Radar Sea Clutter Noise," *Proc. IRE*, **44**, No. 8 (Aug. 1956), pp. 1043–47. [3.5]

———, "Upwind-Downwind Ratio of Radar Return Calculated from Facet Size Statistics of a Wind-Disturbed Water Surface," *Proc. IRE*, **50**, No. 4 (April 1962), pp. 456–61. [3.5]

Schwartz, Mischa, "Effects of Signal Fluctuation on the Detection of Pulse Signals in Noise," *Trans. IRE*, **IT-2**, No. 2 (June 1956), pp. 66–71. [1.2, 11.2]

———, "A Coincidence Procedure for Signal Detection," *Trans. IRE*, **IT-2**, No. 4 (Dec. 1956), pp. 135–39. [1.2, 1.3, 5.3]

———, *Information Transmission, Modulation and Noise*. (New York: McGraw-Hill Book Company, 1959). [1.2, 11.2, 13.1, 13.2]

Sevick, J., "An Experimental Method of Measuring Back-Scattering Cross-Sections of Coupled Antennas," Harvard University, Cruft Lab. Tech. Report 151 (May 1952). [3.1]

Shchukin, A. N., "Dynamic and Fluctuation Errors in Guided Vehicles," *Izdatel'stvo Sovetskoye Radio*, Moscow (1961), as translated by the Foreign Technology Div., Air Force System Command, (Astia Document AD 401, 738). [11.2]

Sherwood, E. M. and E. L. Ginzton, "Reflection Coefficients of Irregular Terrain at 10 Cm," *Proc. IRE*, **43**, No. 7 (July 1955), pp. 877–78. [3.5]

Shucker, Sidney, "Results of Space Tracking and Prediction with Precision Radar," *Proc. 1962 Natl. Symposium on Space Electronics and Telemetry*, IRE-PGSET. [14.3, 17.3]

———, "Error Coefficients Ease Servo Response Analysis," *Control Engineering*, **10**, No. 5 (May 1963), pp. 119–23 [9.4]

Shucker, Sidney and Robert Lieber, "Comparison of Short-Arc Tracking Systems," *Proc. 5th Natl. Conv. on Military Electronics*, IRE-PGMIL, Washington, D.C. (June 1961). [16.2, 17.3]

Siebert, W. M., "A Radar Detection Philosophy," *Trans. IRE*, **IT-2**, No. 3 (Sept. 1956), pp. 204–21. [1.2, 2.1]

Siegel, K. M., "Bistatic Radars and Forward Scattering," *Proc. 1958 Natl. Conf. on Aero. Electronics*, Dayton, Ohio, pp. 286–90. [4.3, 16.1]

Siegel, K. M. and associated authors, "Bistatic Radar Cross-Sections of Surfaces of Revolution," *Journal of Applied Physics*, **26**, No. 3 (March 1955), pp. 297–305. [4.3]

Silver, Samuel, *Microwave Antenna Theory and Design*. (New York: McGraw-Hill Book Company, 1949). [2.2, 4.1, 7.3, 9.2]

Skolnik, Merrill I., "Theoretical Accuracy of Radar Measurements," *Trans. IRE*, ANE-7, No. 4 (Dec. 1960), pp. 123–29. [2.1, 2.2]

——, "An Analysis of Bistatic Radar," *Trans. IRE*, ANE-8, No. 1 (March 1961), pp. 19–27. [4.3, 16.1]

——, *Introduction to Radar Systems* (New York: McGraw-Hill Book Company, 1962). [2.1, 4.4, 7.3, 8.1, 9.1]

Smith, E. K. and S. Weintraub, "The Constants in the Equation for Atmospheric Refractive Index at Radio Frequencies," *Proc. IRE*, 41, No. 8 (Aug. 1953), pp. 1035–37. [15.3]

Spencer, Roy C., "Back Scattering from Conducting Spheres," Air Force Cambridge Research Lab. Report E5070 (April 1951). [3.1]

Stecca, A. J., N. O. O'Neal, and J. J. Freeman, "A Target Simulator," Naval Res. Lab. Report No. 4694 (Feb. 1956). [3.2]

Steinberg, Bernard D., "Signal Enhancement by Linear Filtering in Pulse Radar," Lectures 20–22 of special summer course, University of Pennsylvania (June 1961; to be published). [7.3]

Swerling, Peter, "Probability of Detection for Fluctuating Targets," RAND Corp. Research Memo. RM-1217, March 17, 1954. [This report was reprinted in *Trans. IRE*, IT-6, No. 2 (April 1960).] [1.2, 3.2]

——, "Maximum Angular Accuracy of a Pulsed Search Radar," *Proc. IRE*, Vol. 44, No. 9 (Sept. 1956), pp. 1146–55. [2.2]

Terman, Frederick E., *Electronic and Radio Engineering* 4th ed. (New York: McGraw-Hill Book Company, 1955). [4.1]

Thompson, M. C., H. B. Janes, and R. W. Kirkpatrick, "An Analysis of Time Variations in Tropospheric Refractive Index and Apparent Radio Path Length," *Journal of Geophysics Research*, 65, No. 1 (Jan. 1960), pp. 193–201. [15.3]

Truxal, John G., *Automatic Feedback Control System Synthesis* (New York: McGraw-Hill Book Company, 1955). [9.4]

Van Vleck, J. H., "The Absorption of Microwaves by Oxygen," and "The Absorption of Microwaves by Uncondensed Water Vapor," *Physical Review*, 71, No. 7 (April 1, 1947), pp. 413–33. [15.1]

Van Voorhis, S. N., *Microwave Receivers* (New York: McGraw-Hill Book Company, 1955). [4.1]

Warner, F. L. and R. L. Ford, "Errors Due to Noise in the Measurement of Range, Range Rate and Range Acceleration When Using an Automatic Strobe," *Royal Radar Establishment*, Malvern, England, Tech. Note 657 (January 1960), Astia Document AD 237,223. [11.2]

Wax, Nelson, ed., *Selected Papers on Noise and Stochastic Processes* (New York: Dover Publications, Inc., 1954). (See Rice.)

Wexler, Raymond and Joseph Weinstein, "Rainfall Intensities and Attenuation of Centimeter Electromagnetic Waves," *Proc. IRE*, 36, No. 3 (March 1948), p. 353, [15.1]

Wheeler, M. S., "Antennas and RF Components," Sects. 10.1–10.9 in *Airborne Radar*, by D. J. Povejsil, R. S. Raven, and P. Waterman. (Princeton, N.J.: D. Van Nostrand Company, Inc., 1961.). [4.4]

White, Charles F., "Servo System Theory," Chap. 7 of *Guidance*,, A. S. Locke, ed. (Princeton, N.J.: D. Van Nostrand Company, Inc., 1955). [9.4, 11.2]

White, W. D. and A. E. Ruvin, "Recent Advances in the Synthesis of Comb Filters," *IRE Conv. Record*, 1957, Part 2, pp. 186–200. [7.4]

Wiltse, J. C., S. P. Schlesinger, and C. M. Johnson, "Back-Scattering Characteristics of the Sea in the Region from 10 to 50 KMC," *Proc. IRE*, 45, No. 2 (Feb. 1957), pp. 220–28. [3.5]

Woodward, P. M., *Probability and Information Theory, with Applications to Radar*. (New York: McGraw-Hill Book Company, 1955). [2.1, 2.3]

Zadeh, L. A. and J. R. Ragazzini, "An Extension of Weiner's Theory of Prediction," *Journal of Applied Physics*, 21, No. 7 (July 1950), pp. 645–55. [13.2]

———, "Optimum Filters for the Detection of Signals in Noise," *Proc. IRE*, 40, No. 10 (Oct. 1952), pp. 1223–31. [13.2]

The one-way voltage (or amplitude) pattern of a uniformly illuminated, rectangular aperture is given by the well-known $(\sin x)/x$ function.*

$$f(y) = \frac{\sin\left(\frac{\pi w}{\lambda} \sin y\right)}{\frac{\pi w}{\lambda} \sin y}$$

$$\cong \frac{\sin \frac{\pi w y}{\lambda}}{\frac{\pi w y}{\lambda}} \qquad \text{(C.1)}$$

$$= \frac{\sin x}{x}$$

Here, y represents the off-axis angle in either plane, w is the antenna dimension in that plane, and λ is the wavelength. The parameter x represents $\pi w y/\lambda$. The one-way, half-power beamwidth θ for this type of antenna is given by

$$\theta = 0.885\frac{\lambda}{w} \qquad \text{(C.2)}$$

MATHEMATICAL APPROXIMATIONS FOR RADAR BEAMSHAPE, SPECTRA, AND WAVE FORMS

We may, therefore, express the relative gain function $f(y)$ in terms of the off-axis angle normalized to the beamwidth.

* Samuel Silver, *Microwave Antenna Theory and Design* (New York: McGraw-Hill Book Company, 1949), p. 180.

$$f(y) = \frac{\sin (2.783y/\theta)}{2.783y/\theta} \qquad \textbf{(C.3)}$$

Plots of this function and of the relative power gain (the square of the voltage function) are shown in Figs. C.1 through C.3. The latter two figures are plotted on logarithmic scales.

In the case of a circular aperture of diameter D, the corresponding gain function is

$$f(y) = \frac{2J_1(\pi Dy/\lambda)}{\pi Dy/\lambda} = \frac{2J_1(u)}{u} \qquad \textbf{(C.4)}$$

where J_i is the Bessel function of the first kind, of order one, and the parameter u represents $\pi Dy/\lambda$. Since the half-power width of the beam produced by this antenna is given by

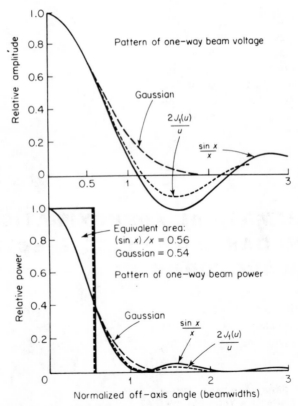

Figure C.1   Comparison of Gaussian, Bessel, and $(\sin x)/x$ functions.

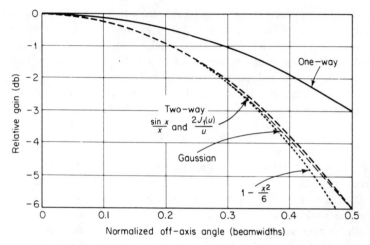

Figure C.2 Comparison of different beam functions near axis.

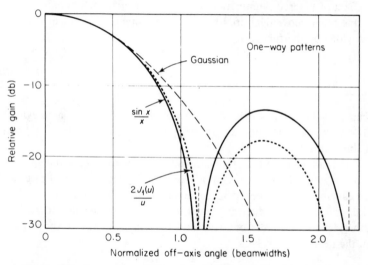

Figure C.3 Comparison of Gaussian, Bessel, and $(\sin x)/x$ functions.

$$\theta = 1.03\frac{\lambda}{D} \qquad\qquad \textbf{(C.5)}$$

we may express the relative gain function $f(y)$ in terms of off-axis angle normalized to this beamwidth.

$$f(y) = \frac{2J_1(3.24y/\theta)}{3.24y/\theta} \qquad\qquad \textbf{(C.6)}$$

The resulting pattern is indistinguishable from that of the rectangular antenna in the region near the axis, and departs from the $(\sin x)/x$ pattern as shown in Figs. C.2 and C.3. The first null, which occurred at $y/\theta = 1.13$ for the rectangular aperture, now occurs at $y/\theta = 1.18$, and the level of the first side lobe is reduced from 13.3 to 17.6 db.

For mathematical convenience, the beam pattern near the axis is often represented by a Gaussian function.

$$f(y) = \exp\left[-1.385(y/\theta)^2\right] \tag{C.7}$$

Curves for this function are also plotted in the figures, showing excellent agreement with the $(\sin x)/x$ and $J_1(u)$ functions, out to the half-power point on the pattern. The Gaussian function, of course, has no side lobe structure, and cannot be generated by any real antenna, since it requires an infinite aperture. However, by tapering the illumination of the aperture it is possible to reduce side-lobe levels by any desired amount. Details on the effect of tapered illumination will be found in Silver's book.

One other convenient approximation for antenna patterns near the axis is provided by the function

$$f(y) = 1 - \frac{x^2}{2} = 1 - \frac{(2.783y/\theta)^2}{6} \tag{C.8}$$

A curve for this function is also shown on Fig. C.2, compared with the two-way voltage (or one-way power) patterns of the $(\sin x)/x$ and Gaussian functions.

Tabulations of the four functions discussed above are given in Tables C.1 through C.4, for values of y/θ out to the second null. These tables were calculated from the standard tables of $(\sin x)/x$ and $J_1(u)$ by interpolation, to provide values at regular intervals within the half-power beamwidth. Relative gain for one-way and two-way patterns is shown, along with the decibel equivalent.

$$(L)_{db} = -20 \log_{10} f \tag{C.9}$$

PULSE SHAPE AND SPECTRAL DENSITY

Most calculations on pulsed systems are made by using the $(\sin x)/x$ form of spectrum generated by a rectangular pulse. The form of this curve is the same as that shown in Fig. C.1 through C.3 for the rectangular aperture, but the parameter used to describe the spectrum is most often the bandwidth $B_t = 1/\tau$, as measured from the center of the spectrum to the first

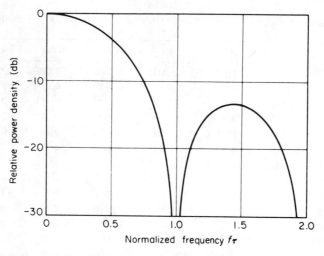

Figure C.4 Spectral density for rectangular pulsed transmission.

null. This is the width of the spectrum between the 40 per cent points (-4 db), rather than the half-power width as used in antenna patterns. The spectrum of Fig. C.4 is normalized to this width, as is Table C.5.

Another type of spectrum which is sometimes used is Gaussian in form, and is applicable to pulse waveforms which also have the Gaussian form. The wave form and spectrum may be represented by the following equations (both giving amplitude functions):

$$a(t) = \exp\left(-\frac{1.385t^2}{\tau^2}\right) \tag{C.10}$$

$$A(f) = \exp\left(-\frac{1.385f^2}{B^2}\right) \tag{C.11}$$

In the above, τ and B represent the half-power pulse width and spectral bandwidth, respectively. These two parameters are related by

$$B\tau = 0.44 \tag{C.12}$$

These Gaussian functions follow the same curves as were used for beam-shape, if the parameters t/τ and f/B are substituted for y/θ [see Eq. (C.7), Figs. C.1 through C.3, and Table C.].

Table C.1 ONE- AND TWO-WAY GAIN FOR RECTANGULAR APERTURE

y/θ	x	f	f^2	f^4	$(L)_{db}$	$2(L)_{db}$
0.0	0.000	1.0000	1.0000	1.0000	0.000	0.000
0.05	0.139	0.9968	0.9936	0.9872	0.028	0.056
0.10	0.278	0.9872	0.9744	0.9494	0.113	0.226
0.15	0.417	0.9712	0.9432	0.8896	0.254	0.508
0.20	0.556	0.9492	0.9014	0.8125	0.451	0.902
0.25	0.695	0.9214	0.8489	0.7200	0.713	1.43
0.30	0.835	0.8879	0.7884	0.6216	1.03	2.06
0.35	0.974	0.8491	0.7210	0.5198	1.42	2.84
0.40	1.113	0.8059	0.6495	0.4218	1.87	3.74
0.45	1.252	0.7585	0.5753	0.3306	2.40	4.80
0.50	1.391	0.7071	0.5000	0.2500	3.01	6.02
0.55	1.531	0.6527	0.4256	0.1811	3.71	7.42
0.60	1.670	0.5959	0.3548	0.1259	4.49	8.99
0.65	1.809	0.5371	0.2884	0.0832	5.39	10.8
0.70	1.948	0.4770	0.2275	0.0516	6.43	12.9
0.75	2.087	0.4164	0.1735	0.0300	7.60	15.2
0.80	2.226	0.3556	0.1264	0.0160	8.97	18.0
0.85	2.366	0.2955	0.0873	0.0076	10.6	21.2
0.90	2.505	0.2372	0.0562	0.0032	12.5	24.9
0.95	2.644	0.1805	0.0326	0.0011	14.9	29.7
1.00	2.783	0.1261	0.0159	0.00025	18.0	36.0
1.05	2.922	0.0745	0.00555	0.00003	22.5	45.0
1.10	3.061	0.0263	0.00069	0.00000	31.6	63.2
1.13	3.142	0.0000	0.00000	0.00000	∞	∞
1.20	3.340	−0.0590	0.00035	0.00000	36.5	73.0
1.30	3.618	−0.1267	0.0160	0.00026	18.0	36.0
1.40	3.896	−0.1757	0.0308	0.00095	15.1	30.2
1.50	4.175	−0.2058	0.0423	0.0018	13.7	27.4
1.60	4.453	−0.2171	0.0471	0.0022	13.3	26.6
1.70	4.731	−0.2113	0.0446	0.0020	13.5	27.0
1.80	5.009	−0.1909	0.0364	0.0013	14.4	28.8
1.90	5.288	−0.1586	0.0251	0.00063	16.0	32.0
2.00	5.566	−0.1180	0.0139	0.00019	18.6	37.1
2.10	5.844	−0.0727	0.0053	0.00003	22.8	45.5
2.20	6.122	−0.0262	0.00069	0.00000	31.6	63.2
2.30	6.400	0.0182	0.00033	0.00000	34.8	69.6

Table C.2 ONE-WAY VOLTAGE AND POWER FOR GAUSSIAN BEAM

y/θ	$(y/\theta)^2$	f	f^2	$(L)_{db}$	$2(L)_{db}$
0.00	0.0000	1.0000	1.0000	0.000	0.000
0.05	0.0025	0.9965	0.9931	0.028	0.056
0.10	0.010	0.986	0.9727	0.118	0.236
0.15	0.0225	0.969	0.9393	0.265	0.530
0.20	0.04	0.946	0.895	0.48	0.96
0.25	0.0625	0.917	0.841	0.75	1.50
0.30	0.090	0.883	0.779	1.08	2.16
0.35	0.1225	0.844	0.712	1.47	2.94
0.40	0.160	0.801	0.642	1.92	3.84
0.45	0.2025	0.755	0.570	2.44	4.88
0.50	0.250	0.707	0.500	3.01	6.02
0.55	0.3025	0.658	0.433	3.63	7.26
0.60	0.360	0.608	0.370	4.31	8.62
0.65	0.4225	0.557	0.310	5.08	10.2
0.70	0.490	0.508	0.258	5.88	11.8
0.75	0.5625	0.459	0.210	6.77	13.5
0.80	0.640	0.412	0.170	7.69	15.4
0.85	0.7225	0.368	0.135	8.70	17.4
0.90	0.810	0.326	0.106	9.75	19.5
0.95	0.9025	0.286	0.082	10.9	21.8
1.00	1.00	0.25	0.063	12.0	24.0
1.10	1.21	0.186	0.035	14.6	29.2
1.20	1.44	0.136	0.018	17.2	34.4
1.30	1.69	0.097	0.0092	20.4	40.8
1.40	1.96	0.066	0.0044	23.6	47.2
1.50	2.25	0.044	0.0020	27.0	54.0
1.60	2.56	0.029	0.0008	30.8	61.6
1.70	2.89	0.018	0.0003	34.8	69.6

Table C.3 ONE- AND TWO-WAY GAIN FOR CIRCULAR APERTURE

y/θ	u	$J_1(u)$	f	f^2	$(L)_{db}$	$2(L)_{db}$
0.00	0.000	0.0000	1.0000	1.0000	0.00	0.00
0.10	0.324	0.1600	0.988	0.976	0.10	0.20
0.20	0.648	0.3073	0.948	0.898	0.46	0.92
0.30	0.972	0.4309	0.886	0.786	1.04	2.08
0.40	1.296	0.5211	0.804	0.648	1.88	3.76
0.50	1.620	0.5718	0.707	0.500	3.01	6.02
0.60	1.944	0.5797	0.596	0.355	4.48	8.96
0.70	2.268	0.5454	0.482	0.232	6.34	12.7
0.80	2.592	0.4731	0.365	0.131	8.70	17.4
0.90	2.916	0.3697	0.254	0.0642	11.9	23.8
1.00	3.240	0.2452	0.151	0.0229	16.4	32.8
1.10	3.564	0.1105	0.0621	0.00385	24.1	48.2
1.20	3.888	−0.0225	−0.0115	0.00013	28.7	57.4
1.30	4.212	−0.1428	−0.0677	0.0046	23.4	46.8
1.40	4.536	−0.2406	−0.1050	0.0110	19.5	39.0
1.50	4.860	−0.3086	−0.1270	0.0161	17.9	35.8
1.60	5.184	−0.3425	−0.132	0.0175	17.6	35.2
1.70	5.508	−0.3410	−0.124	0.0153	18.1	36.2
1.80	5.832	−0.3062	−0.105	0.0110	19.5	39.0
1.90	6.156	−0.2433	−0.079	0.0062	22.0	44.0
2.00	6.480	−0.1595	−0.049	0.0024	26.1	52.2
2.10	6.804	−0.0640	−0.0188	0.00035	34.5	69.0
2.20	7.128	0.0334	0.0094	0.00009	40.6	81.2
2.30	7.452	0.1231	0.033	0.0011	29.6	59.2
2.40	7.776	0.1967	0.0506	0.00255	25.9	51.8
2.50	8.100	0.2476	0.061	0.00375	24.3	48.6

Table C.4 ONE- AND TWO-WAY GAIN APPROXIMATION
(Based on $f = 1 - x^2/6$)

y/θ	x	x^2	$1 - x^2/6$	$(L)_{db}$	$(\triangle L)_{db}$	$2(L)_{db}$
0.00	0.000	0.0000	1.0000	0.000	0.00	0.00
0.05	0.139	0.0194	0.9968	0.028	0.00	0.056
0.10	0.278	0.0773	0.9871	0.113	0.00	0.226
0.15	0.417	0.174	0.9710	0.254	0.00	0.508
0.20	0.556	0.309	0.9483	0.45	0.004	0.90
0.25	0.695	0.483	0.9194	0.73	0.017	1.46
0.30	0.835	0.697	0.884	1.07	0.04	2.14
0.35	0.974	0.950	0.842	1.49	0.07	2.98
0.40	1.113	1.24	0.794	2.00	0.13	4.00
0.45	1.252	1.57	0.739	2.63	0.23	5.26
0.50	1.391	1.94	0.677	3.37	0.36	6.74
0.55	1.531	2.35	0.609	4.30	0.61	8.60
0.60	1.670	2.79	0.536	5.40	0.91	10.8
0.65	1.809	3.26	0.456	6.80	1.41	13.6
0.70	1.948	3.79	0.37	8.60	2.17	17.2

Column showing $(\triangle L)$ represents the error in this approximation, relative to the function $(\sin x)/x$.

Table C.5 AMPLITUDE AND POWER SPECTRA OF RECTANGULAR PULSE

$y\tau$	x	f	f^2	$(L)_{db}$
0.00	0.000	1.0000	1.0000	0.000
0.05	0.157	0.9959	0.9918	0.036
0.10	0.314	0.9836	0.9675	0.130
0.15	0.471	0.9634	0.9320	0.306
0.20	0.628	0.9355	0.8751	0.579
0.25	0.785	0.9004	0.8108	0.911
0.30	0.942	0.8585	0.7370	1.32
0.35	1.100	0.8102	0.6564	1.83
0.40	1.257	0.7566	0.5724	2.42
0.45	1.414	0.6985	0.4879	3.12
0.50	1.571	0.6365	0.4051	3.92
0.55	1.728	0.5715	0.3266	4.86
0.60	1.885	0.5045	0.2545	5.93
0.65	2.042	0.4363	0.1904	7.19
0.70	2.199	0.3679	0.1353	8.69
0.75	2.356	0.3002	0.0901	10.5
0.80	2.513	0.2339	0.0547	12.6
0.85	2.670	0.1701	0.0289	15.4
0.90	2.827	0.1094	0.0120	19.2
0.95	2.985	0.0522	0.0027	25.7
1.00	3.142	0.0000	0.0000	∞
1.10	3.456	-0.0895	0.0080	21.0
1.20	3.770	-0.1560	0.0243	16.1
1.30	4.084	-0.1981	0.0392	14.1
1.40	4.398	-0.2162	0.0467	13.3
1.50	4.712	-0.2122	0.0450	13.5
1.60	5.027	-0.1891	0.0358	14.5
1.70	5.341	-0.1514	0.0229	16.4
1.80	5.655	-0.1039	0.0108	19.7
1.90	5.969	-0.0518	0.0027	25.7
2.00	6.283	0.0000	0.0000	∞

The characteristics of different radar transmissions may be described in a number of ways; time functions, frequency spectra, and ambiguity diagrams may be used. In this appendix, we shall present plots of some of the more common transmission characteristics. For each type of transmission, an equation for the r-f wave form will be given, followed by descriptions of the time resolution, time ambiguity, frequency resolution, and frequency ambiguity of that transmission. Plots of these functions will also be shown.

The transmission types covered here do not include all those used in radar, but they serve to illustrate the characteristics of the most common transmissions, and they may be extended to cover others which become important in practical radar systems. The 12 types given here are as follows:

Figure No.	Transmission
D.1	Pure c-w signal
D.2	c-w signal observed for t_o sec

RESOLUTION AND AMBIGUITY OF VARIOUS RADAR WAVE FORMS

D.3	Two-frequency c-w signal observed for t_o sec
D.4	Amplitude-modulated c-w signal observed for t_o sec
D.5	Single, trapezoidal pulsed signal
D.6	Single, Gaussian pulsed signal

R–F wave form:

$$e_s = E_s \cos (2\pi f_o t + \phi)$$

Time resolution: none

Freq. resolution: infinite

$B = 0$
$\beta_c = 2\pi f_o$
$\beta_e = 0$
$\alpha = \infty$

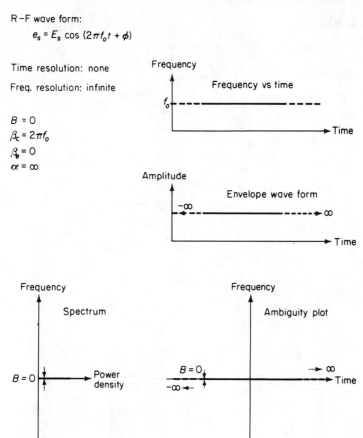

Figure D.1 Characteristics of pure c-w signal.

R-F wave form

$$e_s = E_s \exp\left(-\frac{1.38 t^2}{t_o^2}\right) \cos\left(2\pi f_o t + \phi\right)$$

Time resolution: t_o

Time ambiguity: none

Freq. resolution: $B_o = \dfrac{0.45}{t_o}$

Freq. ambiguity: none

$B_o = \dfrac{0.45}{t_o}$

$\beta_c = 2\pi f_o$

$\beta_e = \dfrac{1.18}{t_o}$

$\alpha = 2.7 t_o$

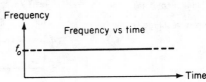

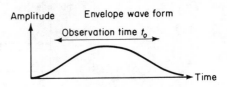

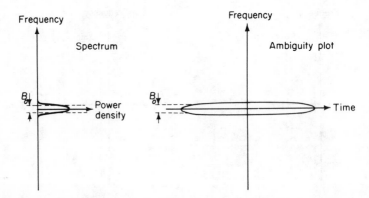

Figure D.2 Characteristics of c-w signal observed for period t_o seconds (Gaussian waveform).

R−F wave form:

$$e_s = E_s \exp\left(-\frac{1.38t^2}{t_o^2}\right)(\cos 2\pi f_1 t + \cos 2\pi f_2 t)$$

$$= 2E_s \exp\left(-\frac{1.38t^2}{t_o^2}\right)\cos \pi(f_2 - f_1)t \, \cos \pi(f_2 + f_1)t$$

Time resolution: t_o

Time ambiguity: $t_p = \dfrac{1}{2\,\Delta f}$

Freq. resolution: B_o

Freq. ambiguity: $\pm \Delta f$

$B_o = \dfrac{0.45}{t_o}$

$\beta_c = \pi(f_2 + f_1)$

$\beta_e = \pi(f_2 - f_1)$

$\alpha = 2.7t_o$

$\Delta f = f_2 - f_1 = \dfrac{1}{2t_p}$

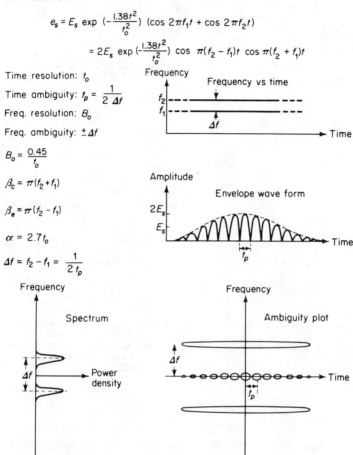

Figure D.3 Characteristics of two-frequency c-w signal observed for period t_o seconds.

R-F wave form

$$e_s = E_s \, \exp \left(\frac{-1.38 t^2}{t_o^2} \right) (1 + m \cos 2\pi f_m t) \cos 2\pi f_o t$$

Time resolution: t_o

Time ambiguity: $t_p = \dfrac{1}{f}$

Freq. resolution: B_o

Freq. ambiguity: $\pm f_m, \, 2f_m$

$B_o = \dfrac{0.45}{t_o}$

$\beta_c = 2\pi f_o$

$\beta_e = 2\pi f_m$

$\alpha = 2.7 t_o$

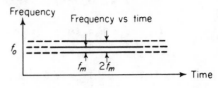

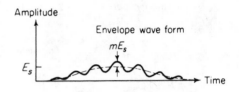

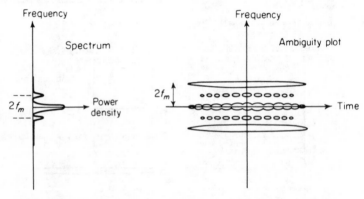

Figure D.4 Characteristics of amplitude modulated c-w signal observed for period t_o seconds.

R-F wave form:

$$e_s = E_s \cos 2\pi f_o t \qquad -\frac{\tau}{2} < t < \frac{\tau}{2}$$

Time resolution: τ

Time ambiguity: none

Freq. resolution: B_f

Freq. ambiguity: side lobes
within. $B_e = \frac{1}{\tau_e}$, centered
at $\pm 3\tau/2$, $5\tau/2$, etc.

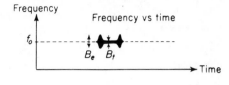

$B_f = \frac{1}{\tau}$

$\beta_c = 2\pi f_o$

$\beta_e \cong \sqrt{2B_e/\tau}$

$\alpha = \frac{\pi\tau}{\sqrt{3}} \cong 1.8\tau$

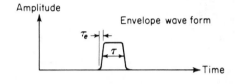

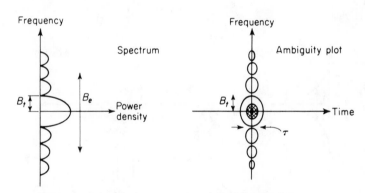

Figure D.5 Characteristics of single, trapezoidal pulsed signal.

R–F wave form:

$$e_s = E_s \exp\left(\frac{-1.38 t^2}{\tau^2}\right) \cos 2\pi f_0 t$$

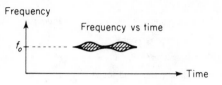

Time resolution: τ

Time ambiguity: none

Freq. resolution: B_f

Freq. ambiguity: none

$B_f = \dfrac{0.45}{\tau}$

$\beta_c = 2\pi f_0$

$\beta_e = \dfrac{1.18}{\tau} = 2.7 B_f$

$\alpha = 2.7\tau$

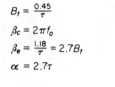

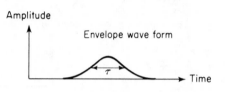

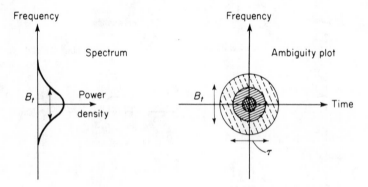

Figure D.6 Characteristics of single, Gaussian pulsed signal.

R–F wave form:

$$e_s = E_s \exp \left(-\frac{1.38\,t^2}{t_o^2}\right) \cos\left(2\pi f_o t + \phi_i\right) \qquad \phi_i = \phi_{i+1}$$

$$\left(it_p - \frac{\tau}{2} < t < it_p + \frac{\tau}{2}\right), \quad i = \pm 0, 1, 2, \ldots$$

Time resolution: τ

Time ambiguity: $\pm it_p$ over interval t_o

Freq. resolution: $B_f = \frac{1}{\tau}$

Freq. ambiguity: side lobes within $B = 1/\tau_e$, centered at $\pm 3\tau/2$, $5\tau/2$, etc,

$B_f = \frac{1}{\tau}$

$\beta_c = 2\pi f_o$

$\beta_e \cong \sqrt{2B_e/\tau}$

$\alpha = 2.7 t_o$

$B_e = \frac{1}{\tau_e}$

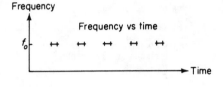

Frequency vs time

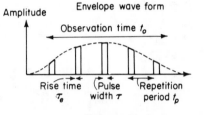

Envelope wave form

Observation time t_o

Rise time τ_e (Pulse width τ)   Repetition period t_p

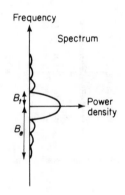

Spectrum

Frequency

B_f

B_e

Power density

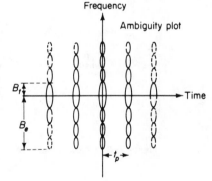

Ambiguity plot

Frequency

B_f

B_e

Time

t_p

Figure D.7 Characteristics of noncoherent pulsed signal train.

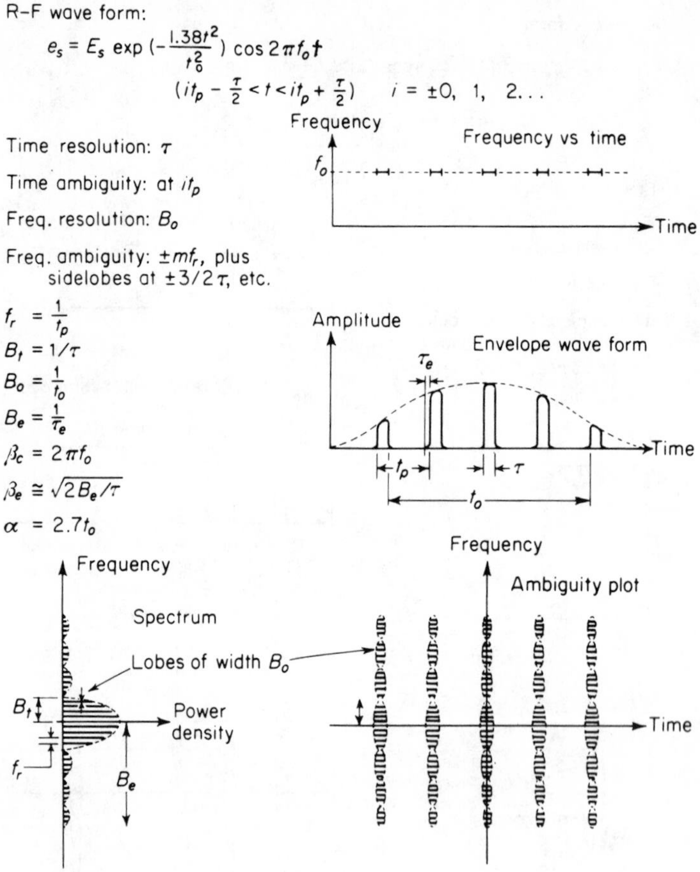

R-F wave form:

$$e_s = E_s \exp\left(-\frac{1.38 t^2}{t_0^2}\right) \cos 2\pi f_0 t$$

$$\left(it_p - \frac{\tau}{2} < t < it_p + \frac{\tau}{2}\right) \qquad i = \pm 0,\ 1,\ 2 \dots$$

Time resolution: τ

Time ambiguity: at it_p

Freq. resolution: B_0

Freq. ambiguity: $\pm m f_r$, plus sidelobes at $\pm 3/2\tau$, etc.

$f_r = \dfrac{1}{t_p}$

$B_f = 1/\tau$

$B_0 = \dfrac{1}{t_0}$

$B_e = \dfrac{1}{\tau_e}$

$\beta_c = 2\pi f_0$

$\beta_e \cong \sqrt{2 B_e / \tau}$

$\alpha = 2.7 t_0$

Figure showing: Frequency vs time plot with f_0 constant over time. Envelope wave form plot showing amplitude with τ_e, t_p, τ, t_0 markings. Spectrum plot showing Frequency, Lobes of width B_0, B_f, f_r, B_e, Power density. Ambiguity plot showing Frequency vs Time.

Figure D.8 Characteristics of coherent pulsed signal train.

R–F wave form:

$$e_s = E_s \cos 2\pi(f_o + k_f t)t \qquad -\frac{\tau}{2} < t < \frac{\tau}{2}$$

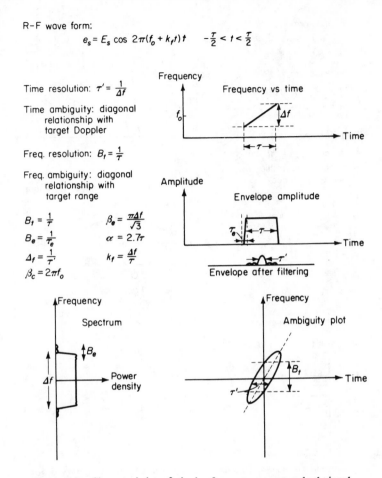

Time resolution: $\tau' = \frac{1}{\Delta f}$

Time ambiguity: diagonal relationship with target Doppler

Freq. resolution: $B_f = \frac{1}{\tau}$

Freq. ambiguity: diagonal relationship with target range

$B_f = \frac{1}{\tau}$ $\beta_e = \frac{\pi \Delta f}{\sqrt{3}}$

$B_e = \frac{1}{\tau_e}$ $\alpha = 2.7\tau$

$\Delta_f = \frac{1}{\tau'}$ $k_f = \frac{\Delta f}{\tau}$

$\beta_c = 2\pi f_o$

Figure D.9 Characteristics of single, frequency-swept pulsed signal.

R-F wave form

$$e_s = E_s \exp\left(\frac{-1.38 t^2}{t_o^2}\right) \cos 2\pi [f_o + k_f (t - it_p)] t \quad i = \pm 0, 1, 2, \ldots$$

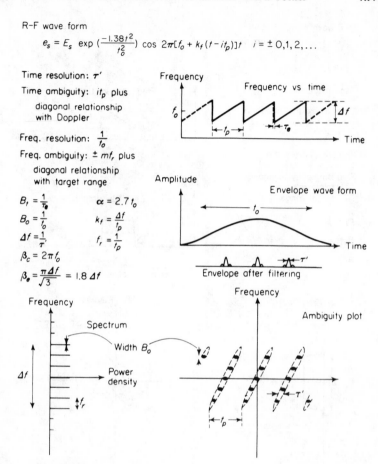

Time resolution: τ'

Time ambiguity: it_p plus

  diagonal relationship
  with Doppler

Freq. resolution: $\frac{1}{t_o}$

Freq. ambiguity: $\pm m f_r$ plus

  diagonal relationship
  with target range

$B_t = \frac{1}{\tau_e}$ $\alpha = 2.7 t_o$

$B_o = \frac{1}{t_o}$ $k_f = \frac{\Delta f}{t_p}$

$\Delta f = \frac{1}{\tau'}$ $f_r = \frac{1}{t_p}$

$\beta_c = 2\pi f_o$

$\beta_e = \frac{\pi \Delta f}{\sqrt{3}} = 1.8 \Delta f$

Figure D.10 Characteristics of linear fm c-w signal observed for period t_o seconds.

R-F wave form:

$$e_s = E_s \exp \left(\frac{-1.38 t^2}{t_0^2}\right) \cos\left\{2\pi\left[f_0 + k_f (t + it_p)\right]t + \phi_i\right\} \quad i = \pm 0, 1, 2, \ldots$$

$$it_p - \frac{T}{2} < t < it_p + \frac{T}{2} \qquad \phi_i \neq \phi_{i+1}$$

Time resolution: τ'

Time ambiguity: it_p plus diagonal
relationship with Doppler

Freq. resolution: B_t

Freq. ambiguity: diagonal relationship
with target range

$B_t = \frac{1}{T}$

$B_e = \frac{1}{T_e}$ $\qquad \alpha = 2.7\tau$

$\Delta_f = \frac{1}{\tau'}$ $\qquad k_f = \frac{\Delta f}{T}$

$\beta_c = 2\pi f_0$ $\qquad \Delta f \tau \gg 1$

$\beta_e = \frac{\pi \Delta f}{\sqrt{3}}$

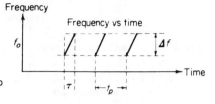

Frequency vs time

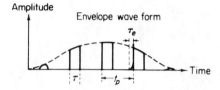

Envelope wave form

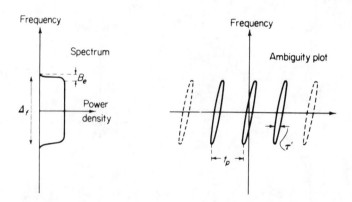

Spectrum

Ambiguity plot

Figure D.11 Characteristics of linear fm pulsed signal train ("chirp")
observed for period t_0 seconds.

R-F wave form:

$$e_s = E_s \exp\left(\frac{-1.38 t^2}{t_o^2}\right) \cos 2\pi(f_o + f)t$$

where f is random with a distribution $P(f) = \exp\left(\frac{-1.38 f^2}{B_f^2}\right)$

Time resolution: $\tau' = \dfrac{0.45}{B_f}$

Time ambiguity: scattered side lobes over t_o

Freq. resolution: B_o

Freq. ambiguity: scattered side lobes over B_f

$B_f = \dfrac{0.45}{\tau'}$

$B_o = \dfrac{0.45}{t_o}$

$\beta_c = 2\pi f_o$

$\beta_e = 2.7 B_f$

$\alpha = 2.7 t_o$

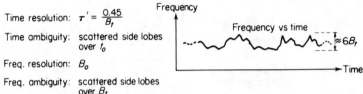

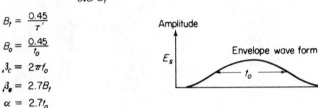

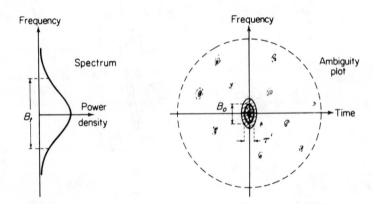

Figure D.12 Characteristics of noise-modulated fm signal observed over period t_0 seconds.

TABLES

Table E.1

TRANSMISSION LINE AND RECEIVER INPUT NOISE TEMPERATURES

Opposite the decimal value of the transmission-line available loss L_r, in the first column, find in the second column the corresponding power-ratio value of L_r. In the third column, find the corresponding value of transmission-line input noise temperature $T_{r(l)}$, assuming that the thermal temperature T_t is approximately equal to $T_o = 290°K$, according to the formula

$$T_{r(l)} = T_o(L_r - 1)$$

If in the actual case T_t has an appreciably different value multiply these values of $T_{r(l)}$ by $T_t/290$.

Opposite the decibel value of receiver noise factor $\overline{NF}$ in the first column, find in the third column the corresponding value of receiver input noise temperature T_e, according to the formula

$$T_e = T_o(\overline{NF} - 1)$$

$\overline{NF}$ L_r db	$\overline{NF}$ L_r power ratios	T_e $T_{r(l)}$ °K	$\overline{NF}$ L_r db	$\overline{NF}$ L_r power ratios	T_e $T_{r(l)}$ °K
0	1.0000	0.00	1.1	1.288	83.5
0.01	1.0023	0.67	1.2	1.318	92.2
0.02	1.0046	1.33	1.3	1.349	101
0.03	1.0069	2.00	1.4	1.380	110
0.04	1.0093	2.70	1.5	1.413	120
0.05	1.0116	3.36	1.6	1.445	129
0.06	1.0139	4.03	1.7	1.479	139
0.07	1.0162	4.70	1.8	1.514	149
0.08	1.0186	5.39	1.9	1.549	159
0.09	1.0209	6.06	2.0	1.585	170
0.10	1.0233	6.76	2.1	1.622	180
0.15	1.0351	10.2	2.2	1.660	191
0.20	1.0471	13.7	2.3	1.698	202
0.25	1.0593	17.2	2.4	1.738	214
0.30	1.0715	20.7	2.5	1.778	226
0.35	1.0839	24.3	2.6	1.820	238
0.40	1.0965	28.0	2.7	1.862	250
0.45	1.1092	31.7	2.8	1.905	262
0.50	1.1220	35.4	2.9	1.950	276
0.55	1.1350	39.2	3.0	1.995	289
0.60	1.1482	43.0	3.1	2.042	302
0.65	1.1614	46.8	3.2	2.089	316
0.70	1.1749	50.7	3.3	2.138	330
0.75	1.1885	54.7	3.4	2.188	345
0.80	1.2023	58.7	3.5	2.239	359
0.85	1.2162	62.7	3.6	2.291	374
0.90	1.2303	66.8	3.7	2.344	390
0.95	1.2445	70.9	3.8	2.399	406
1.00	1.2589	75.1	3.9	2.455	422

Table E.1   cont.

$\overline{NF}$ L_r db	$\overline{NF}$ L_r power ratios	T_e $T_{r(I)}$ °K	$\overline{NF}$ L_r db	$\overline{NF}$ L_r power ratios	T_e $T_{r(I)}$ °K
4.0	2.512	438	7.1	5.129	1197
4.1	2.570	455	7.2	5.248	1232
4.2	2.630	473	7.3	5.370	1267
4.3	2.692	491	7.4	5.495	1304
4.4	2.754	509	7.5	5.623	1341
4.5	2.818	527	7.6	5.754	1379
4.6	2.884	546	7.7	5.888	1418
4.7	2.951	566	7.8	6.026	1458
4.8	3.020	586	7.9	6.166	1498
4.9	3.090	606	8.0	6.310	1540
5.0	3.162	627	8.1	6.457	1583
5.1	3.236	648	8.2	6.607	1626
5.2	3.311	670	8.3	6.761	1671
5.3	3.388	693	8.4	6.918	1716
5.4	3.467	715	8.5	7.079	1763
5.5	3.548	739	8.6	7.244	1811
5.6	3.631	763	8.7	7.413	1860
5.7	3.715	787	8.8	7.586	1910
5.8	3.802	813	8.9	7.762	1961
5.9	3.890	838	9.0	7.943	2013
6.0	3.981	864	9.1	8.128	2067
6.1	4.074	891	9.2	8.318	2122
6.2	4.169	919	9.3	8.511	2178
6.3	4.266	947	9.4	8.710	2236
6.4	4.365	976	9.5	8.913	2295
6.5	4.467	1005	9.6	9.120	2355
6.6	4.571	1036	9.7	9.333	2417
6.7	4.677	1066	9.8	9.550	2480
6.8	4.786	1098	9.9	9.772	2544
6.9	4.898	1130	10.0	10.000	2610
7.0	5.012	1163			

Temperature conversion relations:

$T_{\text{Kelvin}} = 273.16 + T_{\text{Centigrade}} = 255.38 + (\tfrac{5}{9})T_{\text{Fahrenheit}}$

290° Kelvin = 16.84° Centigrade = 62.32° Fahrenheit

(Courtesy L. V. Blake, Naval Res. Lab.)

Table E.2

RADAR RANGE FACTORS FOR SYSTEM POWER
CHANGE FROM 0 TO 40 DECIBELS
(in steps of 0.1 db)

The table is intended for use with an equation of the type

$$R = k\left(\frac{P_t G^2 \lambda^2 \sigma F^4}{P_r L}\right)^{1/4} = kP^{1/4}, \text{ i.e., } R \propto P^{1/4}$$

where R is the radar range and P may be regarded as an equivalent system power variable. The table is based on the relation

$$\frac{R}{R_o} = \text{antilog}\left[\frac{1}{40}\left(10 \log \frac{P}{P_o}\right)\right]$$

where R/R_o is the range factor, and $10 \log P/P_o$ is the power change in decibels. P_t is transmitter power, G antenna gain, λ wave length, σ target cross section, L loss factor, F pattern-propagation factor, and P_r received echo power.

Range factors for power changes greater than 40 db can be obtained from the table by the following procedure: (1) subtract from the absolute value of the power change in decibels the integral multiple of 40 which results in a positive remainder less than 40; (2) look up the range factor corresponding to the remainder; (3) shift the decimal point one place for each 40 db subtracted: for range increase, shift to the right, for decrease shift to the left. For example, the range increase for a power change of 47.3 db is 15.22, and for 87.3 it is 152.2, because for 7.3 db it is 1.522. The decrease factor for 47.3 db is 0.06569, and for 87.3 it is 0.006569, etc.

Power change, db ±	Range increase factor (decimal point)	Range decrease factor (decimal point)	
0.0	1 0000	1 0 000	40.0
0.1	1 0058	9 943	39.9
0.2	1 0116	9 886	39.8
0.3	1 0174	9 829	39.7
0.4	1 0233	9 772	39.6
0.5	1 0292	9 716	39.5
0.6	1 0351	9 661	39.4
0.7	1 0411	9 605	39.3
0.8	1 0471	9 550	39.2
0.9	1 0532	9 495	39.1
1.0	1 0593	9 441	39.0
1.1	1 065	9 386	38.9
1.2	1 072	9 333	38.8
1.3	1 078	9 279	38.7
1.4	1 084	9 226	38.6
1.5	1 090	9 173	38.5
1.6	1 096	9 120	38.4
1.7	1 103	9 068	38.3
1.8	1 109	9 016	38.2
1.9	1 116	8 964	38.1
2.0	1 122	8 913	38.0
2.1	1 129	8 861	37.9
2.2	1 135	8 810	37.8
2.3	1 142	8 760	37.7
2.4	1 148	8 710	37.6
2.5	1 155	8 660	37.5
2.6	1 161	8 610	37.4
2.7	1 168	8 561	37.3
2.8	1 175	8 511	37.2
2.9	1 182	8 463	37.1
3.0	1 189	8 414	37.0
3.1	1 195	8 366	36.9
3.2	1 202	8 318	36.8
3.3	1 209	8 270	36.7
3.4	1 216	8 222	36.6
3.5	1 223	8 175	36.5
3.6	1 230	8 128	36.4
3.7	1 237	8 082	36.3
3.8	1 245	8 035	36.2
3.9	1 252	7 989	36.1
4.0	1 259	7 943	36.0
4.1	1 266	7 898	35.9
4.2	1 274	7 852	35.8
4.3	1 281	7 807	35.7
4.4	1 288	7 763	35.6
4.5	1 296	7 718	35.5
4.6	1 303	7 674	35.4
4.7	1 311	7 630	35.3
4.8	1 318	7 586	35.2
4.9	1 326	7 542	35.1
	Range decrease factor (decimal point)	Range increase factor (decimal point)	± Decibels power change

Power change, db ±	Range increase factor (decimal point)	Range decrease factor (decimal point)	
5.0	1 334	7 499	35.0
5.1	1 341	7 456	34.9
5.2	1 349	7 413	34.8
5.3	1 357	7 371	34.7
5.4	1 365	7 328	34.6
5.5	1 372	7 286	34.5
5.6	1 380	7 244	34.4
5.7	1 388	7 203	34.3
5.8	1 396	7 162	34.2
5.9	1 404	7 120	34.1
6.0	1 413	7 080	34.0
6.1	1 421	7 039	33.9
6.2	1 429	6 998	33.8
6.3	1 437	6 958	33.7
6.4	1 445	6 918	33.6
6.5	1 454	6 879	33.5
6.6	1 462	6 839	33.4
6.7	1 471	6 800	33.3
6.8	1 479	6 761	33.2
6.9	1 488	6 722	33.1
7.0	1 496	6 683	33.0
7.1	1 505	6 645	32.9
7.2	1 514	6 607	32.8
7.3	1 522	6 569	32.7
7.4	1 531	6 531	32.6
7.5	1 540	6 494	32.5
7.6	1 549	6 457	32.4
7.7	1 558	6 420	32.3
7.8	1 567	6 383	32.2
7.9	1 576	6 346	32.1
8.0	1 585	6 310	32.0
8.1	1 594	6 273	31.9
8.2	1 603	6 237	31.8
8.3	1 612	6 202	31.7
8.4	1 622	6 166	31.6
8.5	1 631	6 131	31.5
8.6	1 641	6 095	31.4
8.7	1 650	6 061	31.3
8.8	1 660	6 026	31.2
8.9	1 669	5 991	31.1
9.0	1 679	5 957	31.0
9.1	1 689	5 923	30.9
9.2	1 698	5 888	30.8
9.3	1 708	5 855	30.7
9.4	1 718	5 821	30.6
9.5	1 728	5 788	30.5
9.6	1 738	5 754	30.4
9.7	1 748	5 721	30.3
9.8	1 758	5 689	30.2
9.9	1 768	5 656	30.1
	Range decrease factor (decimal point)	Range increase factor (decimal point)	± Decibels power change

Power change, db ±	Range increase factor (decimal point)	Range decrease factor (decimal point)		Power change, db ±	Range increase factor (decimal point)	Range decrease factor (decimal point)	
10.0	1 778	5 623	30.0	15.0	2 371	4 217	25.0
10.1	1 789	5 591	29.9	15.1	2 385	4 193	24.9
10.2	1 799	5 559	29.8	15.2	2 399	4 169	24.8
10.3	1 809	5 527	29.7	15.3	2 413	4 145	24.7
10.4	1 819	5 495	29.6	15.4	2 427	4 121	24.6
10.5	1 830	5 464	29.5	15.5	2 441	4 097	24.5
10.6	1 841	5 433	29.4	15.6	2 455	4 074	24.4
10.7	1 851	5 401	29.3	15.7	2 469	4 050	24.3
10.8	1 862	5 370	29.2	15.8	2 483	4 027	24.2
10.9	1 873	5 340	29.1	15.9	2 497	4 004	24.1
11.0	1 884	5 309	29.0	16.0	2 512	3 981	24.0
11.1	1 895	5 278	28.9	16.1	2 526	3 958	23.9
11.2	1 905	5 248	28.8	16.2	2 541	3 936	23.8
11.3	1 916	5 218	28.7	16.3	2 556	3 913	23.7
11.4	1 928	5 188	28.6	16.4	2 570	3 890	23.6
11.5	1 939	5 158	28.5	16.5	2 585	3 868	23.5
11.6	1 950	5 129	28.4	16.6	2 600	3 846	23.4
11.7	1 961	5 099	28.3	16.7	2 615	3 824	23.3
11.8	1 972	5 070	28.2	16.8	2 630	3 802	23.2
11.9	1 984	5 041	28.1	16.9	2 645	3 780	23.1
12.0	1 995	5 012	28.0	17.0	2 661	3 758	23.0
12.1	2 007	4 983	27.9	17.1	2 676	3 737	22.8
12.2	2 018	4 954	27.8	17.2	2 692	3 715	22.9
12.3	2 030	4 926	27.7	17.3	2 707	3 694	22.7
12.4	2 042	4 898	27.6	17.4	2 723	3 673	22.6
12.5	2 054	4 870	27.5	17.5	2 738	3 652	22.5
12.6	2 065	4 842	27.4	17.6	2 754	3 631	22.4
12.7	2 077	4 814	27.3	17.7	2 770	3 610	22.3
12.8	2 089	4 786	27.2	17.8	2 786	3 589	22.2
12.9	2 101	4 759	27.1	17.9	2 802	3 569	22.1
13.0	2 113	4 732	27.0	18.0	2 818	3 548	22.0
13.1	2 126	4 704	26.9	18.1	2 835	3 528	21.9
13.2	2 138	4 677	26.8	18.2	2 851	3 508	21.8
13.3	2 150	4 650	26.7	18.3	2 867	3 487	21.7
13.4	2 163	4 624	26.6	18.4	2 884	3 467	21.6
13.5	2 175	4 597	26.5	18.5	2 901	3 447	21.5
13.6	2 188	4 571	26.4	18.6	2 917	3 428	21.4
13.7	2 200	4 545	26.3	18.7	2 934	3 408	21.3
13.8	2 213	4 519	26.2	18.8	2 951	3 388	21.2
13.9	2 226	4 493	26.1	18.9	2 968	3 369	21.1
14.0	2 239	4 467	26.0	19.0	2 985	3 350	21.0
14.1	2 252	4 441	25.9	19.1	3 003	3 330	20.9
14.2	2 265	4 416	25.8	19.2	3 020	3 311	20.8
14.3	2 278	4 390	25.7	19.3	3 037	3 292	20.7
14.4	2 291	4 365	25.6	19.4	3 055	3 273	20.6
14.5	2 304	4 340	25.5	19.5	3 073	3 255	20.5
14.6	2 317	4 315	25.4	19.6	3 090	3 236	20.4
14.7	2 331	4 290	25.3	19.7	3 108	3 217	20.3
14.8	2 344	4 266	25.2	19.8	3 126	3 199	20.2
14.9	2 358	4 241	25.1	19.9	3 144	3 181	20.1
				20.0	3 162	3 162	20.0
	Range decrease factor (decimal point)	Range increase factor (decimal point)	± Decibels power change		Range decrease factor (decimal point)	Range increase factor (decimal point)	± Decibels power change

(Courtesy L. V. Blake, Naval Res. Lab.)

INDEX